Soil Biology

Volume 47

Series Editor
Ajit Varma, Amity Institute of Microbial Technology,
Amity University Uttar Pradesh, Noida, UP, India

More information about this series at http://www.springer.com/series/5138

Alessandra Zambonelli • Mirco Iotti •
Claude Murat
Editors

True Truffle (*Tuber* spp.) in the World

Soil Ecology, Systematics and Biochemistry

Springer

Editors
Alessandra Zambonelli
Department of Agricultural Science
University of Bologna
Bologna, Italy

Mirco Iotti
Department of Life, Health
and Environmental Sciences
University of L'Aquila
L'Aquila, Italy

Claude Murat
INRA, Université de Lorraine
Interactions Arbres-Microorganismes
Lab of Excellence ARBRE
Champenoux, France

ISSN 1613-3382 ISSN 2196-4831 (electronic)
Soil Biology
ISBN 978-3-319-31434-1 ISBN 978-3-319-31436-5 (eBook)
DOI 10.1007/978-3-319-31436-5

Library of Congress Control Number: 2016945975

Printed on acid-free paper

This Springer imprint is published by Springer Nature
The registered company is Springer International Publishing AG Switzerland

Preface

As we were writing this preface, the COP21 international conference on climate change was being held in Paris, highlighting the importance of all initiatives to protect the future of the planet. Forests, and more generally trees, play a key role in carbon sequestration and greenhouse gas mitigation. Many trees live in strict symbiosis with ectomycorrhizal fungi that are important for ecosystems' functioning. Some ectomycorrhizal species, such as boletes and truffles, are also famous because they form edible fructifications, and truffles belonging to the *Tuber* genus, the so-called "true truffles," are gourmet delicacies worldwide. The genus *Tuber* includes around 180 species, most of which are naturally distributed in the northern hemisphere. Some *Tuber* species, such as *Tuber magnatum* (the Italian white truffle), *T. melanosporum* (the Perigord black truffle), *T. aestivum* (the Burgundy truffle), and *T. borchii* (the bianchetto truffle), are the most economically important fungi, but other *Tuber* species are edible and locally appreciated as well. Besides their economic and culinary importance, many truffle species play a key role in forest ecosystems, including disturbed forests, where they are often common ectomycorrhizal symbionts. Moreover, the cultivation of some truffle species such as *T. melanosporum* and *T. aestivum* has spread worldwide in the last two decades and has diversified crops and incomes for local farmers. In this context, many books have been written on truffles, but most of them in French and Italian, or they are focused on a few species or specific aspects.

In this book, we decided to cover much of the taxonomic diversity of the genus *Tuber,* in addition to economically important species, and include information generated from more recent technological innovations (e.g., second-generation DNA sequencing). The book is divided into five parts and comprises chapters written by experienced and internationally recognized scientists. The aim is to provide an inventory of the knowledge on truffle systematics, interactions with abiotic and biotic environments, strategies for spore dispersal, and biochemistry. Such multidisciplinary approach provides a unique insight and a better understanding of the truffle ecology and the role these fungi play in natural and managed ecosystems.

We are grateful to the many scientists who generously assisted us in writing and reviewing the content of this book. It would be too long to cite all the contributors, but we would like to highlight all the corresponding authors of the chapters: Antonella Amicucci, Elena Barbieri, Niccolò Benucci, Gregory Bonito, Gilberto Bragato, Zoltan Bratek, Milan Gryndler, Benoit Jaillard, Chen Juan, Enrico Lancellotti, François Le Tacon, Francis Martin, Cristina Menta, Virginie Molinier, Giovanni Pacioni, Francesco Paolocci, Xavier Parladé, Federica Piattoni, Claudio Ratti, Christophe Robin, Matthew Smith, Richard Splivallo, and Alexander Urban.

Peer review by contributors to this volume and by external internationally recognized scientists helped to maintain the rigor and high quality of material presented. We would like to thank especially all the colleagues who helped us in reviewing the chapters: Antonella Amicucci, Niccolò Benucci, Gilberto Bragato, Aurélie Deveau, Lorenzo Gardin, Milan Gryndler, Ian Hall, Benoit Jaillard, Annegret Kohler, Virginie Molinier, Giovanni Pacioni, Francesco Paolocci, Xavier Parladé, Federica Piattoni, Maria Agnese Sabatini, Elena Salerni, Massimo Turina, Giuliano Vitali, and Yun Wang. We are also grateful to Joey Spatafora who kindly revised this Preface.

We would like also to thank Ajit Varma, series editor, who gave us this great opportunity, Jutta Linderborn, Editor Life Science of Springer, and Sumathy Thanigaivelu, for their help and patience in responding to all the queries regarding the preparation of the book and for giving us the opportunity to include the color pictures provided.

We hope this book will serve as a primary research reference for researchers and research managers interested in mycology, ecology, and soil sciences. Our aim was also to provide a reference book for farmers and foresters who are interested in truffle cultivation worldwide. We are convinced that truffles deserve to be preserved in the context of climate change in order to maintain biodiversity and ecosystem functioning but also to allow future generations to appreciate these unique natural resources.

Bologna, Italy — Alessandra Zambonelli
L'Aquila, Italy — Mirco Iotti
Champenoux, France — Claude Murat
December 2015

Contents

Abbreviation List

1D-SDS-PAGE	One-dimensional sodium dodecyl sulfate polyacrylamide gel electrophoresis
2-DE	Two-dimensional electrophoresis
2D-PAGE	Two-dimensional polyacrylamide gel electrophoresis
AbEV1	*Agaricus bisporus* endornavirus 1
ABV1	*Agaricus bisporus virus 1*
AFLP	Amplified fragment length polymorphism
AFM	Atomic force microscope
AIDS	Acquired immune deficiency syndrome
AMF	Arbuscular mycorrhizal fungi
AMSL	Above mean sea level
AMT	Ammonium transporter
AT	Aminotransferase
ATP	Adenosine triphosphate
BACI	Before-after-control-impact
BCMV	Brown cap mushroom virus
BLAST	Basic local alignment search tool
Cazymes	Carbohydrate active enzymes
CBS	Centraalbureau voor schimmelcultures fungal biodiversity center
cDNA	Complementary deoxyribonucleic acid
CE	Capillary electrophoresis
CEC	Cation exchange capacity
CFB	Cytophaga-Flexibacter-Bacteroides
Cfu	Colony-forming unit
CP	Coat protein
CTAB	Cetyltrimethylammonium bromide
DGGE	Denaturing gradient gel electrophoresis
DNA	Deoxyribonucleic acid
dsRNA	Double-stranded RNA
DTPA	Diethylene triamine pentaacetic acid
ECM	Ectomycorrhiza(l)

EF1	Elongation factor 1
Eno	Enolase
ERM	Ericoid mycorrhiza
ESI	Electrospray ionization
EST	Expressed sequence tags
FF	Fungicolous fungi
FGI	Fungal Genome Initiative
FvBV	Flammulina velutipes browning virus
GAP	GTPase-activating protein
GC	Gas chromatography
GC-MS	Gas chromatography—mass spectrometry
GC-o	Gas chromatography—olfactometry
GDH	Glutamate dehydrogenase
Gdis	GDP-dissociation inhibitor
GDP	Guanosine diphosphate
GEF	Guanine nucleotide exchange factor
GRF1V-M	Glomus sp. strain RF1 virus-like medium dsRNA
GS	Glutamine synthetase
GTP	Guanosine triphosphate
HB	Hydric balance
HPLC	High-performance liquid chromatography
HS-SPME	Head space-solid phase micro-extraction
HS-SPME-GC-MS	Head space-solid phase micro-extraction gas chromatography–mass spectrometry
HXK	Hexokinase
HXT	Hexose transporter
INRA	Institut National de la Recherche Agronomique
ISO	International Standard Organisation
ISRIC	International Soil Reference and Information Centre
ISSR	Inter simple sequence repeat
ITS	Internal transcribed spacer
JGI	Joint Genome Institute
kDa	Kilo dalton
LC	Liquid chromatography
LeSV	Lentinula edodes spherical virus
LeV	*Lentinula edodes* mycovirus
LIV	La France isometric virus
LSU	Large subunit
MALDI	Matrix-assisted laser desorption ionization
MAT	Mating type
MBV	Mushroom bacilliform virus
ME	Embden-Meyerhof
MGI	Mycorrhizal Genome Initiative
MHB	Mycorrhiza helper bacteria

MiSSP	Mycorrhizal induced small secreted protein
MMN	Modified Melin-Norkrans
MRCA	Most recent common ancestor
mRNA	Messenger ribonucleic acid
MS	Mass spectrometry
MTL	Methanethiol
MVOC	Volatile organic compounds produced by microbe
MVX	Mushroom virus X
NADP	Nicotinamide adenine dinucleotide phosphate
NCBI	National Center for Biotechnology Information
NGS	Next-generation sequencing
NIR	Near-infrared spectrometry
NiR	Nitrite reductase
NMR	Nuclear magnetic resonance spectroscopy
NR	Nitrate reductase
Nrt	Nitrate transporter
OM	Organic matter
OMIV	Oyster mushroom isometric virus
OMSV	Oyster mushroom spherical virus
ORC	Orchid mycorrhizas
ORF	Open reading frame
OTU	Operational taxonomic unit
PCWDE	Plant cell wall degrading enzymes
PDA	Potato dextrose agar
pers comm	Personal communication
PKC	Protein kinase C
PoV1	*Pleurotus ostreatus virus 1*
PoV-SN	*Pleurotus* strain Shin-Nong
PP	Pentose phosphate
qPCR	Quantitative polymerase chain reaction
RAPD	Random amplification of polymorphic DNA
rDNA	Ribosomal deoxyribonucleic acid
RdRP	RNA-dependent RNA polymerase
RFLP	Restriction fragment length polymorphism
RNA	Ribonucleic acid
RPB1	RNA polymerase II large subunit
RPB2	RNA polymerase II second largest subunit
RPLC	Reversed-phased liquid chromatography
RPP2	Acidic ribosomal protein P2
rRNA	Ribosomal ribonucleic acid
RT-qPCR	Retrotranscription quantitative polymerase chain reaction
s.l.	Sensu lato
s.s.	Sensu stricto
SCAR	Sequence characterized amplified region

SCIF	Sporocarp-inhabiting fungi
SCL	Sandy clay loam
SD	Standard deviation
SEM	Scanning electron microscope
SGS	Second-generation DNA sequencing
SiCL	Silty clay loam
SiL	Silt loam
SL	Sandy loam
SNP	Single nucleotide polymorphism
SPME-GC-MS	Solid phase micro-extraction-gas chromatography–mass spectrometry
SRP	Signal recognition particle
ssDNA	Single-stranded DNA
SSR	Simple sequence repeat
ssRNA	Single-stranded RNA
TaEV	Tuber aestivum endornavirus
TaMV	*Tuber aestivum* mitovirus
TaV1	*Tuber aestivum virus 1*
TE	Transposable elements
TEF1	Translation elongation factor 1α
TeMV	Tuber excavatum mitovirus
TIF	Truffle-inhabiting fungi
tRNA	Transfer ribonucleic acid
TTGE	Temporal temperature gradient gel electrophoresis
TUNEL	Terminal deoxynucleotidyl transferase dUTP nick end labeling
UPLC	Ultra-performance liquid chromatography
USDA	United States Department of Agriculture
VOC	Volatile organic compound

Part I
Phylogeny

Chapter 1
General Systematic Position of the Truffles: Evolutionary Theories

Gregory M. Bonito and Matthew E. Smith

1.1 Introduction

When the "truffle" concept is evoked, what comes to mind may vary greatly between people and cultural groups. As you read this book, your own concept of what a truffle is may change, as ours has while discovering and learning about these exquisite fungi!

In the very broadest sense, truffles are fungi that sequester their spores within differentiated fruiting structures that are produced below the soil or leaf litter. These fungi have also been referred to in the past as sequestrate fungi or hypogeous fungi, depending on the author and the usage. Hypogeous fungi that belong to the phylum Basidiomycota are sometimes referred to as "false truffles," a name historically used to distinguish these truffles from those in the Ascomycota. We regard truffles as fungi that produce these sequestrate, hypogeous fruiting bodies regardless of their taxonomic or phylogenetic relationships. However, for the purpose of this book, we will use the term truffle in reference to the "true truffles" that belong to the genus *Tuber* (e.g., *Tuber melanosporum* Vittad., *Tuber magnatum* Pico, and related species). Truffles typically fruit on the forest floor just below the leaf litter or sometimes within the mineral horizon. As you will read within this book, we know a lot about the biology and ecology of these organisms, and yet there are still many questions about truffles that remain unanswered.

Truffles often fruit within the rooting zone of forest plants and exhibit a range of variable macroscopic characteristics such as color, shape, size, texture, and aroma.

G.M. Bonito (✉)
Department of Plant, Soil and Microbial Sciences, Michigan State University, East Lansing, MI 48824, USA
e-mail: bonito@msu.edu

M.E. Smith
Department of Plant Pathology, University of Florida, Gainesville, FL 32611-0680, USA

A. Zambonelli et al. (eds.), *True Truffle (*Tuber *spp.) in the World*, Soil Biology 47,
DOI 10.1007/978-3-319-31436-5_1

Microscopic, genomic, and developmental characters also vary widely between truffle species and truffle lineages. Details regarding the natural ecology of the majority of truffle species are still missing. Available evidence suggests that truffles co-diversified with plants and animals and the evolution and distribution of these fungal, plant, and animal symbionts forms a web that overlaps in time and space. Truffle speciation and function in ecosystems are tightly linked to their ectomycorrhizal (ECM) ecology and putative co-diversification with major plant families including Pineaceae (pines), Fagaceae (oak/beech), Myrtaceae (eucalyptus), and Salicaceae (willows/poplar) and also to adaptations for animal dispersal in the Northern and Southern Hemispheres. Because most truffles form ECMs and therefore actively exchange limiting nutrients with plants (truffles usually provide nitrogen and/or phosphorous whereas plants supply carbohydrates), truffle fungi play major roles in the functioning of forest soils and ecosystems as well as the maintenance of Earth's climate and food webs. On the other hand, human-induced climate change appears to be having effects on the distribution and fruiting of truffles and other fungi in Europe and across the globe (Kauserud et al. 2010).

Fungi that form truffle fruiting bodies have evolved independently in at least 13 orders that represent phylogenetically distant fungal lineages (Fig. 1.1) (Smith and Bonito 2013). While there may be some commonalities among these fungi, they are quite diverse in their morphology and ecology, and few generalities can be made about "truffles" at such a coarse level. We do find it interesting that most truffle fungi appear in lineages considered to be plant root-associated mutualists, such as ECM fungi that form a mantel covering the external surface of the root tip and a Hartig net forming laterally between the cortical cells of the root. These structures can be visualized under a microscope or sometimes even with the aid of a hand lens. Some truffle fungi form less evident orchid and ectendomycorrhizal structures, which are only apparent upon staining and visualization under a light microscope.

1.1.1 Loss of Active Spore Discharge in Truffle Fungi

Strong selection for active spore dispersal in fungi has led the evolution of biophysical innovations in forcible spore discharge across the Kingdom. Particularly noteworthy, spores discharged from ascomycetes such as *Podospora curvicolla* (Winter) Niessl can shoot nearly half a meter, and those of *Gibberella zeae* (Schwein.) Petch can reach initial accelerations of 8.5×10^6 m s^{-1} during spore discharge (Yafetto et al. 2008). Such intense force results from a buildup and release of turgor pressure stored in and released from fungal cells, which are critical to dispersal and in maintaining gene flow between populations. However, in truffle fungi that sequester their spores and fruit belowground the ability to actively discharge spores has been lost. This would seem to be detrimental to truffles, yet these fungi are extremely diverse and some species are dominant ECM partners in some ecosystems (Bonito et al. 2011; Smith et al. 2007). Although "self-powered"

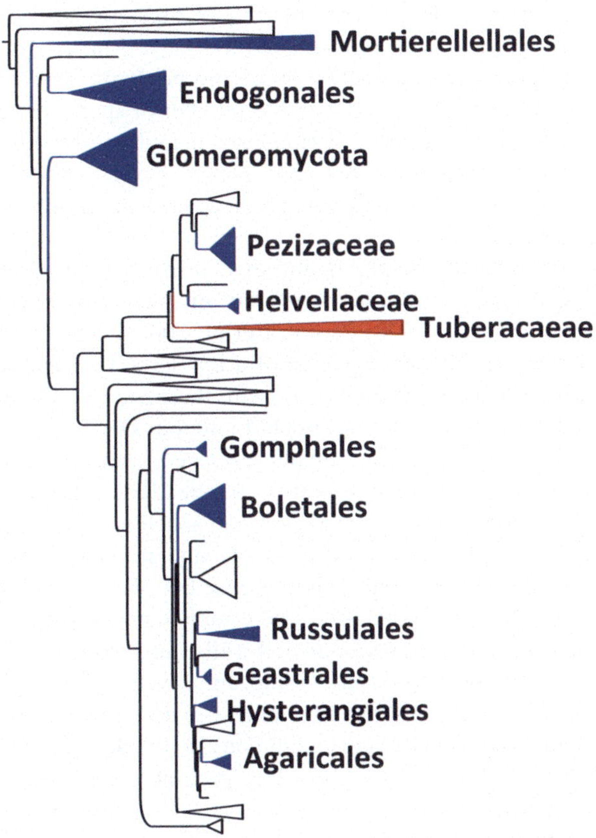

Fig. 1.1 Phylogram showing the distribution of truffle-forming fungi throughout the fungal tree of life. Major fungal orders (and families in the Pezizales) that include truffle taxa are color-coded *blue*. Tuberaceae, the taxonomic family which are the focus of this book, are shown in *red*

active spore discharge has been lost in truffles, novel "passive" mechanisms for spore dispersal have arisen in many truffle lineages. In this scenario the fungi have evolved mechanisms to attract animals through the production of olfactory or visual attractants (Beever and Lebel 2014), coaxing them into the consumption, release, and dispersal of truffle spores. There are a great many instances of coevolution of truffles with mammals in the Northern Hemisphere and with marsupials in the Southern Hemisphere (Claridge et al. 2014). There is also evidence that some birds act as truffle dispersers, such as *Paurocotylis* in New Zealand (Beever and Lebel 2014) and that some insect species could also serve to spread truffle spores (Fogel and Peck 1975).

1.1.2 Enigmatic Truffles and Remaining Mysteries

With their strong aromas, culinary and economic interest, high diversity, and importance to plant and animal nutrition, *Tuber* is a truffle genus that has attracted much interest. However, a number of mysteries still remain concerning its evolution, ecology, and fundamental biology. Mature fruiting bodies of *Tuber* are unusual in their distinctive spore morphology, large spore size (relative to many other truffle fungi), and variable number of spores per ascus. The variable number of spores per ascus is particularly notable since this feature is atypical among Pezizales and varies between *Tuber* species and because the mechanisms for packaging post-meiotic nuclei and nuclear contents into spores are not well understood.

Recently, the sexual nature of *T. melanosporum* and *T. magnatum* was demonstrated using multiple molecular markers and population genetic approaches (Riccioni et al. 2008; Paolocci et al. 2006). Genome sequencing and subsequent studies support the existence of a bipolar sexual mating system in *Tuber* (Martin et al. 2010). In this mating system, there are two idiomorphs, mating loci characterized by large regions of nonhomologous DNA. However, at least some species of *Tuber* also produce mitotically produced (asexual) spores that are hypothesized to function in reproduction or root colonization (Urban et al. 2004; Healy et al. 2013). At least one (currently undescribed) species of *Tuber* belonging to the Puberulum clade is only known from ECMs and masses of these asexual spores (Healy et al. 2013). Although it is possible that these asexually derived spores function as conidia to colonize roots and establish new colonies, these spores are small and abundant and have very thin cell walls, suggesting that instead they may function as spermatia for sexual outcrossing. Improved understanding of the cues and regulation of sexual reproduction and asexual spore production in *Tuber*, aided by population genomic tools, could enable the development of controlled fertilization processes and selective breeding programs for truffles that have so far not been possible.

Truffles evolved from epigeous (mushroom) ancestors, but the specific environments and selective forces leading to truffle evolution in the genus *Tuber* are not clear. The belowground fruiting habit is believed to be adaptive for root-associated fungi, since the spores are produced in closer proximity to roots, but fruiting belowground helps to buffer against environmental fluctuations while the fruiting bodies develops. Further, because the majority of a truffle fruiting body is composed of spore mass, truffle fungi presumably shunt a greater proportion of energy into sexual spore production than do mushroom-producing fungi, which must partition reproduction resources toward the development of sterile cap and stem tissues. Several authors have suggested that sequestrate taxa evolve continually due to chance events, but that sequestrate lineages are selected for when both abiotic environmental conditions (e.g., drought, frequent fires) and biotic interactions (e.g., presence of dispersal agents) are favorable (Albee-Scott 2007; Thiers 1984).

Ancestral biogeographic reconstructions show that *Tuber* most likely had an origin in Eurasia (Bonito et al. 2013; Jeandroz et al. 2008). The most complete phylogenetic treatment of this group indicates that *Tuber* evolved from a lineage of epigeous, cup fungi and then diversified in the Northern Hemisphere throughout the Jurassic and Cretaceous periods (Bonito et al. 2013). What triggered the radiation and high level of *Tuber* diversity is still not completely clear.

In the past, most differences between truffles were considered to be species-specific and strongly influenced by the maturity of the truffle. While maturity is definitely an important factor, there is wide inter- and intraspecific variation in truffle fruiting body shape, size, and aroma and also in the community of bacteria that constitute the truffle microbiome. Evidence from recent studies suggests that the endobiotic bacterial community may be a highly influential and previously underappreciated factor that influences truffle odors and therefore interactions with other organisms (Splivallo et al. 2015; Splivallo and Ebeler 2015). An understanding of how the genomes of endobacteria interact with their fungal hosts and respond to their local environment is one of the grand challenges of truffle ecology, newly invigorated by advances in high-throughput sequencing technologies. Such knowledge will certainly lead to improved strategies for resource management, agricultural production, truffle breeding, and strain development for *Tuber* species.

1.2 Truffle Phylogeny: The Tuberaceae

The family Tuberaceae currently consists of six genera: *Tuber*, *Choiromyces*, *Reddellomyces*, *Labyrinthomyces*, *Dingleya*, and Southern Hemisphere cup fungi *Nothojafnea*. The two Northern Hemisphere genera, *Choiromyces* and *Tuber*, produce sizable and aromatic fruiting bodies that are highly valued in some European countries. It is interesting that species of *Tuber* are incredibly diverse across the Northern Hemisphere, and yet, in contrast, the genus *Choiromyces* includes just a few relatively rare but geographically widespread species.

Cup fungi belonging to the genus *Nothojafnea* have been described from South America and Australia, but DNA sequences are only available for the South American species, *Nothojafnea thaxteri* (E. K. Cash) Gamundí. *N. thaxteri* fruits directly on soil with species of *Nothofagus* and is presumed to form ECMs. Molecular analyses indicate that *N. thaxteri* is phylogenetically nested among Australasian truffles in the genera of *Reddellomyces*, *Labyrinthomyces*, and *Dingleya*. Species in these genera are presumed to be ECM on Australasian Myrtaceae such as *Eucalyptus*, *Corymbia*, *Melaleuca*, and *Leptospermum* as well as with species of *Acacia*, *Nothofagus*, or perhaps other woody plants. These truffle genera are broadly distributed and species rich in Australia, but none of the species are known to have any economic or gastronomic value to humans. Bonito et al. (2010a) found that at least 10 *Tuber* species are also present in Australia and New Zealand, but molecular evidence indicates that these taxa were introduced by humans. Bonito et al. (2010a) also provided molecular data to show that *Tuber clarei* Gilkey is an invalid name erroneously applied to the cosmopolitan and "pioneer" European truffle, *Tuber rapaeodorum* Tul. and Tul. More recently, another *Tuber* collection collected in Australia and deposited in the Melbourne herbarium (MEL2063143) as *Tuber hiromichii* (Imai) Trappe was shown to be *Tuber rapaeodorum* (Bonito unpublished data, GenBank accession KP311464).

1.2.1 Diversity, Ecology, and Distribution of the Genus Tuber

Recently, Bonito et al. (2013) reassessed the diversity, ecology, and historical biogeography of the genus *Tuber* using four genetic loci for inferences (RPP2, TEF1, and ITS and 28S rDNA). They distinguished 11 major clades within the genus *Tuber*. Recent estimates on the number of *Tuber* species are between 180 and 220 species, some of which are known only from "environmental" DNA sequences derived from rhizosphere soil. Characteristics of each of the major *Tuber* clades and exemplars of each are noted below.

1.2.1.1 Aestivum Clade

The Aestivum clade consists of some of the most morphologically diverse *Tuber* species. *Tuber aestivum* Vittad., the type species of the genus *Tuber*, is one of the most widespread and cultivated truffle species and is characterized by a dark warty peridium and aveolate-reticulated ascospores. *Tuber sinoaestivum* Zhang and Liu is an Asian species that is morphologically similar to *T. aestivum*, but *T. sinoaestivum* has ascospores that are more globose and have a shallower reticulum ornamentation (Zhang et al. 2012). In contrast, *Tuber panniferum* Tul. and Tul. is morphologically distinct and characterized by a woolly peridium and very spiny ascospores. *Tuber mesentericum* Vittad. is a species complex composed of at least two species of European truffles (see Chap. 5). Interestingly, the famous and pungent white truffle species *T. magnatum*, with its pale-colored and smooth peridium, also appears to belong within the Aestivum clade despite the fact that it is morphologically quite distinct. Species in this clade form mycorrhizal associations with diverse hosts include angiosperms (e.g., Fagaceae, Betulaceae), gymnosperms (i.e., Pinaceae), and even orchids. The evolutionary history of this clade may be deep and complex.

1.2.1.2 Excavatum Clade

Tuber species belonging in the Excavatum clade are distinguished by having a cavity within the base of their fruiting bodies. Species in this group tend to have a thick and hard peridium and generally have 3–5 coarsely reticulated ascospores per ascus. They are symbionts of angiosperms and are distributed in both Europe and Asia, but this group has never been documented in North America. Often found in association with hardwood tree hosts, several species in this group have also been found as symbionts of orchids (Illyés et al. 2010). The described species that belong to this clade include *Tuber excavatum* Vittad., *Tuber fulgens* Quél from Europe, and *Tuber sinoexcavatum* Fan and Lee from Asia. This group has not been studied as extensively as some *Tuber* clades, yet a number of unique phylogenetic species were detected by Bonito et al. (2010a) indicating the presence of morphologically cryptic species. Truffles in the Excavatum clade can have favorable aromas, but are not typically consumed by humans, likely because of their thick and hard peridium.

1.2.1.3 Gennadii Clade

An early diverging clade within *Tuber* species in the Gennadii group are only thus far known from Europe. Recently, Alvarado et al. (2012) identified two species in this clade [*Tuber gennadii* (Chatin) Pat. and *Tuber lacunosum* Mattir.]. The truffles in this group have only been found in association with the genus *Tuberaria* (Cistaceae) suggesting that these species may have very specific host requirements. Due to the fact that they are relatively rare and restricted in distribution, truffles in the Gennadii clade are generally not consumed by humans.

1.2.1.4 Gibbosum Clade

Endemic to *Pseudotsuga* forests of the Pacific Northwest of the USA, truffles in the *Gibbosum* clade have a light-colored peridium characterized by microscopic beaded hyphae emanating from the surface (Bonito et al. 2010b). The four known species in this group appear to associate exclusively with Pinaceae hosts, particularly *Pseudotsuga* but also occasionally with *Pinus*. Because of their economic value, *Tuber gibbosum* Harkn. and *Tuber oregonense* Trappe, Bonito, and Rawl. are two of the most important species in the Gibbosum clade. These truffles have not yet been cultivated but in the Pacific Northwest of the USA they are wild harvested during winter and spring (Lefevre 2013).

1.2.1.5 Japonicum Clade

Kinoshita et al. (2011) recently discovered a new clade of *Tuber* in Japan. Although species in this group have not yet been officially described, Kinoshita et al. (2011) noted that these species have some unique morphological traits, including pale yellow globose ascospores and fewer spores per ascus than most other *Tuber* species (often only 1 spore per ascus). Internal vein patterning within the gleba of mature truffles in the Japonicum clade tends to be more faint and less conspicuous than in other *Tuber* clades giving them the appearance of unripe truffles. This group is well supported as a monophyletic lineage, but there is still uncertainty regarding the closest relatives of this group (Bonito et al. 2013). We found no information on truffles in the Japonicum clade being consumed by humans.

1.2.1.6 Macrosporum Clade

Truffles in the Macrosporum clade are characterized by the presence of small warts on the outside surface of the peridium and one-, two-, or three-spored asci with relatively large (often >60 μm in length) alveolate-reticulate spores. This group occurs in Asia, Europe, and North America. *Tuber glabrum* Fan and Feng and *Tuber sinomonosporum* Cao and Fan are two new species in the Macrosporum clade that were recently described from China (Fan et al. 2014). Species in this clade tend to be associated with either angiosperm or Pinaceae species. The geographical origin and ancestral host of this clade were not well resolved by Bonito et al. (2013). Paradoxically, truffles in the genus *Paradoxa* actually belong in the Macrosporum clade of *Tuber*. These truffles contain single large ascospores within their asci and had previously been difficult to place phylogenetically without DNA sequence data. Two species in the Macrosporum clade have commercial value. *Tuber macrosporum* Vittad. is found across Italy and eastern Europe and has recently been cultivated in Austria and Hungary (Benucci et al. 2012, 2014). *Tuber canaliculatum* Gilkey is one of the larger and more pungent of the North American *Tuber* species, and this species has a wide distribution in the Eastern USA from the mid-Atlantic states (e.g., North Carolina, Maryland, Virginia) to the upper Midwest (e.g., Michigan) and into Canada. Mycorrhizal synthesis and cultivation trials with *T. canaliculatum* are underway (Benucci et al. 2013).

1.2.1.7 Maculatum Clade

The Maculatum clade produces truffles that have a light-colored peridium with a smooth to cracked texture. The elliptical ascospores of species in this clade tend to have alveolate-reticulate ornamentation. Many species in the Maculatum lineage that have been described from North America and Asia over the past few years and several additional species remain undescribed (Guevara et al. 2013; Su et al. 2013). Truffles in the Maculatum clade are generally not as aromatic as other *Tuber* species. Aside from New Zealand, where *Tuber maculatum* Vittad. has been marketed, species in this clade are not typically consumed by humans but instead are considered undesirable “contaminants” (Amicucci et al. 2000).

1.2.1.8 Melanosporum Clade

Most species belonging to the Melanosporum clade are characterized by a warty outer peridium and spiny ascospore ornamentation, although a few species have spiny-reticulated spores and at least one species (*Tuber pseudoexcavatum* Wang, Moreno, Riousset, Manjón, and Riousset) has spores with alveolate reticulation. Many of the truffle species in this group have pigmented ascospores, giving their gleba a dark color when the spores become mature. There is one currently undescribed species in the Melanosporum clade that has a light-colored outer peridium and gleba (Gregory Bonito, personal observation), putatively ancestral traits that have been fixed in this species. The black truffle *T. melanosporum* is perhaps the most cultivated truffle species internationally, and this species is economically important on several continents. Asian black truffles in the *Tuber indicum* Cooke and Massee complex are also harvested from forests on a massive scale for human consumption, and cultivation trials with this species are underway in China (Wang 2013).

1.2.1.9 Multimaculatum Clade

Tuber multimaculatum Parladé, Trappe, and Alvarez is known only from a few collections in Spain (Alvarez et al. 1992) and is the only species belonging to the Multimaculatum clade. Possibly due to its long branch on the phylogeny, its exact placement within the genus *Tuber* is still not resolved. *Tuber multimaculatum* is characterized by large ellipsoid ascospores with finely meshed alveolate reticulations. Ascospores are produced in one-spored or two-spored asci that have notable apical thickenings in the ascus walls. Because of the rarity of this species, its biology and ecology are not well known.

1.2.1.10 Puberulum Clade

Current data indicate the Puberulum clade has the widest geographic distribution and the most species of any *Tuber* clade. Species in this group are distributed across Europe, Asia, North America, and northern Africa in association with Pinaceae, angiosperms, or both. One species in the Puberulum clade has also been found on the roots of native *Salix humboldtiana* Willd. in South America, suggesting that this may be the only lineage of *Tuber* that has naturally spread to South America with Northern Hemisphere host trees (Bonito et al. 2013). Truffles in the Puberulum clade tend to produce light-colored truffles that have a smooth to cracked peridium, and some species in this clade are known to produce prolific mats of mitospores on soil (Healy et al. 2013). Ascospores of truffles in the Puberulum clade are generally globose to subglobose and are ornamented with alveolate reticulation. Some species in this clade appear to be pioneer ECM species that have been unintentionally introduced into locations in the Southern Hemisphere where they previously did not exist (Guerin-Laguette et al. 2013). Such species could be considered as “weedy” ECM associates. *Tuber borchii* Vittad. is the most important edible truffle species in the Puberulum group and has been shown to produce both ECMs with pine and hardwood species and arbutoid mycorrhizas with *Arbutus unedo* L. (Lancellotti et al. 2014). The list of new species in the Puberulum lineage described from Asia continues to grow, suggesting that there may be many more undescribed taxa in this group (Fan et al. 2012a, b, c). While most species in the Puberulum clade are considered to be undesirable for consumption, one recently described species,

Tuber panzhihuanense Deng and Wang, is reported to have favorable aromatic attributes and commercial potential (Deng et al. 2013).

1.2.1.11 Rufum Clade

The Rufum clade forms a sister group to the Melanosporum clade, and they share some morphological characteristics. For instance, most species in the Rufum clade also have spiny ascospore ornamentation. Some species have a range of spiny-alveolate reticulation, and *Tuber melosporum* (Moreno, Díez, and Manjón) Alvarado, Moreno, Manjón, and Díez in the Rufum clade is the only *Tuber* species known to have smooth ascospores (Alvarado et al. 2012). The Rufum clade is one of the most species-diverse clade in the genus *Tuber*. The peridium of truffles belonging to Rufum clade varies widely; some species may have a verrucose outer peridium covered with small warts, whereas others may have a smooth or cracked peridium. Most species in the Rufum clade have either a faint, unpleasant, or even noxious aroma and are therefore undesirable for human consumption (Iotti et al. 2007). One exception is the North American species *Tuber lyonii* Butters, which is sometimes referred to as the "pecan truffle" because it is commonly found with pecan trees (Bonito et al. 2011). This species has a pleasant aroma and is occasionally harvested and sold in the southeastern USA. Efforts are now underway to cultivate *T. lyonii* and to better understand its biology and ecology (see Chap. 8).

1.3 Biogeography of the Tuberaceae

There is much interest in elucidating the biogeographic origin and evolutionary history of the Tuberaceae. As shown in Fig. 1.2, most of the genera within the Tuberaceae are distributed either in the Southern Hemisphere (*Labyrinthomyces*, *Reddellomyces*, *Dingleya*, *Nothojafnea*) or the Northern Hemisphere (*Choiromyces*, *Tuber*—with the exception of the Puberulum clade), suggesting an ancient phylogeographic split within the family. Based on the work of Bonito et al. (2013), Southern Hemisphere lineages of *Tuber* are more recently diverged than Northern Hemisphere clades, although the Tuberaceae of Australasia have not been thoroughly studied. Based on the large number of undescribed species and blurred generic boundaries, a full systematic revision of the Australasian taxa will be essential to resolve some of these issues (Bonito, Kovacs, Trappe, unpublished).

The geographic origin of *Tuber* has been predicted to be either Europe or Asia. However, even global and multigene datasets assembled by Jeandroz et al. (2008) and by Bonito et al. (2013) were insufficient in reconstructing the geographic center of origin for *Tuber*. Rather than supporting each other, alternate loci mostly gave incongruent results or poorly resolved the branching order among the different lineages. This is especially problematic because several lineages are thus far restricted to only one region (e.g., Japonicum in Asia, Gibbosum in North

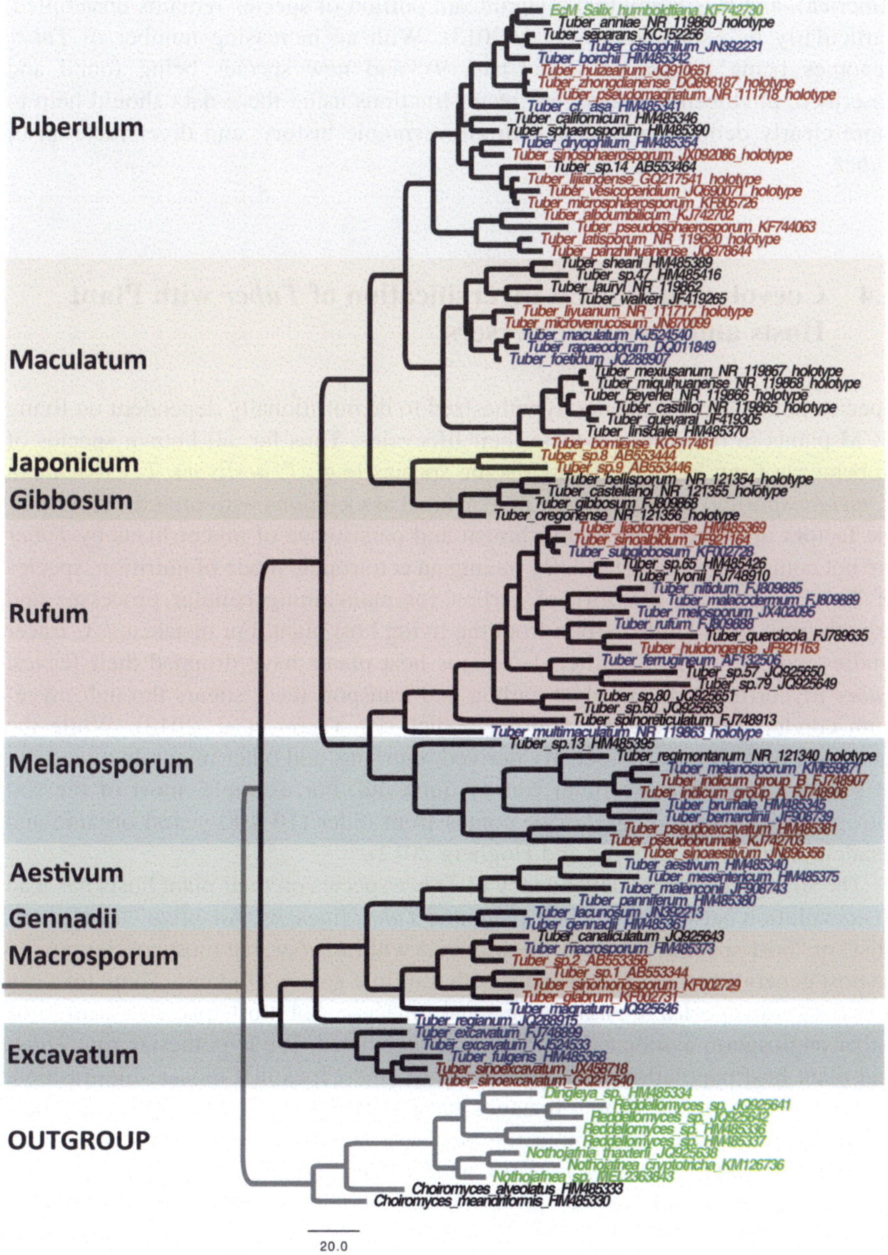

Fig. 1.2 Phylogeny of the Tuberaceae based on ITS rDNA and including all sequence vouchered species and distinct phylotypes. Major clades are distinguished by color and are named on the *left* of the label. Taxon labels are color coded to represent geographic origin of the species: *blue* for Europe, *red* for China, *black* for North America, and *green* for Southern Hemisphere

America), and it is likely that a significant portion of species remains unsampled, particularly in Asia (Bonito et al. 2013). With an increasing number of *Tuber* genomes being sequenced (see Chap. 9), and new species being found and described, phylogenomic network reconstructions using these data should help to more clearly define the center of origin, genomic history, and diversification of *Tuber*.

1.4 Coevolution and Co-diversification of *Tuber* with Plant Hosts and Spore Dispersers

Species in the Tuberaceae are hypothesized to be nutritionally dependent on living ECM plants in order to complete their lifecycles. Thus far, all known species of Tuberaceae form ECMs, although some species (e.g., *T. aestivum*, *T. excavatum*, *T. melanosporum*) may also form mycorrhizal associations with orchids. However, the factors involved in its establishment and persistence of mycorrhizas by *Tuber* are not completely understood. By having an ectotrophic mode of nutrition, species of *Tuber* obtain the majority of carbon for maintaining cellular processes and growth from fixed labile carbon from the living host plant. For instance, ^{13}C tracer studies indicate that even after deciduous host plants have dropped their leaves, *Tuber* mycorrhizas obtain plant carbon and transport these sugars through mycelium conduits to developing fruiting bodies (Le Tacon et al. 2013). While the carbon in truffles may be recently derived, nutrients and other minerals mined out of soil particles by these fungi can by quite old. For example, most of the soil nitrogen obtained by *T. gibbosum* comes from older (10–100 years) organic and recalcitrant factions (Hobbie and Hogberg 2012).

The strong nutritional dependency of *Tuber* species on their plant hosts has lead to coevolution between particular plant and *Tuber* lineages. Adaptive "host-generalist" or "host-specific" strategies could arise within *Tuber*, but most species tend to be host generalists that can associate with multiple genera of plants. Some taxa can associate with species of angiosperms and Pinaceae, but most species appear to be either angiosperm associated or Pineaceae associated. We hypothesize that *Tuber* and plant host populations co-migrate across landscapes in a mosaic-like fashion. Following this model, *Tuber* lineages have putatively migrated with their host symbionts across and throughout the Northern Hemisphere, but even into South America with native species of *Salix* and/or *Alnus* (Bonito et al. 2013). Similar coevolution scenarios have been proposed for other interactions, such as pollination and seed dispersal syndromes. This interplay leads to development of intricate food webs and fascinating complexity in nature (Maser et al. 1978).

1.5 Emergence of *Tuber* Phylogenomics and Molecular Ecology

As the revolution in DNA sequencing technologies continues, with continued interest in truffles, *Tuber* is becoming a model genus for studies of genomics, species diversity, population structure, symbiosis, and evolution at an increasing high resolution (Martin et al. 2010; Rubini et al. 2011; Bonito et al. 2013). There are now genome sequences finished for three *Tuber* species (*T. aestivum*, *T. magnatum*, and *T. melanosporum*), and genomes of four additional *Tuber* species (*T. borchii*, *Tuber brumale* Vittad., *T. indicum*, and *T. lyonii*) are currently being assembled. In total, these taxa represent four of the 11 clades within the genus (Payen et al. 2014; see Chap. 9). These genomic data will help to resolve questions pertaining to truffle growth and development, ecological adaptability, center of origin, and evolutionary history. Molecular tools and understanding arising from these genomic resources will empower a new generation of truffle growers and researchers to tackle age-old questions that have made truffles so perplexing for so long. We expected that genomic approaches will provide streamlined and sensitive protocols for detecting contamination, diseases, genetic diversity, and geographic origin of target truffle strains.

In addition, genomes of eight other fungi in the class Pezizomycetes have been sequenced and are available on the JGI Mycocosm web portal: http://genome.jgi-psf.org/programs/fungi. These data are already providing new perspectives on the evolution of Pezizomycetes and helping to clarify the genomic consequences of different trophic modes in fungi.

1.6 Conclusions

The Tuberaceae continues to be an important family within the Ascomycota for understanding ECM symbiosis, truffle evolution, and fruiting body production. As is evident in the chapters that follow, truffle science has entered into the phylogenomic era. We expect that genomic approaches will bring novel insights and knowledge on truffle development, symbiosis-related genes, molecular crosstalk between fungus and host, genome organization and evolution, and consequences of bacteria on fungal growth, function, and development. It is an exciting time to be studying the biology of *Tuber* and other truffle fungi.

References

Albee-Scott SR (2007) Does secotioid inertia drive the evolution of false-truffles? Mycol Res 111 (9):1030–1039. doi:10.1016/j.mycres.2007.08.008

Alvarado P, Moreno G, Manjon JL (2012) Comparison between *Tuber gennadii* and *T. oligospermum* lineages reveals the existence of the new species *T. cistophilum* (Tuberaceae, Pezizales). Mycologia 104(4):894–910. doi:10.3852/11-254

Alvarez IF, Parlade J, Trappe JM (1992) *Loculotuber gennadii* gen. et comb. nov. and *Tuber multimaculatum* sp. nov. Mycologia 84(6):926–929. doi:10.2307/3760294

Amicucci A, Guidi C, Zambonelli A, Potenza L, Stocchi V (2000) Multiplex PCR for the identification of white *Tuber* species. FEMS Microbiol Lett 189(2):265–269. doi:10.1111/j.1574-6968.2000.tb09241.x

Beever R, Lebel T (2014) Truffles of New Zealand: a discussion of bird dispersal characteristics of fruiting bodies. Auckland Bot Soc 69(2):170–178

Benucci GMN, Csorbai AG, Falini LB, Bencivenga M, Di Massimo G, Donnini D (2012) Mycorrhization of *Quercus robur* L., *Quercus cerris* L. and *Corylus avellana* L. seedlings with *Tuber macrosporum* Vittad. Mycorrhiza 22(8):639–646. doi:10.1007/S00572-012-0441-3

Benucci GMN, Bonito G, Falini LB, Bencivenga M, Donnini D (2013) Truffles, timber, food, and fuel: sustainable approaches for multi-copping truffles and economically important plants. In: Zambonelli A, Bonito G (eds) Edible ectomycorrhizal mushrooms: current knowledge and future prospects, vol 34, Soil biology. Springer, Heidelberg, pp 265–280. doi:10.1007/978-3-642-33823-6_15

Benucci GMN, Raggi L, Albertini E, Csorbai AG, Donnini D (2014) Assessment of ectomycorrhizal biodiversity in *Tuber macrosporum* productive sites. Mycorrhiza 24 (4):281–292. doi:10.1007/S00572-013-0538-3

Bonito GM, Gryganskyi AP, Trappe JM, Vilgalys R (2010a) A global meta-analysis of *Tuber* ITS rDNA sequences: species diversity, host associations and long-distance dispersal. Mol Ecol 19 (22):4994–5008. doi:10.1111/j.1365-294x.2010.04855.x

Bonito G, Trappe JM, Rawlinson P, Vilgalys R (2010b) Improved resolution of major clades within *Tuber* and taxonomy of species within the *Tuber gibbosum* complex. Mycologia 102 (5):1042–1057. doi:10.3852/09-213

Bonito G, Brenneman T, Vilgalys R (2011) Ectomycorrhizal fungal diversity in orchards of cultivated pecan (*Carya illinoinensis*; Juglandaceae). Mycorrhiza 21(7):601–612. doi:10.1007/s00572-011-0368-0

Bonito G, Smith ME, Nowak M, Healy RA, Guevara G, Cazares E, Kinoshita A, Nouhra ER, Dominguez LS, Tedersoo L, Murat C, Wang Y, Moreno BA, Pfister DH, Nara K, Zambonelli A, Trappe JM, Vilgalys R (2013) Historical biogeography and diversification of truffles in the Tuberaceae and their newly identified southern hemisphere sister lineage. PLoS One 8(1):e52765. doi:10.1371/journal.pone.0052765

Claridge AW, Trappe JM, Paul DJ (2014) Ecology and distribution of desert truffles in the Australian outback. In: Kagan-Zur V, Roth-Bejerano N, Sitrit Y, Morte A (eds) Desert truffles: phylogeny, physiology, distribution and domestication, vol 38, Soil biology. Springer, Heidelberg, pp 203–214. doi:10.1007/978-3-642-40096-4_14

Deng XJ, Liu PG, Liu CY, Wang Y (2013) A new white truffle species, *Tuber panzhihuanense* from China. Mycol Prog 12(3):557–561. doi:10.1007/S11557-012-0862-6

Fan L, Cao JZ, Li Y (2012a) *Tuber microsphaerosporum* and *Paradoxa sinensis* spp. nov. Mycotaxon 120:471–475. doi:10.5248/120.471

Fan L, Cao JZ, Li Y (2012b) *Tuber sinosphaerosporum* sp nov from China. Mycotaxon 122:347–353. doi:10.5248/122.347

Fan L, Cao JZ, Yu J (2012c) *Tuber* in China: *T. sinopuberulum* and *T. vesicoperidium* spp. nov. Mycotaxon 121:255–263. doi:10.5248/121.255

Fan L, Feng S, Cao JZ (2014) The phylogenetic position of *Tuber glabrum* sp nov and *T. sinomonosporum* nom. nov., two *Paradoxa*-like truffle species from China. Mycol Prog 13(2):241–246. doi:10.1007/S11557-013-0908-4

Fogel R, Peck SB (1975) Ecological studies of hypogeous fungi. 1. Coleoptera associated with sporocarps. Mycologia 67(4):741–747. doi:10.2307/3758335

Guerin-Laguette A, Cummings N, Hesom-Williams N, Butler R, Wang Y (2013) Mycorrhiza analyses in New Zealand truffieres reveal frequent but variable persistence of Tuber melanosporum in co-existence with other truffle species. Mycorrhiza 23(2):87–98. doi:10.1007/S00572-012-0450-2

Guevara G, Bonito G, Trappe JM, Cazares E, Williams G, Healy RA, Schadt C, Vilgalys R (2013) New North American truffles (*Tuber* spp.) and their ectomycorrhizal associations. Mycologia 105(1):194–209. doi:10.3852/12-087

Healy RA, Smith ME, Bonito GM, Pfister DH, Ge ZW, Guevara GG, Williams G, Stafford K, Kumar L, Lee T, Hobart C, Trappe J, Vilgalys R, Mclaughlin DJ (2013) High diversity and widespread occurrence of mitotic spore mats in ectomycorrhizal Pezizales. Mol Ecol 22 (6):1717–1732. doi:10.1111/Mec.12135

Hobbie EA, Hogberg P (2012) Nitrogen isotopes link mycorrhizal fungi and plants to nitrogen dynamics. New Phytol 196(2):367–382. doi:10.1111/J.1469-8137.2012.04300.X

Illyés Z, Ouanphanivanh N, Rudnoy S, Orczan AK, Bratek Z (2010) The most recent results on orchid mycorrhizal fungi in Hungary. Acta Biol Hung 61(Suppl 1):68–76. doi:10.1556/ABiol.61.2010.Suppl.8

Iotti M, Amicucci A, Bonito G, Bonuso E, Stocchi V, Zambonelli A (2007) Selection of a set of specific primers for the identification of *Tuber rufum*: a truffle species with high genetic variability. FEMS Microbiol Lett 277(2):223–231. doi:10.1111/j.6968.2007.00963.x

Jeandroz S, Murat C, Wang YJ, Bonfante P, Le Tacon F (2008) Molecular phylogeny and historical biogeography of the genus *Tuber*, the 'true truffles'. J Biogeogr 35(5):815–829. doi:10.1111/j.1365-2699.2007.01851.x

Kauserud H, Heegaard E, Semenov MA, Boddy L, Halvorsen R, Stige LC, Sparks TH, Gange AC, Stenseth NC (2010) Climate change and spring-fruiting fungi. Proc R Soc B Biol Sci 277 (1685):1169–1177. doi:10.1098/Rspb.2009.1537

Kinoshita A, Sasaki H, Kazuhide N (2011) Phylogeny and diversity of Japanese truffles (*Tuber* spp.) inferred from sequences of four nuclear loci. Mycologia. doi:10.3852/10-138

Lancellotti E, Iotti M, Zambonelli A, Franceschini A (2014) Characterization of *Tuber borchii* and *Arbutus unedo* mycorrhizas. Mycorrhiza 24(6):481–486. doi:10.1007/S00572-014-0564-9

Le Tacon F, Zeller B, Plain C, Hossann C, Brechet C, Robin C (2013) Carbon transfer from the host to *Tuber melanosporum* mycorrhizas and ascocarps followed using a C-13 pulse-labeling technique. PLoS One 8(5):e64626. doi:10.1371/journal.pone.0064626

Lefevre C (2013) Native and cultivated truffles of North America. In: Zambonelli A, Bonito G (eds) Edible ectomycorrhizal mushrooms: current knowledge and future prospects, vol 34, Soil biology. Springer, Berlin, pp 209–226. doi:10.1007/978-3-642-33823-6_12

Martin F, Kohler A, Murat C, Balestrini R, Coutinho PM, Jaillon O, Montanini B, Morin E, Noel B, Percudani R, Porcel B, Rubini A, Amicucci A, Amselem J, Anthouard V, Arcioni S, Artiguenave F, Aury JM, Ballario P, Bolchi A, Brenna A, Brun A, Buee M, Cantarel B, Chevalier G, Couloux A, Da Silva C, Denoeud F, Duplessis S, Ghignone S, Hilselberger B, Iotti M, Marcais B, Mello A, Miranda M, Pacioni G, Quesneville H, Riccioni C, Ruotolo R, Splivallo R, Stocchi V, Tisserant E, Viscomi AR, Zambonelli A, Zampieri E, Henrissat B, Lebrun MH, Paolocci F, Bonfante P, Ottonello S, Wincker P (2010) Perigord black truffle genome uncovers evolutionary origins and mechanisms of symbiosis. Nature 464 (7291):1033–1038. doi:10.1038/nature08867

Maser CM, Trappe JM, Nussbaum RA (1978) Fungal-small mammal inter-relationships with emphasis on Oregon coniferous forests. Ecology 59(4):799–809. doi:10.2307/1938784

Paolocci F, Rubini A, Riccioni C, Arcioni S (2006) Reevaluation of the life cycle of *Tuber magnatum*. Appl Environ Microb 72(4):2390–2393. doi:10.1128/AEM.72.4.2390-2393.2006

Payen T, Murat C, Bonito G (2014) Truffle phylogenomics: new insights into truffle evolution and truffle life cycle, vol 70. Advances in Botanical Research, Elsevier

Riccioni C, Belfiori B, Rubini A, Passeri V, Arcioni S, Paolocci F (2008) *Tuber melanosporum* outcrosses: analysis of the genetic diversity within and among its natural populations under this new scenario. New Phytol 180(2):466–478. doi:10.1111/j.1469-8137.2008.02560.x

Rubini A, Belfiori B, Riccioni C, Tisserant E, Arcioni S, Martin F, Paolocci F (2011) Isolation and characterization of MAT genes in the symbiotic ascomycete *Tuber melanosporum*. New Phytol 189(3):710–722. doi:10.1111/j.1469-8137.2010.03492.x

Smith ME, Bonito G (2013) Systematics and ecology of edible ectomycorrhizal mushrooms. In: Zambonelli A, Bonito G (eds) Edible ectomycorrhizal mushrooms: current knowledge and future prospects, vol 34, Soil Biology. Springer, Berin, pp 17–39. doi:10.1007/978-3-642-33823-6_2

Smith ME, Douhan GW, Rizzo DM (2007) Ectomycorrhizal community structure in a xeric *Quercus* woodland based on rDNA sequence analysis of sporocarps and pooled roots. New Phytol 174(4):847–863. doi:10.1111/J.1469-8137.2007.02040.x

Splivallo R, Ebeler SE (2015) Sulfur volatiles of microbial origin are key contributors to human-sensed truffle aroma. Appl Microbiol Biotechnol 99(6):2583–2592. doi:10.1007/S00253-014-6360-9

Splivallo R, Deveau A, Valdez N, Kirchhoff N, Frey-Klett P, Karlovsky P (2015) Bacteria associated with truffle-fruiting bodies contribute to truffle aroma. Environ Microbiol 17 (8):2647–2660. doi:10.1111/1462-2920.12521

Su KM, Xiong WP, Wang Y, Li SH, Xie R, Baima D (2013) *Tuber bomiense*, a new truffle species from Tibet, China. Mycotaxon 126:127–132. doi:10.5248/126.127

Thiers HD (1984) The secotioid syndrome. Mycologia 76(1):1–8

Urban A, Neuner-Plattner I, Krisai-Greilhuber I, Haselwandter K (2004) Molecular studies on terricolous microfungi reveal novel anamorphs of two *Tuber* species. Mycol Res 108 (7):749–758. doi:10.1017/S0953756204000553

Wang X (2013) Truffle cultivation in China. In: Zambonelli A, Bonito G (eds) Edible ectomycorrhizal mushrooms: current knowledge and future prospects, vol 34, Soil biology. Springer, Berlin, pp 227–240. doi:10.1007/978-3-642-33823-6_13

Yafetto L, Carroll L, Cui Y, Davis DJ, Fischer MWF, Henterly AC, Kessler JD, Kilroy HA, Shidler JB, Stolze-Rybczynski JL, Sugawara Z, Money NP (2008) The fastest flights in nature: High-speed spore discharge mechanisms among Fungi. PLoS One 3(9):e3237. doi:10.1371/journal.pone.0003237

Zhang JP, Liu PG, Chen J (2012) *Tuber sinoaestivum* sp nov., an edible truffle from southwestern China. Mycotaxon 122:73–82. doi:10.5248/122.73

Chapter 2
The Black Truffles *Tuber melanosporum* and *Tuber indicum*

Juan Chen, Claude Murat, Peter Oviatt, Yongjin Wang, and François Le Tacon

2.1 Introduction

The genus *Tuber* is widespread in Asia (India, China, Mongolia, Japan), Europe and North America, but to date it is almost completely absent in the southern hemisphere. Jeandroz et al. (2008) suggested that the common ancestor of *Tuber* originated in Europe or Eurasia. Considering the relevant genes and molecular clocks, the differentiation of the *Tuber* genus would have taken place between 271 and 140 Mya (Jeandroz et al. 2008) or between 122 and 162 Mya in the early Cretaceous period (Bonito et al. 2013).

The Melanosporum clade, which comprises species in Europe (*Tuber melanosporum* Vittad., *Tuber brumale* Vittad.), Asia (*Tuber indicum* Cooke and Massee and *Tuber pseudoexcavatum* Y. Wang, G. Moreno, Riousset, Manjón and G. Riousset) and America (*Tuber regimontanum* Guevara, Bonito and J. Rodr. and *Tuber* sp. 13; Bonito et al. 2013), could have arisen between 85 and 25 Mya (Jeandroz et al. 2008) or between 65 and 95 Mya (Bonito et al. 2013). According to Jeandroz et al. (2008), the Melanosporum clade may be of Eurasian origin and

J. Chen (✉)
Institute of Medicinal Plant Development, Chinese Academy of Medical Sciences & Peking Union Medical College, 100193 Beijing, China
e-mail: kibchenjuan@126.com

C. Murat • F. Le Tacon
INRA, Université de Lorraine, UMR1136, Interactions Arbres/Micro-organismes, Laboratoire d'Excellence ARBRE, 54280 Champenoux, France

P. Oviatt
Massachusetts Institute of Technology, HASTS Program, 77 Massachusetts Ave. Bldg. E51, Cambridge, MA, USA

Y. Wang
ARL EU International, 132 rue Faubourg Saint Martin, 75010 Paris, France

A. Zambonelli et al. (eds.), *True Truffle (*Tuber *spp.) in the World*, Soil Biology 47,
DOI 10.1007/978-3-319-31436-5_2

could have spread to North America through the Miocene Bering Land Bridge. In another study, America was proposed as the origin of the Melanosporum clade that could have successively migrated towards Asia and Europe (Bonito et al. 2013). For Eurasian species, the differentiation of two subclades (*T. brumale/T. pseudoexcavatum* and *T. melanosporum/T. indicum*) implies two independent dispersals into Asia or Europe, depending on the origin. In each of these two subclades, speciation could have occurred from two vicariance events. The first event, 25 Mya, could have given birth to *T. brumale* in Europe and *T. pseudoexcavatum* in South and Central China, while the second one, 7 Mya, could have given birth to *T. melanosporum* in Europe and *T. indicum* in Asia (Jeandroz et al. 2008).

This clade is particularly important since it comprises two of the most economically important *Tuber* species: *T. melanosporum* and *T. indicum*.

Tuber melanosporum has been harvested and consumed in Europe for several centuries. After the phylloxera crisis at the end of the nineteenth century, the French annual production of *T. melanosporum* may have exceeded 1000 t (Chatin 1869). French production then declined during the twentieth century to roughly 50–100 t per year (Le Tacon et al. 2014). This decline has been explained by sociological factors such as the two world wars, rural desertification, changes in forest use and the modernisation of agriculture (Olivier et al. 2012; Le Tacon et al. 2014). Climate change has also been proposed by Büntgen et al. (2012a, b) to explain the decline of truffle production in Europe since the beginning of the twentieth century. However, the impact of climate change as leading to the decline of truffle production during the past 50 years was not supported by Le Tacon et al. (2014). Nevertheless, annual climatic variations have strongly influenced truffle production in France during the same 50-year period (Le Tacon et al. 1982, 2014).

In the early 1970s, in contrast to previous methods, seedlings were inoculated with *T. melanosporum* in nurseries with well-controlled conditions and then transplanted in truffle orchards (Murat 2015). Today, these new plantations represent at least 80 % of French production (see below).

T. indicum has been traded in the European market since the 1990s. Before the 1990s, in China, *T. indicum* was rarely consumed by local populations. Its commercialisation started when Chinese and European merchants discovered the species. The morphological similarity between *T. melanosporum* and *T. indicum*, and its lower cost, could explain European interest in this species.

The quantity of *T. indicum* production is difficult to estimate. According to Jean-Charles Savignac (President of European Truffle and Truffle Cultivation Association), before 2000, China produced about 1000 t of black truffles per year; during the past few years, however, production has declined to about 300 t. This decline is confirmed by official exportation data from Chinese customs, which show that 200–300 t were exported from 1995 to 2005, less than 50 t in 2013 and at most 30 t in 2014 (http://www.greentimes.com/green/news/fangtan/jbft/content/2014-11/04/content_274881.htm). This decline could be explained by the destruction of the natural habitat of *T. indicum* due to the systematic digging of forest soils, a practice that leads to the disappearance of ectomycorrhizas (ECMs) (Wang et al. 2007).

2.2 *Tuber melanosporum*: The Black Perigord Truffle

2.2.1 *General Characteristics*

Tuber melanosporum, described in Vittadini's (1831) *Monographia Tuberacearum*, is a well-defined species. In 1869, Chatin described *T. hiemalbum* from a batch of *T. melanosporum*. The existence of *T. hiemalbum* as a species, however, is not accepted by all authors. Ceruti et al. (2003) considered *T. hiemalbum* to be a synonym of *T. brumale*, while Pacioni and Pomponi (1989), using chemotaxonomy, and Pacioni et al. (1990), via analysis of the volatile composition of the ascocarps, considered *T. hiemalbum* to be a synonym of *T. melanosporum*. On the contrary, Riousset et al. (2001) maintained that *T. hiemalbum* was an independent species.

Due to the rarity of this species and its low economic interest, *T. hiemalbum* has not been analysed using molecular tools. In the international database GenBank, only two sequences of the internal transcribed spacer (ITS) of the ribosomal DNA are available (FM205570 and FM205571). By comparing these sequences with those of *T. brumale*, *T. indicum*, *T. melanosporum*, *T. regimontanum* and *Tuber magnatum* Pico as an out-group, phylogenetic analysis confirmed that *T. hiemalbum* clustered together with *T. melanosporum* (Fig. 2.1). According to these results, *T. hiemalbum* should be considered synonymous with *T. melanosporum*, confirming previous volatile and chemotaxonomy analyses (Pacioni and Pomponi 1989; Pacioni et al. 1990).

Tuber melanosporum is naturally harvested in Europe between 40° and 48° northern parallel, primarily in France, Italy and Spain (Ceruti et al. 2003). It is found in contrasting climates, including warmer Mediterranean regions such as southern Spain and Italy and colder continental areas such as Alsace (northeastern France). *Tuber melanosporum* has a large number of host species belonging to the genera *Quercus*, *Corylus*, *Populus*, *Tilia*, *Ostrya*, *Carpinus*, *Cistus*, *Pinus* and *Cedrus* (Ceruti et al. 2003). Thanks to inoculated seedlings, *T. melanosporum* has been introduced to countries where it is non-native such as the USA, Australia, Morocco, New Zealand, South Africa, China and Sweden (Wedén et al. 2013). For the moment, *T. melanosporum* ascomata have been harvested in great quantities only in Australia (Reyna and Garcia-Barreda 2014). Nevertheless, its cultivation and the global truffle market are developing rapidly. This is illustrated by the roughly 6 t of *T. melanosporum* produced in Australia in 2013 alone (Lee 2014).

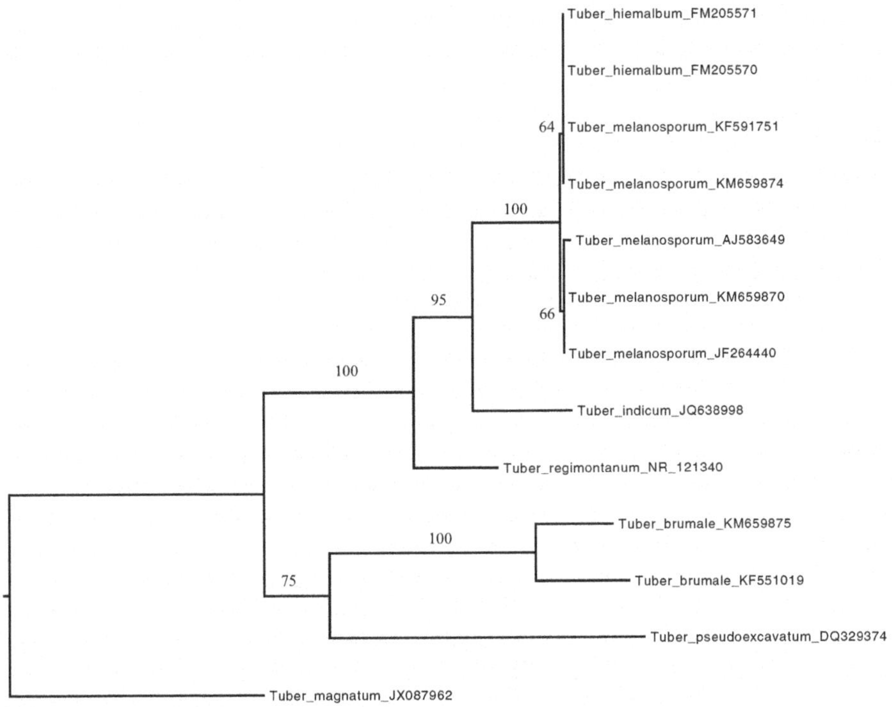

Fig. 2.1 Neighbour-joining phylogenetic tree realised with the ITS regions of the rDNA. The phylogeny was realised with MEGA6. The bootstrap value above 60 is indicated:

2.2.2 Tuber melanosporum *Production in Truffle Orchards*

Currently, *T. melanosporum* production occurs primarily in truffle orchards. The cultivation of this species is based on the planting of seedlings that have been previously inoculated with truffles in controlled conditions, such as nurseries. The inoculation of seedlings with *T. melanosporum* is performed using ascomata as inoculum. Ascomata contain meiotic spores, which, after germination, form ECMs on seedling roots (Paolocci et al. 2006). Seedlings are then grown in greenhouses and sold following a procedure of quality control (Murat 2015). This inoculation technique, developed in Italy and France, is now used worldwide (Chevalier and Grente 1979). In Europe it is estimated that more than 500,000 seedlings inoculated with truffles are sold each year; this compares to the roughly 100,000 inoculated seedlings sold annually on other continents (Murat 2015).

In the twentieth century, the vast majority of truffle production moved from secondary woodlands to planted truffle orchards. Aside from truffle production, truffle orchards help transform agricultural landscapes into sustainable and productive agroforestry ecosystems with high value added; this transition promotes landscape management in fire-prone regions and increases biodiversity compared to that of intensive agriculture (Therville et al. 2013). As a form of agroforestry, truffle orchards may also help mitigate climate change through carbon sequestration (Hamon et al. 2009) and other ecosystem services such as improvements in water and soil quality (Alam et al. 2014). All these aspects should be highlighted together, with truffle production, to convince farmers, foresters and local agencies to invest in *T. melanosporum* cultivation.

Tuber melanosporum production in truffle orchards faces two important bottlenecks: the initiation of sexual reproduction and the growth of ascomata during a period of several months linked to a tree (Fig. 2.2). To complete their life cycle, truffles must associate with a mature tree. It has been shown that the carbon supporting *T. melanosporum* ascomata comes exclusively from tree photosynthesis throughout a continuous association (Le Tacon et al. 2013). Indeed, there is no transfer of carbon from C^{13}-labelled organic matter to fruiting bodies (Le Tacon et al. 2015). Truffle production therefore differs significantly from that of saprotrophic fungi such as button [*Agaricus bisporus* (J. E. Lange) Imbach] or oyster [*Pleurotus ostreatus* (Jacq.) P. Kumm.] mushrooms. Climatic changes such as drought could affect these two bottlenecks in truffle production, and as such, truffle growers have prioritised overcoming the negative effects of drought by applying new management practices (see Chap. 10). Primary interventions realised by truffle growers include soil tilling, tree pruning, irrigation and spore inoculation (Olivier et al. 2012). As indicated in Fig. 2.2, the adaptation capacities of *T. melanosporum* could also be important to overcome stresses.

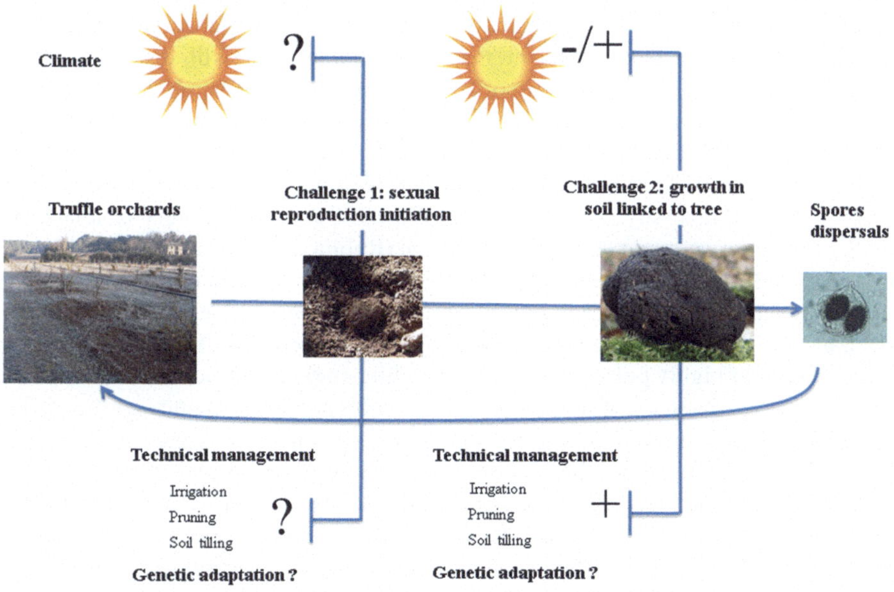

Fig. 2.2 *Tuber melanosporum* production process. The two main bottlenecks are indicated. The unknown and negative effects of drought on bottlenecks 1 and 2 are indicated. To counterbalance the positive effects of cultural practices is indicated, while truffle genetic adaptation capacity is currently unknown

2.2.3 Tuber melanosporum *Production in the Context of Climate Change: The Importance of Adaptation Capacities*

Species can respond to climate change by migration and/or adaptation (Davis and Shaw 2001). In the future, evolutionary responses will likely be critical as habitat fragmentation impedes migration or if suitable habitats are already occupied (Shigesada and Kawasaki 1997). Genetic variability in traits under selection, DNA decay (accumulation of mutations in genes) and reproduction modes are critical for adaptive evolution in response to climate change. Indeed, it seems that genetic variance can enhance the persistence of populations in a changing environment (Etterson 2004). The effect of climate variability on mushroom productivity has been reported for different fungal species, including truffles (Kauserud et al. 2010; Büntgen et al. 2011, 2012a, b). Büntgen and colleagues (2012a) found a boost in ECM fungi harvested in their 30-year survey that may be explained as increases in the growth of mycelium and of host plants, both of which are factors resulting from recent climatic changes. Interestingly, this increase contradicts the decrease of *T. melanosporum* fruiting bodies harvested in the twentieth century. Indeed, it seems that drought could be responsible for the reduction of mycelial growth, as well as the diminution of ascomata formation (Büntgen et al. 2012a, b). Büntgen and colleagues (2012a, b) proposed that *T. melanosporum* would migrate to northern ecosystems to escape the desiccating Mediterranean climate. The authors evoked the possibility that *T. melanosporum* could adapt to these new climatic conditions. Does it mean that *T. melanosporum* will disappear from southern territories? The production of *T. melanosporum* is therefore directly linked to the management of truffle orchards (see above) and adaptation capacities that result from its genetic diversity and reproductive mode.

For many years, *T. melanosporum* was considered a selfing species with a low level of genetic diversity (Bertault et al. 1998). The recent discovery of a heterothallic reproductive mode of *T. melanosporum* (Rubini et al. 2011), and small-scale spatial genetic analyses realised with microsatellites and mating-type (MAT) markers, suggested that *T. melanosporum* invests mainly in sexual reproduction (Murat et al. 2013). Thanks to the sequencing of its genome (Martin et al. 2010), the development of highly polymorphic microsatellite markers provides a better idea of the real genetic diversity of this species. The genotypic diversity index, close to its theoretical maximum of 1, suggested that almost all truffles analysed were genetically different from each other (Murat et al. 2011). A recent investigation of the small-scale genetic structure highlighted the fact that *T. melanosporum* truffle orchards are dynamic ecosystems that can contain up to 13 small genets in 30 m^2 (Murat et al. 2013). Finally, the genome resequencing of six geographic accessions from France, Italy and Spain identified a total of 442,326 single nucleotide polymorphisms (SNPs) corresponding to 3540 SNPs/Mbps, confirming that *T. melanosporum* has a genetic diversity similar to those of other filamentous fungi (Payen et al. 2015). The SNPs were more frequent in repeated sequences

(85 %), but it was possible to identify 4501 SNPs in the coding regions of 2587 genes. This study allowed for the identification of putative genes and genomic regions subjected to positive or negative selection and questioned the adaptation capacities of *T. melanosporum*. Further research is now being conducted at INRA Nancy to specifically address the adaptation of *T. melanosporum* to environmental stresses.

2.2.4 Tuber melanosporum *Biogeography*

Tuber melanosporum biogeography was investigated with different molecular markers such as randomly amplified polymorphism DNA (RAPD), microsatellites, ITS sequencing and inter-simple sequence repeats (ISSR) (Bertault et al. 1998; Murat et al. 2004; Riccioni et al. 2008; García-Cunchillos et al. 2014). These studies highlighted an important genetic structure of *T. melanosporum* in natural populations mainly resulting from the effects of the last glaciation. Two putative postglacial recolonisation routes have been hypothesised for France: one through the west and another through the east (Murat et al. 2004). The existence of glacial refuges was suggested in Italy and Spain (Murat et al. 2004; Riccioni et al. 2008; García-Cunchillos et al. 2014). Recently, Payen and colleagues (2015) estimated that the 60,507 SNPs present in the genomic regions free of selection pressure accumulated between 100,000 and 150,000 years ago. To be sure, this time estimation should be considered with caution; but it still confirms that the last glaciation, which occurred about 120,000 to 11,000 years ago (Van Andel and Tzedakis 1996), has impacted the actual *T. melanosporum* genetic diversity and genetic structure. As a result, the most recent common ancestor (MRCA) of all geographic accession probably occurred before the last glaciation. The sampling was not sufficient to definitively draw a conclusion on *T. melanosporum* history, but it confirms previous conclusions (Murat et al. 2004). This first population genomic analysis highlights the power of genome resequencing to investigate *T. melanosporum* biogeography.

2.3 *Tuber indicum*: A Complex of Cryptic Species?

The Asian black truffle *T. indicum* was first described from a dried sample harvested in January 1892 by Duthie near the Indian city of Mussooree (now Mussoorie), in the northwestern Himalaya at an altitude of about 2000 m AMSL (Cooke and Massee 1892). Much later, other similar specimens have been described and named as different species: *Tuber sinense* (Tao et al. 1989), *Tuber himalayense* (Zhang and Minter 1988), *Tuber pseudohimalayense* (Moreno et al. 1997) and *Tuber formosanum* (Hu 1992). Specimens have also been described in Japan (Kinoshita et al. 2011). The morphological variations supposed to occur between these different specimens, often observed in one or at most a few individuals, could be considered as usual variations within a single species. This was in fact confirmed by molecular analysis, and it is now agreed that these different taxa are synonymous and form a single species, *T. indicum*, with different populations, groups, ecotypes or cryptic species (Zhang et al. 2005; Wang et al. 2006; Chen et al. 2011; Kinoshita et al. 2011). However, Chen et al. (2011) considered that *T. pseudoexcavatum* and *T. pseudohimalayense* are synonymous. This discrepancy is due to taxon sampling or attribution. Samples described as *T. pseudohimalayense* belonged either to *T. pseudoexcavatum* or to *T. indicum*. Based on host plants, geographic distribution and minor morphological differences, Chen et al. (2011) considered that *T. formosanum* is a separate species from *T. indicum*. However, some *Tuber* specimens collected from Japan displayed a close phylogenetic relationship with Chinese *T. indicum* and Taiwanese *T. formosanum* with more than 98 % ITS similarities with both species (Kinoshita et al. 2011), and all of them belong to the Melanosporum group.

In China, *T. indicum* is found mainly in the provinces of Yunnan and Sichuan between 25° and 30° of latitude north. Based on ITS-RFLP or ITS sequences, Roux et al. (1999), Paolocci et al. (1997) and Zhang et al. (2005) distinguished two groups inside the *T. indicum* complex. There were, however, some discrepancies among the three studies. Wang et al. (2006) obtained results congruent with those of Zhang et al. (2005): group I consisted of samples harvested in Huili and Huidong, including those harvested in the South near Chuxiong and Kunming; group II comprised all the samples harvested in Gongshan, Panzhihua, Miyi and Huize. The genetic distances between the Chinese populations based on ITS and β-tubulin sequences confirmed the existence of these two groups. Moreover, there was a significant phylogeographical structure within the Chinese *T. indicum* complex (Wang et al. 2006). The existence of a sinuous limit between the two groups suggested that at least two factors could be involved in their differentiation: a northward migration after the last glaciation and a possible recolonisation from the bottom of the valleys. Chen et al. (2011) also showed that the Chinese *T. indicum* specimens were distributed in two significant clades using the multigene phylogenetic analysis and supposed that they should be at least two cryptic species.

Recently, the mating-type genes of *T. indicum* were described (Belfiori et al. 2013). Similar to *T. melanosporum*, *T. indicum* displays only one mating-type gene per haploid genome, suggesting it is also a heterothallic species. By analysing 115 ascomata imported to Italy from China, Belfiori et al. (2013) found two genetic groups according to ITS sequences called A and B, with the B group being divided into B1 and B2 as in Paolocci et al. (1997).

The sequence and organisation of the mating-type genes and idiomorphs showed significant divergence between *T. indicum* truffles displaying the ITS class A and those displaying classes B1 and B2. This result suggested the presence of at least two cryptic species in *T. indicum* corresponding to *T. indicum*_A and *T. indicum*_B (Belfiori et al. 2013). The use of mating-type genes at a large scale could therefore enhance our understanding of the *T. indicum* complex.

In conclusion, we suggest considering *T. indicum*, *T. himalayense*, *T. sinense*, *T. pseudohimalayense*, *T. formosanum* and the Japanese specimens belonging to the Melanosporum group as forming one morphological species complex, *T. indicum* (Cooke and Massee). Among the Chinese *T. indicum* complex, we suggest the existence of at least two groups, which could be considered as geographical ecotypes or cryptic species. The Taiwanese samples described as *T. formosanum* and the unnamed Japanese samples could form two other ecotypes or cryptic species. It is also probable that the Indian specimens form one or several other ecotypes. Taxonomy and species identification about *T. indicum* complex seems quietly problematic because the holotype *T. indicum*, collected from Indian city of Mussooree one century ago, is too old to DNA extraction. New collection for *Tuber* in this region needs to be done in the near future. Whatever, the Chinese *T. indicum* genome-sequencing project (see Chap. 9) will allow for description of highly polymorphic molecular markers, such as microsatellites that are suitable to analyse population genetics and to identify ecotypes (Molinier et al. 2015).

2.4 *Tuber indicum* and *T. melanosporum*: Friend or Foe?

Phylogenetically, *T. indicum* and *T. melanosporum* are closely related, since they share a common ancestor from a few million years ago (Jeandroz et al. 2008; Bonito et al. 2013). The morphological similarity of *T. indicum* and *T. melanosporum* explains the development of this Chinese species as an export to Europe. However, *T. melanosporum* with ascomata of conspicuous warts, densely spiny ascospores (spine 2–3 μm high), different geological distribution, couple with volatile organic compound (VOC) composition (or more intense and complex aroma) (Culleré et al. 2013) differs from *T. indicum* complex (ascospores with sparse spine 5–7 μm high) and formed a separate clade in the Melanosporum group.

In France, since the 1990s, about 30 t of *T. indicum* were imported each year; according to French customs, data less is imported today. These truffles are preserved in cans, commercialised and sold in supermarkets. *Tuber indicum* can be considered an exotic species and putatively invasive (Murat et al. 2008). Invasive species include organisms that have been introduced, deliberately or not, outside their natural habitats. They can cause significant irreversible environmental and socio-economic impacts on the genetic, species and ecosystem levels (Moore 2000).

Comandini and Pacioni (1997) described the *T. indicum* mycorrhiza with host plant *Quercus pubescens* Willd. ECMs of *T. indicum* with this oak species are monopodial-pinnate, ochraceous-amber and surface smooth with loosely long spiny especially on the very tip. Mantle outer are pseudoparenchymatous and very heterogeneous and inner plectenchymatous. The mycorrhiza of *T. indicum* has very similar morphological character with *T. melanosporum*, presenting both a puzzle-like hypal and spinule-like cystidia sometime with orthogonal branch (Zambonelli et al. 1997). However, the cells in the outer layers of mantle in *T. indicum* are always bigger than those in *T. melanosporum* (Geng et al. 2009).

In addition, the formation of mycorrhizae of *T. indicum* was earlier and more abundant compared with *T. melanosporum* with their host plant (Mabru et al. 2001).

Tuber indicum ECMs were detected in a truffle orchard in Italy (Murat et al. 2008). The same occurred in North America, where *T. indicum* was able to fruit (Bonito et al. 2011). Due to the introduction (intentional or not) of *T. indicum* in *T. melanosporum* habitats, the two species are potentially in contact. It can thus be asked, what ecological consequences might arise? Murat and colleagues (2008) presented different questions regarding the introduction of *T. indicum* in Europe: Are both species able to inbreed? Is *T. indicum* able to spread over long distances? Is *T. indicum* really capable of replacing *T. melanosporum* in truffle grounds? We have no additional information on the last two questions, but the characteristics of mating-type genes for both species highlight the closeness of *T. indicum* and *T. melanosporum*. For both MAT genes, the divergence level between the two *T. indicum* classes (A and B) is similar to that between *T. indicum* and *T. melanosporum* (Belfiori et al. 2013). It is impossible to say with certainty that they are able to inbreed, but the possibility cannot be excluded, especially since interspecies mating has been observed in different fungal taxa, such as *Ophiostoma* (Brasier et al. 1998) or *Heterobasidion* (Chase and Ullrich 1990). It is therefore critical to avoid the introduction of *T. indicum* and to prevent any production of inoculated seedlings with this species outside its natural habitat. To avoid the introduction of *T. indicum* in Europe, new molecular biology high-throughput techniques are now applied to inoculated seedling quality control in France, Italy and Spain (Murat 2015).

2.5 Conclusions

The taxonomic position of *T. melanosporum* is clear, but it is more complex for *T. indicum*. The use of new molecular markers such as MAT genes could therefore be useful to address the complexity of the *T. indicum* clade as suggested by Belfiori et al. (2013). It is also important to inform local agencies that *T. indicum* could represent a risk for European truffles. The progress in *T. melanosporum* cultivation and its recognition by The European Commission as an agricultural product will favour the development of truffle cultivation in Europe. However, scientists have to work closely with the truffle industry to make truffle production more predictable, especially in the context of climate change.

References

Alam M, Olivier A, Paquette A, Dupras J, Revéret JP, Messier C (2014) A general framework for the quantification and valuation of ecosystem services of tree-based intercropping systems. Agrofor Syst 88(4):679–691. doi:10.1007/s10457-014-9681-x

Belfiori B, Riccioni C, Paolocci F, Rubini A (2013) Mating type locus of Chinese black truffles reveals heterothallism and the presence of cryptic species within the *T. indicum* species complex. PLoS One 8(12):e82353. doi:10.1371/journal.pone.0082353

Bertault G, Raymond M, Berthomieu A, Callot G, Fernandez D (1998) Rifling variation in truffles. Nature 394(6695):734. doi:10.1038/29428

Bonito G, Trapp JM, Donovan S, Vilgalys S (2011) The Asian black truffle *Tuber indicum* can form ectomycorrhizas with North American host plants and complete its life cycle in non-native soils. Fungal Ecol 4(1):83–93. doi:10.1016/j.funeco.2010.08.003

Bonito G, Smith ME, Nowak M, Healy RA, Guevara G, Cázares E, Kinoshita A, Nouhra ER, Domínguez LS, Tedersoo L, Murat C, Wang Y, Moreno BA, Pfister DA, Nara K, Zambonelli A, Trappe JM, Vilgalys R (2013) Historical biogeography and diversification of truffles in the *Tuberaceae* and their newly identified southern hemisphere sister lineage. PLos One 81(1):e52765. doi:10.1371/journal.pone.0052765

Brasier CM, Kirk SA, Pipe ND, Buck KW (1998) Rare interspecific hybrids in natural populations of the Dutch elm disease pathogens *Ophiostoma ulmi* and *O. novo-ulmi*. Mycol Res 102 (1):45–57. doi:10.1017/S0953756297004541

Büntgen U, Kauserud H, Egli S (2011) Linking climate variability to mushroom productivity and phenology. Front Ecol Environ 10(1):14–19. doi:10.1890/110064

Büntgen U, Egli S, Camarero JJ, Fischer EM, Stobbe U, Kauserud H, Tegel W, Sproll L, Stenseth NC (2012a) Drought-induced decline in Mediterranean truffle harvest. Nat Clim Change 2:827–829. doi:10.1038/nclimate1733

Büntgen U, Kauserud H, Egli S (2012b) Linking climate variability to mushroom productivity and phenology. Front Ecol Environ 10(1):14–19. doi:10.1890/110064

Ceruti A, Fontana A, Nosenzo C (2003) Le specie europee del genere *Tuber*. Una revision storica. Museo Regionale di Scienze Naturali, Torino

Chase TE, Ullrich RC (1990) Five genes determining intersterility in *Heterobasidion annosum*. Mycologia 82(1):73–81. doi:10.2307/3759965

Chatin AD (1869) La truffe. Bouchard-Huzard, Paris

Chen J, Guo SX, Liu PG (2011) Species recognition and cryptic species in the *Tuber indicum* complex. PLos One 6(1):e14625. doi:10.1371/journal.pone.0014625

Chevalier G, Grente J (1979) Application pratique de la symbiose ectomycorhizienne: production à grande échelle de plants mycorhizés par la truffe. Mushroom Sci 10:483–505

Comandini O, Pacioni G (1997) Mycorrhizae of Asian black truffles, *Tuber himalayense* and *T. indicum*. Mycotaxon 63:77–86

Cooke MC, Massee G (1892) Himalayan truffles. Grevillea 20:67

Culleré L, Ferreira V, Venturini ME, Marco P, Blanco D (2013) Potential aromatic compounds as markers to differentiate between *Tuber melanosporum* and *T. indicum* truffle. Food Chem 141 (1):105–110. doi:10.1016/j.foodchem.2013.03.027

Davis MB, Shaw RG (2001) Range shifts and adaptive responses to quaternary climate change. Science 292(5517):673–679. doi:10.1371/journal.pone.0082353

Etterson JR (2004) Evolutionary potential of *Chamaecrista fasciculata* in relation to climate change. II Genetic architecture of three populations reciprocally planted along an environmental gradient in the great plains. Evolution 58(7):1459–1471. doi:10.1111/j.0014-3820.tb01727.x

García-Cunchillos I, Sänchez S, Barriuso JJ, Pérez-Collazos E (2014) Population genetics of the westernmost distribution of the glaciations-surviving black truffle *Tuber melanosporum*. Mycorrhiza 24(Suppl 1):S89–S100. doi:10.1007/s00572-013-0540-9

Geng LY, Wang XH, Yu FQ, Deng XJ, Tian XF, Shi XF, Xie XD, Liu PG, Shen YY (2009) Mycorrhizal synthesis of *Tuber indicum* with two indigenous hosts, *Castanea mollissima* and *Pinus armandii*. Mycorrhiza 19(7):461–467. doi:10.1007/s00572-009-0247-0

Hamon X, Dupraz C, Liagre F (2009) L'agroforesterie: outils de séquestration du carbone en agroforesterie. Rapport INRA, Agroof, AFAF financé par le CASDAR

Hu H (1992) *Tuber formosanum* sp. Nov and its mycorrhizal associations. J Exp For Natl Taiwan Univ 6:79–86

Jeandroz S, Murat C, Wang Y, Bonfante P, Le Tacon F (2008) Molecular phylogeny and historical biogeography of the genus *Tuber*, the 'true truffles'. J Biogeogr 35(5):815–829. doi:10.1111/j.1365-2699.2007.01851.x

Kauserud H, Heegaard E, Semenov MA, Boddy L, Halvorsen R, Stige LC, Sparks TH, Gange AC, Stenseth NC (2010) Climate change and spring-fruiting fungi. Proc R Soc B 277 (1685):1169–1177. doi:10.1098/rspb.2009.1537

Kinoshita A, Sasaki H, Nara K (2011) Phylogeny and diversity of Japanese truffles (*Tuber* spp.) inferred from sequences of four nuclear loci. Mycologia 103(4):779–794. doi:10.3852/10-138

Le Tacon F, Delmas J, Gleyze R, Bouchard D (1982) Influence du régime hydrique du sol et la fertilisation sur la fructification de la truffe noire du Périgord (*Tuber melanosporum* Vitt.) dans le Sud-Est de la France. Acta Oecol-Oecol Appl 3(4):291–306

Le Tacon F, Zeller B, Plain C, Hossann C, Brechet C, Robin C (2013) Carbon transfer from the host to *Tuber melanosporum* mycorrhizas and ascocarps followed using a ^{13}C pulse-labeling technique. PLoS One 8(5):e64626. doi:10.1371/journal.pone.0064626

Le Tacon F, Marçais B, Courvoisier M, Murat C, Montpied P, Becker M (2014) Climatic variations explain annual fluctuations in French 'Périgord black truffle' wholesale markets but do not explain the decrease in 'black truffle' production over the last 48 years. Mycorrhiza 24(Suppl 1):S115–S125. doi:10.1007/s00572-014-0568-5

Le Tacon F, Zeller B, Plain C, Hossann C, Bréchet C, Martin F, Kohler A, Villerd J, Robin C (2015) Study of nitrogen and carbon transfer from soil organic matter to *Tuber melanosporum* mycorrhizas and ascocarps using ^{15}N and ^{13}C soil labelling and whole-genome oligoarrays. Plant Soil 395(1):351–373. doi:10.1007/s11104-015-2557-7

Lee G (2014) Quality aspects of the Australian black truffle (*T. melanosporum*). In: Abstract of the 1st international conference on truffle research, Vic-Barcelona, 9–12 Mar 2014

Mabru D, Dupre C, Douet JP, Leroy P, Ravel C, Ricard JM, Medina B, Castroviejo M, Chevalier G (2001) Rapid molecular typing method for the reliable detection of Asiatic black truffle (*Tuber indicum*) in commercialized products: fruiting bodies and mycorrhizal seedlings. Mycorrhiza 11(2):89–94. doi:10.1007/s005720100103

Martin F, Kohler A, Murat C, Balestrini R, Coutinho PM, Jaillon O, Montanini B, Morin E, Noel B, Percudani R, Porcel B, Rubini A, Amicucci A, Amselem J, Anthouard V, Arcioni S, Artiguenave F, Aury JM, Ballario P, Bolchi A, Brenna A, Brun A, Buee M, Cantarel B, Chevalier G, Couloux A, Da Silva C, Denoeud F, Duplessis S, Ghignone S, Hilselberger B, Iotti M, Marcais B, Mello A, Miranda M, Pacioni G, Quesneville H, Riccioni C, Ruotolo R, Splivallo R, Stocchi V, Tisserant E, Viscomi AR, Zambonelli A, Zampieri E, Henrissat B, Lebrun MH, Paolocci F, Bonfante P, Ottonello S, Wincker P (2010) Perigord black truffle genome uncovers evolutionary origins and mechanisms of symbiosis. Nature 464 (7291):1033–1038. doi:10.1038/nature08867

Molinier V, Murat C, Peter M, Gollotte A, De la Varga H, Meier B, Egli S, Belfiori B, Paolocci F, Wipf D (2015) SSR-based identification of genetic groups within European populations of *Tuber aestivum* Vittad. Mycorrhiza. doi:10.1007/s00572-015-0649-0

Moore PD (2000) Plant ecology: alien invaders. Nature 403(6769):492–493. doi:10.1038/35000686

Moreno G, Manjòn JL, Diez J, Garcia-Montero LG, Di Massimo G (1997) *Tuber pseudohimalayense* sp. nov. an Asiatic species commercialized in Spain, similar to the "Perigord" truffle. Mycotaxon 63:217–224

Murat C (2015) Forty years of inoculating seedlings with truffle fungi: past and future perspectives. Mycorrhiza 25(1):77–81. doi:10.1007/s00572-014-0593-4

Murat C, Diez J, Luis P, Delaruelle C, Dupré C, Chevalier G, Bonfante P, Martin F (2004) Polymorphism at the ribosomal DNA ITS and its relation to postglacial re-colonization routes of the Perigord truffle *Tuber melanosporum*. New Phytol 164(2):401–411. doi:10.1111/j.1469-8137.2004.01189.x

Murat C, Zampieri E, Vizzini A, Bonfante P (2008) Is the Perigord black truffle threatened by an invasive species? We dreaded it and it has happened! New Phytol 178(4):699–702. doi:10.1111/j.1469-8137.2008.02449.x

Murat C, Riccioni C, Belfiori B, Cichocki N, Labbé J, Morin E, Tisserant E, Paolocci F, Rubini A, Martin F (2011) Distribution and localization of microsatellites in the Perigord black truffle genome and identification of new molecular markers. Fungal Genet Biol 48(6):592–601. doi:10.1016/j.fgb.2010.10.007

Murat C, Rubini A, Riccioni C, De La Varga H, Akroume E, Belfiori B, Guaragno M, Le Tacon F, Robin C, Halkett F, Martin F, Paolocci F (2013) Fine scale spatial genetic structure of the black truffle (*Tuber melanosporum*) investigated with neutral microsatellites and functional mating type genes. New Phytol 199(1):176–187. doi:10.1111/nph.12264

Olivier JM, Savignac JC, Sourzat P (2012) Truffe et trufficulture. Fanlac, Périgueux

Pacioni G, Pomponi G (1989) Chemotaxonomy of some Italian species of *Tuber*. Micol Veg Med 4:63–71

Pacioni G, Bellina-Agostinone C, D'Antonio M (1990) Odour composition of the *Tuber melanosporum* complex. Mycol Res 94(2):201–204. doi:10.1016/S0953-7562(09)80614-X

Paolocci F, Rubini A, Granetti B, Arcioni S (1997) Typing *Tuber melanosporum* and Chinese black truffles by molecular markers. FEMS Microbiol Lett 153(2):255–260. doi:10.1111/j.1574-6968.1997.tb12582.x

Paolocci F, Rubini A, Riccioni C, Arcioni S (2006) Reevaluation of the life cycle of *Tuber magnatum*. Appl Environ Microbiol 72(4):2390–2393. doi:10.1128/AEM.72.4.2390-2393.2006

Payen T, Murat C, Gigant A, Morin E, De Mita S, Martin F (2015) A survey of genome-wide single nucleotide polymorphisms through genome re-sequencing in the Périgord black truffle (*Tuber melanosporum* Vittad.). Mol Ecol Res 15(5):1243–1255. doi:10.1111/1755-0998.12391

Reyna S, Garcia-Barreda S (2014) Black truffle cultivation: a global reality. For Syst 23 (2):317–328. doi:10.5424/fs/2014232-04771

Riccioni C, Belfiori B, Rubini A, Passeri V, Arcioni S, Paolocci F (2008) *Tuber melanosporum* outcrosses: analysis of the genetic diversity within and among its natural populations under this new scenario. New Phytol 180(2):466–478. doi:10.1111/j.1469-8137.2008.02560.x

Riousset G, Riousset L, Chevalier G, Bardet MC (2001) Truffes d'Europe et de Chine. INRA, Paris

Roux C, Sejalon-Delmas N, Martins M, Parguey-Leduc A, Dargent R, Bécard G (1999) Phylogenetic relationships between European and Chinese truffles based on parsimony and distance analysis of ITS sequences. FEMS Microbiol Lett 180(2):147–155. doi:10.1111/j.1574-6968.1999.tb08789.x

Rubini A, Belfiori B, Riccioni C, Tisserant E, Arcioni S, Martin F, Paolocci F (2011) Isolation and characterization of *MAT* genes in the symbiotic ascomycete *Tuber melanosporum*. New Phytol 189(3):710–722. doi:10.1111/j.1469-8137.2010.03492.x

Shigesada N, Kawasaki K (1997) Biological invasions: theory and practice. Oxford University Press, Oxford

Tao K, Liu B, Zhang DC (1989) A new species of the genus *Tuber* from China. J Shanxi Univ 12:215–218

Therville C, Mangenet T, Hinnewinkel C, Guillerme S, de Foresta H (2013) Is truffle growing a response to sustainable development and heritage issues in Mediterranean territories? The case of Uzès, southern France. For Trees Livelihoods 22(4):257–274. doi:10.1080/14728028.2013.859461

Van Andel TH, Tzedakis PC (1996) Palaeolithic landscapes of Europe and environs, 150,000–25,000 years ago: an overview. Q Sci Rev 15(5):481–500. doi:10.1016/0277-3791(96)00028-5

Vittadini C (1831) Monographia Tuberacearum. Ex Tipographia Felicis Rusconi, Milano

Wang YJ, Tan ZM, Zhang DC, Murat C, Jeandroz S, Le Tacon F (2006) Phylogenetic and populational study of the *Tuber indicum* complex. Mycol Res 110(9):1034–1045. doi:10.1016/j.mycres.2006.06.013

Wang Y, Pegui L, Chen J, Hung-Tao H (2007) China – a new emerging truffle-producing nation. In: Reynal B (ed) Actes du colloques la culture de la truffe dans le monde. Imprimerie Georges Lachaise, Brive-La-Gaillarde, pp 35–44

Wedén C, Chevalier G, Backlund A (2013) The first production results of *Tuber melanosporum* in Sweden. In: Abstracts of the 1st international congress of trufficulture, Tuber 2013, Teruel, 5–8 Mar 2013

Zambonelli A, Tibiletti E, Pisi A (1997) Caratterizzazione anatomo-morfologica delle micorrize di *Tuber indicum* Cooke and Massee su *Pinus pinea* L. e *Quercus cerris* L. Micol Ital 26(1):29–36

Zhang BC, Minter DW (1988) *Tuber himalayense* sp. nov. with notes on Himalayan truffles. Trans Br Mycol Soc 91(4):593–597. doi:10.1016/S0007-1536(88)80032-9

Zhang L, Yang ZL, Song DS (2005) A phylogenetic study of commercial Chinese truffles and their allies: taxonomic implications. FEMS Microbiol Lett 245(1):85–92. doi:10.1016/j.femsle.2005.02.028

Chapter 3
The Burgundy Truffle (*Tuber aestivum* syn. *uncinatum*): A Truffle Species with a Wide Habitat Range over Europe

Virginie Molinier, Martina Peter, Ulrich Stobbe, and Simon Egli

3.1 Introduction

Among the at least 180 existing truffle species, *Tuber aestivum* Vittad. (syn. *Tuber uncinatum* Chatin) is one of the most sought after (Bonito et al. 2010). With prices from 200 to 600 € per kg, it is commercially attractive for traders and consumers worldwide. The early season fruiting bodies are called summer truffles, whereas chefs and gourmets make a distinction for the late season fruiting bodies, which are called Burgundy truffles and are of higher quality and value. The confusion around the two varieties and their scientific nomenclature is subject to ongoing discussion and will be clarified in the following chapter. Unlike the even pricier truffle species *Tuber melanosporum* Vittad. and *Tuber magnatum* Pico, which are restricted to Mediterranean regions, *T. aestivum* is distributed all over Europe and occurs in habitats over broad ecological amplitude (Stobbe et al. 2013a). The wide range of possible soils, climates, and host plants, together with its market value and its long harvest season, makes this species interesting for cultivation. Today, Burgundy truffles are harvested on plantations spanning Europe from Italy in the south to Sweden in the north and from Spain in the west to Hungary in the east (De Roman and Boa 2004; Gogan Csorbaine et al. 2007; Hall et al. 2007; Bencivenga et al. 2009; Weden et al. 2009). Particularly in Central and Eastern European countries, until recently excluded from traditional truffle production with *T. melanosporum*, newly emerging cultivation endeavors appear promising (Stobbe et al. 2013a).

V. Molinier (✉) • M. Peter • S. Egli
Swiss Federal Institute for Forest, Snow and Landscape Research (WSL), 8903 Birmensdorf, Switzerland
e-mail: molinier.virginie@gmail.com

U. Stobbe
Deutsche Trüffelbäume, Karl-Bücheler Str. 1, 78315 Radolfzell, Germany

A. Zambonelli et al. (eds.), *True Truffle (*Tuber *spp.) in the World*, Soil Biology 47,
DOI 10.1007/978-3-319-31436-5_3

In this chapter, taxonomic status of *T. aestivum* is presented, morphological characteristics of this species are described, and a summary of geographical distribution including ecological requirements and genetic structure is finally provided.

3.2 Taxonomic Status of *T. aestivum*

3.2.1 First Descriptions of T. aestivum *and the Beginning of Taxonomic Controversy*

In 1831, Carlo Vittadini, a renowned Italian mycologist, described in detail a new *Tuber* species: *T. aestivum*, the summer truffle (Vittadini 1831). This species had already been mentioned in two prior documents: in 1729 by Micheli in "*Nova Plantarum* Genera" and in 1801 by Persoon in "*Synopsis Methodica Fungorum*" (Micheli 1729; Persoon 1801). Although both botanists named the species "*Tuber aestivum*," Vittadini kept the denomination but added his own name and presented himself as the first detailed descriptor. Therefore, "Vittad." became the official authority for the species. The name "*Tuber aestivum*" stems from observations about the maturity period of ascomata (*aestivum* means summer in Latin). This *Tuber* species has a light brown gleba with a black peridium and, according to Vittadini, was present throughout Europe. In 1869, the French botanist Adolphe Chatin mentioned *T. aestivum* in his book "La Truffe" (Chatin 1869). He provided information about the maturity period and indicated that maturity was reached from May-July in the south of France, around August in the north of France, and even during winter near Paris in Charenton and Nogent (Chatin 1869).

A few years later, the same author described a new *Tuber* species, *T. uncinatum* (Chatin 1887). Although very similar to *T. aestivum* in appearance, Chatin considered *T. uncinatum* a new species. From a morphological point of view, fully ripe *T. uncinatum* fruiting bodies have a darker gleba than *T. aestivum* and feature hooks in the spore reticulum (uncinatum means hooked in Latin; Chatin 1887). The presence of these hooks was later determined to be an artifact created by the flexible walls of the spore reticulum bending slightly at the top (Chevalier and Frochot 1997; Fischer 1897).

Since Chatin's work, different opinions have been voiced about whether *T. aestivum* and *T. uncinatum* are two different species, varieties, or merely morphotypes. This discussion was and still is particularly important because of the high commercial value of this precious edible fungus. To differentiate the two taxa, additional morphological criteria have been described, such as the peridium (see Fig. 3.1c). *Tuber uncinatum* has been described as having smaller warts than *T. aestivum* and as having nonstriated warts (Chatin 1887; Riousset et al. 2001). Another important feature is the height of the spore reticulum, which was not mentioned by Vittadini in his first description. Spore reticulum with a 2 μm height is associated with *T. aestivum*, whereas *T. uncinatum* has a 4 μm spore reticulum

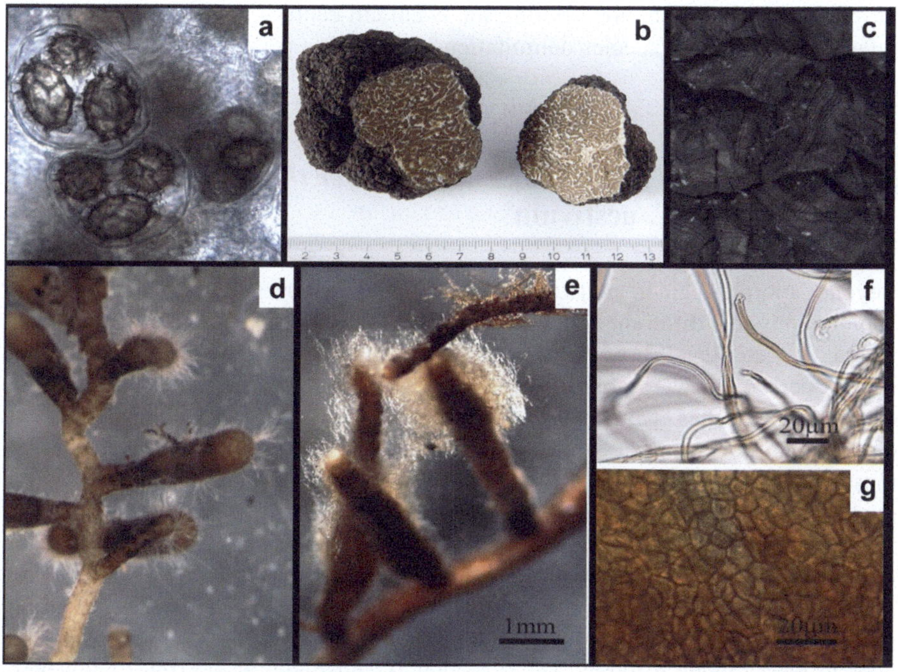

Fig. 3.1 Morphological features of *T. aestivum*. Ascospores of *T. aestivum* (**a**); examples of two *T. aestivum* fruiting bodies (**b**); peridium of *T. aestivum* with *black/gray* striations (**c**); *T. aestivum* mycorrhizas on hazelnut (*C. avellana*) (**d**) and on Norway spruce (*P. abies*) (**e**); *curly, red-brown* cystidia of *T. aestivum* mycorrhiza (**f**); mantle structure with typical angular cells (**g**)

height (Chevalier and Frochot 1997; Riousset et al. 2001). Another study proposed a spore reticulum height limit as a distinguishing feature, with *T. aestivum* smaller than 4 μm and *T. uncinatum* taller than this height threshold (Mello et al. 2002). However, both methods proved to be unreliable and even inappropriate for diagnosis (Mello et al. 2002; Paolocci et al. 2004; Weden et al. 2005). For instance, Weden et al. (2005) found a continuum in the spore reticulum height and this feature was not suitable for discriminating between the two taxa.

Morphological characteristics of other fungal tissues, such as mycorrhizas and in vitro mycelium, appeared to be quite similar (Riousset et al. 2001; Chevalier and Frochot 2002). Besides morphological characteristics, some authors proposed that major differences between the two taxa were their maturity period and their ecological preferences. Botanists determined that *T. uncinatum* is located in the North of France (including the Burgundy region) with a maturity period during autumn, and *T. aestivum* has a Mediterranean distribution with a maturity period during summer (Chevalier et al. 1979; Chevalier and Frochot 1997; Riousset et al. 2001). These first descriptions, as well as the discussion over Chatin's theories, resulted in years of ambiguity and confusion. However, the development of molecular tools in biology during the twentieth century has allowed botanists/scientists to approach species delimitations from a new perspective.

3.2.2 *The Use of Molecular Tools to Infer the Taxonomic Status of* T. aestivum

Beyond morphological and ecological characteristics, diverse studies have explored possible differences between *T. aestivum* and *T. uncinatum* at a biochemical and genetic level. Investigators started by analyzing protein profiles. The first studies performed were able to discriminate the two taxa based on total protein extraction (Mouches et al. 1981; Dupré et al. 1985). Later, several authors could not discriminate the two taxa using the isoenzyme technique or have argued that, at most, *T. uncinatum* is a variety of *T. aestivum* (Pacioni and Pomponi 1991; Pacioni et al. 1993; Gandeboeuf et al. 1994; Dupré 1997; Urbanelli et al. 1998).

From a genetic point of view, one of the first studies using restriction fragment length polymorphism (RFLP) of the internal transcribed spacer (ITS) region could not show differentiation between *T. aestivum* and *T. uncinatum* (Henrion et al. 1994). Later, a random amplified polymorphic DNA (RAPD) approach was used to discriminate different *Tuber* species (Gandeboeuf et al. 1997). However, concerning *T. aestivum* and *T. uncinatum*, no separation could be substantiated. In 2002, Mello and collaborators used the inter-simple sequence repeat (ISSR) method as well as sequencing of the ITS region and discriminated *T. aestivum* from *T. uncinatum* with both methods (Mello et al. 2002). In 2004, Paolocci and collaborators studied the ITS region, as well as sequences of the beta-tubulin and the elongation factor alpha genes, and concluded that *T. aestivum* and *T. uncinatum* belong to a single species (Paolocci ct al. 2004). The samples analyzed by Mello et al. (2002) and Paolocci et al. (2004) originated mostly from Italy. Therefore, it seems unlikely that the whole genetic diversity of the two forms across their geographic range from Sweden to Spain through central Europe was addressed in these studies. In 2005, Weden and collaborators again used the ITS region but studied samples from a broader geographical range to analyze possible correlations between the spore reticulum height and the ITS diversity. In accordance with Paolocci and collaborators (2004), they concluded that the phylogenies of *T. aestivum* and *T. uncinatum* are intermingled. Since these publications, the conspecificity of the two taxa has been accepted for the most part.

The contrasting conclusions obtained in the genetic studies may have been due to which and how many genetic markers and phylogenetic methods were used, but differences in the geographic origin of the samples may also have played a role. Obviously, a limited geographic sampling area can influence results regarding genetic homogeneity, and the use of a small number of genetic markers may not have provided sufficient resolution for these closely related taxa. Moreover, samples were classified as "*T. aestivum*" or "*T. uncinatum*" by different persons, and this point also could have influenced the results. However, combining multigene phylogenies and coalescent analyses of nine regions from five genes, with samples from a broad geographical range across Europe Molinier and collaborators (2013b), supported the conspecificity of *T. aestivum* and *T. uncinatum*. Recently, the development of polymorphic microsatellites (see below) enabled investigation of the gene flow between *T. aestivum* and *T. uncinatum* morphotypes (Molinier et al. 2015a). The existence of samples, described as *T. aestivum* and *T. uncinatum*, belonging to the same multilocus genotype, and the evidence of gene flow between *T. aestivum* and *T. uncinatum* samples excluded their distinction as different species.

Therefore, after a long debate spanning centuries, it is scientifically correct that there is no reason to maintain *T. uncinatum* as a separate taxonomic name. Following the rules of botanical nomenclature (Melbourne 2011), the name *T. aestivum* has priority because it was officially described by Vittadini in 1831, long before *T. uncinatum* was described (Chatin 1887).

Regarding the commercial point of view, truffle traders try hard to maintain the historical species distinction because chefs and gourmets widely accept *T. uncinatum* to be richer in taste than *T. aestivum* and are therefore willing to pay more for it. Moreover, some nurseries make a distinction between seedlings inoculated with *T. aestivum* or *T. uncinatum*. In our opinion, this point contributes to the maintenance of a myth that has no scientific eligibility. In addition, there is no indication about how such nurseries actually discriminate between the two types. Instead of erroneously distinguishing between *T. aestivum* and *T. uncinatum*, it might be useful to indicate the geographical origin of the fungal material used for seedling inoculation. This would enable buyers to know if the inoculated truffle strain is naturally adapted to the precise environmental conditions of their future plantation site.

3.3 Morphological Features of *T. aestivum*

3.3.1 Ascomata

Ascomata are quite variable in size, with a weight from less than 1 g up to more than 1 kg. A typical *T. aestivum* ascoma with ripe gleba is shown in Fig. 3.1b. The shape can be spherical or irregularly lobed, especially when growing in stony ground. The peridium is composed of big, hard, black warts, pyramidal or irregularly polygonal at the base. The structure of the peridium is very variable, especially in the size of the warts. Parts of the ascomata emerging from the soil often have smaller warts than parts under the soil surface. The gleba is firm, fleshy, and marbled, with thin, meandriform veins. Its color can vary from white-ochre to dark-brown, depending on the maturity stage.

The asci contain 1–6 spores (usually 2–4). The spores are brown, ellipsoid to subglobose, about 25–30 × 18–22 μm in size, and reticulate-alveolate and contain irregular, polygonal meshes up to 3–5 μm in height (Fig. 3.1a; Montecchi and Sarasini 2000).

In the center of each verruca, which is often broken up irregularly, brown hyphae emanate from the peridium. They are structurally very similar to the cystidia emanating from the mycorrhizas. The peridium contains distinct striations, as described previously by Callot (1999) (Fig. 3.1c), who hypothesized that these structures are pseudocrystalline deposits rich in calcium. Furthermore, it seems that these structures are synchronized across the individual warts and possibly also among individual fruit bodies. Therefore, they may additionally reflect temporal changes in the growth environment (Büntgen and Egli 2014).

3.3.2 Mycorrhizas

The mycorrhizas of *T. aestivum* show a characteristic woolly surface, due to the presence of many curled, interwoven cystidia (Figs. 3.1d, e). In contrast to vegetative hyphae, cystidia are never ramified, lack septae, and can contain vesicles (Zambonelli et al. 1993). The tips of the cystidia can show irregularly shaped knobs (Fig. 3.1f). Very young mycorrhizas are yellowish-brown to ochre in color and are often just starting to develop cystidia. Older mycorrhizas turn dark-brown and can start to lose their cystidia. The mantle is pseudoparenchymatous with angular cells in the outer and inner layers (Fig. 3.1g). Rhizomorphs are lacking. A detailed morphological and anatomical description was given by Müller et al. (1996).

When comparing *T. aestivum* ectomycorrhizas (ECMs) from different host species (*Quercus pubescens* Willd., Zambonelli et al. 1993; *Pinus pinea* L., Zambonelli et al. 1995; *Corylus avellana* L. and *Carpinus betulus* L., Müller et al. 1996; *Picea abies* L., Stobbe et al. 2013b), the morphological and anatomical features are similar but the shape can vary [e.g., *T. aestivum* on hazelnut (*C. avellana*) and on Norway spruce (*P. abies*); Figs. 3.1d, e, respectively]. Slight differences in mantle thickness as well as in the length and diameter of hyphae are ascribed to natural variability, whereas the presence or absence of cells in a rosette-like arrangement in the inner mantle layers might be a host-specific feature (Müller et al. 1996).

Surprisingly, in 2004, *T. aestivum* was found in orchid roots forming typical endomycorrhizas (Selosse et al. 2004). Such a finding indicates a much broader ability of *T. aestivum* to adapt to varying ecological conditions than previously believed.

3.3.3 Mycelium In Vitro

Isolation of *Tuber* species, including *T. aestivum*, into pure cultures from mycorrhizas or ascomata is feasible but prone to bacterial contamination. Bacterial colonization is often found in truffle ascomata (Mello et al. 2010; see Chap. 18) where they seem to be involved in the maturation process (Antony-Babu et al. 2014). In addition, the propagation of subcultures after isolation is rarely successful, possibly due to difficulties in building up new hyphal contact points on the nutrition medium (Iotti et al. 2002).

Mycelial colonies have a white to brown color, often forming radial growth rings of varying hyphal density. Mycelial radial growth of colonies on Petri dishes is very slow compared to other mycorrhizal fungi. Observations indicated growth rates of 1–2 mm per week on modified Melin-Norkrans (MMN; Marx 1969) and on a *Cenococcum* medium (Trappe 1962). Iotti and colleagues (2002) measured growth rates of 4 mm per week for *T. aestivum* on a modified woody plant medium, which was higher than that for *T. melanosporum* and *Tuber brumale* Vittad. (1.1–1.9 mm per week) but lower than that for *Tuber rufum* Pico (14 mm per week). The hyphae are hyaline to light-yellow-brown with a diameter of around 4 μm. They are septated, granulated, and simple branched and form anastomoses (Iotti et al. 2002). In natural soils, *T. aestivum* seems to produce relatively dense soil mycelium compared with other ECM fungi (Gryndler et al. 2013).

3.4 Geographical Distribution of *T. aestivum*

3.4.1 The Widespread Distribution of T. aestivum

The natural distribution of *T. aestivum* reaches from northern Africa to southern Sweden and from Portugal to the Caucasian region, covering a wide range of suitable habitats throughout the Eurasian Continent (Stobbe et al. 2012, 2013a). The geographical center of the species distribution, with many well-explored truffle regions, can be defined as the temperate regions of Central Europe (Stobbe et al. 2012, 2013a) (Fig. 3.2). The southernmost sites are located in the Atlas Mountains of Morocco (Khabar 2010), and the northernmost populations are located on the Swedish Island of Gotland (Weden et al. 2004a). Descriptions of habitats east of Europe are sparse, with known occurrences in Turkey (Riccioni et al. 2014), Azerbaijan (Bagi and Fekete 2010). *Tuber aestivum* was reported in China (Wang and Liu 2009), but Zambonelli et al. (2012) reported it as a separate sister species that forms a separate clade with respect to the European *T. aestivum* (sensu stricto). Recently, Zhang et al. (2012) named this Chinese species *Tuber sinoaestivum* J. P. Zhang and P. G. Liu. It seems likely that a broad variety of habitats awaits discovery in these regions.

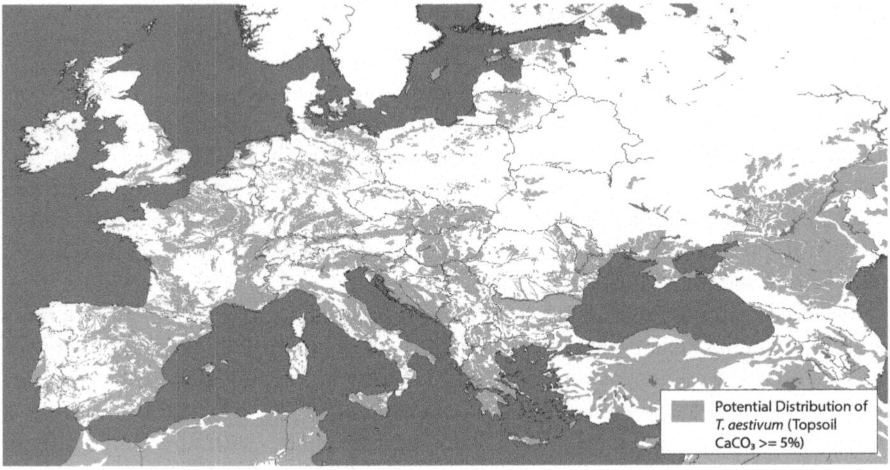

Fig. 3.2 Map of the potential distribution of *Tuber aestivum* in Europe. The *gray* areas indicate potential habitats of *T. aestivum* based on the occurrence of calcium carbonate concentrations greater than 5 % in the topsoil (FAO/IIASA/ISRIC/ISSCAS/JRC, 2012. *Harmonized World Soil Database* (version 1.2); FAO, Rome, Italy and IIASA, Laxenburg, Austria)

3.4.2 Ecological Requirements of T. aestivum

The wide habitat range of *T. aestivum* consists of a specific composition of unique ecological characteristics (Fig. 3.3). Habitats reach high altitudes (Morocco, ~1600 m AMSL) in southern regions with a warm climate, whereas northern habitats are often located near sea level (Sweden, <50 m). The fruiting season has two maxima (July and November) in the temperate habitats of Germany and Switzerland, although observations have been reported throughout the entire year (Stobbe et al. 2012). An earlier fruiting season has been observed in warmer regions and at lower altitudes, e.g., from May to July in Greece (Diamandis and Perlerou 2008), and a shift to later in the year occurs in northern habitats and at higher altitudes, e.g., from August to November in Sweden (Weden et al. 2004a, b). Annual mean temperature varies from 6.8 to 11.5 °C, which is typical for Mediterranean to temperate climates. Of particular importance are the mean temperatures of the coldest and warmest month, which may account for a stop in truffle production due to frost or a decrease of production due to heat-induced drought. The mean temperature for the coldest month is generally >0 °C; significantly lower air temperatures are usually buffered by snow cover, with the corresponding belowground temperatures reaching less distinct depressions (Stobbe et al. 2012). As *T. aestivum* is widespread throughout Europe and tolerant of different climate conditions, and it has an extremely large time range during which fruiting bodies can be harvested. Up to now, no scientific results have shown any qualitative or aromatic differences between truffles harvested in summer and truffles harvested in winter.

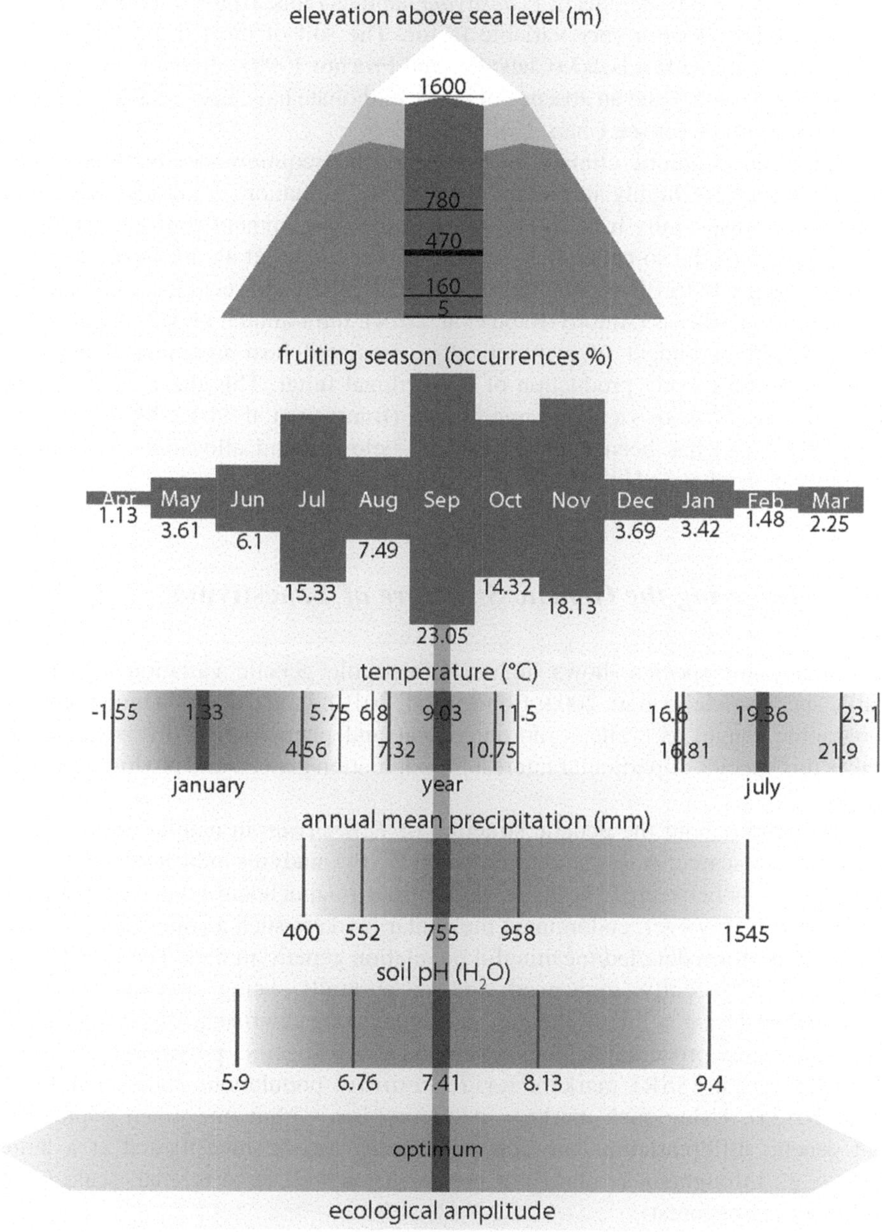

Fig. 3.3 The ecological amplitude of *T. aestivum* including all relevant biotic and abiotic factors. The *thick line* in the *middle* of each partial graphic indicates the factor's optimum, followed outward by the standard deviation and the minimum and maximum, respectively (from Stobbe et al. 2013a)

Annual precipitation totals in *T. aestivum* habitats range from ~400 to 1500 mm, making rainfall alone a very variable factor. The soil of most truffle habitats is calcareous, with pH levels >7 (Chevalier and Frochot 1990), although exceptional sites with a pH of 5.9 and an absence of active carbonate have been reported (Gogan Csorbaine et al. 2012; see Chap. 13).

In addition to abiotic climatic controls and host requirements, biotic environmental factors are highly important for ascoma formation. Soilborne microbial communities, especially bacteria, involved in the development and functioning of ECM symbiosis, the so-called helper bacteria (Frey-Klett et al. 2007), are hypothesized to facilitate truffle production (Mello et al. 2010) and seem to be involved in the maturation process (Antony-Babu et al. 2014). Intra-annual growth patterns and the total carbon budget of symbiotically associated host trees are thought to influence fruiting body production of mycorrhizal fungi. This theory is based on correlation analyses of yield and tree growth (Büntgen et al. 2012; Martínez-Peña et al. 2012) and has been demonstrated by belowground allocation analyses of recent photosynthates (Högberg et al. 2001, 2008).

3.4.3 *Inferring the Genetic Structure of* T. aestivum

In general, this species shows higher intraspecific genetic variation than other truffle species (Mello et al. 2002; Bonito et al. 2010). It also shows a much broader geographic range, as well as morphological and phenological diversity, which makes this species of particular interest for population genetic and phylogeographic studies.

Information about the genetic structure of *T. aestivum* in natural or cultivated truffle sites is scarce. A few studies based on RAPD analyses indicated that genetic diversity exists between *T. aestivum* populations (Gandeboeuf et al. 1997; Weden et al. 2004b). However, codominant molecular markers such as microsatellites are needed to perform detailed/meaningful population genetic studies. The *T. aestivum* genome was recently sequenced and is currently being annotated (Payen et al. 2014). Using a direct shotgun pyrosequencing approach, the first steps of the sequencing process enabled development of specific polymorphic simple sequence repeat (SSR) markers for *T. aestivum* population studies (Molinier et al. 2013a). Using such markers, polymorphism within one given population and genetic differentiation between populations can be investigated at a large scale (e.g., throughout a country or continent) as well as at a small scale (e.g., within a field or forest).

At a large scale, Weden and collaborators (2004b) showed genetic differentiation between samples coming from a population in Gotland (Sweden) and samples coming from France, England, and Italy. Splivallo and collaborators (2012) studied the genetic diversity of *T. aestivum* from nine European populations using amplified fragment length polymorphism (AFLP). They focused on the comparison between genetic and volatile fingerprinting and found that samples coming from the same population tended to cluster together. However, their low number of samples did not lead to a clear conclusion about the genetic structure of *T. aestivum* populations at a European scale. A recent study of 230 ascomata coming from 17 European countries genotyped by SSRs identified four genetic groups (Molinier et al. 2015a). This investigation did not show a clear geographical separation, although one genotype was present exclusively in southeastern regions. This study additionally highlighted that *T. aestivum* is not an endangered species, even in the Mediterranean region, due to the absence of a recent bottleneck. Finally, the existence of ecotypes, i.e., populations adapted to their specific environment, was proposed by Le Tacon (Le Tacon 2011).

At a small scale, even less information is available about the genetic structure of *T. aestivum* in particular locations and over time. The abovementioned study by Splivallo and colleagues (2012) showed that polymorphism was present within a small sampling area, but the study did not focus on the spatial genetic structure but rather on the relationship between the volatile content and genetics. Maximum genet size was recently estimated at 92 m (Molinier et al. 2015b), which is 30–40 times larger than that of *T. melanosporum* (Murat et al. 2013). *Tuber aestivum* therefore has a greater capacity to spread belowground than *T. melanosporum*. As with *T. melanosporum*, a substantial turnover of *T. aestivum* genets was observed over two seasons. Thus, the two species behave similarly in this respect (Molinier et al. 2015b). Finally, both studies highlighted that the genetic background has a pronounced influence on the concentration of C8 and C4 volatiles and on truffle aroma (Splivallo et al. 2012; Molinier et al. 2015b). A very important finding of the ongoing genome sequencing is that *T. aestivum*, like *T. melanosporum* (Martin et al. 2010), is a heterothallic, outcrossing species with a single mating-type gene with two idiomorphs (Murat, pers comm). Finally, small-scale genetic structure analyses in natural and artificial populations, at the level of individuals as well as at the level of sexual mating types, are imperative for improving the understanding of the biology and life history of this commercially important truffle species.

3.5 Conclusions

T. aestivum shows a large distribution area throughout Europe and has a wide amplitude of ecological requirements. It produces ascomata that vary in morphology and matures over a long period of the year. As with *T. melanosporum* and *T. magnatum*, *T. aestivum* is of high economic interest because of its organoleptic properties. Due to its ecological, phenological, and morphological polymorphisms, it was previously divided in two distinct taxa.

Although scientifically proven to be only one species, many producers, traders, and consumers still tend to maintain a differentiation for commercial reasons and continue to name it either "summer truffle" (*T. aestivum*) or "Burgundy truffle"/ "tartufo nero di Fragno" (*T. uncinatum*) according to the geographical region and the harvesting season.

Future studies in population genetics and genomics will be helpful for showing the existence of ecotypes within this species. The establishment of a label for a controlled origin regarding ecotypes could help producers of truffle trees offer high-quality products.

Acknowledgments The authors were supported by the WSL-internal DITREC project and, the Ernst Göhner Stiftung and through the Swiss State Secretariat for Education, Research and Innovation (SERI), the COST Action FP1203. The authors thank Barbara Meier, Ulf Büntgen, and Willy Tegel for their technical help and for their discussions. Authors are grateful for Melissa Dawes for English language editing.

References

Antony-Babu S, Deveau A, Van Nostrand JD, Zhou J, Le Tacon F, Robin C, Frey-Klett P, Uroz S (2014) Black truffle-associated bacterial communities during the development and maturation of *Tuber melanosporum* ascocarps and putative functional roles. Environ Microbiol 16 (9):2831–2847. doi:10.1111/1462-2920.12294

Bagi I, Fekete O (2010) Identification of *Tuber aestivum* habitats in the South Caucasus, Azerbaidjan. In: Communication at the 2nd congress of the *Tuber aestivum/uncinatum*, European Scientific Group, Juva, 20–22 Aug 2010

Bencivenga M, Di Massimo G, Donnini D, Baciarelli Falini L (2009) The cultivation of truffles in Italy. Acta Bot Yunnanica 31(Suppl 16):21–28

Bonito GM, Gryganskyi AP, Trappe JM, Vilgalys R (2010) A global meta-analysis of *Tuber* ITS rDNA sequences: species diversity, host associations and long-distance dispersal. Mol Ecol 19 (22):4994–5008. doi:10.1111/j.1365-294X.2010.04855.x

Büntgen U, Egli S (2014) Breaking new ground at the interface of dendroecology and mycology. Trends Plant Sci 19(10):613–614. doi:10.1016/j.tplants.2014.07.001

Büntgen U, Egli S, Camarero JJ, Fischer EM, Stobbe U, Kauserud H, Tegel W, Sproll L, Stenseth NC (2012) Drought-induced decline in Mediterranean truffle harvest. Nat Clim Change 2 (12):827–829. doi:10.1038/nclimate1733

Callot G (1999) La truffe, la terre, la vie. INRA, Paris

Chatin A (1869) La truffe, 1ère édition. Bouchard-Huzard, Paris

Chatin A (1887) Une nouvelle espèce de truffe. CR Acad Sci 104:1132–1135

Chevalier G, Frochot H (1990) Ecology and possibility of culture in Europe of the Burgundy truffle (*Tuber uncinatum* Chatin). Agr Ecosyst Environ 28(1–4):71–73. doi:10.1016/0167-8809(90) 90016-7

Chevalier G, Frochot H (1997) La truffe de Bourgogne. Pétrarque, Levallois-Perret

Chevalier G, Frochot H (2002) La truffe de Bourgogne. Pétrarque, Levallois-Perret

Chevalier G, Desmas C, Frochot H, Riousset L (1979) L'espèce *Tuber aestivum* Vitt.: I. Définition. Mushroom Sci 10(Part 1):957–975

De Roman M, Boa E (2004) Collection, marketing and cultivation of edible fungi in Spain. Micol Apl Int 16(2):25–33

Diamandis S, Perlerou C (2008) Recent records of hypogeous fungi in Greece. Acta Mycol 43 (2):139–142. doi:10.5586/am.2008.017

Dupré C (1997) De nouveaux outils au service de la trufficulture. Mémoire en vue de l'obtention du diplôme d'ingénieur délivré par l'Etat, ENESAD

Dupré C, Chevalier G, Branlard G (1985) Caractérisation des *Tuber* par électrophorèse de leurs protéines. In: C.R. 1er Colloque national sur les technologies de purification des protéines. DPIC-INPL, Paris, pp 465–467

Fischer E (1897) Ascomyceten: Tuberaceen und Hemiasceen. In: Rabenhorst L (ed) Kryptogamen-flora von Deutschland, Österreich und der Schweiz. Kummer, Leipzig

Frey-Klett P, Garbaye J, Tarkka M (2007) The mycorrhiza helper bacteria revisited. New Phytol 176(1):22–36. doi:10.1111/j.1469-8137.2007.02191.x

Gandeboeuf D, Dupré C, Chevalier G (1994) Use of isoenzyme analysis to differentiate truffles from Europe. Acta Bot Gallica 141(4):455–463

Gandeboeuf D, Dupre C, Roeckel-Drevet P, Nicolas P, Chevalier G (1997) Grouping and identification of *Tuber* species using RAPD markers. Can J Bot 75(1):36–45. doi:10.1139/b97-005

Gogan Csorbaine A, Bratek Z, Dimenyi J (2007) A new tool for rural development: truffle cultivation. Cereal Res Commun 35(2):413–416

Gogan Csorbaine A, Nagy Z, Degi Z, Bagi I, Dimenyi J (2012) Ecological characteristics of a Hungarian summer truffle (*Tuber aestivum* Vittad.) producing area. Acta Mycol 47 (2):133–138. doi:10.5586/am.2012.015

Gryndler M, Trilčová J, Hršelová H, Streiblová E, Gryndlerová H, Jansa J (2013) *Tuber aestivum* Vittad. mycelium quantified: advantages and limitations of a qPCR approach. Mycorrhiza 23 (5):341–348. doi:10.1007/s00572-012-0475-6

Hall IR, Brown GT, Zambonelli A (2007) Taming the truffle, the history, lore and science of the ultimate mushroom. Timber Press, Portland

Henrion B, Chevalier G, Martin F (1994) Typing truffle species by PCR amplification of the ribosomal DNA spacers. Mycol Res 98(1):37–43. doi:10.1016/S0953-7562(09)80333-X

Högberg P, Nordgren A, Buchmann N, Taylor AFS, Ekblad A, Högberg MN, Nyberg G, Ottosson-Lofvenius M, Read DJ (2001) Large-scale forest girdling shows that current photosynthesis drives soil respiration. Nature 411(6839):789–792. doi:10.1038/35081058

Högberg P, Högberg MN, Gottlicher SG, Betson NR, Keel SG, Metcalfe DB, Campbell C, Schindlbacher A, Hurry V, Lundmark T, Linder S, Nasholm T (2008) High temporal resolution tracing of photosynthate carbon from the tree canopy to forest soil microorganisms. New Phytol 177(1):220–228. doi:10.1111/j.1469-8137.2007.02238.x

Iotti M, Amicucci A, Stocchi V, Zambonelli A (2002) Morphological and molecular characterization of mycelia of some *Tuber* species in pure culture. New Phytol 155(3):499–505. doi:10.1046/j.1469-8137.2002.00486.x

Khabar L (2010) Contribution to the study of the ecology of *Tuber aestivum* in Morocco: phytosociological and geological aspects. Österr Z Pilzkunde 19:261–264

Le Tacon F (2011) Point de vue sur *Tuber aestivum* Vittad. Le Trufficulteur 74:13–17

Martin F, Kohler A, Murat C, Balestrini R, Coutinho PM, Jaillon O, Montanini B, Morin E, Noel B, Percudani R, Porcel B, Rubini A, Amicucci A, Amselem J, Anthouard V, Arcioni S, Artiguenave F, Aury JM, Ballario P, Bolchi A, Brenna A, Brun A, Buee M, Cantarel B, Chevalier G, Couloux A, Da Silva C, Denoeud F, Duplessis S, Ghignone S, Hilselberger B, Iotti M, Marcais B, Mello A, Miranda M, Pacioni G, Quesneville H, Riccioni C, Ruotolo R, Splivallo R, Stocchi V, Tisserant E, Viscomi AR, Zambonelli A, Zampieri E, Henrissat B, Lebrun MH, Paolocci F, Bonfante P, Ottonello S, Wincker P (2010) Perigord black truffle genome uncovers evolutionary origins and mechanisms of symbiosis. Nature 464 (7291):1033–1038. doi:10.1038/nature08867

Martínez-Peña F, de Miguel S, Pukkala T, Bonet JA, Ortega-Martínez P, Aldea J, Martínez de Aragón J (2012) Yield models for ectomycorrhizal mushrooms in Pinus sylvestris forests with special focus on *Boletus edulis* and *Lactarius* group deliciosus. For Ecol Manage 282:63–69. doi:10.1016/j.foreco.2012.06.034

Marx D (1969) The influence of ectotrophic mycorrhizal fungi on the resistance of pine roots to pathogenic infection. I Antagonism of mycorrhizal fungi to root pathogenic fungi and soil bacteria. Phytopathology 59:153–163

Mello A, Cantisani A, Vizzini A, Bonfante P (2002) Genetic variability of *Tuber uncinatum* and its relatedness to other black truffles. Environ Microbiol 4(10):584–594. doi:10.1046/j.1462-2920.2002.00343.x

Mello A, Miozzi L, Vizzini A, Napoli C, Kowalchuk G, Bonfante P (2010) Bacterial and fungal communities associated with *Tuber magnatum*-productive niches. Plant Biosyst 144 (2):323–332. doi:10.1080/11263500903374724

Micheli PA (1729) Nova plantarum genera. Bernardi Paperinii, Florence, pp 117–122

Molinier V, Murat C, Morin E, Gollotte A, Wipf D, Martin F (2013a) First identification of polymorphic microsatellite markers in the Burgundy truffle, *Tuber aestivum* (Tuberaceae). Appl Plant Sci 1(2):1200220. doi:10.3732/apps.1200220

Molinier V, van Tuinen D, Chevalier G, Gollotte A, Wipf D, Redecker D (2013b) A multigene phylogeny demonstrates that *Tuber aestivum* and *Tuber uncinatum* are conspecific. Org Divers Evol 13(4):503–512. doi:10.1007/s13127-013-0146-2

Molinier V, Murat C, Peter M, Gollotte A, De la Varga H, Meier B, Egli S, Belfiori B, Paolocci F, Wipf D (2015a) SSR- based identification of four genetic groups within European populations of *Tuber aestivum* Vittad. Mycorrhiza. doi:10.1007/s00572-015-0649-0

Molinier V, Murat C, Frochot H, Wipf D, Splivallo R (2015b) Fine-scale spatial genetic structure analysis of the black truffle *Tuber aestivum* and its link to aroma variability. Environ Microbiol 17(8):3039–3050. doi:10.1111/1462-2920.12910

Montecchi A, Sarasini M (2000) Funghi Ipogei d'Europa. AMB, Brescia

Mouches C, Duthil P, Poitou N, Delmas J, Bove J (1981) Caractérisation des espèces truffières par analyse de leurs protéines en gels de polyacrylamide et application de ces techniques à la taxonomie des champignons. Mushroom Sci 11:819–831

Müller W, Pauscher T, Agerer R (1996) *Tuber aestivum* Vitt + *Corylus avellana*. Descr Ectomyc 1:167–172

Murat C, Rubini A, Riccioni C, De la Varga H, Akroume E, Belfiori B, Guaragno M, Le Tacon F, Robin C, Halkett F, Martin F, Paolocci F (2013) Fine-scale spatial genetic structure of the black truffle (*Tuber melanosporum*) investigated with neutral microsatellites and functional mating type genes. New Phytol 199(1):176–187. doi:10.1111/nph.12264

Pacioni G, Pomponi G (1991) Genotypic patterns of some Italian populations of the *Tuber aestivum-Tuber mesentericum* complex. Mycotaxon 42:171–179

Pacioni G, Frizzi G, Miranda M, Visca C (1993) Genetics of a *Tuber aestivum* population (Ascomycotina, Tubérales). Mycotaxon 47:93–100

Paolocci F, Rubini A, Riccioni C, Topini F, Arcioni S (2004) *Tuber aestivum* and *Tuber uncinatum*: two morphotypes or two species? FEMS Microbiol Lett 235(1):109–115. doi:10.1111/j.1574-6968.2004.tb09574.x

Payen T, Murat C, Bonito G (2014) Truffle phylogenomics: new insights into truffle evolution and truffle life cycle. Adv Bot Res 70:212–234. doi:10.1016/B978-0-12-397940-7.00007-0

Persoon CH (1801) Tuber. In: Synopsis methodica fungorum. Apud H Dieterich, Göttingen, pp 126–129

Riccioni C, Rubini A, Belfiori B, Türkoğlu A, Paolocci F (2014) First evidence of genetic differentiation of *Tuber aestivum/uncinatum* populations from Turkey. In: Khabar L (ed) Abstract of the 5th congress of the *Tuber aestivum/uncinatum*, European Scientific Group, Rabat, 9–13 Apr 2014, p 27

Riousset L, Riousset G, Chevalier G, Bardet MC (2001) Truffes d'Europe et de Chine. INRA, Paris

Selosse MA, Faccio A, Scappaticci G, Bonfante P (2004) Chlorophyllous and achlorophyllous specimens of *Epipactis microphylla* (Neottieae, Orchidaceae) are associated with ectomycorrhizal septomycetes, including truffles. Microb Ecol 47(4):416–426. doi:10.1007/s00248-003-2034-3

Splivallo R, Valdez N, Kirchhoff N, Ona MC, Schmidt JP, Feussner I, Karlovsky P (2012) Intraspecific genotypic variability determines concentrations of key truffle volatiles. New Phytol 194(3):823–835. doi:10.1111/j.1469-8137.2012.04077.x

Stobbe U, Büntgen U, Sproll L, Tegel W, Egli S, Fink S (2012) Spatial distribution and ecological variation of re-discovered German truffle habitats. Fungal Ecol 5(5):591–599. doi:10.1016/j.funeco.2012.02.001

Stobbe U, Egli S, Tegel W, Peter M, Sproll L, Büntgen U (2013a) Potential and limitations of Burgundy truffle cultivation. Appl Microbiol Biotechnol 97(12):5215–5224. doi:10.1007/s00253-013-4956-0

Stobbe U, Stobbe A, Sproll L, Tegel W, Peter M, Büntgen U, Egli S (2013b) New evidence for the symbiosis between *Tuber aestivum* and *Picea abies*. Mycorrhiza 23(8):669–673. doi:10.1007/s00572-013-0508-9

Trappe J (1962) *Cenococcum graniforme* – its distribution, ecology, mycorrhiza formation, and inherent variation. PhD Thesis. University of Washington

Urbanelli S, Sallicandro P, De Vito E, Bullini L, Biocca E (1998) Biochemical systematics of some species in the genus *Tuber*. Mycologia 90(3):537–546. doi:10.2307/3761414

Vittadini C (1831) Monographia tuberacearum. Rusconi, Milano

Wang Y, Liu P (2009) Achievements and challenges of research in truffles in China. Acta Bot Yunnanica 31(Suppl 16):1–9

Weden C, Chevalier G, Danell E (2004a) *Tuber aestivum* (syn. *T. uncinatum*) biotopes and their history on Gotland, Sweden. Mycol Res 108(3):304–310. doi:10.1017/S0953756204009256

Weden C, Danell E, Camacho FJ, Backlund A (2004b) The population of the hypogeous fungus *Tuber aestivum* syn. T uncinatum on the island of Gotland. Mycorrhiza 14(1):19–23. doi:10.1007/s00572-003-0271-4

Weden C, Danell E, Tibell L (2005) Species recognition in the truffle genus *Tuber* – the synonyms *Tuber aestivum* and *Tuber uncinatum*. Environ Microbiol 7(10):1535–1546. doi:10.1111/j.1462-2920.2005.00837.x

Weden C, Pettersson L, Danell E (2009) Truffle cultivation in Sweden: results from *Quercus robur* and *Corylus avellana* field trials on the island of Gotland. Scand J For Res 24(1):37–53. doi:10.1080/02827580802562056

Zambonelli A, Salomoni S, Pisi A (1993) Caratterizzazione anatomo-morfologica delle micorrize di *Tuber* spp su *Quercus pubescens* Willd. Micol Ital 22(3):73–90

Zambonelli A, Salomoni S, Pisi A (1995) Caratterizzazione anatomo-morfologica delle micorrize di *Tuber borchii*, *Tuber aestivum*, *Tuber mesentericum*, *Tuber brumale*, *Tuber melanosporum* su *Pinus pinea*. Micol Ital 24(2):119–137

Zambonelli A, Iotti M, Piattoni F (2012) Chinese *Tuber aestivum* sensu lato in Europe. Open Mycol J 6:22–26

Zhang JP, Liu PG, Chen J (2012) *Tuber sinoaestivum* sp. nov., an edible truffle from southwestern China. Mycotaxon 122:73–82. doi:10.5248/122.73

Chapter 4
Tuber brumale: A Controversial *Tuber* Species

Zsolt Merényi, Torda Varga, and Zoltán Bratek

4.1 Introduction

The winter truffle, also known as black musk truffle (*Tuber brumale* Vittad.), is probably the most controversial species of true truffles. Compared to the estimated 180–230 truffle species worldwide (Bonito et al. 2010), it is one of the few species with remarkable gastronomic importance. In the past, only the Périgord truffle (*Tuber melanosporum* Vittad.) and *T. brumale* were called 'truffles' on French market (Douet et al. 2004). Its gastronomic value is due to its good flavour, which also makes it similar to *T. melanosporum* (Kiss et al. 2011). In the USA, it is the third most valuable truffle species and commands a price of USD660 per kg (Bonito et al. 2013). However, its economic value is lower than those of the Périgord truffle or the Italian white truffle (*Tuber magnatum* Pico), which are 3–5 and 10–15 times more expensive, respectively (Hall et al. 2008). The smaller size of *T. brumale* fruiting bodies makes their collection and processing unfavourable and hinders the demand for it on the market. It is also commercially ignored because of the lack of culinary promotion (Martin-Santafe et al. 2014). All of these features of *T. brumale* have led it to be considered an unwanted, contaminant species in *T. melanosporum* truffle orchards (Benucci et al. 2011; Sourzat 2011; Olivier et al. 2012; Guerin-Laguette et al. 2012; De Miguel et al. 2014). However, the results of studies that examine its effect on the productivity of *T. melanosporum* orchards are ambiguous. References on its aggressivity and ability to displace other truffle species are available (e.g. Chevalier and Frochot 1997; Riousset et al. 2001; Valverde-Asenjo et al. 2009; Chevalier and Sourzat 2012; Olivier et al. 2012; Parladé et al. 2013), although some articles highlighted a positive effect or weak competitiveness of

---

Z. Merényi • T. Varga • Z. Bratek (✉)
Department of Plant Physiology and Molecular Plant Biology, Institute of Biology, Eötvös Loránd University, Budapest, Hungary
e-mail: bratek.zoltan@ttk.elte.hu

A. Zambonelli et al. (eds.), *True Truffle (*Tuber *spp.) in the World*, Soil Biology 47,
DOI 10.1007/978-3-319-31436-5_4

T. brumale versus *T. melanosporum* (Mamoun and Olivier 1993; Belfiori et al. 2012).

Tuber brumale belongs to the six most studied true truffle species (Wang and Marcone 2011), which have been researched for host plant inoculation, optimization of in vitro culture (Iotti et al. 2002) and physiology and biochemistry of mycelium and mycorrhiza (Poitou and Olivier 1990; Amicarelli et al. 1999; Zeppa et al. 2005). It is the most widespread species of Melanosporum clade in Europe and can be frequently found in nature as well as on plantations. In spite of its frequency and the scientific value in truffle research, the genetic variability of *T. brumale* had not been adequately known for a long time.

In this chapter we would like to demonstrate the controversial situation of *T. brumale* in several aspects. There exist relatively few morphological characters in truffle species delimitation; thus the relationship between *T. brumale* and *Tuber moschatum* Bull. (or *T. brumale* f. *moschatum*) was a long debate. However, in addition to the aforementioned two taxa, the phylogenetics and the phylogeographics of the species related to *T. brumale* have been getting clearer. Furthermore in this chapter, we try to summarise the knowledge of the ecology of *T. brumale* and discuss the role of this species on the truffle orchards. Taking everything into account, the aim is to summarise and briefly show the accumulated knowledge on *T. brumale*.

4.2 Phylogenetic Position of *T. brumale*

In many cases, molecular data has remarkably modified former taxa, which were based on the examination of morphological characteristics, indicating serious changes concerning the taxonomy of true truffles. The first molecular and biochemical analyses, including isoenzyme and amino acid diversity, were performed in the 1980s and 1990s and aimed at identifying and differentiating the species (Henrion et al. 1994; Gandeboeuf et al. 1995; Paolocci et al. 1995). Only later, as molecular methods were developing, were the phylogenetic relationships of the species discovered. First, the position of *T. brumale* within genus *Tuber* and its relationship with the other truffle species was determined by random amplified polymorphic DNA (RAPD), where the patterns of *T. melanosporum* indicated remarkable similarity to *T. magnatum*, while *T. brumale* resembled *Tuber rufum* Vittad. (Gandeboeuf et al. 1997). Based on internal transcribed spacer (ITS) phylogenetic tree by Roux et al. (1999), *Tuber* taxa can be divided into three major clades, one of which involves *T. melanosporum*, *Tuber indicum* Cooke and Massee, *Tuber himalayense* B. C. Zhang and Minter and *T. brumale*. All the further molecular phylogenetic works—even applying different loci—uniformly agree that these species belong to the Melanosporum clade (Jeandroz et al. 2008; Bonito et al. 2010). However, the position of *T. brumale* within the Melanosporum clade is not so obvious, where various loci or phylogenetic inference methods contradict each other. Within the Melanosporum clade, *Tuber regimontanum* (Guevara et al. 2008), *T. melanosporum* and the *T. indicum* complex always make up a monophyletic group (Bonito et al. 2013), whereas the position of *Tuber pseudoexcavatum* (Wang et al. 1998) and *T. brumale* is not clear yet. According to most studies, the latter two species establish a new monophyletic genealogy (Zhang et al. 2005; Wang et al. 2006a; Jeandroz et al. 2008; Chen et al. 2011; Bonito et al. 2013). However, in some phylogenetic trees based on the ITS region, the split of *T. brumale* and *T. pseudoexcavatum* is an unsolved polytomy (Jeandroz et al. 2008; Bonito et al. 2013). Other analyses of the ITS and large subunit (LSU) of nuclear rDNA support the split of *T. brumale* (Bonito et al. 2010; Chen et al. 2011; Chen and Liu 2012; Merényi et al. 2014; Fig. 4.1), while, in case of protein kinase C (PKC) locus, it seems obvious that *T. pseudoexcavatum* diverges earlier (Wang et al. 2006a).

The first DNA sequence analysis, in which *T. brumale* were included, revealed that the ITS region of the species is extraordinarily longer (~1000 bp) than that of other truffles (600–700 bp) (Henrion et al. 1994; Gandeboeuf et al. 1995). The difference in length was explained as the duplication of part of the *T. brumale* ITS1 region (Roux et al. 1999), which was later revealed in *T. pseudoexcavatum* too (Wang et al. 2006a). Despite the fact that only 59 % of this insert is similar (Wang et al. 2006a), this ~300 bp insertion, gives evidence to the monophyly of *T. brumale* and *T. pseudoexcavatum* (Wang et al. 2006b). As we discover new species, it lends more and more evidence that this characteristic insertion, beginning at position 73–80 of region ITS1, is not unique to *T. brumale* and *T. pseudoexcavatum* (Fig. 4.1). For example, the first branch of the Melanosporum clade, *Tuber* sp. 13 (Bonito et al. 2013), includes a 169 bp insertion, whereas *T. regimontanum* includes a 51 bp insertion. Members of the pseudoexcavatum lineage, e.g. *Tuber pseudobrumale* (Li et al. 2014) and *Tuber* sp. 5 (Chen et al. 2011), all contain the characteristic insertion, which varies between 53 and 291 bp, depending on the species. This leads us to believe that this is the result of several expansion and reduction events in the ITS1 region, as opposed to a single duplication event (Fig. 4.1).

According to Wang et al. (2006a) the most recent common ancestor (MRCA) of the Melanosporum clade might have lived between Europe and China. First, the genealogy of *T. pseudoexcavatum*/*T. brumale* was split as a result of the ancestor of *T. brumale* migrating towards Europe, while *T. pseudoexcavatum* migrated to China. Later the ancestors of *T. melanosporum* and *T. indicum* migrated following the same route, thus they also formed vicariant species (Jeandroz et al. 2008). The situation became even more complicated with the introduction of species found in North America and Mexico. The radiation of the Melanosporum clade from North America may have started 79 Mya carried by Pinaceae, followed by their migration to Asia and Europe (Bonito et al. 2013). Therefore, during the late Cretaceous age and the Cenozoic era, the Thulean North Atlantic Land Bridge and the Beringia Land Bridge played an important role in the migration of species across the continents (Milne 2006).

Covering the most possible range presently known of *T. brumale*, after phylogenetic and phylogeographic studies, two significantly different clades were proven

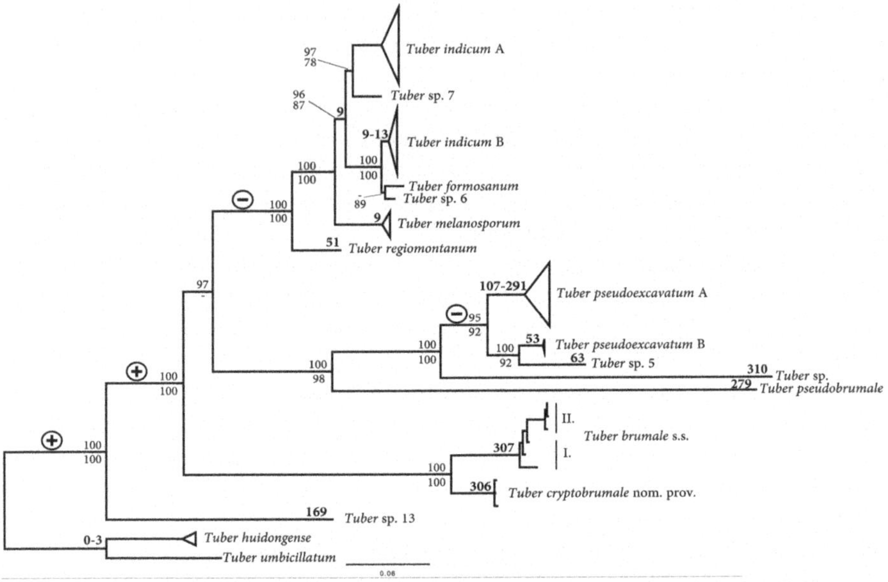

Fig. 4.1 Bayesian phylogenetic tree of Melanosporum clade based on ITS–LSU. Parsimony bootstrap (PB) values are shown above the branches, while the Bayesian posterior probabilities (PP) are below. Only values where PP > 95 % and MP bootstrap support > 70 % are shown. *Tuber umbilicatum* Juan Chen and P. G. Liu and *Tuber huidongense* Y. Wang were served as the outgroup. In the case of *T. brumale* s.s., the numbers I. and II. refer to haplogroups I. and II., respectively. The *bold numbers* indicate the length of a characteristic insertion in the ITS1 locus. In the case of the insertion, the + and – indicate the assumption of the expansion and reduction events, respectively. Scale bar means 0.06 expected nucleotide change per site per branch. Sequence from GenBank are the following (ITS/LSU): *T. brumale* s.s., KF591749/KF581092, KF551063/KF581089, KF551001/KF581087, KF551002/KF581086, KF551005/KF581099, KF551007/KF581095; *T. cryptobrumale*, KF551019/KF581104, KF551049/KF581103, KF551016/KF581106; *T. formosanum*, GU979049/GU979126; *T. huidongense*, FJ797883/GU979099, FJ797881/GU979093; *T. indicum* A, FJ748906/JQ925694, FJ748908/JQ925695, GU979067/GU979127, GU979082/GU979138, GU979061/GU979117, GU979069/GU979122, GU979078/GU979119; *T. indicum* B, GU979051/GU979113, GU979081/GU979137, GU979053/GU979116, GU979059/GU979134, FJ748907/FJ809840, GU979050/GU979112; *T. melanosporum*, KF591751/KF581109, FJ748904/JQ925703, GU979083/GU979139; *T. pseudobrumale*, KJ742703/-; *T. pseudoexcavatum* A, GU979043/GU979105, HM485381/FJ809816, GU979045/GU979106, GU979042/GU979109, GU979041/GU979104, GU979046/GU979107; *T. pseudoexcavatum* B, GU979039/GU979102, GU979040/GU979103; *T. regimontanum*, EU375838/EU375838; *Tuber* sp. (mentioned by the authors as "*T. formosanum*" probably due to a miss identification), GU979047/-; *Tuber* sp. 5, AB553380/AB553516; *Tuber* sp. 6, AB553388/AB553517; *Tuber* sp. 7, AB553414/AB553518; *Tuber* sp. 13, GQ221451/JQ925713; *T. umbilicatum*, FJ797879/GU979096

(Merényi et al. 2014). We can consider that an interspecific relationship exists if the genetic distance in the ITS exceeds 3 % (Nilsson et al. 2008; Bonito et al. 2010), which is true for these two clades (9.3 %), which is also confirmed by LSU and PKC loci, in addition to meeting the compliance of the gene genealogical concordance

(Taylor et al. 2000). Also, based on the molecular clock, the division of the two clades could have happened 17.9 Mya, which nearly concurs with the time of species complex division of *T. melanosporum* and *T. indicum* (Bonito et al. 2013; Merényi et al. 2014). Under these results, the two clades of *T. brumale* can be considered distinct phylogenetic species, one corresponding to the species earlier called and described as *T. brumale* sensu stricto (s.s.), whereas the other clade can be considered as a species under description, called *Tuber cryptobrumale* nom. prov. (see Sect. 4.4). It is worth mentioning that *T. pseudobrumale* from China, described by Li et al. (2014), rather belongs to the lineage of *T. pseudoexcavatum* than the European *T. brumale* aggregate (Fig. 4.1).

4.3 Genetic Variability Within the Species

Already the first studies highlighted the minimal or missing genetic variability within *T. brumale* s.s. It indicated extremely low intraspecific polymorphism in isoenzyme tests (Pacioni and Pomponi 1989) as well as testing DNA with restriction fragment length polymorphism (RFLP) and RAPD methods (Gandeboeuf et al. 1995, 1997). Giomaro et al. (2002) mentions the complete lack of the variability in the ITS region of the species. Later, Pomarico et al. (2007) confirmed, also based on the ITS region, that from the six examined species of truffles, *T. brumale* indicates the lowest heterozygosity (He = 0.09). Wang et al. (2006a) states that the European species indicates far less intraspecific nucleotide divergence (*T. melanosporum* 0–0.3 %, *T. brumale* 0–0.1 %) than the Asian ones (*T. indicum* 3.2 %, *T. pseudoexcavatum* 1 %). The reason for this is due to the bottleneck effect which occurred when the species migrated from its original habitat to Europe. Bonito et al. (2010) found lower than 0.5 % intraspecific ITS variation concerning *T. brumale*. All the studies listed above examined genetic variability of *T. brumale* with a relatively low number of samples and only from Western Europe. After a much more extensive sampling, the structure of *T. brumale* population, and thus its variability, turned out to be more complex than previously assumed. *Tuber brumale* s.s. implies higher variability (0.91 % ± 0.98) of ITS than earlier detected. The high rate of variability is the result of the two major ITS haplotypes (unique DNA variants) which can be differentiated within the species. The ITS variability of the haplogroups I. and II. (*T. brumale* s.s. I. and II.) is 0.032 % ± 0.073 and 0.201 % ± 0.166, respectively. Two ITS haplogroups indicate significant East–West separation in Europe, showing blending only in the Carpathian Basin. The intraspecific ITS variability of *T. cryptobrumale* is 0.07 % ± 0.067, which is still considered extremely low (Merényi et al. 2014). It has to be noted that microsatellite analyses and genome sequencing may point out a larger genetic diversity comparable to other filamentous ascomycetes, as it was shown in case of *T. melanosporum* (Murat et al. 2004; Payen et al. 2015).

When analyzing ITS haplotypes, 119 *T. brumale* s.s. samples were involved in 17 haplotypes, while 17 *T. cryptobrumale* samples were included in merely two haplotypes. Haplotype diversities (Hd; Nei 1987) are as follows: *T. brumale* s.s. I. Hd = 0.29; II. Hd = 0.72; and *T. cryptobrumale* Hd = 0.53 (Merényi et al. 2014; Fig. 4.1). *Tuber brumale* s.s. haplogroup II. is more diverse in the haplotype variance as well as in the structure of the haplotype net than haplogroup I. In addition to these facts, the present distribution of haplogroup II. indicates the existence of an Eastern European refugia during the last glacial. Based on the geographic distribution, the haplotype structure of the ITS region and the tests referring to the neutral evolution, haplogroup I. could have survived the last glacial in a West-European refugium, which is accompanied by a bottleneck effect, as was shown in the case of *T. melanosporum* (see Chap. 2). Certain yet unknown events of the Quaternary period may explain the low ITS diversity and local occurrence of *T. cryptobrumale* and may hint to a Carpathian-Pannonian endemism (Merényi et al. 2014).

4.4 Morphology

The Melanosporum clade includes black warted *Tuber* species with darkly pigmented spiny or, occasionally, spinoreticulate spores (Bonito et al. 2013). Initially based only on morphology, only four species were originally included in this clade (Riousset et al. 2001), but as a result of describing new species, *T. pseudoexcavatum*, *T. regimontanum*, *Tuber formosanum* (Hu 1992) and *T. pseudobrumale* were added as well.

The confusion of certain species of the Melanosporum group, based on morphological similarity, mainly *T. indicum*, *T. melanosporum* and *T. brumale*, is highlighted by several authors (Séjalon-Delmas et al. 2000; Douet et al. 2004; Murat 2015). The determination of the species listed above is possible only by examining micromorphological characters (spores, peridium structure), whose full-grown form can be discovered only on the mature fruiting bodies (Janex-Favre et al. 1996). Still the separation of species is hindered because even though there exist a low number of morphologic characters, they possess an extraordinary amount of variability; as a result, e.g. *T. indicum* is actually a collection of complex cryptic species (see Chap. 2; Chen et al. 2011). An example of the large range of variability can be seen in the spores of *T. pseudoexcavatum* and *T. brumale* s.l. taxa (Hollós 1911; Wang et al. 2006a).

A further example of using highly variable morphological characteristics, we can follow the story of taxa related to *T. brumale*. *Tuber brumale* has been described in Italy by Carlo Vittadini in 1831. Based only on scent and gleba colour variations, which exhibit great variability, *Tuber moschatum* Bonnet (1869) is properly proven separate from the conspecific *T. brumale* s.s. (Pacioni et al. 1990; Montecchi and Sarasini 2000; Mauriello et al. 2004; March et al. 2006). De Ferry de la Bellone (1888) described three further variants named *T. moschatum* var. *vulgare*, *T. moschatum* var. *graveolens* and *T. moschatum* var. *suaveolens*. *Tuber moschatum* was still considered an autonomous species by

Mattirolo's study in 1933. Soon its separation as a species was questioned; according to Chatin (1892) *T. moschatum* was only a variant of *T. melanosporum*, while Fischer (1897) identified it as a subspecies of *T. brumale*. In recent years, several authors considered it as a forma of *T. brumale* (Ceruti 1960; Montecchi and Sarasini 2000; Ceruti et al. 2003). However, the conspecificity of *T. brumale* and *T. moschatum* (or *T. brumale* f. *moschatum*) has been confirmed by the first molecular analyses based on isoenzyme and amino acid polymorphisms (Gandeboeuf et al. 1994), as well as by RAPD (Gandeboeuf et al. 1995, 1997) and ITS sequence similarity (Roux et al. 1999; Merényi et al. 2014). In addition, in the course of history, several species descriptions have been created for *T. brumale*, e.g. *Tuber cibarium* Bull., *Tuber hiemalbum* Chatin, *Tuber renati* H. Bonnet, *Tuber montanum* Chatin and *Oogaster brumalis* (Vittad.) Corda (see Ceruti et al. 2003), which are now considered synonymous to it.

To conclude, we will discuss morphological description of the aforementioned monophyletic sibling species, *T. brumale* s.s. and *T. cryptobrumale*. The diameter of the fruiting bodies of *T. brumale* s.s is 6–41 mm, and their shape varies from subglobose to irregular, frequently with the shape of a kidney, due to the mature base cavity or depression. The surface is covered with easily detachable, penta- or hexagonal flat warts, the average diameter of which is 1.8 ± 0.4 mm. Gleba is rarely penetrated with white veins, and its colour is often sepia, drab (Fig. 4.2). Its asci are subglobose or widely ellipsoid, from which the ones containing 3–5 spores dominate. In four-spored asci, the spore parameters measured without ornamentation are the following: $(25.3) - 28.1 - (33.7) \times (15.7) - 17.4 - (19.1)$ μm^2, $Vol = 4487.4 \pm 541$ μm^3 and $Q = 1.6 \pm 0.1$. The ornamentation of the spores is spiny, the size of which is 3.4 ± 0.38 μm.

Fig. 4.2 Fruit bodies of *T. brumale* aggr. The schematic drawing of fruit bodies shows *T. cryptobrumale* nom. prov. (**a**) and *T. brumale* s.s. (**b**). Bar = 1 cm

Tuber cryptobrumale has never been determined based on morphology, which is not surprising, as traditional morphologic measurements on genotyped fruiting bodies did not reveal obvious differences compared to *T. brumale* s.s. Consequently, their taxonomical consideration was suggested as a cryptic species (Merényi et al. 2014). However, after thorough, detailed morphologic investigations with the addition of new characters, we were able to find differences between the two species, which resulted in their recognition as a pseudocryptic species (Merényi 2014; Merényi et al. in prep.). *Tuber cryptobrumale* only slightly differs from the description of *T. brumale* s.s.; the only way to separate the two species is based on characters that can be only differentiated by statistics such as the size of the warts (1.3 $\pm$ 0.3 mm) and the peridium cells and the proportion of 3- and 5-spored asci, the joint application of which can define the species (Fig. 4.2; Merényi 2014).

The mycelium colony of *T. brumale* s.s. is white and slowly growing (1.1–1.9 mm per week), in which the hyphae usually form chains that are rich in vesicula (Iotti et al. 2002). The ectomycorrhizal (ECM) anatomy of *T. brumale* s.s. may vary, subject to their host plant (Giomaro et al. 2002). Its ECMs belong to the short-distance exploration type (Zambonelli et al. 1995; Agerer 2001). The ECM ends of the roots are mostly 1.0–6.5 mm long; they can be simple, dichotomous or monopodial-pinnate; and their colour is ochre, yellowish brown-dark brown or beige-pale brown. The mycorrhizas of *T. brumale* s.s. have an external mantle with pseudoparenchymatous rounded-epidermoid structure (Giomaro et al. 2002). Cystidia are 20–260 μm long, non-branching, yellowish, without a clamp, having a simple septum, awl-shaped and bristle-like (Fontana and Fasolo-Bonfante 1971; Zambonelli et al. 1993; Giomaro et al. 2002; Fischer et al. 2004).

4.5 Distribution

In general, Ławrynowicz (1993) mentions that *T. brumale* s.l. can be found in most European countries, up to the north latitude 52° and east longitude 19°. Similarly, Ceruti et al. (2003) mention that its range in Europe is between latitudes 40° and 55°. This is further confirmed by Hansen and Knudsen (2000) who recognise nine *Tuber* species in the Northern European countries, but not *T. brumale* s.l. This means that *T. brumale* s.l. is spread throughout most of Europe, except boreal and arctic regions (Ławrynowicz 1993). First, *T. brumale* Vittad. was recorded in Italy in 1831 and then in South England, France and Germany (Vittadini 1831; Hollós 1911; Hawker 1954; Pegler et al. 1993). Later it was considered a species frequent in West and South Europe while rare in North and Central Europe (Hollós 1911; Hawker 1954; Michael and Hennig 1960; Cetto 1979). East of the Alps, its appearance was later found first in the Carpathian Basin (Hollós 1911) and then, in the past decade, in Bulgaria, Greece and Turkey (Dimitrova and Gyosheva 2008; Diamandis and Perlerou 2008; Türkoğlu and Castellano 2014). Today it is considered missing or rare in particular countries (England, Germany, Poland and

Scandinavia), while in others (France, Italy, Hungary, Serbia, Romania, Slovenia) it is considered a common species (Grebenc 2008; Marjanović 2008; Bratek et al. 2013; Merényi et al. 2014). Moreover, its possible existence has been shown from environmental DNA samplings in Iran (Bahram et al. 2012; Merényi et al. 2014). Under the area reconstruction analysis based on the data regarding its already known habitat, extended populations might be found in the Caucasus region (Merényi 2014; Merényi et al. in prep.).

As a result of human-mediated import, it has appeared in New Zealand (Ho et al. 2008), Australia (Linde and Selmes 2012) and North America (Berch and Bonito 2014) in *T. melanosporum* plantations. Guerin-Laguette et al. (2012) emphasise that *T. brumale*, which is assumed to appear in New Zealand in 1995, has only been found in those plantations, brought accidentally via inoculum. Thus the non-native *T. brumale* has not become a spontaneously spreading fungus on the island. Still, by accidentally introducing *T. brumale* to the aforementioned regions, this species now belongs to the 16 *Tuber* species which occur outside of their native range (Bonito et al. 2010).

4.6 Ecology

As it has already become clear, the *T. brumale* aggr. is widely spread, which is reflected by its environmental demands. According to our database and the references, the truffle can be found in areas with an estimated annual rainfall of 240–1400 mm (in average 776 ± 238 mm) that are 70–1000 m AMSL and have a 3.7–15.8 °C annual mean temperature (Gazo et al. 2005; Granetti et al. 2005; Diamandis and Perlerou 2008; Valverde-Asenjo et al. 2009; Merényi et al. 2010; Stobbe et al. 2012). In the case of two species, we amended the data slightly. In the case of *T. brumale* s.s., the annual mean temperature and the annual rainfall range are the same; only the average rainfall is different (786 ± 242 mm). As for *T. cryptobrumale*, the annual rainfall is between 519 and 1311 mm (677 ± 166 mm), and the annual mean temperature is 7.3–11.2 °C. The harvest time for *T. brumale* s.l. is from December to March in France, from November to March in Germany and Italy and from October to January in Turkey (Hall et al. 2007; Stobbe et al. 2012; Türkoğlu and Castellano 2014). Harvest time for both species in the Carpathian Basin may last from August to March, but the maximum amount of the product is expected from November to December.

While Hollós (1911) and Szemere (2005) mentioned that it grows in calcareous soils, Chevalier and Frochot (1997) found, unlike *T. melanosporum*, that it prefers less calcareous and wet soils containing humus. Eighty-one soil samples from habitats in the Carpathian Basin confirmed that it prefers heavy soils of neutral pH, i.e. it prefers humus-rich clay and heavy clay soils. The soil requirement of *T. brumale* aggr., in regard to lime, phosphorous, nitrogen and other micro- and macro-elements, varies a lot (Table 4.1; Olivier et al. 2012). In Northern Spain it was found in soils with an average of 7.9 ± 0.2 pH, where the content of active

Table 4.1 The average values of soil characteristics of 49 and 32 samples of *T. brumale* s.s. and *T. cryptobrumale*, respectively, in the Carpathian Basin

Soil characteristics	*T. brumale* s.s.	*T. cryptobrumale*
Water pH	6.9 ± 0.5	6.8 ± 0.6
Sticky point according to Arany (SPA)	64 ± 10	68 ± 9
Humus content (%)	6.2 ± 1.4	6.9 ± 2.0
Lime-dissolving carbonic acid ($CaCO_3$ %)	2.6 ± 4.3	2.6 ± 6.0
Assimilable phosphorus (P_2O_5; mg kg^{-1})	199 ± 136	165 ± 203
Assimilable nitrite and nitrate-N ($NO_2 + NO_3$-N; mg kg^{-1})	10.1 ± 9.0	14.5 ± 12.4
Assimilable potassium (K_2O; mg kg^{-1})	596 ± 367	306 ± 190
Assimilable sulphate-S (SO_4^{-2}-S; mg kg^{-1})	28.9 ± 15.3	24.3 ± 13.6
Assimilable calcium (Ca; mg kg^{-1})	13,630 ± 11,451	12,910 ± 21,526
Assimilable magnesium (Mg; mg kg^{-1})	365 ± 208	401 ± 154
Assimilable zinc (Zn; mg kg^{-1})	9.6 ± 5.3	7.7 ± 5.6
Assimilable manganese (Mn; mg kg^{-1})	584 ± 441	360 ± 358
Assimilable iron (Fe; mg kg^{-1})	180 ± 115	471 ± 391

carbonate was 124 ± 14 g per kg, while the complete organic carbon was 18.4 ± 2.9 g per kg (Valverde-Asenjo et al. 2009). Based on our data, despite the wide range in soil requirements, *T. brumale* s.s. and *T. cryptobrumale* habitats have only slight soil differences (Table 4.1).

T. brumale aggr., according to the data published in the references (Table 4.2), evidently forms mycorrhizas with nine plant species, but the presence of frequently occurring ECM tree species in its habitat hints at a wider spectrum of host species (Table 4.2). Among them, hornbeam (*Carpinus betulus* L.) is most likely a potential host in the Carpathian Basin. *Tuber brumale* s.l. also grows under Mediterranean tree species in Serbia and Montenegro, e.g. Aleppo pine (*Pinus halepensis* Miller), holm oak (*Quercus ilex* L.) or downy oak (*Quercus pubescens* Willd.). In Greece, it was found under Hungarian oak (*Quercus frainetto* Ten.) and downy oak, while in Turkey, under *Corylus* sp. and Turkish pine (*Pinus brutia* Ten.). It is worth noting that, based on the data from the Carpathian Basin, the habitats of *T. brumale* s.l. involve field maple (*Acer campestre* L.) with a 61 % frequency, which is a species not producing ECMs. Common dogwood (*Cornus sanguinea* L.) and wild privet (*Ligustrum vulgare* L.) are also frequent in the habitat's shrub level (frequency > 39 %). *Viola sylvestris* Lam., *Geum urbanum* L., *Brachypodium sylvaticum* (Huds.) R. et Sch. and *Ajuga reptans* L. are the most frequent in the herb layer (frequency > 40 %). The most frequent and coexisting subterranean fungi in habitats of *T. brumale* aggr. in the Carpathian Basin (>12 %) include *Tuber aestivum* Vittad., *Tuber excavatum* Vittad., *Tuber fulgens* Quél. and *Tuber macrosporum* Vittad.

Table 4.2 The host species of *T. brumale* s.s. and *T. cryptobrumale* inoculation experiments show if successful mycorrhization was made with *T. brumale* aggr

Plant species	Inoculation experiments	Natural co-occurence		References
		T. brumale (%)	*T. cryptobrumale* (%)	
Corylus avellana L.	+	30 (18 ± 21)	35 (2 ± 2)	Palenzona (1969), Chevalier and Frochot (1997), Paolocci et al. (1999), Bencivenga et al. (2009), Stobbe et al. (2012)
Carpinus betulus L.[a]	n.d.	71 (45 ± 21)	91 (36 ± 2 3)	Chevalier and Frochot (1997), Gazo et al. (2005), Marjanović (2008)
Ostrya carpinifolia Scop.	+	n.d.	n.d.	Bencivenga et al. (2009)
Fagus sylvatica L.	n.d.	15 (15 ± 13)	14 (25 ± 18)	Szemere (2005), Stobbe et al. (2012)
Quercus robur L.[a]	n.d.	32 (19 ± 17)	41 (40 ± 17)	Chevalier and Frochot (1997), Gazo et al. (2005), Marjanović (2008)
Quercus cerris L.[a]	+	38 (17 ± 15)	27 (34 ± 12)	Marjanović (2008), Bencivenga et al. (2009)
Quercus petraea (Matt.) Liebl.[a]	n.d.	12 (20 ± 17)	14 (23 ± 19)	
Quercus pubescens Willd.; *Quercus ilex* L.	+	n.d.	n.d.	Zambonelli et al. (1993), Chevalier and Frochot (1997), Paolocci et al. (1999), Fischer et al. (2004), Marjanović (2008), Diamandis and Perlerou (2008), Valverde-Asenjo et al. (2009), Bencivenga et al. (2009)
Tilia cordata Mill.	n.d.	15 (25 ± 16)	9 (10 ± 10)	
Tilia americana Kinné.	+	n.d.	n.d.	Giomaro et al. (2002)
Pinus nigra Arnold	+	2 (14 ± 0)	0 (0)	Fontana and Fasolo-Bonfante (1971), Chevalier and Frochot (1997)

(continued)

Table 4.2 (continued)

Plant species	Inoculation experiments	Natural co-occurence		References
		T. brumale (%)	*T. cryptobrumale* (%)	
Pinus pinea L.	+	n.d.	n.d.	Zambonelli et al. (1995)
Cistus incanus L.; *Cistus salviifolius* L.	+	n.d.	n.d.	Chevalier (1973), Giovannetti and Fontana (1982)
Fumana procumbens (Dun.) Gren. et Godr.	+	n.d.	n.d.	Chevalier (1973)
Populus spp.	n.d.	0 (0)	9 (24 ± 6)	Marjanović (2008)

[a]In many cases the only ECM tree present

The percentage values without and within parentheses indicate the frequency and the cover of tree species, respectively, in 56 habitats of Carpathian Basin

4.7 The Role of *T. brumale* s.l. in Plantations

As mentioned, *T. brumale* s.l. is ranked as a damaging species of high risk factor for man-made truffle plantations (Martin-Santafe et al. 2014), displacing *T. melanosporum* through its aggressive nature as well as reducing its productivity (Chevalier and Frochot 1997; Callot 1999; Lefevre and Hall 2001; Riousset et al. 2001; Ricard 2003; Sourzat 2005; Olivier et al. 2012). Furthermore it is among the three or four most common unwanted mycorrhizal partners in the plantations (De Miguel et al. 2014). Thus, this far less valuable *T. brumale* s.l. species is just contamination, being considered as rather a harmful species to the truffle orchards. In fact, during the saplings' quality control, one of the three most important determining points is checking the roots for mycorrhizas of non-desirable *Tuber* species (e.g. *T. brumale* s.l. or *T. indicum*) (Donnini et al. 2014; Murat 2015). Based on the summary written by De Miguel et al. (2014), *T. brumale* s.l. appears on 78 % of the *T. melanosporum* plantations. Additionally, it was present on all three European *T. aestivum* plantations, though it occurred on only two of the seven *T. magnatum* plantations.

Several factors may promote the presence of *T. brumale* s.l. on plantations. (1) It was observed that *T. brumale* s.l. are more likely to develop on fast-growing host plants (De Miguel et al. 2014). (2) The improper treatment during the initial period of the plantation, such as frequent soil scarification or overwatering, may encourage the appearance of *T. brumale* s.l. (Sourzat 2005, 2011). (3) If *T. melanosporum*'s viability and 'virulence' is reduced, it promotes the spread of *T. brumale* s.l. (Sourzat et al. 2001; Sourzat 2011). (4) If there exists a contamination of the inoculum by *T. brumale* s.l., there exists a higher chance that the plantation will be contaminated as well (Donnini et al. 2014; Murat 2015). (5) Establishing *T. melanosporum* orchards in soils which is more suitable for *T. brumale* s.l. promotes their appearance. (6) The fact that *T. brumale* s.l. and *T. melanosporum* share a common disseminating vector (Hochberg et al. 2003) and production window (winter) increases the chance that a vector carries *T. brumale* s.l. from natural habitats to a *T. melanosporum* plantation. This is contrary to the *Tuber* species with a production window of summer and early autumn.

The lime preference of *T. brumale* s.l. is controversial. Liming of plantations has become a widespread practice in order to control *T. brumale* s.l. (Riousset et al. 2001). However, focused and detailed soil analyses were able to reveal that in the natural habitat and plantation patches of *T. brumale* s.l., the levels of active and full carbonates are much higher than those in the habitats of *T. melanosporum* (Valverde-Asenjo et al. 2009). Moreover, liming does not affect either the mycorrhizas or the production of *T. brumale* s.l. negatively. Additionally, the production of both species *T. brumale* s.l. and *T. melanosporum* is influenced positively by liming, the mechanism of which is summarised in the soil nutrition hypothesis (García-Montero et al. 2008). Finally, based on soil studies done in the Carpathian Basin, *T. brumale* s.l. seems accepting wide range of lime content (Table 4.1).

The succession of hypogeous fungi can be discovered in the brûlé of *T. melanosporum*, in which *T. rufum*, *T. melanosporum* and *T. brumale* follow each other in order (García-Montero et al. 2012; Granetti et al. 2005). On the other hand, according to Sourzat (2011) *T. brumale* s.l. can be found on saplings, which is confirmed by De Miguel et al. (2014).

Scientific articles find that the obviously negative displacing effect of *T. brumale* s.l. doesn't apply to the French truffle (Valverde-Asenjo et al. 2009; Belfiori et al. 2012; Liu et al. 2014; Martin-santafe et al. 2014). The lowest amount of *T. melanosporum* mycelium could be revealed from the production patch of *T. brumale* s.l using DNA-based quantitative methods (Parladé et al. 2013). However, the experiments of Mamoun and Olivier (1993) notice that in some cases, e.g. overwatering *T. brumale* s.l. may help *T. melanosporum* survive and may keep other ECM species away. In other studies, the weaker competitive ability of *T. brumale* s.l. was highlighted, who generates a much bigger ECM diversity on the production area of *T. brumale* s.l. than in the patches of *T. melanosporum* or *T. aestivum* (Belfiori et al. 2012). Also, competition may be assumed between *T. brumale* s.l. and *Tuber borchii* Vittad. as no *T. brumale* s.l. contamination was found on any plants, which reserved *T. borchii* mycorrhizas, of the *T. borchii* plantations (De Miguel et al. 2014). It is worth to mention that this can be due to their different soil demand as well. The competitive role of *T. brumale* s.l. may seriously depend on the current biotic and abiotic factors of the environment, the discovery of which may be implemented only through long-term series of experiments.

Lately the idea that *T. brumale* s.l. should not only be forced out of the plantations has arisen, as they could generate economic benefit if *T. brumale* s.l. plantations were established (Benucci et al. 2012; Liu et al. 2014; Murat 2015). The first greenhouse trials of mycorrhization were often performed using *T. brumale* s.l. spore inocula (Palenzona 1969; Fontana and Fasolo-Bonfante 1971;

Table 4.3 The mycorrhization rate of 4- and 7-year-old *Quercus* spp. trees in Hungarian experimental *T. brumale* s.l. plantations

	ECM colonisation (%)			
	4 years		7 years	
Plantation	Average	SD	Average	SD
I.	36.6	16.2	29.5	13.7
II.	64.2	22.3	24.6	12.9
III.	11.6	14.8	33.4	22.0

SD standard deviation
The average colonisation was calculated according six saplings

Palenzona et al. 1972; Mamoun and Olivier 1993) but direct trials on field plantations have not been published. In Hungary, three microplantations (<20 saplings) have been established with *T. brumale* s.l. mycorrhized plants, going through a mycorrhiza test at the age of four and seven (Table 4.3). In a far more extended trial (>3000 trees), a follow-up *T. brumale* s.l. inoculation has been tried in Spain (Morcillo et al. 2007).

4.8 Conclusions

Tuber brumale s.l. has been considered a harmful, contaminating fungus; however, on the contrary, *T. brumale* s.l. plantations may be economically beneficial (Benucci et al. 2012; Murat 2015; Liu et al. 2014), especially when grown on economically significant host plants (Benucci et al. 2012). Higher long-term market share can be gained only through using the populations of this species that have bigger fruiting bodies, which has already been tested in some of our experiments. Furthermore, to fulfil the demand, the quantity produced on the plantations would need to be supplemented from the natural habitats resulting in overcollection. As for environmental protection, it would be important to understand the effects of the western haplogroup of *T. brumale* s.s., known as a contaminator on the plantations and appearing adventurous on several continents, and the eastern haplogroup, which is much more diverse genetically, as well as *T. cryptobrumale*, which is endemic to the Carpathian Basin. Detailed research of *T. brumale* aggr. has provided a good example for the cryptic/pseudocryptic speciation like it was found in several cases in genus *Tuber* (Bonuso et al. 2010; Chen et al. 2011; Merényi et al. 2014). Additionally, in the near future several experiments are supposed to be concluded on the trial truffle plantations, through the genomic and ecological studies of the genus, which may facilitate the better understanding of the mycorrhizal substitutional abilities and ecological role of the species aggregate.

References

Agerer R (2001) Exploration types of ectomycorrhizae: a proposal to classify ectomycorrhizal mycelial systems according to their patterns of differentiation and putative ecological importance. Mycorrhiza 11(2):107–114

Amicarelli F, Bonfigli A, Colafarina S, Cimini AM, Pruiti B, Cesare P, Cerú MP, Ilio C, Pacioni G, Miranda M (1999) Glutathione dependent enzymes and antioxidant defences in truffles: organisms living in microaerobic environments. Mycol Res 103(12):1643–1648

Bahram M, Polme S, Kõljalg U, Zarre S, Tedersoo L (2012) Regional and local patterns of ectomycorrhizal fungal diversity and community structure along an altitudinal gradient in the Hyrcanian forests of northern Iran. New Phytol 193(2):465–473. doi:10.1111/j.1469-8137.2011.03927

Belfiori B, Riccioni C, Tempesta S, Pasqualetti M, Paolocci F, Rubini A (2012) Comparison of ectomycorrhizal communities in natural and cultivated *Tuber melanosporum* truffle grounds. FEMS Microbiol Ecol 81(1):547–561. doi:10.1111/j.1574-6941.2012.01379.x

Bencivenga M, Di Massimo G, Donnini D, Baciarelli-Falini L (2009) The cultivation of truffle in Italy. Acta Bot Yunnanica 31(Suppl 16):21–28

Benucci GMN, Raggi L, Albertini E, Grebenc T, Bencivenga M, Falcinelli M, Di Massimo G (2011) Ectomycorrhizal communities in a productive *Tuber aestivum* Vittad. orchard: composition, host influence and species replacement. FEMS Microbiol Ecol 76(1):170–184. doi:10.1111/j.1574-6941.2010.01039.x

Benucci GMN, Bonito G, Falini LB, Bencivenga M, Donnini D (2012) Truffles, timber, food, and fuel: sustainable approaches for multi-cropping truffles and economically important plants. In: Zambonelli A, Bonito GM (eds) Edible ectomycorrhizal mushrooms, vol 34. Springer, Berlin, pp 265–280. doi:10.1007/978-3-642-33823-6_15

Berch SM, Bonito G (2014) Cultivation of Mediterranean species of *Tuber* (Tuberaceae) in British Columbia, Canada. Mycorrhiza 24(6):473–479. doi:10.1007/s00572-014-0562-y

Bonito GM, Gryganskyi AP, Trappe JM, Vilgalys R (2010) A global meta-analysis of Tuber ITS rDNA sequences: species diversity, host associations and long-distance dispersal. Mol Ecol 19 (1):4994–5008. doi:10.1111/j.1365-294X.2010.04855.x

Bonito G, Smith ME, Nowak M, Healy RA, Guevara G, Cázares E, Kinoshita NER, Dominguez LS, Tedersoo L, Murat C, Wang Y, Moreno BA, Pfister DH, Nara K, Zambonelli A, Trappe JM, Vilgalys R (2013) Historical biogeography and diversification of truffles in the Tuberaceae and their newly identified southern hemisphere sister lineage. PLoS One 8(1):e52765. doi:10.1371/journal.pone.0052765

Bonnet H (1869) La truffe: études sur les truffes comestibles au point de vue botanique, entomologique, forestier et commercial. A. Delahaye, Paris

Bonuso E, Zambonelli A, Bergemann SE, Iotti M, Garbelotto M (2010) Multilocus phylogenetic and coalescent analyses identify two cryptic species in the Italian bianchetto truffle, *Tuber borchii* Vittad. Conserv Genet 11(1):1453–1466. doi:10.1007/s10592-009-9972-3

Bratek Z, Merényi ZS, Varga T (2013) Changes of hypogeous funga in the Carpathian-Pannonian region in the past centuries. Acta Mycol 48(1):33–39. doi:10.5586/am.2013.005

Callot G (1999) La truffe, la terre, la vie. INRA, Versailles

Ceruti A (1960) Elaphomycetales et Tuberales. In: Bresadola J (ed) Iconographia mycologica, vol 28, Suppl 2. Trento

Ceruti A, Fontana A, Nosenzo C (2003) Le specie europee del genere *Tuber*. Una revisione storica. Museo Regionale di Scienze Naturali, Monografie XXXVII. Regione Piemonte, Torino

Cetto B (1979) Der grosse Pilzführer, vol 3. BLV Verlagsgesellschaft, München

Chatin AD (1892) La truffe. Botanique de la truffe et des plantes truffiéres. J.B. Bailliére et Fils, Paris

Chen J, Liu P (2012) Delimitation of *Tuber pseudohimalayense* and *T. pseudoexcavatum* based on morphological and molecular data. Cryptogam Mycol 32(1):83–93. doi:10.7872/crym.v32.iss1.2012.083

Chen J, Guo SX, Liu PG (2011) Species recognition and cryptic species in the *Tuber indicum* complex. PLoS One 6(1):e14625. doi:10.1371/journal.pone.0014625

Chevalier G (1973) Synthese axenique des mycorhizes de *Tuber brumale* Vitt. a partir de cultures pures du champignon. Ann Phytopathol 5(2):163–182

Chevalier G, Frochot H (1997) La truffe de Bourgogne. Pétrarque, Levallois-Perret

Chevalier G, Sourzat P (2012) Soils and techniques for cultivating *Tuber melanosporum* and *Tuber aestivum* in Europe. In: Zambonelli A, Bonito GM (eds) Edible ectomycorrhizal mushrooms, vol 34, Soil biology. Springer, Berlin, pp 163–189. doi:10.1007/978-3-642-33823-6_10

De Ferry de la Bellone C (1888) La truffe: étude sur les truffes et les truffières. J.B. Baillière et fils, Paris

De Miguel AM, Águeda B, Sánchez S, Parladé J (2014) Ectomycorrhizal fungus diversity and community structure with natural and cultivated truffle hosts: applying lessons learned to future truffle culture. Mycorrhiza 24(1):5–18. doi:10.1007/s00572-013-0554-3

Diamandis S, Perlerou C (2008) Recent records of hypogeous fungi in Greece. Acta Mycol 43 (2):139–142. doi:10.5586/am.2008.017

Dimitrova E, Gyosheva M (2008) Hypogeous ascomycetes in Bulgaria. Phytol Balc 14 (3):309–314

Donnini D, Benucci GMN, Bencivenga M, Baciarelli-Falini L (2014) Quality assessment of truffle-inoculated seedlings in Italy: proposing revised parameters for certification. For Syst 23(2):385–393. doi:10.5424/fs/2014232-05029

Douet JP, Castroviejo M, Mabru D, Chevalier G, Dupré C, Bergougnoux F, Ricard JM, Médina B (2004) Rapid molecular typing of *Tuber melanosporum, T. brumale* and *T. indicum* from tree seedlings and canned truffles. Anal Bioanal Chem 379(1):668–673. doi:10.1007/s00216-004-2643-9

Fischer E (1897) Tuberaceen und Hemiasceen. In: Die Pilze Deutschlands, Österreich und der Schweiz. V. Ascomyceten. L. Rabenhorst's Kryptogamenflora von Deutschland, Österreich und der Schweiz Bd 1, Abt. 5: Ascomyceten 2nd éd. Reprint J. Cramer, Weinheim, Leipzig, pp 37–61

Fischer CR, Suz LM, Martin MP, Colinas C (2004) *Tuber brumale* Vitt. + *Quercus ilex* L. Descr Ectomyc 7(8):135–141

Fontana A, Fasolo-Bonfante P (1971) Sintesi micorrizica di *Tuber brumale* Vitt. con *Pinus nigra* Arnold. Allionia 17:15–18

Gandeboeuf D, Dupré C, Chevalier G (1994) Différenciation des truffes européennes par l'analyse des isoenzymes. Acta Bot Gallica 141(1):455–463. doi:10.1080/12538078.1994.10515183

Gandeboeuf D, Henrion B, Dupre C, Drevet P, Nicolas P, Chevalier G, Martin F (1995) Molecular identification of *Tuber* species and isolates by PCR-based techniques. In: Stocchi V, Bonfante P, Nuti M (eds) Biotechnology of ectomycorrhizae. Plenum Press, New York, pp 151–160

Gandeboeuf D, Dupré C, Chevalier G, Roeckel-Drevet P, Nicolas P (1997) Grouping and identification of Tuber species using RAPD markers. Can J Bot 75(1):36–45. doi:10.1139/b97-005

García-Montero LG, Díaz P, Martín-Fernández S, Casermeiro MA (2008) Soil factors that favour the production of *Tuber melanosporum* carpophores over other truffle species: a multivariate statistical approach. Acta Agric Scand B-S P 58(4):322–329. doi:10.1080/09064710701743286

García-Montero LG, Valverde-Asenjo I, Moreno D, Díaz P, Hernando I, Menta C, Tarasconi K (2012) Influence of edaphic factors on edible ectomycorrhizal mushrooms: new hypotheses on soil nutrition and C sinks associated to ectomycorrhizae and soil fauna using the *Tuber* Brûlé model. In: Zambonelli A, Bonito GM (eds) Edible ectomycorrhizal mushrooms, vol 34. Springer, Berlin, pp 83–104. doi:10.1007/978-3-642-33823-6_6

Gazo J, Miko M, Chevalier G (2005) First results of inventory research on economically important species of truffles (*Tuber*) in the Tribec Mountains. Acta Fytotech Zootech 8(3):66–71

Giomaro G, Sisti D, Zambonelli A, Amicucci A, Cecchini M, Comandini O, Stocchi V (2002) Comparative study and molecular characterization of ectomycorrhizas in *Tilia americana* and *Quercus pubescens* with *Tuber brumale*. FEMS Microbiol Lett 216(1):9–14. doi:10.1111/j.1574-6968.2002.tb11407.x

Giovannetti G, Fontana A (1982) Mycorrhizal synthesis between Cistaceae and Tuberaceae. New Phytol 92(4):533–537. doi:10.1111/j.1469-8137.1982.tb03412.x

Granetti B, De Angelis A, Materozzi G (2005) Umbria terra di tartufi. Regione Umbria, Assessorato regionale Agricoltura, Foreste, Caccia e Pesca

Grebenc T (2008) Research and cultivation of truffes in Slovenia – current status. In: Reynal B (ed) La culture de la truffe dans le monde. Actes du Colloque, Brive-la-Gaillarde. 2. Fevrier 2007, pp 183–191

Guerin-Laguette A, Cummings N, Hesom-Williams N, Butler R, Wang Y (2012) Mycorrhiza analyses in New Zealand truffières reveal frequent but variable persistence of *Tuber melanosporum* in co-existence with other truffle species. Mycorrhiza 23(2):87–98. doi:10.1007/s00572-012-0450-2

Guevara G, Bonito G, Cázares E, Rodríguez J, Vilgalys R, Trappe JM (2008) *Tuber regimontanum*, new species of truffle from Mexico. Rev Mex Micol 26:17–27

Hall IR, Brown G, Zambonelli A (2007) Taming the truffle: the history, lore, and science of the ultimate mushroom. Timber Press, Portland

Hall I, Zambonelli A, Nelson P (2008) Distinguishing the Périgord black truffle from the winter truffle and other look-alikes in the laboratory and field, 3rd edn. Truffles and Mushrooms, Dunedin

Hansen O, Knudsen H (2000) Nordic macromycetes 1: Ascomycetes, vol 1. Nordsvamp, Copenhagen

Hawker LE (1954) British hypogeous fungi. Philos Trans R Soc B 237(650):429–546. doi:10.1098/rstb.1954.0002

Henrion B, Chevalier G, Martin F (1994) Typing truffle species by PCR amplification of the ribosomal DNA spacers. Mycol Res 98(1):37–43. doi:10.1016/S0953-7562(09)80333-X

Ho WH, Anderson S, Guerin-Laguette A, Hesom-Williams N, Wang Y, Braithwaite M, Hill CF, Alexander BJR (2008) *Tuber brumale*, a new truffle in New Zealand. In: Abstract of the 9th international congress of plant pathology, Torino, 24–29 Aug 2008

Hochberg ME, Bertault G, Poitrineau K, Janssen A (2003) Olfactory orientation of the truffle beetle, *Leiodes cinnamomea*. Entomol Exp Appl 109(2):147–153. doi:10.1046/j.1570-7458.2003.00099.x

Hollós L (1911) Magyarország földalatti gombái, szarvasgombaféléi. K.M. Természettudományi Társulat, Budapest

Hu HT (1992) *Tuber formosanum* sp. nov. and its mycorrhizal associations. Q J Exp For Natl Taiwan Univ 6:79–86

Iotti M, Amicucci A, Stocchi V, Zambonelli A (2002) Morphological and molecular characterization of mycelia of some *Tuber* species in pure culture. New Phytol 155(3):499–505. doi:10.1046/j.1469-8137.2002.00486.x

Janex-Favre MC, Parguey-Leduc A, Sejalon-Delmas N, Dargent R, Kulifaj M (1996) Étudê préliminaire de l'ascocarpe de *Tuber indicum*, truffe chinoise récemment introduite en France. CR Acad Sci III-Vie 319(6):517–521

Jeandroz S, Murat C, Wang Y, Bonfante P, Le Tacon F (2008) Molecular phylogeny and historical biogeography of the genus *Tuber*, the "true truffles". J Biogeogr 35(1):815–829. doi:10.1111/j.1365-2699.2007.01851.x

Kiss M, Csóka M, Győrfi J, Korány K (2011) Comparison of the fragrance constituents of *Tuber aestivum* and *Tuber brumale* gathered in Hungary. J Appl Bot Food Qual 84(1):102–110

Ławrynowicz M (1993) Distributional limits of truffles in the northern Europe. Micol Veg Medit 7 (1):31–38

Lefevre CK, Hall IR (2001) The status of truffle cultivation: a global perspective. Acta Hortic 556:513–520. doi:10.17660/ActaHortic.2001.556.75

Li SH, Zheng LY, Liu CY, Wang Y, Li L, Zhao YC, Zhang XL, Yang M, Xiong HK, Qing Y, Wang L, Zhou DQ (2014) Two new truffles species, *Tuber alboumbilicum* and *Tuber pseudobrumale* from China. Mycol Prog 13(4):1157–1163. doi:10.1007/s11557-014-1004-0

Linde CC, Selmes H (2012) Genetic diversity and mating type distribution of *Tuber melanosporum* and their significance to truffle cultivation in artificially planted truffieres in Australia. Appl Environ Microb 78(18):6534–6539. doi:10.1128/AEM.01558-12

Liu B, Fischer C, Bonet JA, Olivera A, Inchusta A, Colinas C (2014) Pattern of *Tuber melanosporum* extramatrical mycelium expansion over a 20-year chronosequence in *Quercus ilex*-truffle orchards. Mycorrhiza 24(1):47–54. doi:10.1007/s00572-014-0559-6

Mamoun M, Olivier JM (1993) Competition between *Tuber melanosporum* and other ectomycorrhizal fungi under two irrigation regimes. Plant Soil 149(1):211–218. doi:10.1007/BF00016611

March RE, Richards DS, Ryan RW (2006) Volatile compounds from six species of truffle – head-space analysis and vapor analysis at high mass resolution. Int J Mass Spectrom 249:60–67. doi:10.1016/j.ijms.2005.12.038

Marjanović Ž (2008) Truffles and possibilities for their cultivation in Serbia-current situation. In: Reynal B (ed) La culture de la truffe dans le monde. Actes du Colloque, Brive-la-Gaillarde. 2 Fevrier 2007, pp 163–172

Martin-Santafe M, Perez-fortea V, Zuriaga P, Barriuso J (2014) Phytosanitary problems detected in truffle cultivation in Spain. For Syst 23(2):307–316. doi:10.5424/fs/2014232-04900

Mattirolo O (1933) I funghi ipogei della Campania, del Lazio e del Molise raccolti dal compianto Prof. Carlo Campbell. Plant Biosyst 40(3):313–326. doi:10.1080/11263503309437906

Mauriello G, Marino R, D'Auria M, Cerone G, Rana GL (2004) Determination of volatile organic compounds from truffles via SPME–GC–MS. J Chromatogr Sci 42(6):299–305. doi:10.1093/chromsci/42.6.299

Merényi Z (2014) Molecular biological and ecological studies on hypogeous fungi. PhD thesis. Eötvös Loránd University, Budapest. doi:10.15476/ELTE.2014.001

Merényi Z, Illyés Z, Völcz G, Bratek Z (2010) Creation database application for development on truffle cultivation methods. In: Proceedings of the first conference on the "European" truffle *Tuber aestivum/uncinatum*, Österreichische Zeitschrift Für Pilzkunde, vol 19, pp 239–244

Merényi Z, Varga T, Geml J, Orczán ÁIK, Chevalier G, Bratek Z (2014) Phylogeny and phylogeography of the *Tuber brumale* aggr. Mycorrhiza 24(1):101–113. doi:10.1007/s00572-014-0566-7

Michael E, Hennig B (1960) Handbuch für Pilzfreunde. Gustav Fischer Verlag, Jena

Milne RI (2006) Northern Hemisphere plant disjunctions: a window on tertiary land bridges and climate change? Ann Bot 98(3):465–472. doi:10.1093/aob/mcl148

Montecchi A, Sarasini M (2000) Funghi ipogei d'Europa. Associazione Micologica Bresadola – Fondacione Centro Studi Micologici, Trento

Morcillo M, Sánchez M, Mateu J, Gracia E (2007) Open field inoculation of adult hazelnut groves with *Tuber brumale & Tuber melanosporum* Vitt. In: Proceedings of the 1st world conference on the conservation and sustainable use of wild fungi, Cordoba, 10–16 Dec 2007, Junta de Andalucia, Consejera de Medio Ambiente, p 322

Murat C (2015) Forty years of inoculating seedlings with truffle fungi: past and future perspectives. Mycorrhiza 25(1):77–81. doi:10.1007/s00572-014-0593-4

Murat C, Díez J, Luis P, Delaruelle C, Dupré C, Chevalier G, Bonfante P, Martin F (2004) Polymorphism at the ribosomal DNA ITS and its relation to postglacial re-colonization routes of the Perigord truffle *Tuber melanosporum*. New Phytol 164(2):401–411. doi:10.1111/j.1469-8137.2004.01189.x

Nei M (1987) Molecular evolutionary genetics. Columbia University Press, New York

Nilsson HR, Kristiansson E, Ryberg M, Hallenberg N, Larsson KH (2008) Intraspecific ITS variability in the kingdom fungi as expressed in the international sequence databases and its implications for molecular species identification. Evol Bioinform 4(1):193–201

Olivier JM, Savignac JC, Sourzat P (2012) Truffe et trufficulture. Éditions Falanc, Périgueux

Pacioni G, Pomponi G (1989) Chemotaxonomy of some Italian species of *Tuber*. Micol Veg Medit 4(1):63–72

Pacioni G, Bellina-Agostinone C, D'Antonio M (1990) Odour composition of the *Tuber melanosporum* complex. Mycol Res 94(2):201–204. doi:10.1016/S0953-7562(09)80614-X

Palenzona M (1969) Sintesi micorrizica tra *Tuber aestivum* Vitt., *Tuber brumale* Vitt., *Tuber melanosporum* Vitt. E semenziali di *Corylus avellana*. Allionia 15:122–131

Palenzona M, Chevalier G, Fontana A (1972) Sintesi micorrizica tra i miceli in coltura di *T. brumale*, *T. melanosporum*, *T. rufum* e semenzali di conifere e latifoglie. Allionia 18:41–52

Paolocci F, Angelini P, Cristofari E, Granetti B, Arcioni S (1995) Identification of *Tuber* spp and corresponding ectomycorrhizae through molecular markers. J Sci Food Agric 69(4):511–517. doi:10.1002/jsfa.2740690416

Paolocci F, Rubini A, Granetti B, Arcioni S (1999) Rapid molecular approach for a reliable identification of *Tuber* spp. FEMS Microbiol Ecol 28(1):23–30. doi:10.1111/j.1574-6941.1999.tb00557.x

Parladé J, De la Varga H, De Miguel AM, Sáez R, Pera J (2013) Quantification of extraradical mycelium of *Tuber melanosporum* in soils from truffle orchards in northern Spain. Mycorrhiza 23(2):99–106. doi:10.1007/s00572-012-0454-y

Payen T, Murat C, Gigant A, Morin E, De Mita S, Martin F (2015) A survey of genome-wide single nucleotide polymorphisms through genome resequencing in the Périgord black truffle (*Tuber melanosporum* Vittad.). Mol Ecol Resour 15(5):1243–1255. doi:10.1111/1755-0998.12391

Pegler DN, Spooner BM, Young TWK (1993) British Truffles – a revision of british hypogeous fungi. Royal Botanic Gardens, Kew

Poitou N, Olivier JM (1990) Effect of copper on mycelium of three edible ectomycorrhizal fungi. Agr Ecosyst Environ 28(1–4):403–408. doi:10.1016/0167-8809(90)90070-T

Pomarico M, Figliuolo G, Rana GL (2007) *Tuber* spp. biodiversity in one of the southernmost European distribution areas. Biodivers Conserv 16(1):3447–3461. doi:10.1007/978-1-4020-6865-2_15

Ricard JM (2003) La truffe. Guide technique de trufficulture. CTIFL, Paris

Riousset L, Riousset G, Chevalier G, Bardet MC (2001) Truffes d'Europe et de Chine. INRA, Paris

Roux C, Séjalon-Delmas N, Martins M, Parguey-Leduc A, Dargent R, Bécard G (1999) Phylogenetic relationships between European and Chinese truffles based on parsimony and distance analysis of ITS sequences. FEMS Microbiol Lett 180(2):147–155. doi:10.1111/j.1574-6968.1999.tb08789.x

Séjalon-Delmas N, Roux C, Martins M, Kulifaj M, Bécard G, Dargent R (2000) Molecular tools for the identification of *Tuber melanosporum* in agroindustry. J Agric Food Chem 48 (6):2608–2613. doi:10.1021/jf9910382

Sourzat P (2005) Pressure of contamination mycorrhizal fungi against black truffle in spontaneous and cultivated truffieres in France. In: Honrubia M, Morte A, Torrente P (eds) Abstracts of the fourth international workshop on edible mycorrhizal mushroom. Universidad de Murcia, Murcia, p 116

Sourzat P (2011) Black truffle cultivation and competing fungi. In: Proceedings of the 7th international conference on mushroom biology and mushroom products (ICMBMP7), pp 516–528

Sourzat P, Sánchez A, Ourzick A, Roux JJ (2001) Analyse du statut mycorhizien de brulés stériles dans trois régions de France: Midi-Pyrenees (Lot), Paca (Var), Charente Poitou (Vienne). In: Proceedings of the 5th international congress: science and cultivation of truffle, Aix-en-Provence, 4–6 Mar 1999, Federation Francaise des trufficulteurs, Paris, pp 8.474–8.479

Stobbe U, Büntgen U, Sproll L, Tegel W, Egli S, Fink S (2012) Spatial distribution and ecological variation of re-discovered German truffle habitats. Fungal Ecol 5(5):591–599. doi:10.1016/j.funeco.2012.02.001

Szemere L (2005) Föld alatti gombavilág. Első Magyar Szarvasgombász Egyesület, Budapest

Taylor JW, Jacobson DJ, Kroken S, Kasuga T, Geiser DM, Hibbett DS, Fisher MC (2000) Phylogenetic species recognition and species concepts in fungi. Fungal Genet Biol 31 (1):21–32. doi:10.1006/fgbi.2000.1228

Türkoğlu A, Castellano MA (2014) New records of some Ascomycete truffle fungi from Turkey. Turk J Bot 38(2):406–416. doi:10.3906/bot-1303-24

Valverde-Asenjo I, García-Montero LG, Quintana A, Velázquez J (2009) Calcareous amendments to soils to eradicate *Tuber brumale* from *T. melanosporum* cultivations: a multivariate statistical approach. Mycorrhiza 19(3):159–165. doi:10.1007/s00572-008-0224-z

Vittadini C (1831) Monographia Tuberacearum ex Typographia. Felicis Rusconi, Milano

Wang S, Marcone MF (2011) The biochemistry and biological properties of the world's most expensive underground edible mushroom: Truffles. Food Res Int 44(9):2567–2581. doi:10.1016/j.foodres.2011.06.008

Wang Y, Moreno G, Riousset LJ, Manjo JL, Riousset G, Fourre G, Di Massimo G, García-Montero LG, Diez J (1998) *Tuber pseudoexcavatum* sp. nov. a new species from China commercialized in Spain, France and Italy with additional comments on Chinese truffles. Cryptogam Mycol 19(1–2):113–120

Wang Y, Tan ZM, Zhang DC, Murat C, Jeandroz S, Le Tacon F (2006a) Phylogenetic relationships between *Tuber pseudoexcavatum*, a Chinese truffle, and other *Tuber* species based on parsimony and distance analysis of four different gene sequences. FEMS Microbiol Lett 259 (2):269–281. doi:10.1111/j.1574-6968.2006.00283.x

Wang Y, Tan ZM, Zhang DC, Murat C, Jeandroz S, Le Tacon F (2006b) Phylogenetic and populational study of the *Tuber indicum* complex. Mycol Res 110(9):1034–1045. doi:10.1016/j.mycres.2006.06.013

Zambonelli A, Salomoni S, Pisi A (1993) Caratterizzazione anatomo-morfologica delle micorrize di *Tuber* spp. su *Quercus pubescens* Willd. Micol Ital 22(3):73–90

Zambonelli A, Salomoni S, Pisi A (1995) Caratterizzazione anatomo-morfologica delle micorrize de *Tuber borchii*, *Tuber aestivum*, *Tuber mesentericum*, *Tuber brumale*, *Tuber melanosporum* su *Pinus pinea*. Micol Ital 24(2):119–137

Zeppa S, Sisti D, Pierleoni R, Potenza L, Guescini M, Vallorani L, Stocchi V (2005) *Tilia platyphyllos* Scop.-*Tuber brumale* Vittad. vs. *T. platyphyllos* Scop.-*T. borchii* Vittad. ectomycorrhizal systems: a comparison of structural and functional traits. Plant Physiol Biochem 43(7):709–716. doi:10.1016/j.plaphy.2005.06.008

Zhang L, Yang ZL, Song DS (2005) A phylogenetic study of commercial Chinese truffles and their allies: taxonomic implications. FEMS Microbiol Lett 245(1):85–92. doi:10.1016/j.femsle.2005.02.028

Chapter 5
Taxonomy, Biology and Ecology of *Tuber macrosporum* Vittad. and *Tuber mesentericum* Vittad.

Gian Maria Niccolò Benucci, Andrea Gógán Csorbai, Leonardo Baciarelli Falini, Giorgio Marozzi, Edoardo Suriano, Nicola Sitta, and Domizia Donnini

5.1 Introduction

Nowadays, some truffle species such as *Tuber melanosporum* Vittad., *Tuber aestivum* Vittad., *Tuber borchii* Vittad. and *Tuber brumale* Vittad. have a flourishing market and are routinely and successfully cultivated. Plantations are established with mycorrhized seedlings produced by commercial nurseries giving important extra income to farmers, especially in marginal lands or disadvantaged areas (Chevalier 1998; Olivier 2000; Hall and Yun 2002; Bencivenga and Baciarelli Falini 2012). In addition to those previously mentioned species, for some other less famous truffles, locally collected by truffle hunters, national and international markets are struggling but recently starting to develop.

G.M.N. Benucci (✉)
Department of Agricultural, Food and Environmental Sciences, University of Perugia, Borgo XX Giugno 74, 06121 Perugia, Italy

Department of Plant, Soil and Microbial Sciences, Michigan State University, East Lansing, MI 48824, USA
e-mail: gian.benucci@gmail.com

A.G. Csorbai
Department of Horticulture, Szent István University, Páter Károly str. 1, 2100 Gödöllő, Hungary

L.B. Falini • G. Marozzi • D. Donnini
Department of Agricultural, Food and Environmental Sciences, University of Perugia, Borgo XX Giugno 74, 06121 Perugia, Italy

E. Suriano
Via F.lli Mazzocchi 23, 00133 Rome, Italy

N. Sitta
Loc. Farné 32, 40042 Lizzano in Belvedere, Bologna, Italy

A. Zambonelli et al. (eds.), *True Truffle (*Tuber *spp.) in the World*, Soil Biology 47,
DOI 10.1007/978-3-319-31436-5_5

In particular, the black smooth truffle *Tuber macrosporum* Vittad. (also known as "garlic truffle" for its aromatic resemblance of garlic) has an aroma that is vaguely similar to the esteemed white truffle *Tuber magnatum* Pico (Vittadini 1831; Hollós 1911; Ceruti 1968; Montecchi and Lazzari 1993; Iotti et al. 2002; Riousset et al. 2012).

Despite its attractive aromatic traits, *T. macrosporum* is only accidentally collected in Italy by truffle hunters who are trying to find *T. magnatum*, growing in the same environment; hence, it has no commercial interest and does not enjoy privileged hunting. Small ascoma size, insecure, strongly weather-dependent yields in natural habitats, the common practice for traders to mix it with *T. aestivum* or immature *T. melanosporum*, are all the reasons of its limited reputation. Notwithstanding the above facts, *T. macrosporum* is certainly an attractive species and can merit more attention because of its enticing organoleptic features and its wide distribution across Europe. Moreover, its successful cultivation on experimental orchards (Vezzola 2005, 2010) has resulted in the production of seedlings inoculated with *T. macrosporum* spores by commercial nurseries with the aim of expanding cultivation.

Another species of a certain market which deserves specific attention is *Tuber mesentericum* Vittad., also called "black truffle of Bagnoli Irpino", from the name of a city in the Province of Avellino (Italy), where this truffle has a long history and tradition of harvest and gastronomic use (Garofoli 1906). It is also famous in northeast of France where it is more expensive than *T. aestivum* [often about 450 € per kg in detail markets, e.g. Pulnoy township (Claude Murat, pers comm)]. This species often emits a strong and very distinct phenolic-like aroma that makes it not so appreciated outside the traditional area (Vittadini 1831; Granetti et al. 2005; Riousset et al. 2012). Opinions on the gastronomic and consequently economic value of this truffle are anyway controversial, subjective and often linked to local traditions or customs. Immature fruiting bodies of *T. mesentericum* can be easily confused with the morphologically very similar *T. aestivum* and *T. aestivum* var. *uncinatum* (Chatin) I. R. Hall, P. K. Buchanan, Y. Wang and Cole, which now are considered conspecific (Paolocci et al. 2004; Wedén et al. 2005; Molinier et al. 2013) even if the identity of "uncinatum" survives as a commercial type of *T. aestivum* mainly in Italy and in France (see Chap. 3).

In this chapter, morphological characteristics of *T. macrosporum* and *T. mesentericum* fruiting bodies and ectomycorrhizas (ECMs) are described; taxonomic controversies about macro- and microscopic traits necessary for specific identification are disclosed; phylogenetic findings are highlighted in order to clarify the position of *T. macrosporum* and *T. mesentericum* in the *Tuber* phylogenetic tree, and a summary of their ecological requirements (including soil, climate and host trees) is finally provided.

5.2 Methodology

From fresh/dried *T. macrosporum* ascomata collected in Italy but also in some locations in Europe, genomic DNA was directly amplified with ITS1 and ITS4 universal primers (White et al. 1990) for fungal barcoding and then sequenced on both strands following the methodology reported by Bonito (2009). Forward and reverse sequences were then edited into contigs in BioEdit (Hall 1999). Corrected sequences were submitted to GenBank (Benson et al. 2013) with the following accession numbers: KP738345 to KP738396. Additional available ITS sequences of *T. macrosporum* were also downloaded from GenBank (www.ncbi.nlm.nih.gov) to improve the phylogenetic resolution of major clades. For *T. mesentericum*, only public sequences were used for the molecular analyses. Sequence accessions and details are reported in the phylogenetic trees (Figs. 5.2 and 5.5).

Before building phylogenetic trees, all the obtained sequences labelled as *Tuber macrosporum* and *Tuber mesentericum* downloaded from GenBank were aligned in MEGA6 to detect mislabelled sequences or misidentifications. Sequences that did not align correctly were then compared to others in GenBank using the BLAST algorithm to verify their identity (Altschul et al. 1990).

Phylogenetic tree reconstructions were performed using the maximum likelihood method (Tamura et al. 2011). The trees were drawn to scale, with branch lengths in the same units as those of the evolutionary distances used to infer the phylogenetic tree, and bootstrap values (999 replicates) are shown next to the branches (Felsenstein 1985). Evolutionary analyses were conducted in MEGA6 (Tamura et al. 2013).

5.3 Characteristics of *Tuber macrosporum*

5.3.1 Morphology of Tuber macrosporum *Ascomata*

Ascoma maturation usually ranges from August to December, in the same period of *T. magnatum*, with the peak of mature specimens during the autumn. Typical ascomata of *T. macrosporum* have generally an irregular shape, lobed, but also regular and/or subglobose, with a diameter of 2–5 cm, exceptionally bigger (Fig. 5.1a). The blackish peridium is irregularly stained by reddish-brown, very short and flat warts of variable shape and size. Gleba varies from grey brown to brown lilac and purple brown when mature, with thick, branching and winding white veins. The asci of $90-120 \times 60-80$ µm size contain 1–3(4), generally three yellowish-brown spores. The ellipsoid spores are considered definitely the biggest between the main truffle species. These spores are $40-70(-80) \times 30-55(-60)$ µm (Fig. 5.1b, c), covered with reticulate-alveolate, polygonal, 2–4 µm high, dense, closed and small meshes.

Fig. 5.1 *Tuber macrosporum* characteristics: (**a**) mature ascomata; (**b**) gleba with spores; (**c**) a two spores ascus (25 μm); (**d**) natural productive site with *Q. cerris*, *Q. pubescens* and *O. carpinifolia*; (**e** and **f**) ECMs with cystidia (0.4 mm); (**g**) ramification of cystidia (25 μm); (**h**) outer mantle layer (25 μm)

5.3.2 Tuber macrosporum *Ectomycorrhizal Synthesis and Morphology*

The first mycorrhizal synthesis of *T. macrosporum* with hornbeam seedlings was published by Giovannetti and Fontana (1980–1981) with the description of the ECMs and their distinctive traits. Some subsequent works expanded the topics: oaks and hazel seedlings were inoculated by *T. macrosporum* spore slurry and the obtained ECMs were photographed and described (Zambonelli et al. 1993; Granetti 1995; Vezzola 2005; Agerer and Rambold 2004–2008). Nevertheless, those descriptions are controversial and do not really focus on simple and valuable morphological traits that are fundamental for a correct species identification. Moreover, no molecular confirmation for ECMs belonging to *T. macrosporum* was reported in literature before 2012 when Benucci and colleagues (2012) described morphologically *T. macrosporum* ECMs on *Quercus robur* L., *Quercus cerris* L. and *Corylus avellana* L. and identified its DNA through the use of species-specific primers (Benucci et al. 2011). The same authors also described and characterized the ECM communities of cultivated and natural *T. macrosporum* sites (Benucci et al. 2014).

Tuber macrosporum ECMs on *Q. robur* and *C. avellana* are simple or ramified in a monopodial-pinnate or monopodial-pyramidal pattern (Fig. 5.1e, f). Simple ECM tips are almost straight, cylindrical or club shaped with rounded ends. The colour of the ECMs varies considerably: the youngest are light yellow with pale grey shades and cystidia are sinuous and septate, with very thick walls and branched at various angles (frequently with sharp angles). The colour of the cystidia varies from light yellow when young (sometimes with a greyish shade) to ochre at maturity. They are ramified (Fig. 5.1g) and tend to merge, creating anastomoses that form an abundant web of mycelium around the ECM that is typically orange in colour (Fig. 5.1e, f). Formation of needle-shaped cystidia reported by Granetti et al. (2005) is never found in any of the *T. macrosporum* ECMs examined by Benucci et al. (2012).

The mantle is pseudoparenchymatous and composed of four to six cell layers. The Hartig net penetrates into the first two to three cell layers of the root parenchyma (Benucci et al. 2012). The outer mantle surface is either covered densely by mycelium (cottony) or it is smooth to loosely grainy. In both cases, it is composed of angular (type L according to Agerer and Rambold 2004–2008) and epidermoid (type M) cells that form an uneven, regular puzzle-like pattern (Fig. 5.1h). Benucci et al. (2012) showed that ECM mantle might differ among the apex, middle part and base of the ECM with the middle part being the most variable.

5.3.3 Tuber macrosporum *Taxonomy and Phylogeny*

According to Index Fungorum, the global nomenclator of fungal taxonomic names (www.indexfungorum.org), the correct name of this species is *Tuber macrosporum* Vittad.

Recent phylogenetic studies on the *Tuber* genus, based on the ITS (internal transcribed spacer) region and LSU (large subunit) of the nuclear rDNA (ribosomal DNA), show that the Macrosporum clade is one of the ancestral lineages and includes two species: *T. macrosporum* and *Tuber canaliculatum* Gilkey (Jeandroz et al. 2008; Bonito et al. 2013). In addition, molecular evidence of truffles belonging to the Macrosporum group has been reported also for Japan (Kinoshita et al. 2011).

It is worth noting that some sequences downloaded from GenBank have been misidentified or mislabelled and do not belong to *T. macrosporum*. In particular, FJ809838, FJ809839, JN392325 and HE601929 show the highest similarity with *T. canaliculatum*; JQ288921 shows the highest similarity with *Tuber malenconii* Donadini, Riousset, G. Riousset and G. Chev; and HE602584 shows the highest similarity with *Tuber pseudoexcavatum* Y. Wang, G. Moreno, Riousset, Manjón and G. Riousset.

The maximum likelihood phylogenetic reconstruction based on the ITS region shows *T. macrosporum* position in the Macrosporum clade that includes the North American species *T. canaliculatum* (Fig. 5.2) (Bonito et al. 2010). Two bootstrap-supported distinct clades are present in the tree: most of the Italian sequences cluster in the clade II, while clade I comprises many samples from Central and Eastern Europe. Interesting to note that in the clade I two sequences from ECM tips (JX474822 and JX474809) clustered together and with sequences obtained from fruiting bodies (e.g. KP738346) which were collected in Sigillo (Italy) and previously analysed by Benucci et al. (2014) in a fungal community analysis study. Even if with high divergence, the sample AB553344 from a truffle fruiting body collected in Japan showed to be close to *T. canaliculatum*, suggesting the possible presence of a new Asiatic species belonging to the Macrosporum clade.

5.3.4 **Tuber macrosporum** *Geographic Distribution and Ecological Demand*

The truffle *T. macrosporum* has a wide distribution in Europe, being considered common in Serbia, Hungary and Romania, less frequent in Italy and rare in France and Great Britain, but also occurs in Switzerland, Germany, Ukraine, Croatia and Slovenia and has been recently reported from Slovakia, Poland and Turkey (Ceruti et al. 2003; Miko et al. 2006; Hall et al. 2007; Marjanović et al. 2010; Piltaver and Ratoŝa 2010; Benucci et al. 2012; Stobbe et al. 2012; Hilszczańska et al. 2013). Mature ascomata can be found as early as June (Vezzola 2010), but more often from September to December. *Tuber macrosporum* is generally collected from plain sites or from foothills to low mountains, often found on north-oriented slopes, lowlands or floodplains of watercourses (Vittadini 1831; Milenkovic and Marjanović 2001). Although annual rainfall in *T. macrosporum* sites was reported variable

Fig. 5.2 *Tuber macrosporum* maximum likelihood phylogenetic tree based on the Jukes–Cantor model (Jukes and Cantor 1969): bootstrap values >65 % are shown next to branching nodes. A discrete gamma distribution [+*I*] was used to model evolutionary rate differences among sites. The analysis involved 44 nucleotide sequences. All positions containing gaps and missing data were eliminated. There were a total of 337 positions in the final dataset. Sequences produced in this study have accessions starting with KP

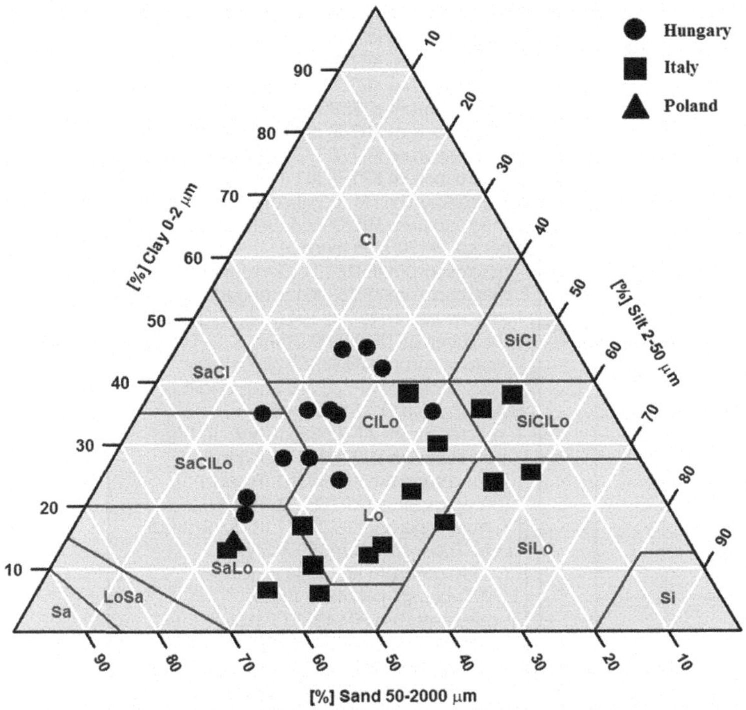

Fig. 5.3 Soil textures of different *T. macrosporum* habitats (Benucci 2011 modified; Gógán Csorbai 2011; Hilszczańska et al. 2014)

(520–850 mm), water dependence of the species is undoubted and soil moisture is very often complemented by arriving waters (subsurface water, flooding, etc.).

Soil genetic types include chernozems, luvisols, and planosols but also rendzic leptosols (Gógán Csorbai 2011; Hilszczańska et al. 2014). The species regularly shares habitats with *T. magnatum*, resulting in similar characteristics of lime-rich, neutral or slightly alkaline soil with both good aeration and humid environment. However, recent findings revealed that soil compaction in *T. macrosporum*-inhabited soils is very common. Compacted layers are typical in 30–60 cm depth, but in some cases, they occur close to the surface (5–10 cm). Due to compacted layers and the presence of water, gleys and ferric precipitations are frequent (Gógán Csorbai 2011). Soil granulometry of samples coming from different geographical origins represents slight variability in soil textures but without extreme patterns (Fig. 5.3). The most common soil types are clay loam, loam and sandy loam.

Some findings cite *T. macrosporum* from neutral or alkaline soils of pH around 7.5 with various lime contents (Djurdjevic et al. 1999; Miko et al. 2006; Hilszczańska et al. 2014). Researches focusing on the ecological demands of *T. macrosporum* affirm its preference to the above-mentioned characteristics; however, lime-free, slightly acidic environment cannot be considered as limiting factor for the species (Gógán Csorbai et al. 2010). Results also reveal high organic

Table 5.1 Main chemical characteristics of soil samples from 88 *T. macrosporum* habitats (Benucci et al. 2014, modified; Gógán Csorbai et al. 2010; Gógán Csorbai 2011)

	Range	Median	Average	SD
pH H_2O	5.7–8.2	6.87	6.89	0.59
pH KCl	4.6–7.3	6.54	6.48	0.59
$CaCO_3$ (%)	0–59	1.95	6.55	9.74
Organic matter (%)	3.2–13.4	7.32	6.95	2.16
AL-P (ppm)	5–512	49	103.26	123.59
AL-K (ppm)	85–1790	330.5	441.96	295.59

Range, median, average and standard deviation (SD) value are reported

matter and variable content of phosphorus and potassium in natural habitats (Table 5.1).

The developing environment of *T. macrosporum* is also characterized of different symbiotic partners involved in the life cycle of this truffle. Mixed deciduous, closed-canopy forests are considered as very suitable habitats for *T. macrosporum*. The most common host trees of the species are oaks (*Quercus pubescens* Willd., *Q. robur*, *Quercus petraea* Liebl., *Q. cerris*), hazelnut (*C. avellana*), hornbeams (*Ostrya carpinifolia* Scop., *Carpinus betulus* L.), willows (*Salix viminalis* L., *Salix alba* L., *Salix vitellina* L., *Salix caprea* L.), lindens (*Tilia cordata* Miller, *Tilia platyphyllos* Scop.), beeches (*Fagus sylvatica* L.) and poplars (*Populus nigra* L., *Populus tremula* L., *Populus alba* L.) (Ceruti et al. 2003; Miko et al. 2006; Marjanović et al. 2010; Gógán Csorbai 2011).

5.4 Characteristics of *Tuber mesentericum*

5.4.1 Morphology of Tuber mesentericum *Ascomata*

The morphology of *T. mesentericum* ascomata is very similar to that of *T. aestivum*; the overlapping features of the two species make some authors designate them as "*Tuber aestivum-mesentericum* complex" (Pacioni and Pomponi 1991). Ascomata of *T. mesentericum* are rounded or subglobose, can reach the size of 10 cm in diameter and are covered with brown-black pyramidal warts (Fig. 5.4a, on the right). Transverse streaks (Fig. 5.4e) are present in the peridium warts in *T. aestivum* and in *T. mesentericum* as well, even if less evident and frequent in the latter. Even if basal depression or cavity has been considered a distinguishing feature for differentiating ascomata of *T. mesentericum* from ones of *T. aestivum* (Montecchi and Sarasini 2000; Ceruti et al. 2003), in our experience, this characteristic is not a valid taxonomic trait, as it is possible to find ascomata of *T. aestivum* with the basal cavity as well as *T. mesentericum* ascomata without it (Fig. 5.4a). Gleba colour is typically dark grey brown, often with violet shades (Fig. 5.4a, on the right) with numerous white intensively winding veins at full maturity, in contrast to the gleba

Fig. 5.4 *Tuber mesentericum* characteristics: (**a**) mature *T. aestivum* var. *uncinatum* (on the *left*) and *T. mesentericum* (on the *right*) ascomata; (**b–d**) different *T. mesentericum* ascospores (25 μm); (**e**) transverse streaks in *T. mesentericum* peridium warts (*red arrows*); (**f**) *T. mesentericum* natural productive site with *F. sylvatica*

of *T. aestivum* (Fig. 5.4a, on the left) that ranges from yellowish or light brown to ochre, but never dark brown with violet shades.

Globose or subglobose and pedunculate asci contain (1)2–4(6), yellowish-brown, ellipsoid spores (Figs. 5.4b–f) of $28 - 33 \times 20 - 23$ μm size according to Montecchi and Sarasini (2000) and Ceruti et al. (2003). It is worth noting that in our experience spore shape can vary from ellipsoid to perfectly globose (Figs. 5.4b–d), and this variability (more or less important) can be detected even in the same ascoma. Spore surface is reticulate-alveolate with irregular polygonal meshes of 3–5 μm height. Meshes are typically incomplete and often with a crest in the inside, but it can also happen, even in the same ascoma, to find complete meshes similar to those of *T. aestivum* spores (Figs. 5.4b–d).

Tuber mesentericum scent is generally strong, with frequent unpleasant note reminding of phenol, tar and/or iodine. This note, highly variable, can be immediately perceivable in the specimens when freshly harvested or can reveal itself only some days after, especially if the truffles are conserved at low temperatures in the fridge. It can be absent in mature specimens, as when freshly harvested, as some days after; on the contrary, sometimes it can be present even in immature ascomata. It has been showed for *T. aestivum* that the variability in the truffle aroma caused by volatile organic compounds (VOCs) can have a genotype basis (Splivallo et al. 2012; see Chap 3). Besides it can be also influenced by soil and ascoma-associated microbes (Buzzini et al. 2005; Splivallo et al. 2014). Anyway, the phenolic unpleasant note of *T. mesentericum* can be present or absent even in the same area of harvesting, and this variation has not been studied in details yet. In our experience, it is possible to find freshly harvested truffles with *T. mesentericum* morphological characteristics, but with a pleasant aroma, complex and deep, similar to those of *T. aestivum* or even *T. melanosporum*. A collection of mature *T. mesentericum* specimens (in the order of some tens of kilos) absolutely free of phenolic aromatic component were observed in Mediterranean habitats of Salento (Southern Italy) and in the province of Rome, under *Q. ilex* and in flat areas with reforestations of *Quercus* spp., in late spring and early summer. These particular *T. mesentericum* ascoma collections are at present under study.

5.4.2 Tuber mesentericum *Ectomycorrhizal Synthesis and Morphology*

The ECMs of *T. mesentericum* have been mentioned as early as 1988 (Giraud 1988), collected in a natural truffle field. Detailed description of the ECMs on nursery plants revealed monopodial-pyramidal ramification type on *C. avellana* (Rauscher et al. 1995), dichotomous on *Pinus pinea* L. (Zambonelli et al. 1995) and monopodial-pinnate and monopodial-pyramidal on *Q. pubescens* seedlings (Zambonelli et al. 1993). Ectomycorrhizas are reported as densely wolly, their color can vary from ochre, to yellowish brown and to red. The surface of the mantle layer is plectenchymatous and pseudoparenchymatous with angular (type L, according to Agerer and Rambold 2004–2008) mantle cells of 3 – 11 × 6 – 20 μm size. Cystidia are awl shaped, bristle-like (type A) with proximal ramification, although no ramification was also reported (Zambonelli et al. 1995). Brownish cystidia were measured of 1.9–5 μm of diameter and 130–1520 μm long. Despite the morphological descriptions of the ECM, no molecular evidence is present in literature regarding isolation of DNA from *T. mesentericum* ECMs so far.

5.4.3 **Tuber mesentericum** ***Taxonomy and Phylogeny***

In Index Fungorum, the names *Tuber mesentericum* Vittad., with its variety (var. *mesentericum* Vittad.), and *Tuber mesentericum* var. *tesserulatum* Zobel are reported. *Tuber bituminatum* Berk. and Broome is considered a synonym of *T. mesentericum* by several authors (Montecchi and Sarasini 2000; Granetti et al. 2005), but in Index Fungorum, this name refers to a holotype of *T. aestivum* (Kew Royal Botanic Gardens—Accession n. 30594). The species *Tuber bellonae* Quél. (synonym of *Tuber bituminatum* var. *sphaerosporum* Ferry de la Bellone) is reported to be close to *T. mesentericum* with some distinctive morphological features, in particular the globose spores and the higher spore volume (Pacioni and Fantini 1997). In our opinion, also *T. bellonae*, from the morphological point of view, can be included into the variability of *T. mesentericum*, even if some authors, without any molecular evidence, continue to consider it a separate species (Ławrynowicz et al. 2008). A study in progress will add molecular to morphological data to investigate the intraspecific diversity of *T. mesentericum*.

According to Bonito and colleagues (2013), *T. mesentericum* belongs to the Aestivum clade of the *Tuber* genus phylogeny, together with *T. aestivum*, *T. panniferum* Tul. and C. Tul. and *T. magnatum*. The maximum likelihood phylogenetic tree based on *T. mesentericum* ITS sequences downloaded from GenBank shows the presence of three distinct clades (Fig. 5.5). The clade I includes mainly sequences coming from Central-North Europe, comprising sequences from Sweden, and Gotland Island (Wedén et al. 2005) which likely went through a reproductive isolation. In the clades II and III, only sequences from Italian *T. mesentericum* ascomata are present, with the exception of two sequences from Spain (FM205536 and FM205535) and one from France (JQ348414). The phylogenetic reconstruction includes also *T. aestivum* sequences, which are close in the basal lineage with the clades of *T. mesentericum* and together are separated from the out-group (Fig. 5.5). The data reported here are consistent with the finding of Sica et al. (2007) showing a strong genetic structuring of the samples in different geographical areas, with the Italian clade very well distinguishable. In this instance, it may therefore be assumed that *T. mesentericum* is a species complex, but wider sampling campaign and higher genetic support (e.g. multiple gene phylogenies, population studies) are needed to confirm this hypothesis.

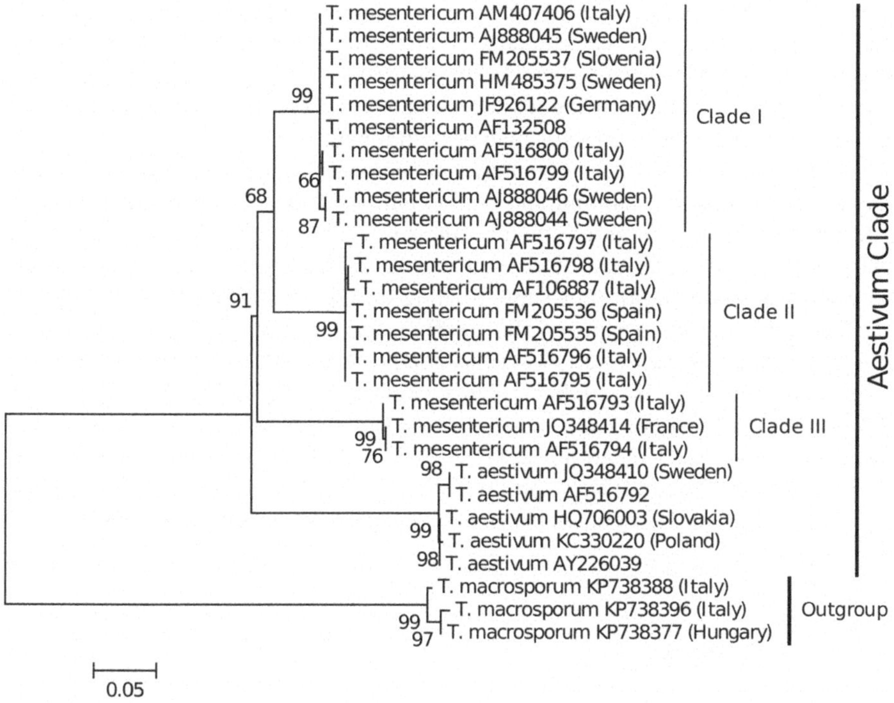

Fig. 5.5 *Tuber mesentericum* maximum likelihood phylogenetic tree based on the Tamura 3-parameter model (Tamura 1992). Bootstrap values >65 % are shown next to branching nodes. A discrete gamma distribution [+*I*] was used to model evolutionary rate differences among sites. The analysis involved 28 nucleotide sequences. All positions containing gaps and missing data were eliminated. There were a total of 440 positions in the final dataset

5.4.4 Tuber mesentericum *Geographic Distribution and Ecology*

Tuber mesentericum is a well-distributed species in Europe, occurring from Spain to Turkey and from Sweden to Italy (Pegler et al. 1993; Montecchi and Sarasini 2000; Wedén et al. 2001; Ceruti et al. 2003; Granetti et al. 2005; Sica et al. 2007; Castellano and Türkoğlu 2012).

Tuber mesentericum can grow from hilly to mountainous areas, in the former frequently associated with *C. avellana* and *O. carpinifolia*, whereas in the latter principally in relationship with *F. sylvatica*; the truffle can be common up to the altitudinal limit of the beech, that means 1800 m AMSL and more. *Tuber mesentericum* prefers north-oriented slopes and humid environment, with well-shaded, closed forests, but orientation seems to be less determinant on lower elevations (Granetti et al. 2005; Miko et al. 2006). Soil parameters proved to be variable citing mostly high lime content with a pH of 7–7.7 but also as low as 6–6.5, agreeing in well-aerated, wet soils rich in organic matter and also in volcanic subacid soils (Granetti et al. 2005). Able to share the habitat with all the black truffle species, but only occasionally in the growth areas of *T. aestivum* and *T. melanosporum*, *T. mesentericum* becomes predominant species in the mountain belt, especially in the calcareous beechwoods, where only *T. aestivum* var. *uncinatum*, and rarely *T. brumale*, can be also found.

The ripening season of *T. mesentericum* can begin in the summer, with a peak between the end of the autumn and the beginning of the winter. In the mountain belt, the harvest is often interrupted by the snowfalls; nevertheless in Italy, fresh and mature *T. mesentericum* ascomata can be found (sometimes along with *T. aestivum* var. *uncinatum*) after snow melting during the spring, even in April or May at higher altitudes (approx.1500–1800 m AMSL). These specimens are often in a perfect state of preservation, but this condition can last only for few days.

A more complete list of host plants comprises oaks (*Q. robur*, *Q. cerris*, *Q. petraea* and *Q. pubescens*), beeches (*F. sylvatica*), pines (*Pinus nigra* Arnold and *Pinus sylvestris* L.), hornbeams (*C. betulus* and *O. carpinifolia*), hazelnuts (*C. avellana*) and lindens (*Tilia* spp.) (Ceruti et al. 2003; Granetti et al. 2005; Miko et al. 2006; Marjanović et al. 2010; Bencivenga and Baciarelli-Falini 2012).

5.5 Conclusions

Tuber macrosporum and *T. mesentericum* are two minor truffles with a promising market and this perspective fosters researchers to disclose their taxonomy, biology and ecology. In this chapter, we showed that *T. macrosporum* is an adaptable species growing in soils with variable pH and lime content, and at the same time, it prefers fresh environment characterized by the constant presence of water. The unreliable, strongly water-dependent natural yields could be one of the main reasons for low quantities of *T. macrosporum* in the international market, which hinder its success. The establishment of plantations could resolve the problem. In recent years, due to its outstanding organoleptic characteristics, some Italian nurseries started producing mycorrhized plants with *T. macrosporum* and some orchards were realized. In spite of some successful initiatives of already productive *T. macrosporum* plantations, the species is still considered as marginal; therefore, further research is necessary to enhance *T. macrosporum* cultivation. The species has low intraspecific variability, and according to the available data, the relation between its genetic diversity and geographic distribution remains unclear. Concerning *T. mesentericum*, its ecology overlaps the one of *T. aestivum*, both prefer calcareous soils with sub-alkaline or neutral pH, but the former can in some cases be found in volcanic subacid soils. Unlike *T. macrosporum*, *T. mesentericum* shows evident patterns of genetic diversity in relation to geographic origin. The identification of five clusters of sequences, well-distinguished one from the other, suggested the existence of cryptic species within *T. mesentericum*. A study in progress aims to investigate morphological and genetic intraspecific diversity of *T. mesentericum* species complex and its relationships with its sister species *T. aestivum*.

Acknowledgements We would like to thank Facchini A, Grebenc T, Gazo J and Marjanovic Z for the provision of part of the *T. macrosporum* ascomata used in this study. We are grateful to Reilly Wilkinson for the English revision of the manuscript.

References

Agerer R, Rambold G (2004–2008) DEEMY – an information system for characterization and determination of ectomycorrhizae. www.deemy.de, München

Altschul SF, Gish W, Miller W, Myers EW, Lipman DJ (1990) Basic local alignment search tool. J Mol Biol 215(3):403–410

Bencivenga M, Baciarelli Falini L (2012) Manuale di Tartuficoltura. Esperienze di coltivazione dei tartufi in Umbria. Regione Umbria, Assessorato Regionale Agricoltura e Foreste, Artegraf, Citta di Castello

Benson DA, Cavanaugh M, Clark K, Karsch-Mizrachi I, Lipman DJ, Ostell J, Sayers EW (2013) GenBank. Nucleic Acids Res 41(Database issue):D36–D42. doi:10.1093/nar/gks1195

Benucci GMN (2011) Truffles: from individuals to communities and their structure. Ectomycorrhizal ecology and new cultivation possibilities for potential hosts and *Tuber* species. PhD thesis, University of Perugia. doi:10.1007/s00572-012-0441-3

Benucci GMN, Raggi L, Di Massimo G, Baciarelli Falini L, Bencivenga M, Falcinelli M, Albertini E (2011) Species-specific primers for the identification of the ectomycorrhizal fungus *Tuber macrosporum* Vittad. Mol Ecol Resour 11(2):378–381. doi:10.1111/j.1755-0998.2010.02915.x

Benucci GMN, Gógán Csorbai A, Baciarelli Falini L, Bencivenga M, Di Massimo G, Donnini D (2012) Mycorrhization of *Quercus robur* L., *Quercus cerris* L. and *Corylus avellana* L. seedlings with *Tuber macrosporum* Vittad. Mycorrhiza 22(8):639–646. doi:10.1007/s00572-012-0441-3

Benucci GM, Raggi L, Albertini E, Csorbai AG, Donnini D (2014) Assessment of ectomycorrhizal biodiversity in *Tuber macrosporum* productive sites. Mycorrhiza 24(4):281–292. doi:10.1007/s00572-013-0538-3

Bonito G (2009) Fast DNA-based identification of the black truffle *Tuber melanosporum* with direct PCR and species-specific primers. FEMS Microbiol Lett 301(2):171–175. doi:10.1111/j.1574-6968.2009.01812.x

Bonito G, Gryganskyi AP, Trappe JM, Vilgalys R (2010) A global meta-analysis of *Tuber* ITS rDNA sequences: species diversity, host associations and long-distance dispersal. Mol Ecol 19 (22):4994–5008. doi:10.1111/j.1365-294X.2010.04855.x

Bonito G, Smith ME, Nowak M, Healy RA, Guevara G, Cazares E, Kinoshita A, Nouhra ER, Dominguez LS, Tedersoo L, Murat C, Wang Y, Arroyo Moreno B, Pfister DH, Nara K, Zambonelli A, Trappe JM, Vilgalys R (2013) Historical biogeography and diversification of truffles in the Tuberaceae and their newly identified Southern hemisphere sister lineage. PLoS One 8(1):e52765. doi:10.1371/journal.pone.0052765

Buzzini P, Gasparetti C, Turchetti B, Cramarossa MR, Vaughan-Martini A, Martini A, Pagnoni UM, Forti L (2005) Production of volatile organic compounds (VOCs) by yeasts isolated from the ascocarps of black (*Tuber melanosporum* Vitt.) and white (*Tuber magnatum* Pico) truffles. Arch Microbiol 184(3):187–193. doi:10.1007/s00203-005-0043-y

Castellano MA, Türkoğlu A (2012) New records of truffle taxa in *Tuber* and *Terfezia* from Turkey. Turk J Bot 36:295–298. doi:10.3906/bot-1106-10

Ceruti A (1968) Biologia e possibilità di coltivazione dei tartufi. In: Atti del 1° congresso internazionale sul tartufo, Ente Rocca di Spoleto, Spoleto, 24–25 May 1968, pp 29–42
Ceruti A, Fontana A, Nosenzo C (2003) Le specie del genere *Tuber*. Una revisione storica. Museo Regionale di Scienze Naturali di Torino, Regione Piemonte, Torino
Chevalier G (1998) The truffle cultivation in France: assessment of the situation after 25 years of intensive use of mycorrhizal seedlings. In: Abstract of the second international conference on mycorrhiza, Uppsala, 5–10 July 1998, p 43
Djurdjevic L, Milenkovic M, Pavlovic P, Kostic O (1999) Allelopathic investigations in the Fraxino angustifoliae-Quercetum roboris (Jov. et Tomić 1979) forest community with the autumnal truffle (*Tuber macrosporum* Vitt.). Arch Biol Sci 50:27–33
Felsenstein J (1985) Confidence limits on phylogenies: an approach using the bootstrap. Evolution 39(4):783–791. doi:10.2307/2408678
Garofoli A (1906) Funghi e tartufi. Cassone, Casale Monferrato
Giovannetti G, Fontana A (1980–1981) Simbiosi micorrizica di *Tuber macrosporum* Vitt. con alcune Fagales. Allionia 24:13–17
Giraud M (1988) Prélèvement et analyse de mycorhizes. In: CTIFL (ed), La truffe, Bull FNPT, vol 10, pp 49–63
Gógán Csorbai A (2011) Studies on cultivation possibilities of summer truffle (*Tuber aestivum* Vittad.) and smooth black truffle (*Tuber macrosporum* Vittad.) in Hungary. PhD thesis, Szent István University, Gödöllő
Gógán Csorbai A, Bratek Z, Illyés Z, Dimény J (2010) Studies on *Tuber macrosporum* Vittad. natural truffle habitats in the Carpatho-Pannon region. In: Donnini D (ed) Atti del 3° congresso internazionale di Spoleto sul tartufo. Comunità Montana dei Monti Martani, Serano e Subasio, pp 182–188
Granetti B (1995) Caratteristiche morfologiche, biometriche e strutturali delle micorrize di *Tuber* di interesse economico. Micol Ital 24(2):101–117
Granetti B, De Angelis A, Materozzi G (2005) Umbria terra di tartufi. Regione Umbria, Terni
Hall TA (1999) BioEdit: a user-friendly biological sequence alignment editor and analysis program for Windows 95/98/NT. Nucleic Acids Symp Ser 41:95–98
Hall IR, Yun W (2002) Truffles and other edible mycorrhizal mushrooms - some new crops for the southern hemisphere. In: Hall IR, Yun W, Danell E, Zambonelli A (eds) Edible mycorrhizal mushrooms and their cultivation, Proceedings of the 2nd international conference on edible mycorrhizal mushrooms. Crop & Food Research, Christchurch
Hall IR, Zambonelli A, Brown G (2007) Taming the truffle. Timber Press, Portland
Hilszczańska D, Rosa-Gruszecka A, Sikora K, Szmidla H (2013) First report of *Tuber macrosporum* occurrence in Poland. Sci Res Essays 8(23):1096–1099. doi:10.5897/SRE2013.5529
Hilszczańska D, Rosa-gruszecka A, Szmidla H (2014) Characteristic of *Tuber* spp. localities in natural stands with emphasis on plant species composition. Acta Mycol 49(2):267–277. doi:10.5586/am.2014.024
Hollós L (1911) Magyarország földalatti gombái, szarvasgombafélei. KM Természettudományi Társulat, Budapest
Iotti M, Amicucci A, Stocchi V, Zambonelli A (2002) Morphological and molecular characterization of mycelia of some *Tuber* species in pure culture. New Phytol 155(3):499–505. doi:10.1046/j.1469-8137.2002.00486.x
Jeandroz S, Murat C, Wang Y, Bonfante P, Le Tacon F (2008) Molecular phylogeny and historical biogeography of the genus *Tuber*, the true truffles. J Biogeogr 35(5):815–829. doi:10.1111/j.1365-2699.2007.01851.x
Jukes TH, Cantor CR (1969) Evolution of protein molecules. In: Munro HN (ed) Mammalian protein metabolism. Academic, New York, pp 21–132
Kinoshita A, Sasaki H, Nara K (2011) Phylogeny and diversity of Japanese truffles (*Tuber* spp.) inferred from sequences of four nuclear loci. Mycologia 103(4):779–794. doi:10.3852/10-138

Ławrynowicz M, Krzyszczyk T, Fałdzinski M (2008) Occurrence of black truffles in Poland. Acta Mycol 43(2):143–151

Marjanović Ž, Grebenc T, Markovic M, Glisic A, Milenkovic M (2010) Ecological specificity and molecular diversity of truffles (genus *Tuber*) originating from mid-west of the Balkan Peninsula. Sydowia 62(1):67–87

Miko M, Jan G, Bratek Z (2006) *Tuber macrosporum* Vitt. and *Tuber mesentericum* Vitt. one hundred years neglected hypogeous fungi species in Slovak Republic. Acta Fytotechn Zootech 4:85–90. doi:10.5897/SRE2013.5529

Milenkovic M, Marjanović Ž (2001) Current results on *Tuber* spp. research in Yugoslavia. In: Actes du Ve congrès international science et culture de la truffe et des autres champignons hypogés comestibles, Aix-en-Provence, 4–6 Mar 1999. Federation Française des Trufficulteurs, Paris, pp 218–222

Molinier V, Van Tuinen D, Chevalier G, Gollotte A, Wipf D, Redecker D (2013) A multigene phylogeny demonstrates that tuber aestivum and tuber uncinatum are conspecific. Org Divers Evol 13(4):503–512. doi:10.1007/s13127-013-0146-2

Montecchi A, Lazzari G (1993) Atlante fotografico dei funghi ipogei. Associazione Micologica Bresadola, Trento

Montecchi M, Sarasini A (2000) Funghi Ipogei d'Europa. Associazione Micologica Bresadola, Trento

Olivier JM (2000) Progress in the cultivation of truffles. In: Van Griensven LJLD (ed) Science and cultivation of edible fungi, vol 2. Proceedings of the 15th international congress on the science and cultivation of edible fungi. A. A. Balkema, Rotterdam, pp 937–942

Pacioni G, Fantini P (1997) *Tuber bellonae* un tartufo mediterraneo del complesso *Tuber aestivum–Tuber mesentericum*. Micol Veg Medit 12(1):15–20

Pacioni G, Pomponi G (1991) Genotypic patterns of some Italian populations of the *Tuber aestivum-mesentericum* complex. Mycotaxon 42:171–179

Paolocci F, Rubini A, Riccioni C, Topini F, Arcioni S (2004) *Tuber aestivum* and *Tuber uncinatum*: two morphotypes or two species? FEMS Microbiol Lett 235(1):109–115. doi:10.1016/j.femsle.2004.04.029

Pegler DN, Spooner BM, Young TWK (1993) British truffles: a revision of British hypogeous fungi. Royal Botanic Gardens, Kew

Piltaver A, Ratoŝa I (2010) Hypogeous fungi from Slovenia. In: Donnini D (ed) Atti del 3° congresso internazionale di Spoleto sul tartufo, Spoleto, 25–28 Nov 2008. Comunità Montana dei Monti Martani, Serano e Subasio, Spoleto, pp 333–345

Rauscher T, Agerer R, Chevalier G (1995) Ektomykorrhizen von *Tuber melanosporum*, *Tuber mesentericum* und *Tuber rufum* (Tuberales) on *Corylus avellana*. Nova Hedwig 61(3):281–322

Riousset L, Riousset G, Chevalier G, Bardet MC (2012) Truffes d'Europe et de Chine. Quae, Paris

Sica M, Gaudio L, Aceto S (2007) Genetic structure of *Tuber mesentericum* Vitt. based on polymorphisms at the ribosomal DNA ITS. Mycorrhiza 17(5):405–414. doi:10.1007/s00572-007-0115-8

Splivallo R, Valdez N, Kirchhoff N, Ona MC, Schmidt JP, Feussner I, Karlovsky P (2012) Intraspecific genotypic variability determines concentrations of key truffle volatiles. New Phytol 194(3):823–835. doi:10.1111/j.1469-8137.2012.04077.x

Splivallo R, Deveau A, Valdez N, Kirchhoff N, Frey-Klett P, Karlovsky P (2014) Bacteria associated with truffle-fruiting bodies contribute to truffle aroma. Environ Microbiol 17 (8):2647–2660. doi:10.1111/1462-2920.12521

Stobbe U, Büntgen U, Sproll L, Tegel W, Egli S, Fink S (2012) Spatial distribution and ecological variation of re-discovered German truffle habitats. Fungal Ecol 5(5):591–599. doi:10.1016/j.funeco.2012.02.001

Tamura K (1992) Estimation of the number of nucleotide substitutions when there are strong transition-transversion and G + C-content biases. Mol Biol Evol 9(4):678–687

Tamura K, Peterson D, Peterson N, Stecher G, Nei M, Kumar S (2011) MEGA5: molecular evolutionary genetics analysis using maximum likelihood, evolutionary distance, and maximum parsimony methods. Mol Biol Evol 28(10):2731–2739. doi:10.1093/molbev/msr121

Tamura K, Stecher G, Peterson D, Filipski A, Kumar S (2013) MEGA6: molecular evolutionary genetics analysis version 6.0. Mol Biol Evol 30(12):2725–2729. doi:10.1093/molbev/mst19

Vezzola V (2005) Primi risultati produttivi con piante micorrizate da *T. macrosporum* Vittad. In: Atti del seminario sullo stato attuale della tartuficoltura italiana. Grafiche Millefiorini, Spoleto-Norcia, pp 51–55

Vezzola V (2010) La coltivazione di *Tuber macrosporum*. In: Donnini D (ed) Atti del 3° congresso internazionale di Spoleto sul tartufo. Comunità Montana dei Monti Martani, Serano e Subasio, pp 694–697

Vittadini C (1831) Monographia Tuberacearum. Ex Typographia Felicis Rusconi, Milano

Wedén C, Ericsson L, Danell E (2001) Research on *Tuber aestivum* syn. *T. uncinatum*, and *T. mesentericum* reported from Sweden for the first time. Sven Bot Tidskr 95(4):205–211

Wedén C, Danell E, Tibell L (2005) Species recognition in the truffle genus *Tuber* – the synonyms *Tuber aestivum* and *Tuber uncinatum*. Environ Microbiol 7(10):1535–1546. doi:10.1111/j.1462-2920.2005.00837.x

White TM, Bruns T, Lee S, Taylor J (1990) Amplification and direct sequencing of fungal ribosomal RNA for phylogenetics. In: Innis MA, Gelfand DH, Sninsky JJ, White JJ (eds) PCR Protocols: a guide to methods and applications. Academic, San Diego, pp 315–322

Zambonelli A, Salomoni S, Pisi A (1993) Caratterizzazione anatomo-morfologica delle micorrize di *Tuber* spp. su *Quercus pubescens* Willd. Micol Ital 22(3):73–90

Zambonelli A, Salomoni S, Pisi A (1995) Caratterizzazione anatomo-morfologica delle micorrize di *Tuber borchii*, *Tuber aestivum*, *Tuber mesentericum*, *Tuber brumale*, *Tuber melanosporum* su *Pinus pinea*. Micol Ital 24(2):119–137

Chapter 6
Tuber magnatum: The Special One. What Makes It so Different from the Other *Tuber* spp.?

Claudia Riccioni, Andrea Rubini, Beatrice Belfiori, Gianluigi Gregori, and Francesco Paolocci

6.1 Introduction

The Italian physician Vittorio Pico was the first who described *Tuber magnatum* (1788). This truffle species produces white ascomata endowed with a distinctive aroma. Thus, among the *Tuber* spp. of economical relevance, *T. magnatum* is considered by most people as the premium one. Truffles of this species, however, are very hard to find, and for centuries, they have been considered as an icon made in Italy, so that this species is known worldwide as "Italian white truffle" or even more strictly as "Piedmont truffle" or "Tartufo bianco di Alba". In the 1700s, the princes of Savoy used *tartufi* as a promotional gift in diplomatic relations (Nowak 2015), and according to Rittersma (2011), by using white truffles in this manner, they generated a real "*truffle mania*" throughout Europe.

The onset of the first truffle market that was established in Alba at the beginning of the twentieth century (1930) contributed to reinforce the link between white truffles and Piedmont region. Nevertheless, we are now aware that there are other territories and small cities, scattered throughout several Italian regions from north to south, whose names are tightly linked to the production of this delicacy. Better still, it became clear only around the late 1990s of the last century that *T. magnatum* can be also found in the Balkan Peninsula and countries nearby (Marjanovic et al. 2010). More marginally, since 2011, it is harvested in southeast of France

Claudia Riccioni and Andrea Rubini contributed equally to this work

C. Riccioni • A. Rubini • B. Belfiori • F. Paolocci (✉)
Institute of Biosciences and Bioresources – Perugia Division, National Research Council, via Madonna Alta 130, 06128 Perugia, Italy
e-mail: francesco.paolocci@ibbr.cnr.it

G. Gregori
Centro Sperimentale Tartuficoltura ASSAM – Regione Marche, via Macina 1, 61048 Sant'Angelo in Vado, Pesaro e Urbino, Italy

A. Zambonelli et al. (eds.), *True Truffle (*Tuber *spp.) in the World*, Soil Biology 47,
DOI 10.1007/978-3-319-31436-5_6

(Tabouret 2011) and also sporadically in Switzerland (Tabouret 2012). Notwithstanding, with respect to the other *Tuber* spp., the geographical range of *T. magnatum* remains far more limited.

Information relative to annual gross production of white truffles is very limited because neither in Italy nor in the Balkan areas are there reliable historical records of this production, although different estimations have been done. In this respect, Hall and colleagues (2003) reported that the estimated world production of *T. magnatum* ascomata ranged between 50 and 200 t per year. Brun and Mosso (2010) estimated that the *T. magnatum* production in Italy was around 124, 19, and 12 t in the years 2002, 2006, and 2007, respectively, and reported that the regions that likely contributed the most to this production were Umbria and Abruzzo, with about 29 and 20 %, with only a modest contribution from Piedmont (3 %).

The annual production is generally insufficient to meet the rising worldwide demand for truffles, despite the fact that some of them, such as *Tuber melanosporum* Vittad. and *Tuber aestivum* Vittad. truffles, can be successfully cultivated (see Chaps. 2 and 3; Hall et al. 2003). This supply/demand gap becomes even larger if one considers *T. magnatum* because, as we shall detail later, examples of successful cultivation of this fungus are very few, and the contribution in terms of overall quantity of truffles produced by these plantations remains *de facto* negligible. Given the insufficient provision of this highly appreciated species, it comes therefore as no surprise that *T. magnatum* is the most expensive truffle. Because of the drought, in 2007, its production fell dramatically and, accordingly, the price paid to hunters peaked up to 7000 € kg^{-1}; conversely, the last harvesting season (fall-winter 2014) was a good one, with prices ranging between 300 and 400 € kg^{-1}. However, on average, the prices of *T. magnatum* truffles paid to hunters ran over the last 10 years between 1000 and 1500 € kg^{-1}.

Considering the aspects discussed above, it is clearly evident that *T. magnatum* is a special truffle with distinctive features in relation to all the other *Tuber* spp. Stemming from problems related to its cultivation, here, we will review the recent breakthroughs on the life cycle and population genetics of this fungus. Perspectives and approaches that might be of help to deepen our understanding on ecology and biology of this species will be also discussed.

6.2 Problems that Might Have Affected *T. magnatum* Cultivation

Truffle production results from a complex interplay between the fungus, its host, and the habitat. The past century has seen a marked decline in harvests of some edible mycorrhizal mushrooms, and among them are truffles (Hall et al. 2003). The drop in natural truffle production, coupled with the possibility of generating an income from marginal areas, have in turn fuelled intense interest in truffle cultivation over the last five to six decades. Man-made truffle plantations can be established by planting host plants nursery inoculated with *Tuber* spp. The synthesis of *T. magnatum* mycorrhizas on a wide range of host plants has become a routine process in several nurseries in Italy since the early 1980s. Records from the public nursery located in Sant'Angelo in Vado (Italy) documented that approximately 167,000 host plants, produced and certified on the basis of morphological analysis by this structure, as colonized by the white truffle, have been outplanted all over Italy between 1984 and 2004. Overall, by considering the number of public and private nurseries operating in that period in Italy, it can be estimated that at least 500,000 *T. magnatum*-inoculated and certified seedlings have been outplanted in this country. This estimation can be likely doubled if one considers the production coming from private nurseries that used to sell plants without any certification. In face of this massive introduction of inoculated host plants in regions and habitats where this fungus grows spontaneously, in the late 1990s, truffle growers and scientists started to realize that the cultivation of this fungus presents specific drawbacks that cannot be overcome by following the notions and practices developed through the successful cultivation of other *Tuber* spp. As matter of the fact, the overall number of *T. magnatum* fields that turned out to be productive after two or more decades from their onset were and are still less than a few dozen. Also, not infrequently, nursery-inoculated *T. magnatum* plants have unexpectedly produced *Tuber borchii* Vittad. or *Tuber maculatum* Vittad. ascomata in the field (Gregori 2002; Gregori et al. 2010). In stark contrast to *T. melanosporum* and *T. aestivum* orchards, the average production of *T. magnatum* plantations has never exceeded, in good years, 10–15 kg ha^{-1}, while on average the harvest from these orchards is around 4–5 kg ha^{-1}. This scenario becomes even worse if one considers that the fructification was never detected earlier than 15–20 years post implantation. This evidence has led to the suspicion that the nursery mycorrhizal plants were of bad quality in terms of the mycorrhization level with the target fungal species. In turn, the hypothesis that the late fructification of these orchards is imputable to the occasional colonization of the introduced host plants by native *T. magnatum* strains cannot be ruled out. Alternatively, or in addition, orchard management practices were erroneous.

Despite a plethora of studies that have been carried out over the last decades to decipher the pedological and climatic conditions that should favor the fructification of this fungus (also reviewed in this book), the cultivation of *T. magnatum* has never reached an acceptable rate of success.

6.2.1 The Morphology of T. magnatum *Mycorrhizas: A Dilemma Solved by a Molecular-Assisted Analysis*

As we reported above, a critical analysis of the reasons underlying the difficulties in cultivating *T. magnatum* started more than 15 years ago. Around that period, papers have been published that questioned the reliability of procedures and morphological criteria used for obtaining and certifying the presence of *T. magnatum* ectomycorrhizas (ECMs) on nursery-inoculated host plants, respectively (Gandeboeuf et al. 1997). Although the possibility of typing *T. magnatum* ascomata and ECMs from those of other *Tuber* spp. by molecular markers (i.e., on the basis of the length of the ITS rDNA) was reported since 1995 (Paolocci et al. 1995a, b), the certification of *T. magnatum* mycorrhizal plants has relied for decades on morphological methods only (Bencivenga et al. 1987). Diagnostic and/or species-specific *T. magnatum* morphological traits have been in fact coded by a number of different publications that compared the morphology of ECMs from different *Tuber* species (Bencivenga et al. 1987; Fontana et al. 1992; Zambonelli et al. 1993; Granetti 1995). Yet, morphology of *Tuber* ECMs is influenced by age, host plant, and soil conditions, and in some species, such as *Tuber oligospermum* Tul. and C. Tul., and *T. borchii* and *T. magnatum*, the morpho-anatomical characters are quite similar and overlapping (Rubini et al. 2001; Tedersoo et al. 2006; Boutahir et al. 2013). Thus, when molecular markers were used to ascertain the identity of ECMs obtained after host plant inoculation with white truffles, it turned out that most of them were indeed to be ascribed to other "minor" white *Tuber* spp. such as *T. borchii* and *T. maculatum* or to other still undefined species (Gandeboeuf et al. 1997; Amicucci et al. 1998). Hence, not only have these studies demonstrated the unreliability of morphological analysis *per se* when certifying the presence of *T. magnatum* on host plants, they also pointed to the need of developing innovative procedures for inoculating and growing host plants treated with the spores of this species. In fact, very often species such as *Sphaerosporella brunnea* (Alb. and Schwein.) Svrček and Kubička and *Pulvinula constellatio* (Berk. and Broome) Boud. that compete with the truffle mycelium for the host roots both in greenhouse and later in the field can be found on *T. magnatum*-inoculated host plants (Danielson 1984; Amicucci et al. 2001).

Bearing all these points in mind, Rubini and colleagues (2001) developed a protocol to synthesize ECMs in a controlled environment and reported for the first time the morphological description of *T. magnatum* ECMs assisted by molecular marker characterization. Thus, the work by Rubini and colleagues (2001) enabled the redefinition of the morpho-anatomical traits specific to *T. magnatum* ECMs (Table 6.1). The parallel study carried out by Mello and colleagues (2001) basically reached the same conclusions.

Thus, by adopting erroneous criteria of species attribution, it is more than likely that seedlings harboring truffle species of less value than *T. magnatum* have passed the quality control of the different certification agencies until the early 2000s. Consequently, mistyped nursery mycorrhizal plants might have been used for establishing white truffle orchards. Ironically, those seedlings that indeed sustained *T. magnatum* ECMs might have been conversely discarded.

Overall, the works by Rubini and colleagues (2001) and Mello and collaborators (2001) represented the point of no return in the commercialization of white truffle-inoculated seedlings. Since these works, at least in Italy where most of the nurseries have been operating, the production and market of *T. magnatum* mycorrhizal plants has come to a standstill. Subsidies by many regional governments to white truffle cultivation have been also stopped, with both local governments and truffle growers

Table 6.1 Comparison between diagnostic morphological traits ascribed to *T. magnatum* ECMs before and after the use of molecular tools for their species attribution

Diagnostic features	Pre-molecular validation	Post-molecular validation[a]
Color	Young ECMs: amber color with whitish apex	Young ECMs: uniform color tending to grayish
	Old ECMs: brown to brown reddish	Old ECMs: dark gray or grayish-amber[b]
Cystidia	Short spinule-like cystidia with uniform length (from 40 to 120 μm), never ramified	Cystidia variable in length but often very long (up to 400 μm), colorless, pluriseptate, frequently ramified with perpendicularly oriented branches

[a]Only the simultaneous presence of simple and/or ramified cystidia, puzzle-like pattern of the mycoclena and gray color makes the morphotyping of *T. magnatum* ECMs reliable (Rubini et al. 2001)

[b]Mycorrhizal pigmentation is due to the presence of small speckles probably caused by polyphenol accumulation, which turn amber in old ECMs

awaiting for the development of large-scale nursery production of genuine *T. magnatum* colonized seedlings as well as for the development of agro-silvicultural practices specifically tailored for this species.

In conclusion, it is reasonable that the mischaracterization of *T. magnatum*-inoculated seedlings has contributed to make the cultivation of this fungus unsuccessful. Therefore, the next logical step would be the large-scale production of *T. magnatum*-inoculated seedlings validated by molecular analysis. In this context, we note that only recently the French nursery Robin has started to commercialize *T. magnatum* mycorrhizal plants produced under INRA know-how license (Murat 2015). Each plant is individually controlled by INRA for the presence of white truffle ECMs on its roots using either classical PCR or real-time PCR with species-specific primers (Rubini et al. 2001; Iotti et al. 2012). However, this plant by plant control concurs to elevate the cost of each single plant in the market up to 100 € each, which has no comparison with the cost of seedlings inoculated with *T. melanosporum* (ranging from 10 to 20 €). Whether outplanting of genuine *T. magnatum* mycorrhizal plants in areas with proper ecological conditions is sufficient to ensure white truffle production is the next challenge.

6.2.2 *The Scarcity of Mycorrhizas in the Field: A Different Ecological Strategy of* T. magnatum *with Respect to Other* Tuber *spp.?*

As we learned above, the synthesis of ECMs between *T. magnatum* and its host plants is not a trivial issue: even academic and/or highly specialized labs and nurseries have experienced problems in producing genuine *T. magnatum* mycorrhizal plants. Curiously, but in keeping with this evidence, recent studies have documented the rarity of *T. magnatum* ECMs in nature, also at productive sites. Bertini and colleagues (2006) used molecular methods to investigate the ECM communities in a productive *T. magnatum* truffle ground. This screening revealed that a high percentage of ECMs belonged to as diverse *Tuber* species as *Tuber dryophilum* Tul. and C. Tul., *Tuber rufum* Pico, and *T. maculatum*, whereas the percentage of *T. magnatum* ECMs was less than 4.4 %. Similar results have also been obtained in other *T. magnatum* areas where either *T. magnatum* ECMs were at low frequency, if not totally absent, or their distribution was not superimposed to the collection sites of white ascomata (Murat et al. 2005; Leonardi et al. 2013). Hence, these are additional aspects that mark the difference between the ecology and life cycle of this fungus with respect to the other *Tuber* spp. which, conversely, extensively colonize the root apparatus of their hosts (Baciarelli-Falini et al. 2006; Belfiori et al. 2012; Napoli et al. 2010; De Miguel et al. 2014).

The peculiar distribution of *T. magnatum* ECMs with respect to that of other truffle species has prompted scientists to investigate the presence of this fungus directly in the soil. By a PCR analysis based on β-tubulin species-specific primers, Zampieri and collaborators (2010) monitored the presence of *T. magnatum* mycelium in soil samples collected in a natural site in Piedmont. These authors found that *T. magnatum* "propagules were much more widespread than could be inferred from the distribution of truffles." *Tuber magnatum* was in fact also present in soil samples collected in nonproductive areas. Incidentally, soil samples displaying *T. magnatum*-specific signals have been detected as far as 100 m away from the nearest productive tree. Other studies have investigated the amount of *T. magnatum* mycelia or propagules in the soil by qPCR. This is an innovative and species-targeted approach that has been used to detect a large number of soil fungi (Landeweert et al. 2003; Bidartondo and Gardes 2005; Smith and Osborn 2008). Iotti and colleagues (2012, 2014) have developed a *T. magnatum*-specific real-time assay based on the ITS rDNA region to evaluate the spatial and temporal distribution of its mycelium in soil samples from four different natural grounds in Italy. While confirming a patchy distribution of *T. magnatum* in the soil, they also reported that the amount of this mycelium inside the productive patches undergoes seasonal fluctuations. In fact, in early spring, the amount of *T. magnatum* mycelium is higher than in summer, to increase again in fall when it concentrates all around the fruiting points. Incidentally, it has also been pointed out that *T. magnatum* might be more sensitive to summer drought and temperature peaks than black truffle species *T. melanosporum* and *T. aestivum* (Iotti et al. 2014). This view becomes even more likely if one embraces the hypothesis that in *T. magnatum* the saprophytic phase is prevalent with respect to the symbiotic one. Indeed, the presence of abundant extra-radical *T. magnatum* mycelia/propagules in the field, but the rarity of its ECMs prompted some authors to argue that in *T. magnatum* mycelia rather than ECMs play an essential role in ascomata formation and development (Iotti et al. 2014). This runs contrary to what reported for *T. melanosporum* (Rubini et al. 2011a; Murat et al. 2013; Le Tacon et al. 2013, 2015).

Although tantalizing, the abovementioned hypothesis has yet to be tested. The sequencing of the *T. melanosporum* genome has shown that this fungus presents a limited repertoire of genes that are able to degrade dead organic matter (Martin et al. 2010a), and later studies conducted by monitoring the natural abundance of ^{15}N and ^{13}C (Zeller et al. 2008) and by labeling a single host plant with $^{13}CO_2$ (Le Tacon et al. 2013), have proven, for the same *Tuber* species, a dependency of the ascomata on their hosts throughout their development. On the other hand, the physical link between ECMs and ascomata has been directly observed neither in *T. melanosporum* nor in *T. magnatum*. In *T. magnatum* ascomata, however, Barbieri et al. (2010) detected the expression of the nitrogenase gene *nifH* from *Bradyrhizobium* and found that the nitrogenase activity was comparable to that of young legume nodules. Despite the presence of bacterial *nif* genes inside the *T. melanosporum* ascomata, N_2 fixation was neither detected in immature nor in mature ascomata, to suggest that these bacteria would participate very little or at all in the nitrogen nutrition of *T. melanosporum* ascomata, in contrast to what occurs in *T. magnatum* (Le Tacon et al. 2015).

Alternatively, the abundance of extra-radical mycelium and the rarity of *T. magnatum* ECMs can also be explained with mutualistic relationships of this species with plant roots that do not result in the differentiation of "canonical" ECM structures. To this regard, recent studies have shown that some *Tuber* spp. can indeed colonize the roots of species generally recognized as nonhost, by developing structures that differ from ECMs *sensu stricto*. More specifically, Selosse and colleagues (2004) showed the presence of mycelia of *T. aestivum* associated with roots of orchids belonging to the genus *Epipactis*, whereas Lancellotti and colleagues (2014) showed that *T. borchii* is able to colonize the roots of *Arbutus unedo* L. forming arbutoid-like ECMs. Pacioni and colleagues (2014) reported that mycelia of some *Tuber* spp. such as *T. aestivum* and *T. melanosporum* can form different types of mycorrhizal structures on genetically transformed roots of *Cistus incanus* L., depending on the characteristics of the culture medium. Finally, Gryndler and collaborators (2014), using qPCR analyses, revealed that within the "brûlé," the concentration of *T. aestivum* mycelium in contact with roots of nonhost plants is higher than that found in the surrounding soil. Overall, these observations suggest that *Tuber* spp. can interact with a wider range of plant species than previously thought and that the functional plant fungus relationship results in the differentiation of different exchange structures that might or might not be structured as canonical ECMs. Embracing this hypothesis, it is plausible that several hosts might represent the reservoir of *T. magnatum* mycelia under unfavorable environmental conditions and that they eventually spread into the soil patches nearby only when favorable environmental conditions are met. Finally, we note that the difficulties of obtaining genuine *T. magnatum* ECMs on nursery-inoculated host plants might result from the scarce attitude of this fungus in competing with other fungi and/or in perpetrating these exchange structures on their hosts over several months and/or environmental conditions.

6.3 *Tuber magnatum* Life Cycle and Reproductive Biology

The study of the reproductive mode of truffles is a subject that has been fascinating scientists, truffle hunters, seekers, and growers for decades. First studies on life cycle and reproductive biology of *Tuber* spp. were based on morphological analyses. Several authors described the presence of reproductive structures in *Tuber* fruiting bodies (Marchisio 1964; Callot 1999). Such observations however were only occasional, and formation of these structures cannot be obtained in controlled conditions for any *Tuber* species. This limitation has represented a major obstacle in understanding the life cycle and reproductive system of these fungi. Other studies have shown the presence of hyphae with paired nuclei in mycelia forming the ECMs and in in vitro cultured mycelia of different *Tuber* spp. In virtue of this karyological pattern, it was suggested that the life cycle of these ascomycetes is sustained by a secondary mycelium with dikaryotic hyphae (Lanfranco et al. 1995). The advent of analyses based on molecular markers has then provided the most powerful tools to investigate the life cycle and reproductive modes of truffles. To make a long story short (for a specific review, see Rubini et al. 2014), a breakthrough on these aspects has been gained when codominant SSR markers were applied to study population genetics in *T. magnatum*. By this approach, the occurrence of gene flow and the absence of significant linkage disequilibrium among individuals within the same population or neighboring populations were provided for the first time (Rubini et al. 2005). These results mined the view, prevalent at that time, that truffles were strictly selfing species (Bertault et al. 1998). Rather they neatly pointed to a *T. magnatum* reproduction system based on outcrossing. Additionally, SSR markers have been used to genotype single *T. magnatum* ECMs obtained under controlled conditions as reported in Rubini et al. (2001) and provided the unequivocal evidence that truffle ECMs are formed by primary (haploid) mycelia derived from thc germination of single ascospores. Taken together, these results have led to a reevaluation of *T. magnatum* reproductive modes, in that they provided the clear-cut evidence that this species can outcross and that its life cycle is predominantly haploid (Paolocci et al. 2006; Rubini et al. 2007). In turn, the approaches pursued and results obtained in *T. magnatum* have paved the way for studies aimed at assessing the life cycle and reproductive mode in *T. melanosporum* and reach the same conclusion, i.e., this species can also outcross (Riccioni et al. 2008). Finally, the sequencing of the *T. melanosporum* genome led scientists to conclude that *T. melanosporum* is a heterothallic (i.e., obligate outcrossing) fungus since the two mating type genes, namely, *MAT1-1-1* and *MAT1-2-1*, that control sexual reproduction in Ascomycetes, are harbored by different strains (Martin et al. 2010a; Rubini et al. 2011b). In turn, sequence information of the two *T. melanosporum* mating type genes has been seminal to clone their orthologs in *T. magnatum* and prove that also the prestigious white truffles result from sexual crossing between strains harboring the opposite mating type genes (Martin et al. 2010b).

The distribution of sexually compatible truffle strains has been investigated recently by taking into account both ECMs and extra-radical mycelium in natural and man-made *T. melanosporum* grounds. Briefly, these studies have documented an uneven distribution of ECMs of different mating type in truffle grounds that has been interpreted as the result of a competition between strains of opposite mating type for their persistence on host plants (Rubini et al. 2011a; Linde and Selmes 2012; Murat et al. 2013). Although the genetic determinants underlying this phenomenon have yet to be elucidated (for more information, see Rubini et al. 2011a, 2014; Selosse et al. 2013; Le Tacon et al. 2015), such pattern of ECM distribution might represent a factor limiting fructification. In fact, in a heterothallic fungus, mycelia of opposite mating type must come into contact for fertilization and fructification to occur. A different scenario was found considering the distribution of extra-radical mycelia of the two mating types in the soil. The analyses of DNA from soil samples have shown that often both mating types are present in the same sample. However, the source of this DNA in the soil remains to be understood, whether it comes from ascospores, mitotic spores (Urban et al. 2004), free-living mycelia, or mycelia connected to ECMs and the possible role of these elements in the fertilization process (Rubini et al. 2011a, 2014; Murat et al. 2013; Le Tacon et al. 2015). But what about the distributional patterns of *T. magnatum* strains of opposite mating type? As we learned above, the vegetative propagation pattern of *T. magnatum* seems not to rely on the formation of stable ECM structures on host plants. Consequently, investigating the distribution of strains with different mating types on host plants is not only very challenging, but it might be also not so informative. Notwithstanding, it is crucial to monitor the spatial and temporal dynamics of sexually compatible *T. magnatum* strains, regardless of the structures/organs and propagules that they might form. Thus, in order to glean a leap forward into the sexual propagation pattern of this fungus, the next logical step will be to develop real-time PCR assays based on the two *T. magnatum* *MAT* genes to monitor, in addition to the presence and abundance, also the mating type of *T. magnatum* mycelia and/or propagules present in soil samples and likely on plant partners.

6.4 Genetic Variability and Population Genetics of *T. magnatum*

Different from other truffle species, *T. magnatum* has a particularly narrow distributional range, being patchily found in some Italian regions from Piedmont to Sicily and in some countries in the Balkan Peninsula (Frizzi et al. 2001; Rubini et al. 2005), with a few samples recently found in the south of France (Tabouret 2011).

Within a given truffle species, the geographic provenance of truffles is much taken into account to determine the price. As we highlighted previously,

T. magnatum is traditionally thought as "Tartufo bianco di Alba", and thus white truffles marketed in Alba generally command a higher price. Indeed, across their geographic range, truffles of a given species generally exhibit a significant variability in several key traits, one of the most salient being the aroma (Mauriello et al. 2004; Splivallo et al. 2011). Whether or not genetic determinants concur with environmental factors in driving organoleptic differences among truffles of different provenance is a long-lasting debate among scientists and truffle stakeholders. However, at least for *T. aestivum*, a correlation between genetic and aroma variability was reported (Splivallo et al. 2012; Molinier et al. 2015). In light of these considerations, it is of pivotal importance to evaluate and map, by means of large-scale population genetics studies, the extent of genetic diversity within each single *Tuber* species of economical relevance.

The first studies dealing with the genetic variability of *T. magnatum* showed a very limited intraspecific polymorphism. No genetic diversity was evidenced by assessing RAPD (random amplification of polymorphic DNA) and ITS-RFLP (restriction fragment length polymorphism of the internal transcribed spacer of rDNA) markers over a low number of Italian ascomata (Gandeboeuf et al. 1997; Mello et al. 2001). Later on, the presence of single nucleotide polymorphism (SNPs) in the ITS region, in the β-tubulin gene, and in a sequence-characterized amplified region (SCAR) was investigated across *T. magnatum* ascomata from different Italian regions and from Croatia (Mello et al. 2005). Because of the exiguous number of SNPs detected, the number of different haplotypes found was very low. Similarly, when a broader investigation was performed by analyzing the polymorphism of isoenzymes on 139 *T. magnatum* samples distributed in 13 Italian populations, only two of the 11 enzymes tested were polymorphic. Consequently, no interpretable evolutionary processes correlated with geological or climatic phenomena emerged (Frizzi et al. 2001).

Parallel studies conducted on *T. melanosporum* by Bertault and colleagues (1998) also showed a scant variation, within and among Italian and French populations of this species. These authors explained such very low levels of polymorphism as a result of a population bottleneck that occurred during the last and coldest glaciation in the Quaternary, when the deciduous forest of Europe was restricted to the Mediterranean coastal zone (Bennett et al. 1991). On these premises, it was concluded that "morphological and organoleptic differences seen over the geographical range of the black truffles can probably be explained by environmental variation rather than by genetic factors" (Bertault et al. 1998). In a following paper, the same group suggested that such a bottleneck could have been also experienced by other truffle species, such as *T. magnatum* that, similarly to *T. melanosporum*, have a rather sparsely wooded habitat that makes them susceptible to mortality due to freezing (Bertault et al. 2001). Overall, under these assumptions, the genetic diversity in species such as *T. melanosporum* and *T. magnatum* was expected to be low.

However, for both *T. magnatum* and *T. melanosporum*, a genetic differentiation over their habitats was disclosed afterwards when a larger number of SSRs, amplified fragment length polymorphism (AFLP), and/or ITS-SNP markers

(Rubini et al 2004, 2005; Riccioni et al. 2008; Murat et al. 2004, 2011) was employed. SSRs are the markers of choice for population genetics studies also in fungi (Dutech et al. 2007; Douhan et al. 2011). Rubini and colleagues (2004) isolated and characterized seven polymorphic SSR loci in *T. magnatum* and assessed their variability in 316 individuals grouped in 26 populations, collected from the entire distributional range of this species (Rubini et al. 2005). The different parameters evaluated concurred to show a consistent intraspecific genetic variation. Some populations exhibited private alleles as well as single highly frequent alleles. For example, ascomata from three places of the Piedmont region, including the well-known Alba, had four private SSR alleles. Similarly, the southern Italian regions Molise, Campania, and Basilicata showed very high frequencies of another specific allele. A high number of haplotypes was found across all samples (265 among the 316 individuals) and many different haplotypes were generally present within single populations as well as within single truffle grounds. Nevertheless, significant signals of a genetic and phylogeographic structure emerged. Finally, the southernmost (Campania and Basilicata regions) and the north-westernmost (Piedmont) *T. magnatum* populations resulted to be clearly distinguishable from all the others.

The glacial refugia of *T. magnatum* host species (e.g., *Quercus*, *Corylus*, *Tilia*, and *Carpinus* spp.) were located in central and southern Italy from where they spread northward as ice receded (Brewer 2002; Petit et al. 2003). Data on *T. magnatum* suggest that the refugia for this species were located in central Italy and that populations of this species spread northward and southward when ice receded (Rubini et al. 2005).

Going back to the issue of what factors concur in shaping the aromatic profiles of truffles, we note that there are many, spanning from pedoclimatic conditions of the collection sites (Gioacchini et al. 2008), host plant species, microbiome (yeasts, bacteria, and filamentous fungi) inhabiting fruiting bodies (Buzzini et al. 2005; Pacioni et al. 2007; Splivallo et al. 2014; Splivallo and Ebeler 2015), down to the genetic background of the fungal strain (Splivallo et al. 2012; Molinier et al. 2015; see also Chap. 23). As we are now aware that *T. magnatum* populations are genetically structured, their genetic background should be taken into account as a potential driver of the aroma variability across white truffles from different areas. To this purpose, metabolic profiling of truffles should be paralleled by high-throughput genotyping. Along this line of reasoning, it will be interesting to assess whether the *T. magnatum* truffles that show significant differences in the proportion of volatile organic compound (VOCs), as those from Marche, Tuscany, and Emilia Romagna regions (Gioacchini et al. 2008), can also be differentiated by molecular markers.

In summary, it should be possible to genetically discriminate white truffles according to their provenance. To this end, the ongoing *T. magnatum* genome sequencing project will provide scientists in the near future with a plethora of molecular markers to test even further this hypothesis. In turn, the successful differentiation of *T. magnatum* strains and populations is the premise for actions aimed at (a) tracing genetically the origin of local production as an effort to promote the social and economic development of specific rural areas and (b) preserving *T. magnatum* biodiversity. In this regard, if programs of *T. magnatum* cultivation are relaunched, then local fungal strains adapted to the areas of cultivation should be, whenever possible, used for the inoculation of host plants.

6.5 Conclusions

The last years have witnessed a major breakthrough in several aspects of the life cycle and population genetics of *T. magnatum*. These studies have highlighted that *T. magnatum* populations are genetically structured and in turn that its biodiversity across the species range has to be preserved. Also, these studies have questioned the ability of *T. magnatum* in establishing a mutualistic symbiosis with its hosts via canonical ECM structures. In turn, the recent progresses made in producing and characterizing genuine *T. magnatum* mycorrhizal plants represent a fundamental but still not sufficient step to manage productive man-made plantations. Thus, more efforts need to be made to trace the fate of this fungus in the field and disclose the structures and/or propagules that this fungus differentiates to accomplish its vegetative and sexual propagation phases. The development of molecular tools for monitoring the spatial and temporal dynamics of sexually compatible *T. magnatum* strains in the field will be the next frontier. These innovative molecular tools will be of invaluable help for investigating the effects of soil practices on persistence, spreading, and fructification of this fungus in the field and thus to set new guidelines for the correct management of both natural and man-made truffle grounds.

References

Amicucci A, Zambonelli A, Giomaro G, Potenza L, Stocchi V (1998) Identification of ectomycorrhizal fungi of the genus *Tuber* by species-specific ITS primers. Mol Ecol 7 (3):273–277. doi:10.1046/j.1365-294X.1998.00357.x

Amicucci A, Zambonelli A, Guidi C, Stocchi V (2001) Morphological and molecular characterisation of *Pulvinula constellatio* ectomycorrhizae. FEMS Microbiol Lett 194(2):121–125. doi:10.1111/j.1574-6968.2001.tb09456.x

Baciarelli-Falini L, Rubini A, Riccioni C, Paolocci F (2006) Morphological and molecular analyses of ectomycorrhizal diversity in a man-made *T. melanosporum* plantation: description of novel truffle-like morphotypes. Mycorrhiza 16(7):475–484. doi:10.1007/s00572-006-0066-5

Barbieri E, Ceccaroli P, Saltarelli R, Guidi C, Potenza L, Basaglia M, Fontana F, Baldan E, Casella S, Ryahi O, Zambonelli A, Stocchi V (2010) New evidence for nitrogen fixation within the Italian white truffle *Tuber magnatum*. Fungal Biol 114(11):936–942. doi:10.1016/j.funbio.2010.09.001

Belfiori B, Riccioni C, Tempesta S, Pasqualetti M, Paolocci F, Rubini A (2012) Comparison of ectomycorrhizal communities in natural and cultivated *Tuber melanosporum* truffle grounds. FEMS Microbiol Ecol 81(3):547–561. doi:10.1111/j.1574-6941.2012.01379.x

Bencivenga M, Granetti B, Gregori G, Rebaudengo E, Tocci A, Zambonelli A (1987) Valutazione dello stato di micorrizazione di piantine tartufigene – proposta di un metodo. Ministero Agricoltura e Foreste Corpo Forestale dello Stato, Genzano, Rome

Bennett KD, Tzedakis PC, Willis KJ (1991) Quaternary refugia of North European trees. J Biogeogr 18(1):103–115. doi:10.2307/2845248

Bertault G, Raymond M, Berthomieu A, Callot G, Fernandez D (1998) Trifling variation in truffles. Nature 394:734. doi:10.1038/29428

Bertault G, Roussett F, Fernandez D, Berthomieu A, Hochberg ME, Callot G, Raymond M (2001) Population genetics and dynamics of the black truffle in a man-made truffle field. Heredity 86 (4):451–458. doi:10.1046/j.1365-2540.2001.00855.x

Bertini L, Rossi I, Zambonelli A, Amicucci A, Sacchi A, Cecchini M, Gregori G, Stocchi V (2006) Molecular identification of *Tuber magnatum* ectomycorrhizae in the field. Microbiol Res 161 (1):59–64. doi:10.1016/j.micres.2005.06.003

Bidartondo MI, Gardes M (2005) Fungal diversity in molecular terms: profiling, identification, and quantification in the environment. In: Dighton J, White JF, Oudemans P (eds) The fungal community, its organization and role in the ecosystem, 3rd edn. CRC Press/Taylor & Francis, Boca Raton, pp 215–239

Boutahir S, Iotti M, Piattoni F, Zambonelli A (2013) Morphological and molecular characterization of *Tuber oligospermum* mycorrhizas. Afr J Agric Res 8(29):4081–4087. doi:10.5897/AJAR2013.7354

Brewer S (2002) Recolonisation postglaciaire de quelques taxons tempérés en Europe: une approche spatiale et temporelle. PhD thesis. Institut Méditerranéen d'Ecologie et de Paléoécologie, Marseille

Brun F, Mosso A (2010) Studio delle filiere dei prodotti trasformati a base di tartufo in Piemonte. Mercato, produzioni di qualità e valorizzazione del territorio. Regione Piemonte Direzione Opere Pubbliche, Difesa del suolo, Economia Montana e Foreste, Dipartimento di Economia e Ingegneria agraria forestale e ambientale dell'Università degli Studi di Torino, Grugliasco

Buzzini P, Gasparetti C, Turchetti B, Cramarossa MR, Vaughan-Martini A, Martini A, Pagnoni UM, Forti L (2005) Production of volatile organic compounds (VOCs) by yeasts isolated from the ascocarps of black (*Tuber melanosporum* Vitt.) and white (*Tuber magnatum* Pico) truffles. Arch Microbiol 184(3):187–193. doi:10.1007/s00203-005-0043-y

Callot G (1999) La Truffe, la Terre, la Vie. INRA, Paris

Danielson RN (1984) Ectomycorrhiza formation by the operculate discomycetes *Sphaerosporella brunnea* (Pezizales). Mycologia 76(3):454–461. doi:10.2307/3793328

De Miguel AM, Águeda B, Sánchez S, Parladé J (2014) Ectomycorrhizal fungus diversity and community structure with natural and cultivated truffle hosts: applying lessons learned to future truffle culture. Mycorrhiza 24(Suppl 1):S5–S18. doi:10.1007/s00572-013-0554-3

Douhan GW, Vincenot L, Gryta H, Selosse MA (2011) Population genetics of ectomycorrhizal fungi: from current knowledge to emerging directions. Fungal Biol 115(7):569–597. doi:10.1016/j.funbio.2011.03.005

Dutech C, Enjalbert J, Fournier E, Delmotte F, Barres B, Carlier J, Tharreau D, Giraud T (2007) Challenges of microsatellite isolation in fungi. Fungal Genet Biol 44(10):933–949. doi:10.1016/j.fgb.2007.05.003

Fontana A, Ceruti A, Meotto F (1992) Criteri istologici per il riconoscimento delle micorrize di *Tuber albidum*. Micol Veg Medit 7:121–136

Frizzi G, Lalli G, Miranda M, Pacioni G (2001) Intraspecific isozyme variability in Italian populations of the white truffle *Tuber magnatum*. Mycol Res 105(3):365–369. doi:10.1017/S0953756201003513

Gandeboeuf D, Dupré C, Roeckel-Drévet P, Nicolas P, Chevalier G (1997) Grouping and identification of *Tuber* species using RAPD markers. Can J Bot 75(1):36–45. doi:10.1139/b97-005

Gioacchini AM, Menotta M, Guescini M, Saltarelli R, Ceccaroli P, Amicucci A, Barbieri E, Giomaro G, Stocchi V (2008) Geographical traceability of Italian white truffle (*Tuber magnatum* Pico) by the analysis of volatile organic compounds. Rapid Commun Mass Spectrom 22(20):3147–3153. doi:10.1002/rcm.3714

Granetti B (1995) Caratteristiche morfologiche, biometriche e strutturali delle micorrize di *Tuber* di interesse economico. Micol Ital 24(2):101–117

Gregori G (2002) Problems and expectations with the cultivation of *Tuber magnatum*. In: Hall IR, Wang Y, Dannell E, Zambonelli A (eds) Edible mycorrhizal mushroom and their cultivation. Proceedings of the 2nd conference on edible mycorrhizal mushroom, Christchurch, 2001

Gregori G, Donnini D, Bencivenga M (2010) *Tuber magnatum*: alcuni esempi produttivi di tartufaie coltivate in Italia. In: Donnini D (ed) Proceedings of the 3rd international conference on truffles, Spoleto, 25–28 Nov 2008. Comunità Montana dei Monti Martani, Serano e Subasio, Spoleto, pp 741–749

Gryndler M, Černá L, Bukovská P, Hršelová H, Jansa J (2014) *Tuber aestivum* association with non-host roots. Mycorrhiza 24(8):603–610. doi:10.1007/s00572-014-0580-9

Hall IR, Yun W, Amicucci A (2003) Cultivation of edible mushrooms. Trends Biotechnol 21(10):433–438. doi:10.1016/S0167-7799(03)00204-X

Iotti M, Leonardi M, Oddis M, Salerni E, Baraldi E, Zambonelli A (2012) Development and validation of a real-time PCR assay for detection and quantification of *Tuber magnatum* in soil. BMC Microbiol 12:93. doi:10.1186/1471-2180-12-93

Iotti M, Leonardi M, Lancellotti E, Salerni E, Oddis M, Leonardi P, Perini C, Pacioni G, Zambonelli A (2014) Spatio-temporal dynamic of *Tuber magnatum* mycelium in natural truffle grounds. PLoS One 9(12):e115921. doi:10.1371/journal.pone.0115921

Lancellotti E, Iotti M, Zambonelli A, Franceschini A (2014) Characterization of *Tuber borchii* and *Arbutus unedo* mycorrhizas. Mycorrhiza 24(6):481–486. doi:10.1007/s00572-014-0564-9

Landeweert R, Veenman C, Kuyper TW, Fritze H, Wernars K, Smit E (2003) Quantification of ectomycorrhizal mycelium in soil by real-time PCR compared to conventional quantification techniques. FEMS Microbiol Ecol 45(3):283–292. doi:10.1016/S0168-6496(03)00163-6

Lanfranco L, Arlorio M, Matteucci A, Bonfante P (1995) Truffles: their life cycle and molecular characterization. In: Stocchi V, Bonfante P, Nuti P (eds) Biotechnology of ectomycorrhizae. Plenum Press, New York, pp 139–149. doi:10.1007/978-1-4615-1889-1_12

Le Tacon F, Zeller B, Plain C, Hossann C, Bréchet C, Robin C (2013) Carbon transfer from the host to *Tuber melanosporum* mycorrhizas and ascocarps followed using a ^{13}C pulse-labeling technique. PLos One 8(5):e64626. doi:10.1371/journal.pone.0064626

Le Tacon F, Rubini A, Murat C, Riccioni C, Robin C, Belfiori B, Zeller B, De la Varga H, Akroume E, Deveau A, Martin F, Paolocci F (2015) Certainties and uncertainties about the life cycle of the Périgord black truffle (*Tuber melanosporum* Vittad.). Ann For Sci 1–13. doi:10.1007/s13595-015-0461-1

Leonardi M, Iotti M, Oddis M, Lalli G, Pacioni G, Leonardi P, Mccherini C, Perini C, Salerni E, Zambonelli A (2013) Assessment of ectomycorrhizal fungal communities in the natural habitats of *Tuber magnatum* (Ascomycota, Pezizales). Mycorrhiza 23(5):349–358. doi:10.1007/s00572-012-0474-7

Linde CC, Selmes H (2012) Genetic diversity and mating type distribution of *Tuber melanosporum* and their significance to truffle cultivation in artificially planted truffieres in Australia. Appl Environ Microbiol 78(18):6534–6539. doi:10.1128/AEM.01558-12

Marchisio V (1964) Sulla cariologia degli aschi e delle spore di *Tuber maculatum* Vitt. Allionia 10:105–113

Marjanovic Z, Saljnikov E, Milenkovic M, Grebenc T (2010) Ecological features of *Tuber magnatum* Pico in the conditions of west Balkans – soil characterization. In: Donnini D (ed) Proceedings of the 3rd international conference on truffles, Spoleto, 25–28 Nov 2008. Comunità Montana dei Monti Martani, Serano e Subasio, Spoleto, pp 190–195

Martin F, Kohler A, Murat C, Balestrini R, Coutinho PM, Jaillon O, Montanini B, Morin E, Noel B, Percudani R, Porcel B, Rubini A, Amicucci A, Amselem J, Anthouard V, Arcioni S, Artiguenave F, Aury JM, Ballario P, Bolchi A, Brenna A, Brun A, Buée M, Cantarel B, Chevalier G, Couloux A, Da Silva C, Denoeud F, Duplessis S, Ghignone S, Hilselberger B, Iotti M, Mello M, Miranda M, Pacioni G, Quesneville H, Riccioni C, Ruotolo R, Splivallo R, Stocchi V, Tisserant E, Viscomi AR, Zambonelli A, Zampieri E, Henrissat B, Lebrun MH, Paolocci F, Bonfante P, Ottonello S, Wincker P (2010a) Périgord black truffle genome uncovers evolutionary origins and mechanisms of symbiosis. Nature 464(7291):1033–1038. doi:10.1038/nature08867

Martin F, Murat-Furminieux C, Paolocci F, Rubini A, Riccioni C, Belfiori B, Arcioni S (2010b) Molecular method for the identification of mating type genes of truffles species. Patent EP2426215A1

Mauriello G, Marino R, D'Auria M, Cerone G, Rana GL (2004) Determination of volatile organic compounds from truffles via SPME-GC-MS. J Chromatogr Sci 42(6):299–305. doi:10.1093/chromsci/42.6.299

Mello A, Fontana A, Meotto F, Comandini O, Bonfante P (2001) Molecular and morphological characterization of *T. magnatum* mycorrhizas in a long-term survey. Microbiol Res 155 (4):279–284. doi:10.1016/S0944-5013(01)80005-7

Mello A, Murat C, Vizzini A, Gavazza V, Bonfante P (2005) *Tuber magnatum* Pico, a species of limited geographical distribution: its genetic diversity inside and outside a truffle ground. Environ Microbiol 7(1):55–65. doi:10.1111/j.1462-2920.2004.00678.x

Molinier V, Murat C, Frochot H, Wipf D, Splivallo R (2015) Fine-scale spatial genetic structure analysis of the black truffle *Tuber aestivum* and its link to aroma variability. Environ Microbiol 17(8):3039–3050. doi:10.1111/1462-2920.12910

Murat C (2015) Forty years of inoculating seedlings with truffle fungi: past and future perspectives. Micorrhiza 25(1):77–81. doi:10.1007/s00572-014-0593-4

Murat C, Diez J, Luis P, Delaruelle C, Dupre C, Chevalier G, Bonfante P, Martin F (2004) Polymorphism at the ribosomal DNA ITS and its relation to postglacial re-colonization routes of the Perigord truffle *Tuber melanosporum*. New Phytol 164(2):401–411. doi:10.1111/j.1469-8137.2004.01189.x

Murat C, Vizzini A, Bonfante P, Mello A (2005) Morphological and molecular typing of the below-ground fungal community in a natural *Tuber magnatum* truffle-ground. FEMS Microbiol Lett 245(2):307–313. doi:10.1016/j.femsle.2005.03.019

Murat C, Riccioni C, Belfiori B, Cichocki N, Labbe J, Morin E, Tisserant E, Paolocci F, Rubini A, Martin F (2011) Distribution and localization of microsatellites in the Perigord black truffle genome and identification of new molecular markers. Fungal Genet Biol 48(6):592–601. doi:10.1016/j.fgb.2010.10.007

Murat C, Rubini A, Riccioni C, De la Varga H, Akroume E, Belfiori B, Guaragno M, Le Tacon F, Robin C, Halkett F, Martin F, Paolocci F (2013) Fine scale spatial genetic structure of the black truffle (*Tuber melanosporum*) investigated with neutral microsatellites and functional mating type genes. New Phytol 199(1):176–187. doi:10.1111/nph.12264

Napoli C, Mello A, Borra A, Vizzini A, Sourzat P, Bonfante P (2010) *Tuber melanosporum*, when dominant, affect fungal dynamics in truffle grounds. New Phytol 185(1):237–247. doi:10.1111/j.1469-8137.2009.03053.x

Nowak Z (2015) Truffle: a global history. In: Smith AF (ed) Reaktion Books-Edible, London

Pacioni G, Leonardi M, Aimola P, Ragnelli AM, Rubini A, Paolocci F (2007) Isolation and characterization of some mycelia inhabiting *Tuber* ascomata. Mycol Res 111(12):1450–1460. doi:10.1016/j.mycres.2007.08.016

Pacioni G, Ragnelli AM, Aimola P, Leonardi M, Marinucci D, Roth-Bejerano N, Kagan-Zur V (2014) Endomycorrhiza and pseudo-ectomycorrhiza produced in vitro by two species of *Tuber* on transformed *Cistus incanus* roots. Int J Plant Biol Res 2(4):1021

Paolocci F, Angelini P, Cristofari E, Granetti B, Arcioni S (1995a) Identification of *Tuber* spp. and corresponding ectomycorrhizae through molecular markers. J Food Sci Agric 69(4):511–517. doi:10.1002/jsfa.2740690416

Paolocci F, Cristofari E, Angelini P, Granetti B, Arcioni S (1995b) The polymorphism of the rDNA region in typing ascocarps and ectomycorrhizae of truffle species. In: Stocchi V, Bonfante P, Nuti M (eds) Biotechnology of ectomycorrhizae. Plenum Press, New York, pp 171–184. doi:10.1007/978-1-4615-1889-1_15

Paolocci F, Rubini A, Riccioni C, Arcioni S (2006) Re-evaluation of the life cycle of *Tuber magnatum*. Appl Environ Microbiol 72(4):2390–2393. doi:10.1128/AEM.72.4.2390-2393. 2006

Petit RJ, Aguinagalde I, de Beaulieu JL, Bittkau C et al (2003) Glacial refugia: hotspots but not melting pots of genetic diversity. Science 300(5625):1563–1565. doi:10.1126/science. 1083264

Riccioni C, Belfiori B, Rubini A, Passeri V, Arcioni S, Paolocci F (2008) *Tuber melanosporum* outcrosses: analysis of the genetic diversity within and among its natural populations under this new scenario. New Phytol 180(2):466–478. doi:10.1111/j.1469-8137.2008.02560.x

Rittersma RC (2011) A culinary Captatio Benevolentiae; the use of the truffles as a promotional gift by the Savoy dynasty in the eighteenth century. In: De Vooght D (ed) Royal taste: food, power and status at the European courts after 1789. Ashgate, Farnham, pp 31–55

Rubini A, Paolocci F, Granetti B, Arcioni S (2001) Morphological characterization of molecular typed *Tuber magnatum* ectomycorrhizae. Mycorrhiza 11(4):179–185. doi:10.1007/s005720100116

Rubini A, Topini F, Riccioni C, Paolocci F, Arcioni S (2004) Isolation and characterization of polymorphic microsatellite loci in white truffle (*Tuber magnatum*). Mol Ecol Notes 4 (1):116–118. doi:10.1111/j.1471-8286.2004.00587.x

Rubini A, Paolocci F, Riccioni C, Vendramin GG, Arcioni S (2005) Genetic and phylogeographic structures of the symbiotic fungus *Tuber magnatum*. Appl Environ Microbiol 71 (11):6584–6589. doi:10.1128/AEM.71.11.6584-6589.2005

Rubini A, Riccioni C, Arcioni S, Paolocci F (2007) Troubles with truffles: unveiling more of their biology. New Phytol 174(2):256–259. doi:10.1111/j.1469-8137.2007.01976.x

Rubini A, Belfiori B, Riccioni C, Arcioni S, Martin F, Paolocci F (2011a) *Tuber melanosporum*: mating type distribution in a natural plantation and dynamics of strains of different mating types on the roots of nursery-inoculated host plants. New Phytol 189(3):723–735. doi:10.1111/j.1469-8137.2010.03493.x

Rubini A, Belfiori B, Riccioni C, Tisserant E, Arcioni S, Martin F, Paolocci F (2011b) Isolation and characterization of MAT genes in the symbiotic ascomycete *Tuber melanosporum*. New Phytol 189(3):710–722. doi:10.1111/j.1469-8137.2010.03492.x

Rubini A, Riccioni C, Belfiori B, Paolocci F (2014) Impact of the competition between mating types on the cultivation of *Tuber melanosporum*: Romeo and Juliet and the matter of space and time. Mycorrhiza 24(1):19–27. doi:10.1007/s00572-013-0551-6

Selosse MA, Faccio A, Scappaticci G, Bonfante P (2004) Chlorophyllous and achlorophyllous specimens of *Epipactis microphylla (Neottieae, Orchidaceae)* are associated with ectomycorrhizal septomycetes, including truffles. Microb Ecol 47(4):416–426. doi:10.1007/s00248-003-2034-3

Selosse MA, Taschen E, Giraud T (2013) Do black truffles avoid sexual harassment by linking mating type and vegetative incompatibility? New Phytol 199(1):10–13. doi:10.1111/nph.12329

Smith CJ, Osborn AM (2008) Advantages and limitations of quantitative PCR (Q-PCR)-based approaches in microbial ecology. FEMS Microbiol Ecol 67(1):6–20. doi:10.1111/j.1574-6941.2008.00629.x

Splivallo R, Ebeler SE (2015) Sulfur volatiles of microbial origin are key contributors to human-sensed truffle aroma. Appl Microbiol Biotechnol 99(6):2583–2592. doi:10.1007/s00253-014-6360-9

Splivallo R, Ottonello S, Mello A, Karlovsky P (2011) Truffle volatiles: from chemical ecology to aroma biosynthesis. New Phytol 189(3):688–699. doi:10.1111/j.1469-8137.2010.03523.x

Splivallo R, Valdez N, Kirchhoff N, Ona MC, Schmidt JP, Feussner I, Karlovsky P (2012) Intraspecific genotypic variability determines concentrations of key truffle volatiles. New Phytol 194(3):823–835. doi:10.1111/j.1469-8137.2012.04077.x

Splivallo R, Deveau A, Valdez N, Kirchhoff N, Frey-Klett P, Karlovsky P (2014) Bacteria associated with truffle-fruiting bodies contribute to truffle aroma. Environ Microbiol 17 (8):2647–2660. doi:10.1111/1462-2920.12521

Tabouret P (2011) Exclusif! Description d'un site français producteur de truffe blanche d'Italie. Le Trufficulteur 78:18

Tabouret P (2012) Constitution d'une fédération Suisse de la truffe et de la trufficulture. Le Trufficulteur 81:25

Tedersoo L, Hansen K, Brian AP, Kjøller R (2006) Molecular and morphological diversity of pezizalean ectomycorrhiza. New Phytol 170(3):581–596. doi:10.1111/j.1469-8137.2006.01678.x

Urban A, Neuner-Plattner I, Krisai-Greilhuber I, Haselwandter K (2004) Molecular studies on terricolous microfungi reveal novel anamorphs of two *Tuber* species. Mycol Res 108 (7):749–758. doi:10.1017/S0953756204000553

Zambonelli A, Salomoni S, Pisi A (1993) Caratterizzazione anatomo-morfologica delle micorrize di *Tuber* spp. su *Quercus pubescens* Willd. Micol Ital 3:73–90

Zampieri E, Murat C, Cagnasso M, Bonfante P, Mello A (2010) Soil analysis reveals the presence of an extended mycelial network in a *Tuber magnatum* truffle ground. FEMS Microbiol Ecol 71(1):43–49. doi:10.1111/j.1574-6941.2009.00783.x

Zeller B, Bréchet C, Maurice JP, Le Tacon F (2008) Saprotrophic versus symbiotic strategy during truffle ascocarp development under holm oak. A response based on ^{13}C and ^{15}N natural abundance. Ann For Sci 65(6):607–614. doi:10.1051/forest:2008037

Chapter 7
The Puberulum Group Sensu Lato (Whitish Truffles)

Enrico Lancellotti, Mirco Iotti, Alessandra Zambonelli, and Antonio Franceschini

7.1 Introduction

Tuber species in the Puberulum group sensu lato (s.l.) produce small and light-colored ascomata with alveolate-reticulated ascospores. Members of this group are commonly called "whitish truffles" (Zambonelli et al. 2000a), in order to distinguish them from the most precious species *Tuber magnatum* Pico (the Italian white truffle). Puberulum group s.l. is the most widely distributed group and has the highest species richness within *Tuber* genus (Bonito et al. 2010a). Its members have been found all over the Northern Hemisphere associated with angiosperm and/or gymnosperm in a wide range of habitats: from the northern boreal forests of Europe and North America to the semiarid environments of Mexico and North Africa. Moreover, unclassified members of the Puberulum group were also found as ectomycorrhizas (ECMs) in the Southern Hemisphere (Bonito et al. 2010b). The Puberulum group s.l. includes commercially valuable species which are becoming increasingly popular in the marketplace. The most noteworthy species is *Tuber borchii* Vittad. ("bianchetto" truffle) which, actually, is cultivated outside its native areas and the extent of its plantations is increasing around the world. Actually, it is successfully cultivated in New Zealand and Australia, and, more recently, it has also been introduced in the USA (Zambonelli et al. 2015). *Tuber gibbosum* Harkn.

E. Lancellotti (✉) • A. Franceschini
Department of Agriculture, University of Sassari, via Enrico De Nicola 1, 07100 Sassari, Italy
e-mail: lance@uniss.it

M. Iotti
Department of Life, Health and Environmental Sciences, University of L'Aquila, via Vetoio, Coppito 1, I-67100 L'Aquila, Italy
e-mail: mirco.iotti@univaq.it

A. Zambonelli
Department of Agricultural Science, University of Bologna, viale Fanin 46, 40127 Bologna, Italy

A. Zambonelli et al. (eds.), *True Truffle (*Tuber *spp.) in the World*, Soil Biology 47,
DOI 10.1007/978-3-319-31436-5_7

(Oregon spring white truffle) and *Tuber oregonense* Trappe, Bonito and P. Rawl. (Oregon winter white truffle) are well appreciated in North America, but the efforts to cultivate them are in their infancy (Lefevre 2012). The potential culinary value of all the other whitish truffles has to be investigated yet.

The term "Puberulum" was used for the first time by Knapp (1950) to group together the *Tuber* species forming ascomata with smooth, papillate or rough surface, soft texture, and globose asci bearing an early-disappearing stalk and reticulate ascospores. Furthermore, the author define a Puberulum group A including nine species (*Tuber puberulum* Berk. & Broome, *Tuber albidum* Pico, *Tuber michailowskianum* Bucholtz, *Tuber rapaeodorum* Tul. & C. Tul., *T. borchii*, *Tuber maculatum* Vittad., *Tuber mougeotii* Quél., *Tuber asa-foetida* Lesp., and *Tuber lacunosum* Mattir.) with small-meshed (<10 μm) and regularly reticulate ascospores and a Puberulum group B consisting of three species (*Tuber dryophilum* Tul. & C. Tul., *Tuber foetidum* Vittad., and *T. magnatum*) with large-meshed (>10 μm) and irregularly reticulate ascospores. Thereafter, taxonomic relationships of the whitish truffle species have been debated for decades by many other European mycologists. Based on the spore dimension and morphology, Gross (1987) identified a Puberulum "cluster" within the white truffle species with small-sized ascomata, which included *T. borchii* var. *sphaerosperma* Malençon (=*T. puberulum* var. c), *T. puberulum* (=var. b), *Tuber murinum* R. Hesse (=*T. puberulum* var. a), *Tuber exiguum* R. Hesse, and *T. borchii*. Montecchi and Lazzari (1987) reported a redundant increase of species within the Puberulum group and accepted as valid only *T. borchii*, *T. rapaeodorum*, *T. dryophylum*, *T. maculatum*, *T. puberulum*, and *T. foetidum*. As time passes, the validity of some of these species was questioned, and new European species were included within the Puberulum group such as *Tuber oligospermum* (Tul. & C. Tul.) Trappe (Riousset et al. 2001) and *Tuber cistophilum* P. Alvarado, G. Moreno, Manjón, Gelpi & J. Muñoz (Alvarado et al. 2012).

The recent studies on molecular phylogeny of truffles have significantly improved the taxonomy of Puberulum group s.l., unraveling most of the phylogenetic affiliations within it. For example, Jeandroz et al. (2008) demonstrated that *T. magnatum* is phylogenetically well differentiated from the whitish truffles and belongs to the Aestivum group. On the contrary, many Asiatic and American species have been recently added to the Puberulum group s.l. (Wang et al. 2007, 2013; Jeandroz et al. 2008; Bonito et al. 2010a; Guevara et al. 2013).

Based on recent insight, we can consider as whitish truffles the *Tuber* species forming fruiting bodies with the following general features (Fig. 7.1):

Ascomata: hypogeous, small (commonly <5 cm in diameter), solid, and firm; globose or subglobose, sometimes lobed; whitish at first becoming pale yellowish to yellow-brown or reddish-brown; surface smooth or finely pubescent, rarely with minute and flat warts (*T. foetidum*)

Peridium: pseudoparenchymatous or plectenchymatous; yellow or darker toward the surface, otherwise white/hyaline

Dermatocystidia: cylindric or needle with septa when present; hyaline or yellowish

Gleba: whitish at first becoming grayish-brown to yellowish-brown, marbled with branching white veins arising from different points on the peridium

Asci: globose to ovate with 1–4 (−5) spores, thin-walled, sessile or short stipitate

Ascospores: hyaline at first becoming yellow-brown or reddish-brown at maturity, bearing an alveolate reticulum 3–6 μm deep

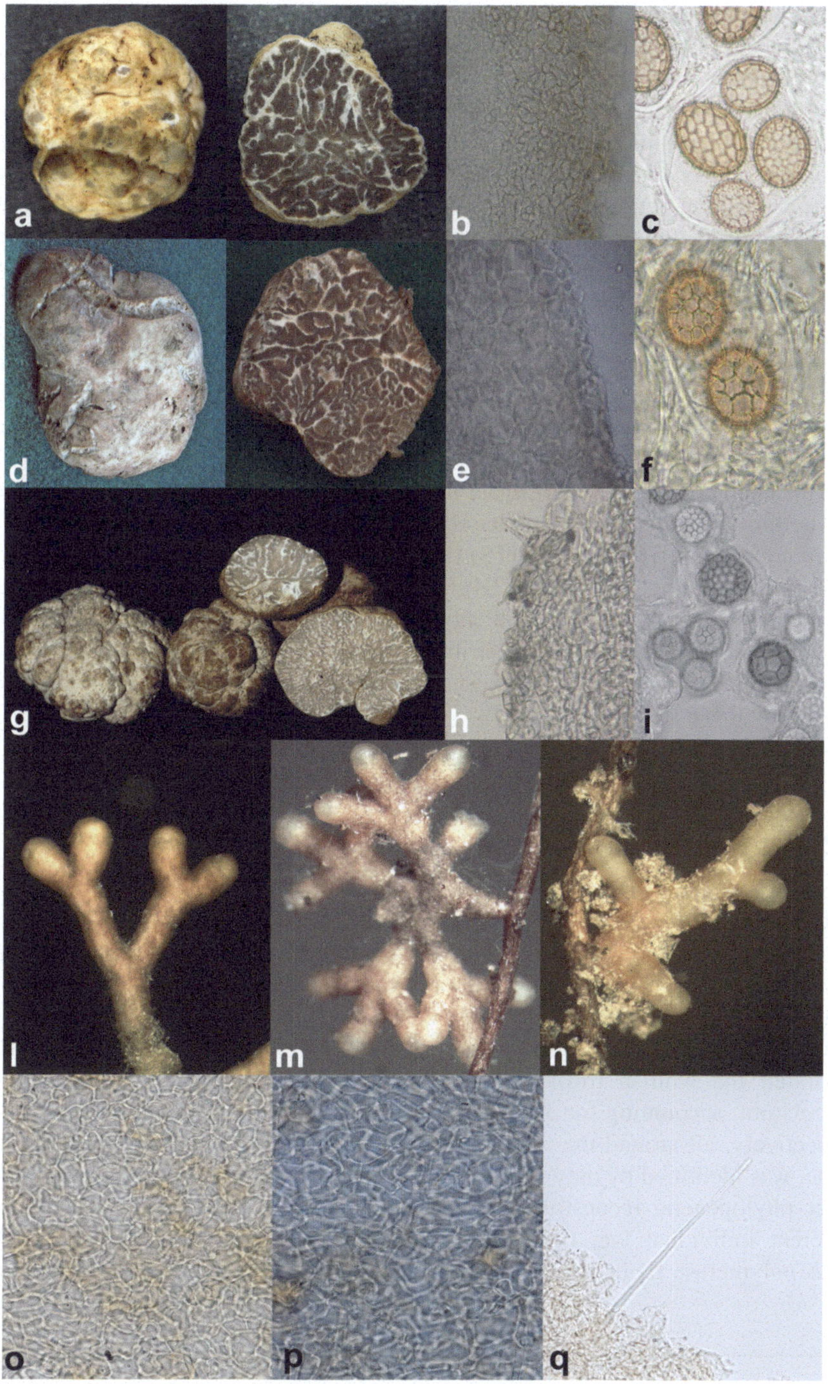

Fig. 7.1 Ascomata, peridium, and spores of *T. borchii* (**a**, **b**, **c**), *T. dryophilum* (**d**, **e**, **f**), and *T. oligospermum* (**g**, **h**, **i**). Mycorrhizas of whitish truffles with *P. nigra*, *A. unedo*, and *Q. robur*

The large amount of molecular data deposited in the public sequence repositories have represented a great resource to study the relationships among whitish truffle species but their analysis also revealed inconsistencies and contradictions. Indeed, part of the misidentified entries have been generated from the difficulties in identifying the whitish truffle ascomata because only few taxonomic informative traits of peridium and spores are available for species identification. However, most of the conflicting entries are from European species described before the twentieth century because the poor quality of the original descriptions led to disagreement among taxonomists and a different application of the species concepts (Jeandroz et al. 2008; Zambonelli et al. 2012). Then, the poor quality of the oldest holotypes in herbarium collections made often difficult to obtain reference DNA sequences useful to solve these controversies. Considering also that Puberulum group s.l. contains the highest number of insufficiently identified taxa and most of the diversity within *Tuber* genus (Bonito et al. 2010a), it remains the most controversial group of truffle species.

This chapter aimed to investigate the phylogenetic relationships and the diversity within Puberulum group s.l. based on the recent findings and the screening of the internal transcribed spacer (ITS) rDNA sequences available in GenBank database. We attempted to select an ITS reference sequence and, consequently, to assess the current extent of misidentified entries for each whitish truffle species. Finally, we reported the geographical distribution and intraspecific variability of each member of the Puberulum group s.l. as well as the description of mycorrhizas formed by these fungi.

7.2 Phylogeny of the Puberulum Group s.l.

Halász et al. (2005) were the first to explore the phylogenetic relationships within Puberulum group, validating the distinction among *T. rapaeodorum*, *T. foetidum*, *T. maculatum*, *T. borchii*, and *T. puberulum*. Wang et al. (2007) and Jeandroz et al. (2008) sorted members of this group in four subclades, including new species from Europe, Asia, and North America. Later, Bonito and colleagues (2010a) separated the whitish truffle species in two different clades, Puberulum and Maculatum, accounting for 37 (25 undescribed) and 19 (12 undescribed) species, respectively, all around the world. A third closely related lineage, the Gibbosum clade, was identified by only four species from North America. More recently, this latter phylogenetic reconstruction was also confirmed by the analysis of three different individual loci (28S large subunit rDNA, elongation factor 1-alpha, RNA polymerase II) in addition to the common ITS rDNA-based phylogeny, and

---

Fig. 7.1 (continued) (**l**, **m**, **n**). Puzzle-like (**o**) and puzzle-like bearing a hyphal net (**p**) mantles and cystidia (**q**) of whitish truffles it was estimated that the most recent common ancestors of Puberulum, Maculatum, and Gibbosum clades diverged 65, 67, and 27 Mya, respectively (Bonito et al. 2013).

Actually (date of accession 18 January 2015), about 800 ITS1-5.8S-ITS2 rDNA complete sequences (>400 bp) attributable to the Puberulum group s.l. are deposited in GenBank. To retrieve all the sequences (identified and insufficiently identified) belonging to this group of truffles, three reference sequences were selected for each clade (Puberulum, Maculatum, and Gibbosum) and BLAST against GenBank nucleotide database adjusting to 1000 the "max target sequences" parameter. The output sequences of each clade were aligned together by Muscle in MEGA6 (http://www.megasoftware.net/, Tamura et al. 2013), and those of poor quality and short length as well as the redundant entries were removed by the datasets. A provisional neighbor-joining tree was then generated for each of the three dataset for identifying and removing the sequences belonging to other *Tuber* lineages after Bonito et al. (2010a). We assumed 97 % ITS sequence similarity as threshold for phylotype definition and species approximation. Phylotypes were defined based on a similarity matrix constructed using MEGA6. Sequences with a p-distance < 3 % were grouped into the same phylotype. The 3 % is commonly accepted as global cutoff values for intraspecific variation in fungi (Nilsson et al. 2008; Smith et al. 2007) and in *Tuber* spp. (Bonito et al. 2013) although it is unable to discriminate cryptic species of whitish truffles (Bonuso et al. 2010; Wang et al. 2013). For some species of the Maculatum clade, we adopted a more stringent threshold for species definition after the work of Guevara et al. (2013) on North American truffles. We defined a reference ITS sequence for each described species of the Puberulum group s.l. according to one of the following criteria (listed in order of priority) (Table 7.1, second column): (1) sequences from holotypes considered by the "ITS RefSeq Target Loci project" (Schoch et al. 2014); (2) sequences from holotypes characterized in various scientific works; and (3) sequences from specimens in which characters are congruent with those reported by Pegler et al. (1993), Astier (1998), Zambonelli et al. (2000b), Montecchi and Sarasini (2000), Riousset et al. (2001), and Ceruti et al. (2003).

The validity of the reference sequences selected for *T. borchii* and *T. maculatum* is supported by the analysis carried out by Mello et al. (2000) which amplified some neotypes of "Vittadini original" collections with the species-specific primer pairs TBA–TBB (Mello et al. 1999) and TmacI–TmacII (Amicucci et al. 1998), respectively.

The ITS sequence divergence of each phylotype was measured by p-distance in MEGA6, using either the correctly identified or misidentified sequences as well as the unidentified sequences. Mean and max values of p-distance are reported in Table 7.1.

It has not been possible to assign reference ITS sequences to *T. asa* and *T. sphaerospermum* because, in our opinion, their genetic and morphological identity is still controversial. All sequences from GenBank identified as *T. asa* (1) and *T. sphaerospermum* (5) clustered within our *T. oligospermum* phylotype (<3 % diversity).

Table 7.1 List of the described species within the Puberulum group sensu lato (Puberulum, Maculatum, and Gibbosum clades)

Tuber species (reference)	Ref. Seq.	GB Seq.	Id Seq.	Un Seq.	≠Species name	*p*-distance max/mean	≠ITS
Puberulum clade							
T. anniae (Colgan and Trappe 1997)	NR 119860 (1)	46	9	29	*T. pacificum* (7), *T. puberulum* (1)	0.034/0.013	
T. borchii (Vittadini 1831)	FJ554490 (3)	126	87	26	*T. maculatum* (1), *T. puberulum* (12)	0.036/0.012	13
T. californicum (Harkness 1899)	HM485351 (3)	7	5	2		0.004/0.001	
T. cistophilum (Alvarado et al. 2012)	NR119983 (1)	16	12	4		0.015/0.005	
T. dryophilum (Tulasne and Tulasne 1845)	AF003917 (3)	44	4	17	*T. melanosporum* (1), *T. puberulum* (11), *T. borchii* (9), *T. oligospermum* (2)	0.024/0.011	2
T. latisporum (Chen and Liu 2007)	NR119620 (1)	4	1	3		0.005/0.003	
T. lijiangense (Fan et al. 2011b)	KF805727 (2)	7	1	4	*T. microsphaerosporum* (1), *T. borchii* var. *sphaerospermum* (1)	0.013/0.008	
T. liui (Xu 1999)	DQ898182 (3)	1	1	0			
T. oligospermum (Trappe 1979)	KF021624 (3)	14	8	1	*T. sphaerospermum* (4), *T. asa* (1)	0.024/0.012	28
T. pacificum (Trappe and Castellano 2000)	HM485378 (3)	9	5	3	*Choiromyces alveolatus* (1)	0.024/0.014	5
T. pseudosphaerosporum (Fan and Yue 2013)	KF744063 (2)	1	1	0			
T. zhongdianense (He et al. 2004)	NR119621 (1)	3	3	0		0.005/0.003	
T. panzhihuanense (Deng et al. 2013)	NR120126 (1)	13	12	1		0.001/0.000	
T. sphaerosporum (Gilkey 1939)	HM485390 (3)	4	4	0		0.008/0.006	
T. huizeanum (Fan et al. 2012b)	JQ910651 (2)	2	2	0		0.003/0.003	
T. alboumbilicum (Li et al. 2014)	KJ742702 (2)	1	1	0			
T. puberulum (Berkeley and Broome 1846)	AJ969626 (2)	25	8	17		0.009/0.004	31
T. separans (Gilkey 1916)	HM485387 (3)	20	5	15		0.029/0.002	0
T. pseudomagnatum (Fan and Cao 2012)	NR111718 (1)	2	1	0	*T. liyuanum* (1)		
T. sinosphaerosporum (Fan et al. 2012c)	JX092086 (2)	2	0	0			

Maculatum clade							
T. maculatum (Vittadini 1831)	AF003919 (3)	90	13	53	*T. rapaeodorum* (21), *T. borchii* (2), *T. dryophilum* (1)	0.016/0.007	14
T. scruposum (Hesse 1891)	DQ011845 (3)	14	4	10		0.018/0.007	12
T. lauryi (Guevara et al. 2013)	NR119862 (1)	13	6	7		0.020/0.008	
T. rapaeodorum (Tulasne and Tulasne 1843)	DQ011850 (3)	8	3	5		0.007/0.003	22
T. shearii (Murrill 1920)	HM485389 (3)	3	2	1			
T. foetidum (Vittadini 1831)	AJ557544 (3)	15	5	10		0.003/0.009	
T. whetstonense (Frank et al. 2006)	NR119864 (1)	12	5	7		0.026/0.013	
T. walkeri (Guevara et al. 2013)	JF419265 (2)	9	7	2		0.021/0.015	
T. beyerlei (Guevara et al. 2013)	NR119866 (1)	2	0	2		0.005/0.005	
T. guevarai (Guevara et al. 2013)	JF419251 (3)	5	3	2		0.011/0.006	
T. castilloi (Guevara et al. 2013)	NR119865 (1)	19	6	13		0.031/0.014	
T. miquihuanense (Guevara et al. 2013)	NR119868 (1)	26	2	19	*T. rapaeodorum* (1), *T. scruposum* (3), *T. mexiusanum* (1)	0.030/0.014	
T. mexiusanum (Guevara et al. 2013)	NR119867 (1)	13	7	5	*T. miquihuanense* (1)	0.026/0.014	
T. bomiense (Su et al. 2013)	KC517480 (2)	3	0	3		0.00/0.00	
T. microverrucosum (Fan et al. 2012b)	JN870099 (3)	2	0	1	*T. liyuanum* (1)		
T. linsdalei (Gilkey 1954)	HM85370 (2)	2	1	1			
Gibbosum clade							
T. bellisporum (Bonito et al. 2010b)	NR121354 (1)	3	3	0		0.004/0.004	
T. castellanoi (Bonito et al. 2010b)	NR121355 (1)	4	4	0		0.002/0.001	
T. gibbosum (Harkness 1899)	FJ809868 (3)	13	11	2		0.014/0.003	
T. oregonense (Bonito et al. 2010b)	NR121356 (1)	14	14	0		0.002/0.001	

Reference sequence (Ref. Seq.), number of sequences in GenBank: total (GB Seq., date of accession 18/01/2015), correctly identified (Id. Seq.), unidentified (Un. Seq.), misidentified [≠ species name, number (between brackets) of entries with same ITS phylotype but differently named; ≠ITS, number of entries with same name but different ITS phylotype] and intraspecific diversity (*p*-distance, mean, and max values calculated in MEGA6 software) are reported for each species

T. asa was considered synonym of *T. oligospermum* by Alvarado et al. (2012), but they were regarded as true species by other authors which, instead, considered *T. lacunosum* synonym of *T. asa* (Ceruti 1960; Montecchi and Sarasini 2000; Riousset et al. 2001; Ceruti et al. 2003).

Tuber sphaerospermum has been promoted to the rank of species only recently by Roux (2006) but it was firstly identified as *T. borchii* var. *sphaerosperma* by Malençon (1973). The original description of its ascomata reported regularly spherical spores and a pseudoparenchymatous peridium. *Tuber oligospermum* shows the same spore morphology but a plectenchimatous peridium as described by Ceruti et al. (2003) who examined a syntype of the Mattirolo herbarium.

No ITS sequences are available in GenBank for the American truffle species *Tuber guzmanii* Trappe and Cázares, *Tuber irradians* Gilkey, and *Tuber levissimum* Gilkey which were included within the Puberulum group by Bonito et al. (2010a) based on the sequence of their LSU rDNA gene.

7.3 Unreliability of the ITS Accessions and Unknown Species

Several studies have questioned the taxonomic reliability of the entries in public sequence databases, demonstrating that more than 10 % of the identified ITS sequences from fungi have been incorrectly annotated (Bridge et al. 2003; Nilsson et al. 2006). Despite the efforts recently made by the scientific community to improve the quality and the interpretation of database submissions (Tedersoo et al. 2011; Nilsson et al. 2012), compromised annotations still hamper the correct interpretation of the molecular data of some fungal taxa. Among these, whitish truffles account for a significant number of incorrect annotations.

GenBank ambiguities have been revealed in two different ways with respect to each reference entry: (1) entries with the same ITS sequence but differently named (Table 7.1, column 6) and (2) entries with the same name but with ITS sequence clustering in a different phylotype (Table 7.1, column 8). According to our criteria of analysis, we found 83 ambiguous entries (31 % of total identified sequences belonging to the Puberulum group s.l.) in the first case and 127 ambiguous entries (47 % of total) in the second case. Erroneous annotations were found in 10 out of 38 described whitish truffle species represented by at least one GenBank entry (Table 7.1). The highest number of misidentified entries has been found for *T. puberulum*, *T. maculatum*, *T. oligospermum*, *T. rapaeodorum*, *T. dryophilum*, and *T. borchii* (Table 7.1). More than 75 % of the sequences identified as *T. puberulum* (31 out of 39), *T. oligospermum* (28 out of 36), and *T. rapaeodorum* (22 out of 25) fall into other phylotypes in addition to that of the respective reference sequence (Table 7.1, columns 4 and 8). On the contrary, *T. maculatum* and *T. dryophilum* show the highest number of sequences (24 and 23, respectively) erroneously accessioned under other whitish truffle species (Table 7.1, column 6).

The reliability of the recent described Asiatic species *Tuber polyspermum* (Fan et al. 2011a) and *Tuber microsphaerosporum* (Fan et al. 2012a) must be verified because they are both represented by a single ITS sequence clustering within the *Tuber lijiangense* L. Fan and J. Z. Cao phylotype. Likewise, the unique sequence identified as *T. murinum* (JF261371) is similar to those of the *Tuber beyerlei* Trappe, Bonito, and Guevara holotype. The validity of *T. murinum* was debated for a long time, and, recently, it has been ignored by taxonomists (Ceruti et al. 2003).

Also the number of insufficiently identified sequences attributable to described species of the Puberulum group s.l. is remarkable. In fact, 42 % (249 out of 594) of the sequences of these species are mainly from ECMs or soil clones with the highest values registered in *T. maculatum* (53 out of 90) and *Tuber anniae* W. Colgan & Trappe (29 out of 46).

Based on the sequences from mycorrhiza, soil clone and unidentified ascomata deposited in GenBank, Bonito et al. (2010a) added 25 and 12 putative undescribed species to the Puberulum and Maculatum clades, respectively. A number of these unknown phylotypes have been recently identified and their ascomata characterized (Guevara et al. 2013). Our analysis found a total of 53 unidentified phylotypes in the Puberulum (43) and Maculatum (10) groups, whereas no unidentified species were revealed for the Gibbosum group. Most of these unidentified phylotypes appeared only recently in GenBank (later than Bonito et al. 2010a), and 18 out of 53 are only represented by one sequence.

7.4 Phylogeography

Whitish truffles represent the most widely distributed group of species in the world although many of them appear to have a native distribution range restricted to continental or subcontinental scale. Indeed, truffle cultivation, forestry, and nursery trade have contributed to introduce a number of whitish truffle species into novel habitats (Barroetaveña et al. 2005; Hall et al. 2007; Bulman et al. 2010; Wang et al. 2013). However, to date, the number of species shared between continents is still low (six for both Puberulum and Maculatum groups).

7.4.1 Puberulum Clade

Puberulum clade is the most abundant and ubiquitous clade of whitish truffle species, with members distributed across Europe, Asia, North America, South America, and northern Africa (Fig. 7.2). A new unidentified species belonging to the Puberulum group was also found in Argentina analyzing ITS sequences from ECMs of *Nothofagus* spp. and *Salix humboldtiana* Willd. natural stands (Bonito et al. 2013). Based on the sequence metadata, all described species from Asia and America have been found only in China (11 species), the USA (*Tuber californicum* Harkn.; *Tuber pacificum* Trappe, Castellano, and Bushnell; *Tuber sphaerosporum*

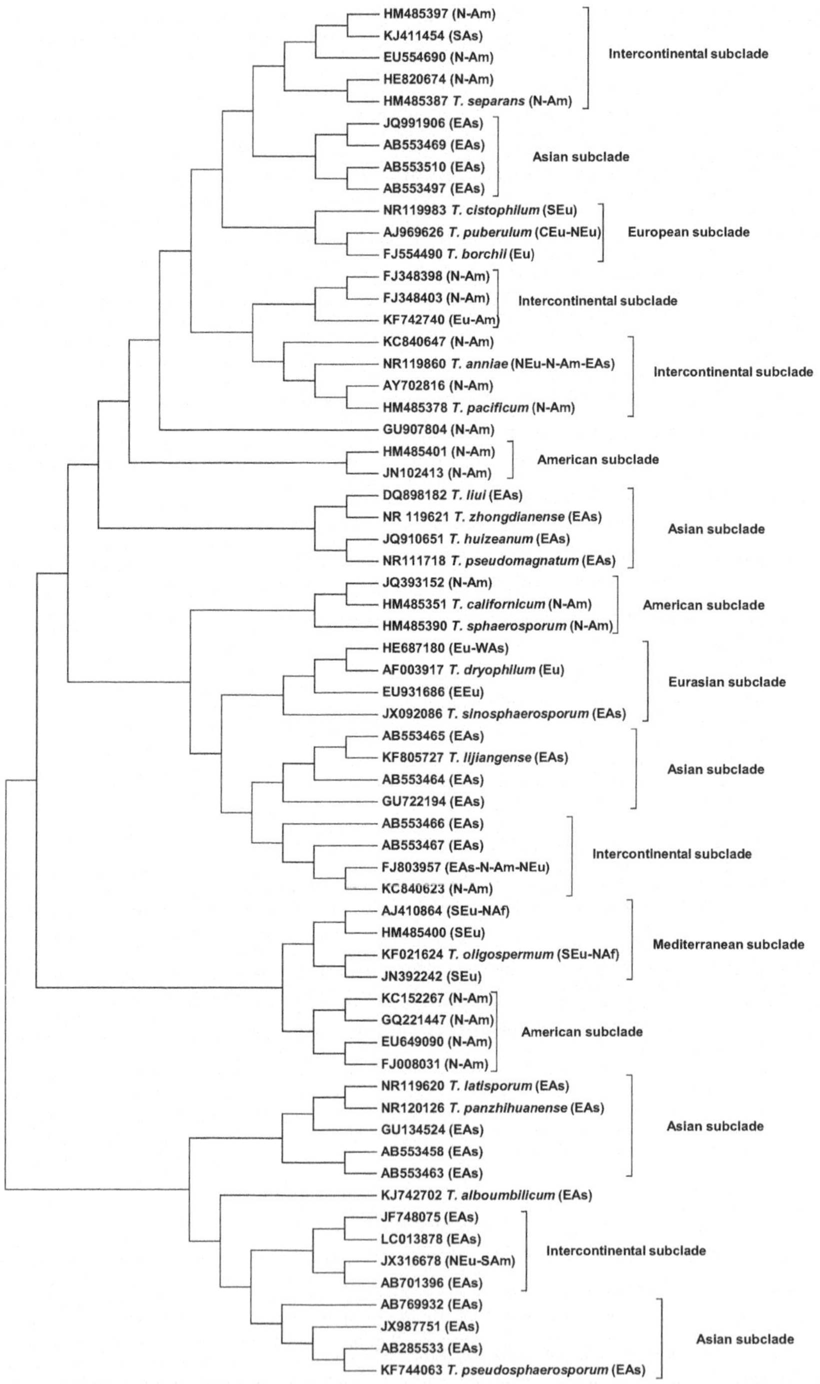

Fig. 7.2 Neighbor-joining tree (*p*-distance) of described and undescribed species of Puberulum clade. Accessions without the species name are referred to undescribed *Tuber* species. The bootstrap consensus tree was inferred from 100 replicates. Branches corresponding to partitions

Gilkey), or both the USA and Mexico (*Tuber separans* Gilkey). However, the number of entries attributable to these species is probably still too low to determine their entire range of distribution. On the contrary, four out of five species are widespread in Europe and have been also found in Iran (*T. dryophilum* and *T. puberulum*) and Morocco (*T. oligospermum*) or are cultivated around the world (*T. borchii*). According to the current metadata, *T. cistophilum* is only present in the Iberian Peninsula, and as *T. oligospermum*, it might be a species confined to the Western Mediterranean areas, with a preference for acid soils (Alvarado et al. 2012). The higher intraspecific diversity within *T. oligospermum* than *T. cistophilum* is probably due to the different range of collection sites. *Tuber oligospermum* has also been found outside the Iberian Peninsula, and its ITS sequences are grouped in two subclades, the first distributed in Morocco, Portugal, and Spain and the second in Sardinia (Italy) and Spain. On the contrary to the latter two species, the phylotype identified by us as *T. puberulum* almost grows in the northern temperate areas of middle Europe, from France and England to Poland and Czech Republic. To date, no molecular evidence of this species in Italian and Iberian Peninsulas as well as in the south Balkans or Scandinavia has been found.

Tuber borchii and *T. dryophilum* are ubiquitous in Europe and share the same distribution range and, often, also the same habitat (Iotti et al. 2010). Both species grow in cold-temperate to Mediterranean regions, from Iberian to Balkan Peninsulas (Hall et al. 2007). They are commonly found in calcareous sub-alkaline soils but can also colonize host species in acid soils (Peintner et al. 2007; Lancellotti and Franceschini 2013). ITS sequences of *T. borchii* are also from nonnative countries (the USA, British Columbia, New Zealand) probably due to increased cultivation of this truffle or introduced host plants. *Tuber borchii* is the whitish truffle mostly represented in GenBank with more than 120 entries, and our analysis revealed highest diversity within Puberulum group s.l. Its ITS variation is distributed between two main subclades, considered as cryptic species by Bonuso et al. (2010). Collections of the first group (clade I of Bonuso's work) are distributed mainly throughout Italy, whereas the second group (clade II) is ubiquitous in Europe. A significant genetic diversity was also found in *T. dryophilum* and the analysis of ITS sequence from 44 entries distinguished 6 subclades without any geographic structure. However, most of the diversity was found in Italy which might represent the area of diversification for this species.

Tuber anniae is the species of the Puberulum group with the largest geographical distribution. Its ascomata and ECMs have been mainly found in the coldest region of the Northern Hemisphere, but there is evidence for introduction of this species also in New Zealand (Bulman et al. 2010) and Hawaii (Hynson et al. 2013). Wang et al. (2013) considered *T. anniae* a species complex composed by three distinct

Fig. 7.2 (continued) reproduced in less than 50 % bootstrap replicates are collapsed. Evolutionary analyses were conducted in MEGA6. *Am* America, *N-Am* North America, *SAm* South America, *EAs* East Asia, *SAs* South Asia, *WAs* West Asia, *Eu* Europe, *NEu* North Europe, *SEu* South Europe, *CEu* Central Europe, *EEu* East Europe clades not supported by distinct morphological differences of their ascomata. Members of the clade I have been only found in North America, whereas clades II and III represent geographically disjunct phylotypes, which both include sequences from Europe and North America (Wang et al. 2013). *Tuber anniae* has been recently found also in northern Japan (Hashimoto et al. 2012) to form symbiosis with *Betula ermanii* Cham., a tree species widespread also in Russian, Korea, and China (Li and Skvortsov 1999). So, this species complex could have a wider distribution range than previously supposed.

With respect to Bonito et al. (2010a), we found 23 additional unidentified phylotypes within Puberulum clade, mostly from East Asia. Two unidentified phylotypes (*Tuber* sp. 33 and *Tuber* sp. 29 from Bonito's work) have been recently described as *Tuber pseudosphaerosporum* L. Fan (Fan and Yue 2013) and *T. lijiangense* (Fan et al. 2011b).

7.4.2 Maculatum Clade

Maculatum clade includes ten described species found exclusively in North America (*Tuber shearii* Harkn.; *Tuber lauryi* Trappe, Bonito & Guevara; *Tuber whetstonense* J. L. Frank, D. Southw. & Trappe; *Tuber walkeri* Healy, Bonito & Guevara; *T. beyerlei*; *Tuber guevarai* Bonito & Trappe; *Tuber castilloi* Guevara, Bonito & Trappe; *Tuber miquihuanense* Guevara, Bonito & Cázares; *Tuber mexiusanum* Guevara, Bonito & Cázares, and *Tuber linsdalei* Gilkey) and four described species from Europe (*T. maculatum*, *Tuber scruposum* R. Hesse, *T. rapaeodorum*, *T. foetidum*) rarely found also in Asia (Fig. 7.3). In general, intraspecific ITS diversity of European species is lower than American species.

Within Maculatum group, *T. maculatum* was the phylotype with the highest number of ITS entries which are grouped in six main subclades. Members of these subclades share a similar geographic range of distribution, from the Netherlands to Iran and from Italy to Finland. A unique sequence from China forms an additional basal subclade which might represent a native Asiatic cryptic species of *T. maculatum* (GU134516). Our analysis revealed also the introduction of *T. maculatum* in Canada, the USA, and New Zealand.

Tuber maculatum ascomata can be easily confused with those of *T. rapaeodorum* and our analysis confirmed the taxonomic difficulties in distinguishing between these species (Table 7.1). According on our criteria for species selection, *T. rapaeodorum* should be only present in East Europe and South Caspian region.

Tuber scruposum was found in the same areas where *T. rapaeodorum* grow as well as in central Europe (Italy, Germany, and Austria). Its ITS sequences are grouped in two subclades which do not show any geographic structure. The reference sequence selected for *T. scruposum* correspond to the *Tuber* sp. 37 (scruposum 1) phylotype after Bonito et al. (2010a) which is phylogenetically distinct from *Tuber* sp. 40 (scruposum 2), found both in North America and Europe. ITS diversity between these two phylotypes (>7 %) does not seem

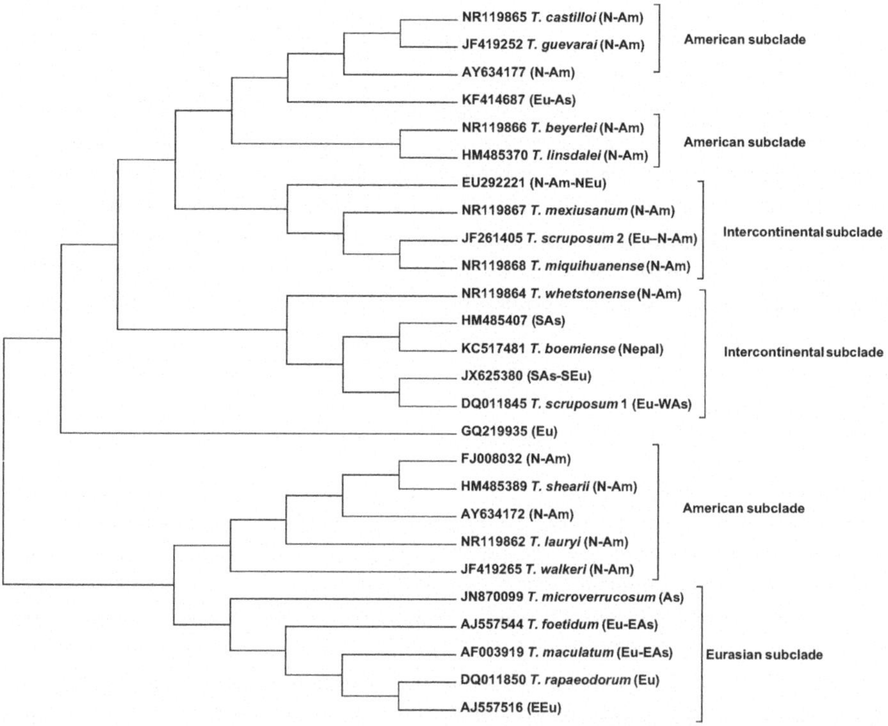

Fig. 7.3 Neighbor-joining tree (*p*-distance) of described and undescribed species of Maculatum clade. Accessions without the species name are referred to undescribed *Tuber* species. The bootstrap consensus tree was inferred from 100 replicates. Branches corresponding to partitions reproduced in less than 50 % bootstrap replicates are collapsed. Evolutionary analyses were conducted in MEGA6. *N-Am* North America, *As* Asia, *EAs* East Asia, *SAs* South Asia, *WAs* West Asia, *Eu* Europe, *NEu* North Europe, *SEu* South Europe, *EEu* East Europe

supported by evident differences in morphology of their ascomata. In fact, Badalyan et al. (2005) classified as *T. scruposum* the ascomata of both species collected in an oak-hornbeam forest near Dilijan (Armenia).

With respect to Bonito et al. (2010a), our analysis revealed only one new unidentified phylotype from ECMs collected in a *T. magnatum* truffle ground located in South Italy (Leonardi et al. 2013). Two unidentified phylotypes (*Tuber* sp. 39 and *Tuber* sp. 45 from Bonito's work) and other four new species have been recently described by Guevara et al. (2013).

7.4.3 Gibbosum Clade

Gibbosum clade includes only four described species (*T. gibbosum*, *Tuber bellisporum* Bonito & Trappe, *Tuber castellanoi* Bonito & Trappe, and *T. oregonense*). All 34 analyzed ITS sequences have been obtained by ascomata or

ECMs collected in western North America. Montecchi and Sarasini (2000) and Pomarico et al. (2007) reported *T. gibbosum* ascomata from Emilia-Romagna (northern Italy) and Basilicata (Southern Italy), respectively, but evidences for these collections are not available in GenBank database. This truffle species was probably introduced in Italy together with *Pseudotsuga menziesii* (Mirb.) Franco, a nonnative European tree present in both collection sites. Species of the Gibbosum clade are characterized by a low intraspecific ITS variation (Table 7.1), and unlike the other *Tuber* lineages, they appear to be exclusively associated with Pinaceae hosts (Bonito et al. 2010b; Bonito et al. 2013). Five years after Bonito et al. (2010a), also our analysis did not reveal unidentified phylotypes within Gibbosum clade.

7.5 Mycorrhizas

Many species of the Puberulum group s.l. have been proven to form ECMs with a wide range of host plants either in greenhouse or in the field. Mycorrhizas of *T. borchii* have been synthesized and described on a wide range of host plants: *Pinus strobus* L. (Scannerini and Palenzona 1967), *Pinus pinea* L. (Zambonelli et al. 1995), *Pinus radiata* D. Don (Duñabeitia et al. 1996), *Quercus robur* L. (Fontana et al. 1992; Boutahir et al. 2013), *Quercus pubescens* Willd. (Zambonelli et al. 1993), *Quercus suber* L. (Zambonelli and Branzanti 1989), *Tilia platyphyllos* Scop. (Granetti et al. 1995; Sisti et al. 1998), *Corylus avellana* L. (Zambonelli and Branzanti 1984; Fontana et al. 1992; Rauscher et al. 1996), *Populus* sp. (Fontana and Palenzona 1969), *Castanea sativa* Miller (Zambonelli and Branzanti 1989), *Alnus cordata* (Loisel.) Desf. (Zambonelli and Branzanti 1989), *Carya illinoinensis* (Wangenh.) K. Koch (Benucci et al. 2012), and *Arbutus unedo* L. (Lancellotti et al. 2014). ECMs of *T. puberulum* on *Picea abies* Karst. (Blaschke 1988), *T. maculatum* on *Ostrya carpinifolia* Scop. (Zambonelli et al. 1999), *T. oligospermum* on *Quercus cerris* L. (Bencivenga et al. 1997), and *Q. robur* (Boutahir et al. 2013) have also been synthesized. In natural condition whitish truffles have been found in association with a broad host range, including both gymnosperm and angiosperm (Kovács and Jakucs 2006; Iotti et al. 2010; Leonardi et al. 2013). Whitish truffles are also susceptible to orchid colonization (Tešitelová et al. 2012).

ECMs established by the whitish truffle species share a common morphology and only little differences could be detected in some anatomical characters (Fig. 7.1). However, these differences are usually wide even within the same species and partially overlap between species of whitish truffles. Unramified ends are commonly club-shaped, straight, yellow ochre (younger tissues) to brown (older tissues), with a smooth or spiny surface. The mantel is pseudoparenchymatous with roundish to epidermoid cells, sometimes covered by a hyphal network, and the cystidia, when present, are needle-shaped, straight, or slightly bent, hyaline with a blunt tip and one basal septa (sometimes 2). Emanating hyphae are rare, septate, hyaline, without clamps, and not frequently ramified. Rhizomorphs have never been found.

Characters useful to discriminate ECMs of different whitish truffles can be only found in cystidia and mantle cells although the shape of mantle cells can slightly vary depending on the strain and the host plant (Giomaro et al. 2000). Kovács and Jakucs (2006) confirmed this statement analyzing ECMs formed by different species of the *Puberulum* group s.l. in natural conditions. For example, cystidia of *T. oligospermum* show a frequent basal inflation which can be used to discriminate ECMs of this species (Bencivenga et al. 1997; Boutahir et al. 2013). *Tuber maculatum* ECM differs from those of *T. borchii* for the shorter cystidia (Zambonelli et al. 1999).

7.6 Conclusions

The Puberulum group s.l. is the phylogenetic group inside the genus *Tuber* having the greatest genetic and species diversity, not supported by a similar variability in morphology of ascomata, ECM, or mycelia. Most of the whitish truffle species are difficult to differentiate morphologically and the number of cryptic species within this group is probably high. As a consequence, these species are often misidentified or insufficiently identified in GenBank, and, so, caution has to be exercised in interpreting the BLAST results for the molecular identification.

This context led to confusion among researchers who are concerned with the study of these truffles. Moreover, being some whitish truffles of high economic importance, this confusion may have negative effects on truffle market and cultivation. In fact, in many parts of Italy and throughout Europe, the precious species *T. borchii* is very often confused with other *Tuber* species in the Puberulum group s.l. such as *T. maculatum*, *T. dryophilum*, and *T. puberulum* which have poor culinary quality (Montecchi and Sarasini 2000). Misidentification of *T. borchii* ascomata may produce ecological and economic problems when alien or nontarget species are used to inoculate the seedlings for truffle cultivation.

We aim that reference sequences and morphological considerations reported in this chapter will help in research and practical studies on whitish truffles and will be a starting point for developing shared species concepts.

References

Alvarado P, Moreno G, Manjón JL (2012) Comparison between *Tuber gennadii* and *T. oligospermum* lineages reveals the existence of the new species *T. cistophilum* (Tuberaceae, Pezizales). Mycologia 104(4):894–910. doi:10.3852/11-254

Amicucci A, Zambonelli A, Giomaro G, Potenza L, Stocchi V (1998) Identification of ectomycorrhizal fungi of the genus *Tuber* by species-specific ITS primers. Mol Ecol 7 (3):273–277. doi:10.1046/j.1365-294X.1998.00357.x

Astier J (1998) Truffes blanche et noires (Tuberaceae et Terfeziaceae). Louis-Jean, Gap

Badalyan SM, Hovsepyan RA, Iotti M, Zambonelli A (2005) On the presence of truffles in Armenia. Flora Medit 15:683–692

Barroetaveña C, Rajchenberg M, Cázares E (2005) Mycorrhizal fungi in *Pinus ponderosa* introduced in central Patagonia (Argentina). Nova Hedwigia 80(3–4):453–464. doi:10.1127/0029-5035/2005/0080-0453

Bencivenga M, Tanfulli M, Donnini D (1997) Micorrizazione di *Quercus cerris* L. con *Tuber oligospermum* (Tul. et Tul.) Trappe. Micol Ital 26(3):89–93

Benucci GMN, Bonito G, Baciarelli Falini L, Bencivenga M (2012) Mycorrhization of pecan trees (*Carya illinoinensis*) with commercial truffle species: *Tuber aestivum* Vittad. and *Tuber borchii* Vittad. Mycorrhiza 22(5):383–392. doi:10.1007/s00572-011-0413-z

Berkeley MJ, Broome CE (1846) Notices of British hypogaeous fungi. Ann Mag Nat Hist 18:73–82

Blaschke H (1988) *Tuber puberulum*. In: Agerer R (ed) Colour Atlas of ectomycorrhizae. Einhorn, Schwabisch Gmund, p 22

Bonito GM, Gryganskyi AP, Trappe JM, Vilgalys R (2010a) A global meta-analysis of *Tuber* ITS rDNA sequences: species diversity, host associations and long-distance dispersal. Mol Ecol 19 (22):4994–5008. doi:10.1111/j.1365-294X.2010.04855.x

Bonito G, Trappe JM, Rawlinson P, Vilgalys R (2010b) Improved resolution of major clades within *Tuber* and taxonomy of species within the *Tuber gibbosum* complex. Mycologia 102 (5):1042–1057. doi:10.3852/09-213

Bonito G, Smith ME, Nowak M, Healy RA, Guevara G, Cazares E, Kinoshita A, Nouhra ER, Dominguez LS, Tedersoo L, Murat C, Wang Y, Moreno BA, Pfister DH, Nara K, Zambonelli A, Trappe JM, Vilgalys R (2013) Historical biogeography and diversification of truffles in the Tuberaceae and their newly identified southern hemisphere sister lineage. PLoS One 8(1), e52765. doi:10.1371/journal.pone.0052765

Bonuso E, Zambonelli A, Bergemann SE, Iotti M, Garbelotto M (2010) Multilocus phylogenetic and coalescent analyses identify two cryptic species in the Italian bianchetto truffle, *Tuber borchii* Vittad. Conserv Genet 11(4):1453–1466. doi:10.1007/s10592-009-9972-3

Boutahir S, Iotti M, Piattoni F, Zambonelli A (2013) Morphological and molecular characterization of *Tuber oligospermum* mycorrhizas. Afr J Agric Res 8(29):4081–4087. doi:10.5897/AJAR2013.7354

Bridge PD, Roberts PJ, Spooner BM, Panchal G (2003) On the unreliability of published DNA sequences. New Phytol 160(1):43–48. doi:10.1046/j.1469-8137.2003.00861.x

Bulman SR, Visnovsky SB, Hall IR, Guerin-Laguette A, Wang Y (2010) Molecular and morphological identification of truffle-producing *Tuber* species in New Zealand. Mycol Prog 9 (2):205–214. doi:10.1007/s11557-009-0626-0

Ceruti A (1960) Elaphomycetales et tuberales. In: Bresadola J (ed) Iconographia mycologica, vol 28 (Suppl 2). Trento

Ceruti A, Fontana A, Nosenzo C (2003) Le specie Europee del genere *Tuber*, una revisione storica. Museo Regionale di Scienze Naturali, Monographie XXXVII. Regione Piemonte, Torino

Chen J, Liu PG (2007) *Tuber latisporum sp. nov*. and related taxa, based on morphology and DNA sequence data. Mycologia 99(3):475–481. doi:10.3852/mycologia.99.3.475

Colgan W, Trappe JM (1997) NATS tuffle and truffle-like fungi 7: *Tuber anniae sp. nov.* (Ascomycota). Mycotaxon 64:437–441. doi:10.1007/s11557-012-0862-6

Deng XJ, Liu PG, Liu CY, Wang Y (2013) A new white truffle species, *Tuber panzhihuanense* from China. Mycol Prog 12(3):557–561. doi:10.1007/s11557-012-0862-6

Duñabeitia MK, Hormilla S, Salcedo I, Peña JI (1996) Ectomycorrhizae synthesized between *Pinus radiata* and eight fungi associated with *Pinus* spp. Mycologia 88(6):897–908. doi:10.2307/3761052

Fan L, Cao JZ (2012) Two new species of white truffle from China. Mycotaxon 121:297–304. doi:10.5248/121.297

Fan L, Yue SF (2013) Phylogenetic divergence of three morphologically similar truffles: *Tuber sphaerosporum*, *T. sinosphaerosporum*, and *T. pseudosphaerosporum* sp. nov. Mycotaxon 125:283–288. doi:10.5248/125.283

Fan L, Hou CL, Cao JZ (2011a) *Tuber sinoalbidum* and *T. polyspermum*—new species from China. Mycotaxon 118:403–410. doi:10.5248/118.403
Fan L, Cao JZ, Liu Y, Li Y (2011b) Two new species of the genus *Tuber* from China. Mycotaxon 116:349–354. doi:10.5248/116.349
Fan L, Cao JZ, Li Y (2012a) *Tuber microsphaerosporum* and *Paradoxa sinensis spp. nov*. Mycotaxon 120:471–475. doi:10.5248/120.471
Fan L, Hou C, Li Y (2012b) *Tuber microverrucosum* and *T. huizeanum* – two new species from China with reticolate ascospores. Mycotaxon 122:161–169. doi:10.5248/122.161
Fan L, Cao JZ, Li Y (2012c) *Tuber sinosphaerosporum sp. nov*. from China. Mycotaxon 122:347–353. doi:10.5248/122.347
Fontana A, Palenzona M (1969) Sintesi micorrizica di *Tuber albidum* in coltura pura con *Pinus strobus* e pioppo euroamericano. Allionia 15:99–104
Fontana A, Ceruti A, Meotto F (1992) Criteri istologici per il riconoscimento delle micorrize di *Tuber albidum*. Micol Veg Medit 7(1):121–136
Frank JL, Southworth D, Trappe JM (2006) NATS truffle and truffle-like fungi 13: *Tuber quercicola* and *T. whetstonense*, new species from Oregon, and *T. candidum* redescribed. Mycotaxon 95:229–240
Gilkey HM (1916) A revision of the Tuberales of California. Univ Calif Publ Bot 6:275–356
Gilkey HM (1939) Tuberales of North America. Or State Monogr Stud Bot 1:1–63
Gilkey HM (1954) Tuberales. N Am Flora 1:1–36
Giomaro G, Zambonelli A, Cecchini D, Stocchi V, Saffi V (2000) Anatomical and morphological characterization of mycorrhizas of five strains of *Tuber borchii* Vittad. Mycorrhiza 10 (3):107–114. doi:10.1007/s005720000065
Granetti B, Angelini P, Rubini A (1995) Morfologia e struttura delle micorrize di *Tuber magnatum* Pico e di *Tuber borchii* Vitt. con *Tilia platyphyllos*. Micol Ital 24(2):27–34
Gross G (1987) Zu den europäischen Sippen der Gattung *Tuber*. In: Derbsch H, Schmitt JA (eds) Atlas der Pilze des Saarlandes, Teil 2, Nachweise, Okologie, Vorkommen und Beschreibungen. Derlattinia, Saarbruken, pp 79–99
Guevara G, Bonito G, Trappe JM, Cázares E, Williams G, Healy RA, Schadt C, Vilgalys R (2013) New North American truffles (*Tuber* spp.) and their ectomycorrhizal associations. Mycologia 105(1):194–209. doi:10.3852/12-087
Halász K, Bratek Z, Szego D, Rudnoy S, Racz I, Lasztity D, Trappe JM (2005) Tests of species concepts of the small, white, European group of Tuber spp based on morphology and rDNA ITS sequences with special reference to Tuber rapaeodorum. Mycol Prog 4(4):281–290. doi:10.1007/s11557-006-0132-6
Hall IR, Brown GT, Zambonelli A (2007) Taming the truffle: the history, lore and science of the ultimate mushroom. Timber Press, Portland
Harkness HW (1899) Californian hypogeous fungi. Proc Calif Acad Sci 3(1):241–292
Hashimoto Y, Fukukawa S, Kunishi A, Suga H, Richard F, Sauve M, Selosse MA (2012) Mycoheterotrophic germination of *Pyrola asarifolia* dust seeds reveals convergences with germination in orchids. New Phytol 195(3):620–630. doi:10.1111/j.1469-8137.2012.04174.x
He XY, Li HM, Wang Y (2004) *Tuber zhongdianense sp. nov*. from China. Mycotaxon 90 (1):213–216
Hesse R (1891) Die Hypogaeen Deutschlands, Band 1: Die Hymenogastreen. Verlag von Ludw/ Hofstetter, Halle
Hynson NA, Merckx VS, Perry BA, Treseder KK (2013) Identities and distributions of the co-invading ectomycorrhizal fungal symbionts of exotic pines in the Hawaiian Islands. Biol Invasions 15(11):2373–2385. doi:10.1007/s10530-013-0458-3
Iotti M, Lancellotti E, Hall I, Zambonelli A (2010) The ectomycorrhizal community in natural *Tuber borchii* grounds. FEMS Microbiol Ecol 72(2):250–260. doi:10.1111/j.1574-6941.2010.00844.x

Jeandroz S, Murat C, Wang Y, Bonfante P, Le Tacon F (2008) Molecular phylogeny and historical biogeography of the genus *Tuber*, the 'true truffles'. J Biogeogr 35(5):815–829. doi:10.1111/j.1365-2699.2007.01851.x

Knapp A (1950) Die Europaischen Hypogaeen Gattungen und ihre Gattungstypen. Schwelz Zeits Pilzk 28:29–42, 101–118, 153–179

Kovács GM, Jakucs E (2006) Morphological and molecular comparison of white truffle ectomycorrhizae. Mycorrhiza 16(8):567–574. doi:10.1007/s00572-006-0071-8

Lancellotti E, Franceschini A (2013) Studies on the ectomycorrhizal community in a declining *Quercus suber* L. stand. Mycorrhiza 23(7):533–542. doi:10.1007/s00572-013-0493-z

Lancellotti E, Iotti M, Zambonelli A, Franceschini A (2014) Characterization of *Tuber borchii* and *Arbutus unedo* mycorrhizas. Mycorrhiza 24(6):481–486. doi:10.1007/s00572-014-0564-9

Lefevre C (2012) Native and cultivated truffles of North America. In: Zambonelli A, Bonito GM (eds) Edible ectomycorrhizal mushrooms, current knowledge and future prospects, vol 34, Soil biology. Springer, Berlin, pp 209–226. doi:10.1007/978-3-642-33823-6_12

Leonardi M, Iotti M, Oddis M, Lalli G, Pacioni G, Leonardi P, Maccherini S, Perini C, Salerni E, Zambonelli A (2013) Assessment of ectomycorrhizal fungal communities in the natural habitats of *Tuber magnatum* (Ascomycota, Pezizales). Mycorrhiza 23(5):349–358. doi:10.1007/s00572-012-0474-7

Li PC, Skvortsov AK (1999) *Betula ermanii*. In: Flora of China, vol 4, Betulaceae, p 310. http://www.efloras.org/florataxon.aspx?flora_id=2&taxon_id=200006146

Li SH, Zheng LY, Liu CY, Wang Y, Li L, Zhao YC, Zhang XL, Yang M, Xiong HK, Qing Y, Wang L, Zhou DQ (2014) Two new truffle species, *Tuber alboumbilicum* and *Tuber pseudobrumale* from China. Mycol Prog 13:1157–1163. doi:10.1007/s11557-014-1004-0

Malençon G (1973) Champignons hypogés du Nord de l'Afrique. Persoonia 7(2):261–288

Mello A, Garnero L, Bonfante P (1999) Specific PCR-primers as a reliable tool for the detection of white truffle in mycorrhizal roots. New Phytol 141(3):511–516. doi:10.1046/j.1469-8137.1999.00356.x

Mello A, Vizzini A, Longato S, Rollo F, Bonfante P, Trappe JM (2000) *Tuber borchii* versus *Tuber maculatum*: neotype studies and DNA analyses. Mycologia 92(2):326–331. doi:10.2307/3761569

Montecchi A, Lazzari G (1987) Un nuovo tartufo di montagna: *Tuber regianum* sp. nov. Riv Micol 30:3–11

Montecchi A, Sarasini M (2000) Atlante fotografico di funghi ipogei. Associazione Micologica Bresadola, Vicenza

Murrill WA (1920) Another new truffle. Mycologia 12(3):157–158. doi:10.2307/3753258

Nilsson RH, Ryberg M, Kristiansson E, Abarenkov K, Larsson KH, Kõljalg U (2006) Taxonomic reliability of DNA sequences in public sequence databases: a fungal perspective. PLoS One 1 (1):e59. doi:10.1371/journal.pone.0000059

Nilsson RH, Kristiansson E, Ryberg M, Hallenberg N (2008) Intraspecific ITS variability in the kingdom fungi as expressed in the international sequence databases and its implications for molecular species identification. Evol Bioinform 4:193–201

Nilsson RH, Tedersoo L, Abarenkov K, Martin Ryberg M, Kristiansson E, Hartmann M, Schoch CL, Nylander JAA, Bergsten J, Porter TM, Jumpponen A, Vaishampayan P, Ovaskainen O, Hallenberg N, Bengtsson-Palme J, Eriksson KM, Larsson KH, Larsson E, Koljalg U (2012) Five simple guidelines for establishing basic authenticity and reliability of newly generated fungal ITS sequences. MycoKeys 4:37–63. doi:10.3897/mycokeys.4.3606

Pegler DN, Spooner BM, Young TWK (1993) British truffles, a revision of British hypogeous fungi. Royal Botanic Gardens, Kew

Peintner U, Iotti M, Klotz P, Bonuso E, Zambonelli A (2007) Soil fungal communities in a *Castanea sativa* (chestnut) forest producing large quantities of *Boletus edulis* sensu lato (porcini): where is the mycelium of porcini? Environ Microbiol 9(4):880–889. doi:10.1111/j.1462-2920.2006.01208.x

Pomarico M, Figliuolo G, Rana GL (2007) *Tuber* spp. biodiversity in one of the southernmost European distribution areas. Biodivers Conserv 16:3447–3461. doi:10.1007/s10531-006-9013-1

Rauscher T, Muller WR, Agerer R, Chevalier G (1996) *T. borchii* Vitt. + *Corylus avellana* L. In: Agerer R, Danielson RM, Egli S, Ingleby K, Luoma D, Treu R (eds) Description of ectomycorrhizae, vol 1. Einhorn, Schwabisch-Gmund, pp 173–178

Riousset G, Riousset L, Chevalier G, Bardet MC (2001) Truffes d'Europe et de Chine. INRA, Paris

Roux P (2006) Mille et un champignons. Edition Roux

Scannerini S, Palenzona M (1967) Ricerche sulle ectomicorrize di *Pinus strobus* in vivaio, III Micorrize di *Tuber albidum* Pico. Allionia 13:187–194

Schoch CL, Robbertse B, Robert V, Vu D, Cardinali G, Irinyi L, Meyer M, Nilsson RH, Hughes K, Miller AN, Kirk PM, Abarenkov K, Aime MC, Ariyawansa HA, Bidartondo M, Boekhout T, Buyck B, Cai Q, Chen J, Crespo A, Crous PW, Damm U, De Beer ZW, Dentinger BTM, Divakar PK, Dueñas M, Feau N, Fliegerova K, García MA, Ge ZW, Griffith GW, Groenewald JZ, Groenewald M, Grube M, Gryzenhout M, Gueidan C, Guo L, Hambleton S, Hamelin R, Hansen K, Hofstetter V, Hong SB, Houbraken J, Hyde KD, Inderbitzin P, Johnston PR, Karunarathna SC, Kõljalg U, Kovács GM, Kraichak E, Krizsan K, Kurtzman CP, Larsson KH, Leavitt S, Letcher PM, Liimatainen K, Liu JK, Lodge DJ, Luangsa-ard JJ, Lumbsch HT, Maharachchikumbura SSN, Manamgoda D, Martín MP, Minnis AM, Moncalvo JM, Mulè G, Nakasone KK, Niskanen T, Olariaga I, Papp T, Petkovits T, Pino-Bodas R, Powell MJ, Raja HA, Redecker D, Sarmiento-Ramirez JM, Seifert KA, Shrestha B, Stenroos S, Stielow B, Suh SO, Tanaka K, Tedersoo L, Telleria MT, Udayanga D, Untereiner WA, Uribeondo JD, Subbarao KV, Lgyi CV, Visagie C, Voigt K, Walker DM, Weir BS, Weiß M, Wijayawardene NN, Wingfield MJ, Xu JP, Yang ZL, Zhang N, Zhuang WY, Federhen S (2014). Finding needles in haystacks: linking scientific names, reference specimens and molecular data for fungi. Database 1–21. doi:10.1093/database/bau061

Sisti D, Giomaro G, Zambonelli A, Rossi I, Ceccaroli P, Citterio B, Stocchi V, Benedetti PA (1998) *In vitro* mycorrhizal synthesis of micropropagated *Tilia platyphyllos* Scop. plantlets with *Tuber borchii* Vittad. mycelium in pure culture. Acta Hortic 457:379–387. doi:10.17660/ActaHortic.1998.457.47

Smith ME, Douhan GW, Rizzo DM (2007) Intra-specific and intra-sporocarp ITS variation of ectomycorrhizal fungi as assessed by rDNA sequencing of sporocarps and pooled ectomycorrhizal roots from a *Quercus* woodland. Mycorrhiza 18(1):15–22. doi:10.1007/s00572-007-0148-z

Su KM, Xiong WP, Wang Y, Li SH, Xie R, Baima D (2013) *Tuber bomiense* a new truffle species from Tibet, China. Mycotaxon 126:127–132. doi:10.5248/126.127

Tamura K, Stecher G, Peterson D, Filipski A, Kumar S (2013) MEGA6: molecular evolutionary genetics analysis version 6.0. Mol Biol Evol 30(12):2725–2729. doi:10.1093/molbev/mst197

Tedersoo L, Abarenkov K, Nilsson RH, Schussler A, Grelet GA, Kohout P, Oja J, Bonito GM, Veldre V, Jairus T, Ryberg M, Larsson KH, Kõljalg U (2011) Tidying up international nucleotide sequence databases: ecological, geographical, and sequence quality annotation of ITS sequences of mycorrhizal fungi. PLoS One 6(9):e24940. doi:10.1371/journal.pone.0024940

Tešitelová T, Tešitel J, Jersáková J, Ríhová G, Selosse MA (2012) Symbiotic germination capability of four *Epipactis* species (Orchidaceae) is broader than expected from adult ecology. Am J Bot 99(6):1020–1032. doi:10.3732/ajb.1100503

Trappe JM (1979) The orders, families and genera of hypogeous Ascomycotina (truffles and their relatives). Mycotaxon 9(1):297–340

Trappe JM, Castellano MA (2000) New sequestrate Ascomycota and Basidiomycota covered by the Northwest Forest Plan. Mycotaxon 75:153–179

Tulasne LR, Tulasne C (1843) Champignons hypogés de la famille des Lycoperdacées. Ann Sci Nat Bot 19:373–381

Tulasne LR, Tulasne C (1845) Fungi nonulli hypogaei, novi v. minus cogniti. G Bot Ital 1 (2):55–63
Vittadini C (1831) Monographia Tuberacearum. Ex Typographia Felicis Rusconi, Milano
Wang YJ, Tan ZM, Murat C, Jeandroz S, Le Tacon F (2007) Molecular taxonomy of Chinese truffles belonging to *Tuber rufum* and *Tuber puberulum* groups. Fungal Divers 24:301–328
Wang XH, Benucci GMN, Xie XD, BonitoG LM, Liu PG, Shamekh S (2013) Morphological, mycorrhizal and molecular characterization of Finnish truffles belonging to the *Tuber anniae* species-complex. Fungal Ecol 6(4):269–280. doi:10.1016/j.funeco.2013.03.002
Xu AS (1999) Notes on the genus of *Tuber* from Tibet. Mycosystema 18(4):361–365
Zambonelli A, Branzanti B (1984) Prove di micorrizzazione del nocciolo con *Tuber aestivum* e *Tuber albidum*. Micol Ital 13(1):47–52
Zambonelli A, Branzanti MB (1989) Mycorrhizal synthesis of *Tuber albidum* Pico with *Castanea sativa* Mill., *Quercus suber* L. and *Alnus cordata* Loisel. Agric Ecosyst Environ 28 (1):563–567. doi:10.1016/0167-8809(90)90099-Y
Zambonelli A, Salomoni S, Pisi A (1993) Caratterizzazione anatomo-morfologica delle micorrize di *Tuber* spp. su *Quercus pubescens* Willd. Micol Ital 22(3):73–90
Zambonelli A, Salomoni S, Pisi A (1995) Caratterizzazione anatomo-morfologica delle micorrize di *Tuber borchii*, *Tuber aestivum*, *Tuber mesentericum*, *Tuber brumale*, *Tuber melanosporum* su *Pinus pinea*. Micol Ital 24(2):119–137
Zambonelli A, Iotti M, Amicucci A, Pisi A (1999) Caratterizzazione anatomo-morfologica delle micorrize di *Tuber maculatum* Vittad. su *Ostrya carpinifolia* Scop. Micol Ital 28(3):29–35
Zambonelli A, Iotti M, Rossi I, Hall I (2000a) Interaction between *Tuber borchii* and other ectomycorrhizal fungi in a field plantation. Mycol Res 104(6):698–702. doi:10.1017/S0953756299001811
Zambonelli A, Rivetti C, Percurdani R, Ottonello S (2000b) TuberKey: a DELTA-based tool for the description and interactive identification of truffles. Mycotaxon 74:57–76
Zambonelli A, Iotti M, Boutahir S, Lancellotti E, Perini C, Pacioni G (2012) Ectomycorrhizal fungal communities of edible ectomycorrhizal mushrooms. In: Zambonelli A, Bonito GM (eds) Edible ectomycorrhizal mushrooms, current knowledge and future prospects, vol 34, Soil biology. Springer, Berlin, pp 105–124. doi:10.1007/978-3-642-33823-6_7
Zambonelli A, Iotti M, Hall IR (2015) Current status of truffle cultivation: recent results and future perspectives. Micol Ital 44:31–40

Chapter 8
A Brief Overview of the Systematics, Taxonomy, and Ecology of the *Tuber rufum* Clade

Rosanne Healy, Gregory M. Bonito, and Matthew E. Smith

8.1 Introduction

The most comprehensive assessment of *Tuber* phylogeny to date showed that the Rufum clade is one of the three most speciose in the Tuberaceae (Bonito et al. 2010). Perhaps not surprisingly, the three most species-rich clades (the Rufum clade, the Maculatum clade, and the Puberulum clade) also contain the highest number of insufficiently identified sequences and undescribed taxa, illustrating the fact that these groups have not been adequately studied (see also Chap. 7). In this chapter we provide an overview of what is known about the Rufum clade of *Tuber*.

8.2 The *Tuber rufum* Species Concept

Tuber rufum Pico is a well-known, if not well-understood, species. It was described more than two centuries ago (1788) and was one of the original *Tuber* species treated by Elias Fries in his Systema Mycologicum (1823). *Tuber rufum* was described as globose, smooth, with a dirty white gleba that later becomes rufescent with white veins. We do not know the etymology for certain, but the epithet "rufum" means "red," and we infer that "rufum" refers to the reddish color of the mature gleba, although the peridium is also reddish. This species is thought of as

R. Healy (✉) • M.E. Smith
Department of Plant Pathology, University of Florida, Gainesville, FL 32611-0680, USA
e-mail: rosanne.healy@gmail.com

G.M. Bonito
Department of Plant, Soil and Microbial Sciences, Michigan State University, East Lansing, MI 48824, USA

A. Zambonelli et al. (eds.), *True Truffle (*Tuber *spp.) in the World*, Soil Biology 47,
DOI 10.1007/978-3-319-31436-5_8

morphologically variable and common in Europe wherever *Tuber magnatum* Pico and *Tuber melanosporum* Vittad. are found. Perhaps due to the high level of morphological variability, other described species have been synonymized with *T. rufum*, and a number of forms, varieties, and subspecies have been described for *T. rufum*. Corda (1854) recognized the species with spiny spores as being sufficiently different from the species with alveolate-reticulate spores to transfer the former, including *T. rufum*, to his new genus *Oogaster*. However, *Oogaster* was not much accepted and fell into disuse. Fischer (1923) subdivided *T. rufum* into subspecies *Tuber rutilum* and *T. rufum*, and he also described three varieties (*apiculatum*, *brevisporum*, and *oblongisporum*). *Tuber ferrugineum* Vittad. and *Tuber nitidum* Vittad. were considered forms of *T. rufum* by Montecchi and Lazzari (1993).

Knapp (1950) attempted to order the genus *Tuber* by delineating groups based on morphology. He designated the "rufum" group because *T. rufum* was the earliest described of the species in the group. Riousset et al. (2001) also adopted this convention. Besides *T. rufum*, this group included *T. nitidum*, *Tuber panniferum* Tul. and C. Tul., *Tuber requienii* Tul. and C. Tul., and *T. rufum* f. *lucidum* (H. Bonnet) Montecchi and Lazzari. Jeandroz et al. (2008) were the first to use phylogenetic inference to delimit *Tuber* based on molecular characters. In their analyses based on 5.8S and ITS2, they found that *T. panniferum* (another spiny-spored *Tuber* species) belonged in the Aestivum clade and that the following species were well supported in a monophyletic Rufum clade: *T. rufum*, *Tuber candidum* Harkn., *T. ferrugineum*, *Tuber huidongense* Y. Wang, *Tuber liaotongense* Y. Wang, *Tuber quercicola* J. L. Frank, D. Southw., and Trappe, *Tuber taiyuanense* B. Liu, and Tuber sp. 1 (Wang et al. 2007). They also found that *Tuber spinoreticulatum* Uecker and Burds. represents the earliest diverging lineage and is sister to the rest of the Rufum clade. Later analyses of a much larger ITS data set by Bonito et al. (2010) agreed overall with Jeandroz et al. (2008), but their improved taxon sampling showed that *T. candidum* was a species complex, and they added the following to the Rufum clade: *Tuber lucidum* H. Bonnet, *Tuber lyonii* Butters, *Tuber malacodermum* E. Fisch., *T. nitidum*, and *Tuber texense* Heimsch, along with 25 undescribed species. The undescribed taxa were delimited based on ITS clustering at 96 % similarity. These numerical designations are species hypotheses (Kõljalg et al. 2013) intended to stabilize a species concept until the species can be fully described. Bonito et al. (2013) conducted additional multilocus analyses that also supported the monophyly of the Rufum clade and confirmed that *T. spinoreticulatum* is sister to all other known taxa. Bonito et al. (2010, 2013) supported the validity of *T. ferrugineum* and *T. nitidum*, whereas Bonito et al. (2010) also supported the distinctness of *T. lucidum* (which was not included in the 2013 study). No sequence data are available for *T. requienii*, but putatively undescribed species close to *T. ferrugineum* and *T. rufum* should be morphologically compared with the description of *T. requienii* and also compared with the varieties described by Fischer (1923).

8.3 Taxonomy and Species Diversity Within the Rufum Clade

In our most recent reconstruction of the Rufum clade (Fig. 8.1), we have removed erroneous species names in order to reduce confusion and clarify species limits. We included described species names for 18 types or verified specimen-based sequences. For all other taxa, we used species hypotheses designations based on the phylogenetic inferences of Jeandroz et al. (2008) and Bonito et al. (2010, 2013). Since the first major molecular phylogenetic studies of *Tuber*, the number of species in the Rufum clade has grown from 8 (Jeandroz et al. 2008) to 39 (Bonito et al. 2010) to our current hypothesis of approximately 43 species, including five species complexes that require further study. The number of species in the Rufum clade will likely continue to increase as investigators sequence additional ectomycorrhizal (ECM) root tips and truffles from diverse locations. Here we include 18 described species and 25 undescribed species or species hypotheses, including five taxa known only from ECM root tips (*Tuber* sp. 51, 58, 61, 63, 67). *Tuber furfuraceum* H. T. Hu and Y. I. Wang was synonymized with *T. huidongense* by Deng et al. (2009), but in agreement with Huang et al. (2009), our analysis suggests that *T. furfuraceum* is a unique taxon.

8.4 Morphological Characterization of the Rufum Clade

Prior to the application of molecular tools, Trappe defined the *T. rufum* group for taxa with (1) smooth to minutely verrucose peridia (Fig. 8.2a), (2) asci with three-layered walls and a stem (Fig. 8.2b), and (3) spiny spores (Trappe et al. 1996). The current phylogenetic reconstruction of the Rufum clade suggests a more diverse morphology. Although most species have spiny spores, this character appears to exist along a continuum of elaboration of spore ornamentation. Toward one end of the continuum are taxa with isolated spines [e.g., *T. rufum*, *T. candidum* complex, *T. quercicola*, and *T. lucidum* complex (Fig. 8.1)], in the middle of the continuum are taxa with spines that are sometimes connected to adjacent spines by low ridges to form an incomplete reticulum [e.g., members of the *T. lyonii* complex and many undescribed North American species (Fig. 8.2c)], and at the other end of the continuum are taxa with fully connected spines that form a full reticulum [e.g., *T. spinoreticulatum* and many of the Chinese species (Figs. 8.1 and 8.2d)]. Character state reconstruction analysis in *Tuber* indicates that the ancestral spore ornamentation was most likely alveolate-reticulate and that spiny-reticulate ornamentation evolved independently by the loss of tissue along the ridges connecting junctions of the alveolae in the Aestivum clade and in the common ancestor of the Melanosporum and Rufum lineages (Bonito et al. 2013). The reduction of ornamentation within the Rufum lineage is at its most extreme in *Tuber melosporum* (G. Moreno, J. Díez, and Manjón) P. Alvarado, G. Moreno, J. L. Manjón, and Díez,

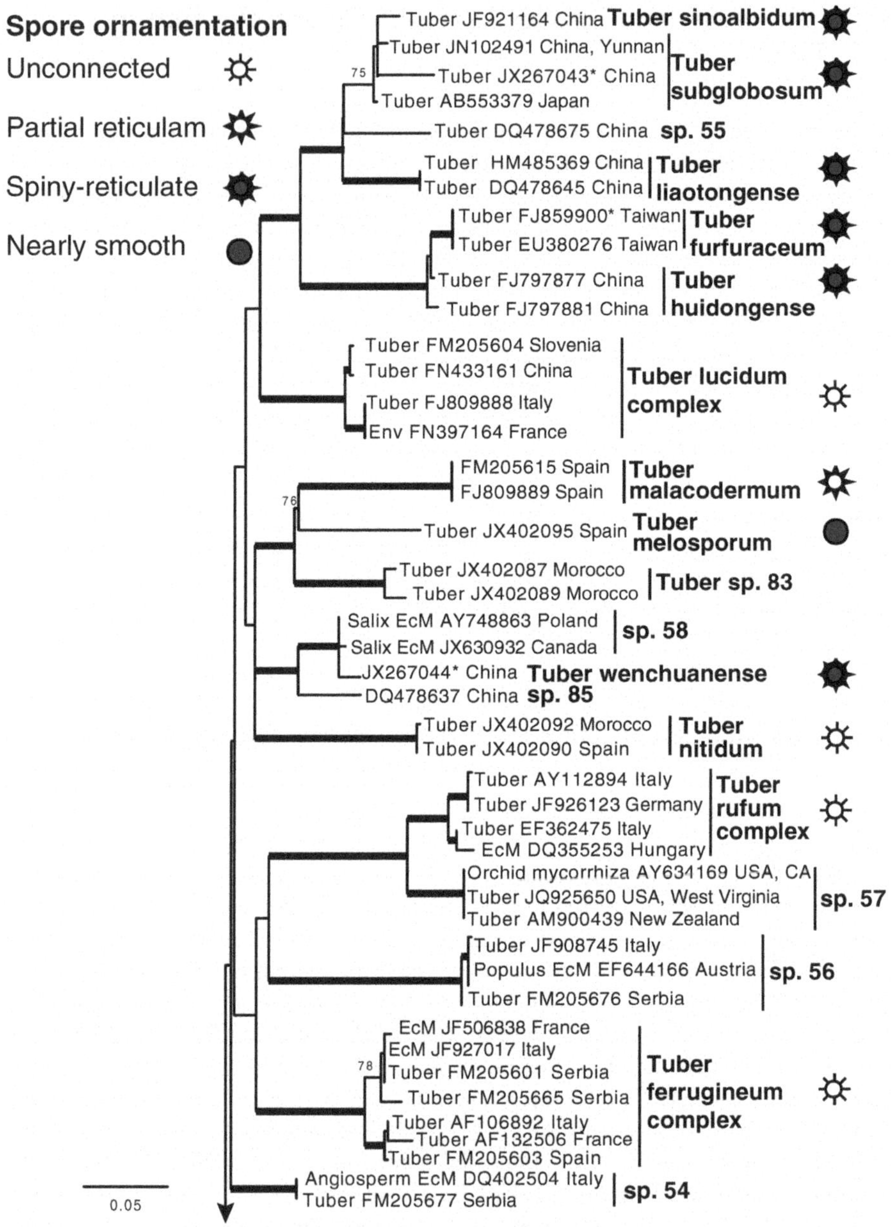

Fig. 8.1 Maximum likelihood phylogram of *T. rufum* clade based on publicly available ITS rDNA sequences (with GenBank numbers) and unpublished data (with collector numbers). Ambiguous regions within the nucleotide alignment were excluded in Gblocks under the least stringent settings (Castresana 2000), and RAxML using a GTR + gamma model of nucleotide substitution (Stamatakis 2014) was used to generate the phylogeny. Based on the analysis of Bonito et al. (2013), we used *T. spinoreticulatum* as the out-group taxon. Thickened branches represent bootstrap support greater than 84 % based on 1000 iterations. *Asterisks* denote sequences from type collections. Species hypotheses for undescribed species (*Tuber* sp. followed by numbers) follow the conventions of Bonito et al. (2010)

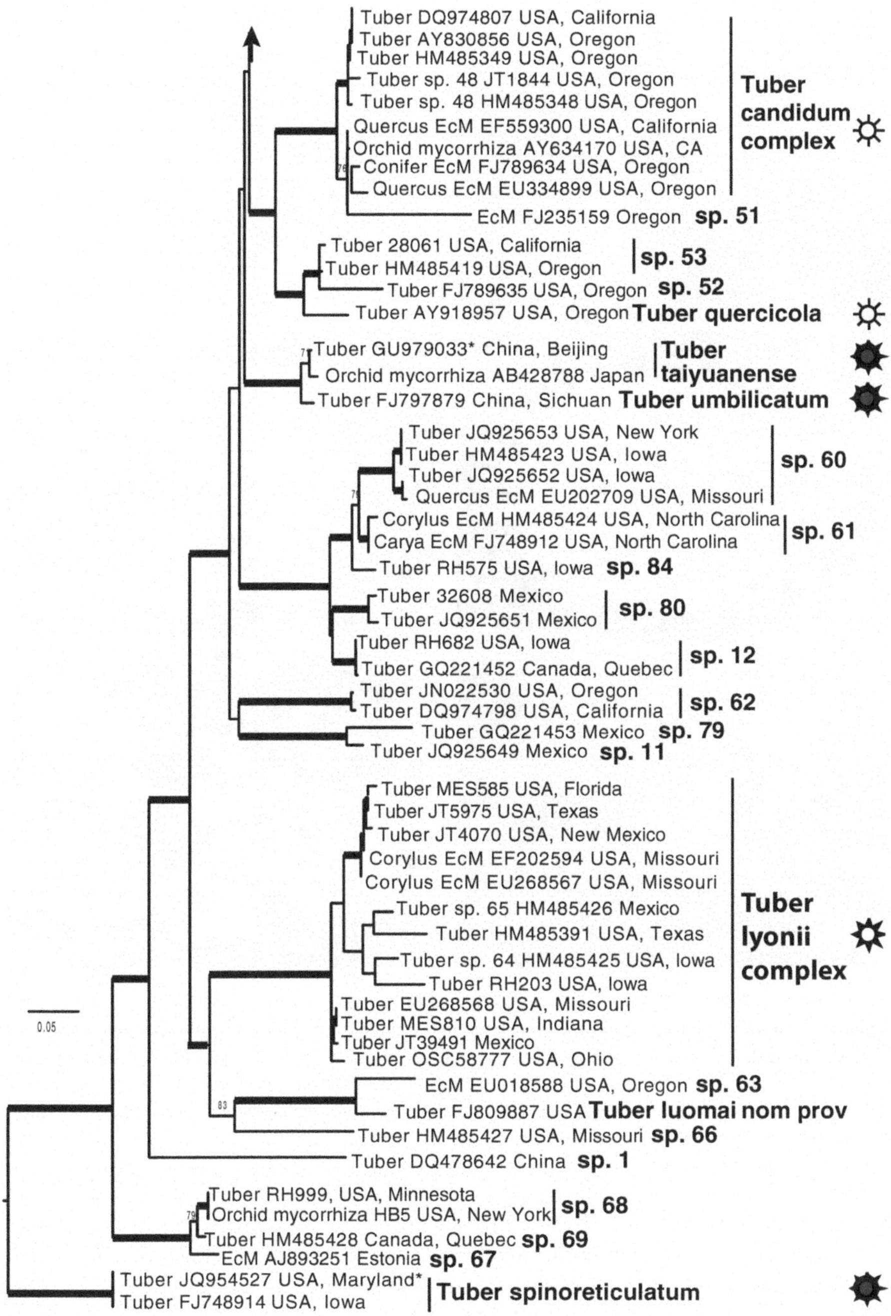

Fig. 8.1 (continued)

Fig. 8.2 (**a**) Smooth to slightly verrucose peridium of *T. lyonii* complex species (bar = 2 mm). (**b**) Stalked ascus of *T. spinoreticulatum*, with spiny-reticulate spores (bar = 5 μm). (**c**) Spores of *T. lyonii* complex species showing low-lying connections between spines that create a partial reticulum (bar = 5 μm). (**d**) Young spores of *T. spinoreticulatum* developing with full reticulum and spines in an ascus (bar = 10 μm). (**e**) ECM roots of *C. illinoensis* and *T. lyonii* (bar = 500 μm). (**f**) *T. lyonii* harvest in pecan orchard (bar = 5 cm)

where the spores have lost all ornamentation and are smooth. *Tuber polyspermum* L. Fan and J. Z. Cao was thought to belong to the Rufum clade (Fan et al. 2011) but the ITS sequence shows affinities to the Puberulum clade (see Chap. 5). This suggests a third, putatively independent transition toward spiny spores, this time in the Puberulum lineage.

While many species in the Rufum clade have a smooth to minutely verrucose peridium, *T. huidongense*, *T. melosporum*, and *Tuber subglobosum* L. Fan and C. L.

Hou are more obviously verrucose or warted. Most species are pale to more deeply yellow-brown to reddish-brown, but *T. quercicola* and *T. lucidum* are much darker reddish-brown to purple-black than the others.

Most Rufum clade species are described with a two-layered peridium where the layers are sometimes difficult to distinguish: an outer layer of textura intricata to angularis (with occasional swollen cells in some species) and an inner layer of textura intricata. However, the peridium of *T. quercicola* is composed of textura intricata in both layers (Frank et al. 2006). Globose spores have not been found in any Rufum clade species; all species described so far have spores that are broadly or narrowly ellipsoid to subglobose with one spore pole often having an acute angle.

ECMs of species in the Rufum clade have been morphologically described for *T. rufum* (Iotti et al. 2007 and references therein), *T. candidum* (Frank et al. 2006), and *T. furfuraceum* (Huang et al. 2009), among others. The ECM mantle is similar to those of other Pezizales in being thin, pseudoparenchymatous, and with limited emanating hyphae (Fig. 8.2e). They tend to lack the bristlelike cystidia or other cystidia types that are characteristic of other *Tuber* clades.

8.5 Biogeography and Distribution of Species Within the Rufum Clade

Species in the Rufum clade are mostly found in North America, Europe, and Asia. One species (sp. 57) was recovered from a *Tilia* root in New Zealand and an herbarium specimen of "*Tuber rufum*" was collected from New Zealand in 1941, but this specimen has not been sequenced. These undoubtedly represent an introduction to New Zealand with foreign soil or plant material (Bonito et al. 2010; Bulman et al. 2010).

There have been two molecular clock analyses of *Tuber* origin and biogeography. For the diversification of the Melanosporum and Rufum clades from a common ancestor, Jeandroz et al. (2008) hypothesized a European origin around 70 Mya, while Bonito et al. (2013) hypothesized a North American origin around 86 Mya. Differences between the estimates may be attributed to the greater taxon sampling and model complexity used by Bonito et al. (2013). Bonito et al. (2013) incorporated topological and branch length uncertainty into their Bayesian model and had a larger, more diverse taxon sampling set that included Southern Hemisphere Tuberaceae that were not available to Jeandroz et al. (2008). Half of the Rufum clade taxa delimited thus far are from North America which is consistent with a hypothesized initial species radiation in North America. Dispersal from North America to Europe and Asia was putatively via the Thulean North Atlantic Land Bridge (Bonito et al. 2013).

8.6 Mycorrhizal Ecology of *Tuber* Species Belonging to the Rufum Clade

Most species in the *T. rufum* clade are ECM associates of angiosperms, particularly of *Quercus*, but they are also found on *Populus*, *Salix*, *Corylus*, and *Carya* (Frank et al. 2006; Huang et al. 2009; Bonito et al. 2010). Members of the clade are widespread in many plant communities but are most dominant and diverse in angiosperm-dominated forests and woodlands with neutral to high pH soils (e.g., Bertini et al. 2006; Jumpponen et al. 2010; Smith et al. 2007). There are a few ECM root sequences from pines in North America (Gladish et al. 2010; Glassman et al. 2015) and China (Deng et al. 2009) and reports of truffles from the Rufum clade fruiting with Pinaceae in China (Cao et al. 2011) and Europe (Stobbe et al. 2012). A number of species have also been found from the roots of *Epipactis* (Bidartondo et al. 2004). These are mycoheterotrophic terrestrial orchids that parasitize ECM fungi and some species of *Epipactis* are known to regularly parasitize *Tuber* species (e.g., Selosse et al. 2004). An ecological consequence is that *Tuber* and other Pezizalean ECM fungi appear to provide a favorable environment for *Epipactis* orchids to become established and may thus facilitate the invasion of this orchid into nonnative regions (Ogura-Tsujita and Yukawa 2008).

8.7 Commercial Value and Edibility of *Tuber* Species in the Rufum Clade

Tuber rufum itself is considered to have little flavor, and the taste was said to be insipid and undelightful ("insipidum and gusto injucundo," attributed to Borch by Fries 1823). It has no commercial value, but since it grows where commercially important species grow, it is inadvertently harvested along with target species and its spores make it into root inoculum. Thus, it is regarded as an accidental contaminant or weed on inoculated tree roots in truffle orchards (Iotti et al. 2007). However, one member of the Rufum clade, *T. lyonii* (Fig. 8.2f), is among the most commercially important *Tuber* species in North America, along with *Tuber gibbosum* Harkn. and *Tuber oregonense* Trappe, Bonito, and P. Rawl. (Bonito et al. 2011, 2013). The value of *T. lyonii* is currently around \$160–\$200 per lb or 314–393 € per kg (Smith et al. 2012; Sharma et al. 2012). Members of the *T. lyonii* complex can be found with a wide array of host plants across the Eastern USA, including species of *Quercus*, *Corylus*, and *Carya*. However, *T. lyonii* is most commonly collected in commercial orchards of pecan [*Carya illinoensis* (Wangenh.) K. Koch] in the Southeastern USA where truffle orchards can take advantage of the association of *T. lyonii* with pecan trees to form a dual crop (Bonito et al. 2009, 2011; Sharma et al. 2012; Smith et al. 2012). *Tuber lyonii* was first described from Minnesota in early spring (Butters 1903), an unusual time to find Pezizales truffles in the Upper Midwest USA and north of the range limit for

C. illinoensis. The type collection is in good condition and has been examined morphologically but DNA sequencing has not been possible. To date, none of the spiny-spored truffles collected more recently in Minnesota fit the morphological description of *T. lyonii*, and the exact identity of this taxon still needs to be resolved. The identity of *T. texense* is also uncertain, and the possibility that it is a synonym of *T. lyonii* prevents us from trying to match its description to any recently sequenced specimens pending further phylogenetic investigation. Due to the current state of taxonomic uncertainty, the practical identification of *T. lyonii* is confused by numerous morphologically similar species.

There is no available data that compares the organoleptic properties among species within the *T. lyonii* complex, so whether parsing out the species is important from a commercial standpoint is unknown. Collections identified as *T. lyonii* are from Canada (Quebec) south to Florida and Mexico and west to New Mexico (Trappe et al. 1996; Bonito et al. 2010). Members of the *T. lyonii* species complex have not been found west of the Rocky Mountains, suggesting that this group is bound to the eastern part of North America by the continental divide formed by the Rocky Mountains. *Tuber lyonii* is not only important as a crop truffle, but is ecologically significant as well. Pyrosequencing of *Quercus* roots in Kansas revealed that species from the *T. lyonii* complex are among the most abundant recovered (Jumpponen et al. 2010). Similarly, *T. lyonii* is abundant on the roots of pecan trees in many orchards (Bonito et al. 2011; Smith and Ge, unpublished).

8.8 Conclusions

Ecological studies and commercial strategies are only as sound as the reference information that is available to them. Verified species with ITS reference sequences are vital for these endeavors. In this regard, the most immediate task for those who work with the Rufum clade is species delimitation, which includes tackling species complexes. The number of described species and undescribed species in the Rufum clade has increased fivefold since this group was first molecularly delimited by Jeandroz et al. (2008). More than half of the undescribed species are from North America, where there is a need for basic taxonomic work on the species that are native there. This is time-consuming, but not difficult. A more difficult task resides in the untangling of the five species complexes and potentially delimiting an even greater number of species for the Rufum clade. Three of the species complexes are in Europe (*T. ferrugineum*, *T. lucidum*, and *T. rufum*) and two are in North America (*T. lyonii* and *T. candidum*). For this work, the addition of RPB1, RPB2, and EF1a markers across the clade would be helpful, although the identification of more informative markers would be ideal.

For greater insight into many aspects of the Rufum clade, comparative genomics is a possibility. To date there is no published genome for any member of the Rufum clade. However, some species such as those in the *T. lyonii* complex are culturable which makes them good candidates for sequencing their genomes.

More work can also be done on the biology of the Rufum clade. As yet, no mitosporic forms are known for this group. Do they exist and have not yet been recognized, or do members of the Rufum clade lack this stage? If the latter case, their life cycle may differ from the Puberulum and Maculatum clades where mitosporic forms are increasingly being identified (Urban et al. 2004; Healy et al. 2013).

References

Bertini L, Rossi I, Zambonelli A, Amicucci A, Sacchi A, Cecchini M, Gregori G, Stocchi V (2006) Molecular identification of *Tuber magnatum* ectomycorrhizae in the field. Microbiol Res 161 (1):59–64. doi:10.1016/j.micres.2005.06.003

Bidartondo MI, Burghardt B, Gebauer G, Bruns TD, Read DJ (2004) Changing partners in the dark: isotopic and molecular evidence of ectomycorrhizal liaisons between forest orchids and trees. Proc R Soc Lond B 271(1550):1799–1806. doi:10.1098/rspb.2004.2807

Bonito G, Trappe JM, Vilgalys R (2009) North American truffles in the Tuberaceae: molecular and morphological perspectives. Acta Bot Yunnanica Suppl 31(Suppl 16):39–51

Bonito GM, Gryganskyi AP, Trappe JM, Vilgalys R (2010) A global meta-analysis of *Tuber* ITS rDNA sequences: species diversity, host associations and long-distance dispersal. Mol Ecol 19 (22):4994–5008. doi:10.1111/j.1365-294X.2010.04855.x

Bonito GM, Brenneman T, Vilgalys R (2011) Ectomycorrhizal fungal diversity in orchards of cultivated pecan (*Carya illinoinensis*; *Juglandaceae*). Mycorrhiza 21(7):601–612. doi:10.1007/s00572-011-0368-0

Bonito G, Smith ME, Nowak M, Healy RA, Guevara G, Cázares E, Kinoshita A, Nouhra E, Domínuez LS, Tedersoo L, Murat C, Wang Y, Moreno BA, Pfister DH, Nara K, Zambonelli A, Trappe JM, Vilgalys R (2013) Historical biogeography and diversification of truffles in the Tuberaceae and their newly identified Southern Hemisphere sister lineage. PLoS One 8(1): e52765. doi:10.1371/journal.pone.0052765

Bulman SR, Visnovsky SB, Hall IR, Guein-Laguette A, Wang Y (2010) Molecular and morphological identification of truffle-producing *Tuber* species in New Zealand. Mol Prog 9 (2):205–214. doi:10.1007/s11557-009-0626-0

Butters FK (1903) A Minnesota species of *Tuber*. Bot Gaz 35(6):427–431. doi:10.1086/328364

Cao JZ, Chi WD, Fan L, Li Y (2011) Notes on the neotype of *Tuber taiyuanense*. Mycotaxon 116:7–11. doi:10.5248/116.7

Castresana J (2000) Selection of conserved blocks from multiple alignments for their use in phylogenetic analysis. Mol Biol Evol 17(4):540–552

Corda ACJ (1854) Icones fungorumhucusque cognitorum, vol 6. JG Calve, Prague, p 70

Deng XJ, Chen J, Yu FQ, Liu PG (2009) Notes on *Tuber huidongense* (Tuberaceae, Ascomycota), an endemic species from China. Mycotaxon 109:189–199. doi:10.5248/109.189

Fan L, Hou CL, Cao JZ (2011) *Tuber sino-albidum* and *T. polyspermum* – new species from China. Mycotaxon 118:403–410. doi:10.5248/118.403

Fischer E (1923) Verhandlungen der naturforscheenden gesellschaft in Basel. Festband Hermann Christ 35(1):48

Frank JL, Southworth D, Trappe JM (2006) NATS truffle and truffle-like fungi 13. *Tuber quercicola* and *T. whetstonense*, new species from Oregon, and *T. candidum* redescribed. Mycotaxon 95:229–240

Fries EM (1823) Systema mycologicum, vol 2. Sumtibus Ernesti Mauritti, p 293. doi:10.5962/bhl.title.5378

Gladish S, Frank J, Southworth D (2010) The serpentine syndrome below ground: ectomycorrhizas and hypogeous fungi associated with conifers. Can J Forest Res 40 (8):1671–1679. doi:10.1139/x10-092

Glassman SI, Peay KG, Talbot JM, Smith DP, Chung JA, Taylor JW, Vilgalys R, Bruns TD (2015) A continental view of pine-associated ectomycorrhizal fungal spore banks: a quiescent functional guild with a strong biogeographic pattern. New Phytol 205(4):1619–1631. doi:10.1111/nph.13240

Healy RA, Smith ME, Bonito GM, Pfister DH, Ge ZW, Guevara GG, Williams G, Stafford K, Kumar L, Lee T, Hobart C, Trappe J, Vilgalys R, McLaughlin DJ (2013) High diversity and widespread occurrence of mitotic spore mats in ectomycorrhizal *Pezizales*. Mol Ecol 22 (6):1717–1732. doi:10.1111/mec.12135

Huang JY, Hu HT, Shen WC (2009) Phylogenetic study of two truffles, *Tuber formosanum* and *Tuber furfuraceum*, identified from Taiwan. FEMS Micobiol Lett 294(2):157–171. doi:10.1111/j.1574-6968.2009.01571.x

Iotti M, Amicucci A, Bonito G, Bonuso E, Stocchi V, Zambonelli A (2007) Selection of a set of specific primers for the identification of *Tuber rufum*: a truffle species with high genetic variability. FEMS Microbiol Lett 277(2):223–231. doi:10.1111/j.1574-6968.2007.00963.x

Jeandroz S, Murat C, Wang Y, Bonfante P, Tacon FL (2008) Molecular phylogeny and historical biogeography of the genus *Tuber*, the 'true truffles'. J Biogeogr 35(5):815–829. doi:10.1111/j.1365-2699.2007.01851.x

Jumpponen A, Jones KL, Mattox D, Yaege C (2010) Massively parallel 454-sequencing of fungal communities in *Quercus* spp. ectomycorrhizas indicates seasonal dynamics in urban and rural sites. Mol Ecol 19(Suppl 1):41–53. doi:10.1111/j.1365-294x.2009.04483.x

Knapp A (1950) Die europaischen Hypogaeen-Gattungen und ihre Gattungstypen. Schweiz Z Pilzk 28:29–42, 101–118, 153–179

Kõljalg U, Nilsson RH, Abarenkov K, Tedersoo L, Taylor AFS, Bahram M, Bates ST, Bruns TD, Bengtsson-Palme J, Callaghan TM, Douglas B, Drenkhan T, Eberhardt U, Dueñas M, Grebenc T, Griffith GW, Hartmann M, Kirk PM, Kohout P, Larsson E, Lindahl BD, Lücking R, Martín MP, Matheny B, Nguyen NH, Niskanen T, Oja J, Peay KG, Peinter U, Peterson M, Põldmaa K, Saag L, Saar I, Schüßler A, Scott JA, Senés C, Smith ME, Suija A, Taylor DL, Telleria MT, Weiss M, Larsson KH (2013) Towards a unified paradigm for sequence-based identification of fungi. Mol Ecol 22(21):5271–5277. doi:10.1111/mec.12481

Montecchi A, Lazzari G (1993) Atlante fotografico di funghi ipogei. Associazione Micologica Bresadola, Vicenza

Ogura-Tsujita Y, Yukawa T (2008) *Epipactis helleborine* shows strong mycorrhizal preference towards ectomycorrhizal fungi with contrasting geographic distributions in Japan. Mycorrhiza 18(6):331–338. doi:10.1007/s00572-008-0187-0

Riousset G, Riousset L, Chevalier G, Bardet MC (2001) Truffes d'Europe et de Chine. INRA, Paris

Selosse MA, Faccio A, Scappaticci G, Bonfante P (2004) Chlorophyllous and achlorophyllous specimens of *Epipactis microphylla*, (Neottieae, Orchidaceae) are associated with ectomycorrhizal septomycetes, including truffles. Microb Ecol 47(4):416–426. doi:10.1007/s00248-003-2034-3

Sharma J, Trela B, Wang S, Smith M, Bonito G (2012) Pecan truffle (*Tuber lyonii*) in Texas. Pecan South 2012:16–24

Smith ME, Douhan GW, Rizzo DM (2007) Ectomycorrhizal community structure in a xeric *Quercus* woodland based on rDNA sequence analysis of sporocarps and pooled roots. New Phytol 174(4):847–863. doi:10.1111/j.1469-8137.2007.02040.x

Smith ME, Bonito G, Sharma J, Long J, Davis-Long B, Brenneman T (2012) Pecan truffles (*Tuber lyonii*): what we know and what we need to know. Georgia Pecan Magazine, 2012 (Spring):52–59

Stamatakis A (2014) RAxML Version 8: a tool for phylogenetic analysis and post-analysis of large phylogenies. Bioinformatics 30(9):1312–1313. doi:10.1093/bioinformatics/btu033

Stobbe U, Büntgen U, Sproll L, Tegel W, Egli S, Fink S (2012) Spatial distribution and ecological variation of re-discovered German truffle habitats. Fungal Ecol 5(5):591–599. doi:10.1016/j.funeco.2012.02.001

Trappe JM, Jumpponen AM, Cázares E (1996) NATS truffle and truffle-like fungi 5: *Tuber lyonii* (=*T. texense*) with a key to the spiny-spored *Tuber* species groups. Mycotaxon 60:365–372

Urban A, Neuner-Plattner I, Krisai-Greilhuber I, Haselwandter K (2004) Molecular studies on terricolous microfungi reveal novel anamorphs of two *Tuber* species. Mycol Res 108 (7):749–758. doi:10.1017/s0953756204000553

Wang Y, Tan ZM, Murat C, Jeandroz S, Le Tacon F (2007) Molecular taxonomy of Chinese truffles belonging to the *Tuber rufum* and *Tuber puberulum* groups. Fungal Divers 24:301–328

Chapter 9
Truffle Genomics: Investigating an Early Diverging Lineage of *Pezizomycotina*

Claude Murat and Francis Martin

9.1 Introduction

In the 1980s, the term genomics came into common use to describe the emerging scientific field of sequencing and analyzing genomes, but genomics started in the second half of the twentieth century, thanks to numerous discoveries, such as the first sequencing of the alanine tRNA of *Saccharomyces cerevisiae* Meyen ex E.C. Hansen (Holley et al. 1965), the sequencing of the bacteriophage MS21 (Fiers et al. 1976), and the development of the Sanger DNA sequencing method (Sanger et al. 1977). The Sanger sequencing method was used until the mid-2000s and allowed the sequencing of the first bacterial and eukaryotic genomes, *Haemophilus influenzae* (Fleischmann et al. 1995) and *S. cerevisiae* (Goffeau et al. 1996), respectively.

In 2003, the first filamentous fungal genome was sequenced (*Neurospora crassa* Shear and B. O. Dodge; Galagan et al. 2003), and afterward the number of sequenced fungal genomes began to increase rapidly (Galagan et al. 2005). Fungal genomics benefits from different large-scale programs, such as the Fungal Genome Initiative (FGI), a consortium that aims to sequence fungal genomes, initiated in the year 2000 at the Broad Institute (http://www.broadinstitute.org/science/projects/fungal-genome-initiative/fungal-genome-initiative). The FGI aimed to sequence the fungal species that are important in medicine, agriculture, and industry. In the last decade, large-scale projects that aimed to sequence the genomes of fungi from environmental settings were launched at the Joint Genome Institute (JGI), the sequencing center of the US Department of Energy (http://jgi.doe.gov/).

C. Murat (✉) • F. Martin
INRA, Université de Lorraine, UMR1136, Interactions Arbres/Micro-organismes, Laboratoire d'Excellence ARBRE, 54280 Champenoux, France
e-mail: claude.murat@nancy.inra.fr

A. Zambonelli et al. (eds.), *True Truffle (*Tuber *spp.) in the World*, Soil Biology 47,
DOI 10.1007/978-3-319-31436-5_9

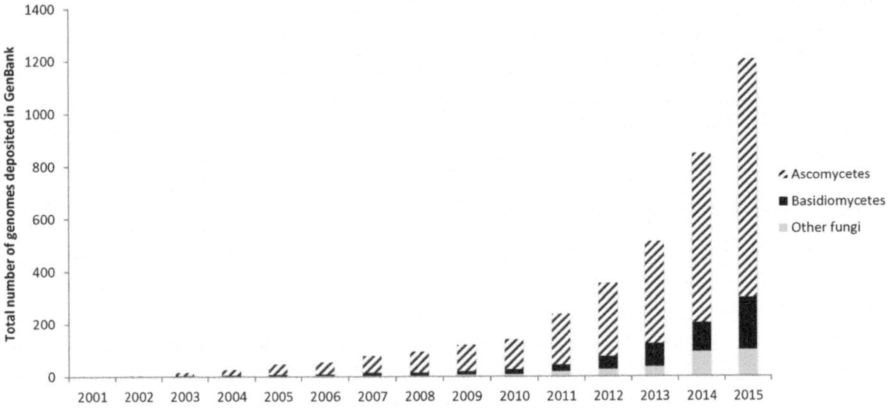

Fig. 9.1 Number of partial or complete fungal genomes deposited in GenBank from 2001 to December 10, 2015 (http://www.ncbi.nlm.nih.gov/genome/browse/)

The 1000 Fungal Genomes project (http://genome.jgi-psf.org/programs/fungi/1000fungalgenomes.jsf), which aimed to sequence at least two reference genomes from the more than 500 recognized families of fungi, started to fill in the gaps in the Fungal Tree of Life (Spatafora 2011). Additional projects, such as the Mycorrhizal Genomics Initiative (MGI; http://genome.jgi.doe.gov/Mycorrhizal_fungi/Mycorrhizal_fungi.info.html), which focuses on sequencing specific ecologically relevant groups of fungi, participated and increased the number of released fungal genomes (van der Heijden et al. 2015). All these projects rely on second-generation DNA sequencing (SGS) technologies. Indeed, until the mid-2000s, most sequencing projects were limited to one or a few genomes due to the cost of Sanger sequencing. In 2005, a revolution in genome sequencing started with SGS (Margulies et al. 2005), which dramatically dropped the cost of sequencing (Metzker 2010). To date, 1204 fungal genomes have been deposited in the GenBank database, belonging to the Ascomycetes (903), the Basidiomycetes (200), and early diverging fungi (101) (Fig. 9.1). The activity of the international consortia mentioned above explains the exponential increase in the number of fungal genomes deposited in the GenBank in the last few years (Fig. 9.1).

9.2 The Evolution of Mycorrhizal Symbiosis Unraveled by Comparative Genomics

Almost all vascular plants interact with mycorrhizal fungi, but woody shrub and tree species in particular rely on symbiotic associations with ectomycorrhizal (ECM) fungi to generate large amounts of biomass and store carbon (Smith and

Read 2010). The ability to establish ECM symbioses is a widespread characteristic of various soil Ascomycetes and Basidiomycetes; interactions with these soil symbionts are essential for efficient acquisition of growth-limiting nutrients, such as phosphorus and nitrogen. The ECM species have great potential as a versatile ecological model due to their involvement in symbioses with economically important conifer and hardwood tree species worldwide. Several ECM species have been disseminated globally via exotic eucalypt, poplar, and pine plantations. Reforestation and large-scale plantations on marginal soils introduced ECM symbionts to be successful; as a consequence, this mutualistic partnership is being exploited as a strategy to bolster biomass production for next-generation biofuels in the USA, Europe, Brazil, and Australasia. In addition, the fruiting bodies of ~400 ECM fungi are appreciated worldwide as edible mushrooms, e.g., boletes, chanterelles, and truffles, and they represent a global market evaluated at more than US$23 billion (Boa 2004).

According to Hibbett and colleagues (2000), ECM symbiosis arose several times in the evolution from saprotrophic species. In phylogenetic analyses, however, it is impossible to know if all ECM fungi use the same "tool kit" to establish symbiosis or if multiple ECM species developed their own tool kits. To gain information on mycorrhizal evolution, we sequenced the genome of the ECM species *Laccaria bicolor* (Maire) P. D. Orton (Martin et al. 2008) and *Tuber melanosporum* Vittad. (Martin et al. 2010). In the frame of the MGI project (see above), an additional 40 mycorrhizal genomes were sequenced, among them ECM, arbuscular (AMF), orchids (ORC), and ericoid (ERM) mycorrhizal species (http://genome.jgi.doe.gov/Mycorrhizal_fungi/Mycorrhizal_fungi.info.html). The questions addressed in the frame of the MGI project are:

1. Do all ECM lineages arise from similar saprotrophic ancestors?
2. Are there a common set of mycorrhiza-related genes that interact with the host plants and differentiate the symbiotic structures: a putative *symbiosis tool kit*?
3. Do the different types of mycorrhizal symbioses (ECM, AMF, ORC, ERM) use similar gene networks to differentiate the mutualistic interaction?

The first large-scale comparative analysis of the genomes from mycorrhizal species included 11 ECM (10 Basidiomycetes and one Ascomycete), 2 ORC, and 1 ERM. These genomes were compared to those of 33 white and brown rots, soil/litter decayers, and pathogenic fungi (Kohler et al. 2015). The first conclusion is that ECM symbiosis arose several times during the *Agaricomycotina* evolution from ancestors of either white-rot, brown-rot, or litter decayers. On average, ECM lineages have a reduced complement of genes encoding plant cell-wall degrading enzymes (PCWDEs), compared to ancestral white-rot wood decayers, suggesting convergent genome erosion. This decrease in PCWDE can be explained by two hypotheses: (1) being fed by their host plant, ECM fungi do not need to maintain PCWDEs in their genome to degrade the soil's organic matter or (2) their reduced number of PCWDEs allows them to avoid eliciting the host plant's defenses as a result of the oligosaccharide elicitors. Transcriptomic data allowed the identification, for each basidiomycete ECM species, of a specific set of mycorrhizal-induced small secreted proteins (MiSSP; Kohler et al. 2015). One of these proteins, MiSSP7, is an effector protein that contributes to the dialogue between *L. bicolor* and the plant (Plett et al. 2011, 2014); most of these MiSSPs are species specific (Kohler et al. 2015). Finally, a different picture was observed for ORC and ERM, since the genomes of these fungi are very rich in PCWDEs and some PCWDEs are expressed during symbiosis (Kohler et al. 2015). In conclusion, the major types of mycorrhizal symbioses (ECM, ORC, and ERM) evolved different molecular mechanisms to interact with their host plants.

In this large-scale comparative genomic analysis of mycorrhizal genomes, one *Ascomycete* species (i.e., *T. melanosporum*) was included; can one generalize the conclusions reached with basidiomycete ECM fungi to *Ascomycete* ECM fungi?

9.3 Pezizomycetes Pan-Genome Project

True truffles belong to the genus *Tuber* and form ECM symbiosis with trees and shrubs. The *Tuber* genus evolved in the last 140 Mya in 11 phylogenetic groups (Bonito et al. 2013) and 180 species (Bonito et al. 2010; see Chap. 1). *Tuber* belongs to the Pezizomycetes class, which constitutes an early diverging lineage in *Pezizomycotina* (Spatafora et al. 2006). The Pezizomycetes class is composed of 200 genera and 1683 described species (Kirk et al. 2008) that are saprophytic, mycorrhizal, or pathogenic. The formation of apothecia that contain operculate asci with forcible spore discharge (Laessøe and Hansen 2007) characterizes Pezizomycetes, although other forms exist, such as hypogeous fungi. Fungi with subterranean fruiting bodies are called truffles, and they comprise many taxonomically unrelated species (i.e., Ascomycetes and Basidiomycetes) that show remarkable features of phenotypic convergent evolution as a result of adaptation to this specialized habitat (i.e., fruiting belowground). Truffle ascomata contain asci with passive spore dispersal, which have multiple evolutionary origins in Pezizomycetes (O'Donnell et al. 1997; Laessøe and Hansen 2007). They probably derive from epigeous fruiting bodies which evolved as an adaptation to animal grazing or water stress (Thiers 1984; Bruns et al. 1989). Although truffle lifestyles have evolved in nearly every major group of fleshy fungi and over 100 independent instances within the Ascomycota, Basidiomycota, and Mucoromycotina (Tedersoo et al. 2010), the majority of transitions to a truffle form occur in ECM fungal lineages (Trappe et al. 2009). This pattern suggests that the symbiotic association with plants may be an important driver in the evolution of the truffle fruiting body phenotype. Although protected from desiccation and other environmental stresses, like grazing, by their belowground location, truffle ascomata grow slower than those of most epigeous mushrooms, e.g., *T. melanosporum* ascomata growth during the 6–9 months underground during summer and fall (Olivier et al. 2012). It is tempting to speculate that this slow hypogeous differentiation of the truffle ascomata is only possible, thanks to the constant flux of carbohydrates from the host plant (Le Tacon et al. 2013).

Comparing *Tuber* spp. with other Pezizomycetes, therefore, allowed the investigation of the different fungal life strategies in Ascomycetes (mycorrhizal vs. saprotroph), the development of hypogeous versus epigeous ascocarps, and the evolution of the particular organoleptic qualities of some species (e.g., truffles and morels). To date, the *T. melanosporum* genome is the only truffle genome that has been published (Martin et al. 2010). Several projects to sequence additional truffle genomes are ongoing (Payen et al. 2014), and the genomes of fourteen additional *Tuber* species will be released in the next few years (Table 9.1). These genomes are sequenced in the frame of the TuberEvol project (Comparative Genomics of Truffle Species and the Evolution Ectomycorrhizal Symbiotic Genomes), involving scientists in France, Italy, the USA, and both the JGI and Genoscope sequencing centers (Payen et al. 2014). The haploid genomes of *Tuber aestivum* Vittad. and *Tuber magnatum* Pico have been sequenced and assembled, and gene annotations are now completed. The genomes of *Tuber brumale* Vittad., *Tuber indicum* Cooke and Massee, and *Tuber lyonii* Butters have been sequenced and assembled, and the gene annotation is underway. For *Tuber borchii* Vittad., *Tuber canaliculatum* Gilkey, *Tuber dryophilum* Tul. and C. Tul., *Tuber excavatum* Vittad., *Tuber gibbosum* Harkn., *Tuber macrosporum* Vittad., *Tuber maculatum* Vittad., *Tuber oregonense* Trappe, Bonito, and P. Rawl., and *Tuber rufum* Pico, genome sequencing will be carried out in 2016–2017. These 14 species belong to 8 out of the 11 clades identified in the *Tuber* genus (Table 9.1, Bonito et al. 2013); indeed, only species of the multimaculatum, japonicum, and gennadii clades are not yet represented in this genomic project.

Additional Pezizomycetes genomes are now available (http://genome.jgi.doe.gov/pezizomycetes/pezizomycetes.info.html), including:

1. The saprotrophic species: *Pyronema confluens* Tu. and C. Tul. (Traeger et al. 2013), *Ascobolus immersus* Pers., *Ascodesmis nigricans* Tiegh., *Morchella conica* Pers., *Morchella importuna* M. Kuo, O'Donnell, and T. J. Volk, and *Sarcoscypha coccinea* (Gray) Boud
2. The symbiotic species: *Choiromyces venosus* (Fr.) Th. Fr., *Terfezia boudieri* Chatin, and *Wilcoxina mikolae* (Chin S. Yang and H. E. Wilcox) Chin S. Yang and Korf

A first characteristic of truffle genomes is their large size (>125 Mbp) and their high repeat sequence content (Table 9.1). The Tuberaceae member *C. venosus* has a genome that shares features with *Tuber* species (i.e., a large size with a high repeat sequence content), in agreement with some phylogenetic studies (Percudani et al. 1999), although recent phylogenetic analyses clearly separated both genera (Bonito et al. 2013). Moreover, *W. mikolae* (Pyronemataceae) also has a large genome (111 Mbp), similar to other ECM Pezizomycetes (i.e., *Tuber* spp. and *Choiromyces*), confirming that the genomes of ECM species are generally larger than those of saprotrophic species (Payen et al. 2014). The comparative analysis of these Pezizomycetes genomes is currently underway and should provide new insights into truffle evolution, but also into ECM evolution by comparing Basidiomycetes ECM and Pezizomycetes ECM fungi.

Table 9.1 List of *Tuber* species for which the genome is sequenced or being sequenced

Species	Clade[a]	Genome size (Mb)	% of repeated sequences	Material	Statut of the project	Natural habitat	Project
T. aestivum	Aestivum	145	50	Ascomata	Complete	North Africa, Europe	T. Evolve (Génoscope-INRA)
T. magnatum	Aestivum	192	60	Ascomata	Complete	Italy, South-Central Europe	INRA-UNITO
T. excavatum	Excavatum	na	na	Mycelium	Start in 2016	Europe	1K Fungal Genome (JGI-DOE)
T. gibbosum	Gibbosum	na	na	Mycelium	Start in 2016	North America, Europe	1K Fungal Genome (JGI-DOE)
T. oregonense	Gibbosum	na	na	Ascomata	Start in 2016	North America	1K Fungal Genome (JGI-DOE)
T. canaliculatum	Macrosporum	na	na	Mycelium	Start in 2016	North America	1K Fungal Genome (JGI-DOE)
T. macrosporum	Macrosporum	na	na	Mycelium	Start in 2016	Europe	1K Fungal Genome (JGI-DOE)
T. maculatum	Maculatum	na	na	Mycelium	Start in 2016	Europe	1K Fungal Genome (JGI-DOE)
T. brumale	Melanosporum	180[b]	na	Ascomata	Annotation pending	Europe	INRA
T. indicum	Melanosporum	147[b]	na	Ascomata	Annotation pending	China	INRA-University Kunming
T. melanosporum	Melanosporum	125	58	Mycelium	Published	South-West Europe	Martin et al. (2010)
T. borchii	Puberulum	na	na	Mycelium	Pending	Europe	Metatranscriptomic of soil forest (JGI-DOE)

T. dryophilum	Puberulum	na	na	Mycelium	Start in 2016	Europe	1K Fungal Genome (JGI-DOE)
T. lyonii	Rufum	na	na	Mycelium	Pending	North America	Duke University
T. rufum	Rufum	na	na	Mycelium	Start in 2016	Europe	1K Fungal Genome (JGI-DOE)

[a]Clade in the *Tuber* phylogeny according to Bonito et al. (2013)
[b]Genome size estimation obtained by Kmer approach in AllPaths-LG genome assembler

9.4 Population Genomic: Resequencing of Geographic Accessions

The increase of genomic resources in the last few decades provided new genetic tools for population geneticists. For example, using whole genome sequences, it was possible to characterize highly polymorphic microsatellite markers for investigating the genetic structures of *T. aestivum* (Molinier et al. 2013) and *T. melanosporum* (Murat et al. 2011) populations (see Chaps. 2 and 3). Traditional population genetic studies, however, use a limited set of molecular markers (typically a dozen or less), impeding the identification of genomic evolution and adaptation signatures. Thanks to the decrease in DNA sequencing costs (see above), it is now possible to move from population genetics to the population genomics of mycorrhizal symbionts by resequencing the genome of several geographic accessions (Branco et al. 2015; Payen et al. 2015). In the latter studies, single nucleotide polymorphisms (SNP) were characterized; SNPs are more informative than microsatellites due to their distribution throughout the genome, bi-allelic nature, and high potential for automation (Brumfield et al. 2003). SNP locations in coding sequences increase the probability of identifying the signatures of adaptation to environmental cues. Such an approach was used, for example, to identify the genomic signature of adaptation to temperature in *N. crassa* (Ellison et al. 2011). Recently, Branco and colleagues (2015) re-sequenced 28 individual strains of the ECM basidiomycete *Suillus brevipes* (Peck) Kuntze from coastal and montane sites in California. Reduced nucleotide diversity was observed among coastal individuals for the *Nha1-like* gene, a membrane Na^{+}/H^{+} exchanger known to enhance salt tolerance in plants and yeast, suggesting an adaptation to saline soils by *S. brevipes*.

The first population genomics analysis of *T. melanosporum* was recently published (Payen et al. 2015). More than 440,000 SNPs were identified by comparing seven genomes. These SNPs were used to detect genomic regions putatively under selection (see Chap. 2). Among these SNPs, 60,507 present in the intergenic regions free of selective pressure were selected to reconstruct a phylogeny with the seven geographic accessions. Interestingly, the phylogenetic tree clustered samples according to their geographical origin, with a cluster comprising the northern France samples, another with those of southeastern France and Italy, and a last one with the Spanish samples (Fig. S1B in Payen et al. 2015). The use of SNPs is now facilitated by SGS and high-throughput SNP arrays (Davey et al. 2011). Medium- to high-throughput technologies, such as the competitive allele-specific PCR (KASPar) assay from KBiosciences (Hertfordshire, UK; http://www.kbioscience.co.uk) or the Affymetrix Axiom SNP microarrays, are now available. The KASPar assay is commonly used for genotyping up to 1000–2000 SNPs, whereas the Axiom SNP microarrays allow genotyping of 1500 to several million SNPs. We are now developing an array based on the 60,507 SNPs for analyzing the population genetic structure throughout the natural regions of *T. melanosporum* production.

9.5 Conclusions

Truffle genomics began with the sequencing of the *T. melanosporum* genome (Martin et al. 2010). The genomes of 14 additional truffle species (*T. aestivum*, *T. borchii*, *T. brumale*, *T. canaliculatum*, *T. dryophilum*, *T. excavatum*, *T. gibbosum*, *T. indicum*, *T. lyonii*, *T. macrosporum*, *T. maculatum*, *T. magnatum*, *T. oregonense*, and *T. rufum*) should be released in the next 2 years. In the framework of the 1000 Fungal Genomes Project, we recently proposed the genome sequencing of about 20 Pezizomycetes belonging to the *Balsamia*, *Barssia*, *Discina*, *Helvella*, *Tuber*, *Underwoodia*, and *Verpa* genera.

Based on the ongoing studies of the *Tuber* and other Pezizomycetes genomes, we identified several key questions that future analyses can help resolve, presented below in the form of six currently unanswered questions, rather than an exhaustive list:

1. How did the different lifestyles (e.g., mutualism vs. saprotrophism) evolve in Ascomycetes?
2. Which are the key developmental genes explaining the shift from epigeous to hypogeous ascomata?
3. Are the sex-related pathways in *Tuber* species similar to those characterized in other Pezizomycetes, such as the genetic model *A. immersus*?
4. Which enzymatic pathways are at the origin of the particular organoleptic volatiles of truffle species and are these pathways species or genus specific?
5. Are truffles able to adapt to environmental stresses, such as drought or frost?
6. Is it possible to genotype the geographic origin(s) of truffles?

In summary, our knowledge of the truffle life cycle, evolution, and population dynamics have increased, thanks to the availability of genomic resources. In addition to answering fundamental questions, genomic resources could also help us respond to truffle industry requests, since the truffle industry, for several decades, has sought innovative tools to identify the geographic origin of the truffles, mainly to valorize local territories. Protected designation of origin certification was developed for boletes, i.e., "Fungo di Borgotaro" (http://www.fungodiborgotaro.com/ita/igp.jsp), although molecular markers allowing us to certify bolete origins do not exist yet. In a population genomic study, using SNPs, the seven geographic accessions are clustered according to their geographic origin (Payen et al. 2015). Using SNPs to identify the harvesting region could have many applications for the truffle industry regarding local geographic certification.

The development of truffle genome sequencing will therefore provide scientists and the truffle industry new innovative tools or molecular markers to investigate not only population genetics but also taxonomy. Indeed, the recent discovery and characterization of mating-type genes for *T. indicum* confirmed that such functional markers could also be used for taxonomic purposes (Belfiori et al. 2013; see Chap. 2).

Acknowledgments The UMR1136 is supported by a grant overseen by the French National Research Agency (ANR) as part of the "Investissements d'Avenir" program (ANR-11-LABX-0002-01, Lab of Excellence ARBRE). Most of the genome sequencing was financed by the Joint Genome Institute and the Genoscope. Discussion with our colleagues from the Mycorrhizal Genomics Initiative and the Pezizomycete Pan-Genome consortium contributed to this chapter.

References

Belfiori B, Riccioni C, Paolocci F, Rubini A (2013) Mating type locus of Chinese black truffles reveals heterothallism and the presence of cryptic species within the *T. indicum* species complex. PLoS One 8(12), e82353. doi:10.1371/journal.pone.0082353

Boa ER (2004) Wild edible fungi: a global overview of their use and importance to people. FAO, Rome

Bonito GM, Gryganskyi AP, Trappe JM, Vilgalys R (2010) A global meta-analysis of *Tuber* ITS rDNA sequences: species diversity, host associations and long-distance dispersal. Mol Ecol 19 (22):4994–5008. doi:10.1111/j.1365-294x.2010.04855.x

Bonito G, Smith ME, Nowak M, Healy RA, Guevara G, Cazares E, Kinoshita A, Nouhra ER, Dominguez LS, Tedersoo L, Murat C, Wang Y, Moreno BA, Pfister DH, Nara K, Zambonelli A, Trappe JM, Vilgalys R (2013) Historical biogeography and diversification of truffles in the Tuberaceae and their newly identified southern hemisphere sister lineage. PLoS One 8(1):e52765. doi:10.1371/journal.pone.0052765

Branco S, Gladieux P, Ellison CE, Kuo A, LaButti K, Lipzen A, Grigoriev IV, Liao HL, Vilgalys R, Peay KG, Taylor JW, Bruns TD (2015) Genetic isolation between two recently diverged populations of a symbiotic fungus. Mol Ecol 24(11):2747–2758. doi:10.1111/mec.13132

Brumfield RT, Beerli P, Nickerson DA, Edwards SV (2003) The utility of single nucleotide polymorphisms in inferences of population history. Trends Ecol Evol 18(5):249–256. doi:10.1016/S0169-5347(03)00018-1

Bruns TD, Fogel R, White TJ, Palmer JD (1989) Accelerated evolution of a false-truffle from a mushroom ancestor. Nature 339(6220):140–142. doi:10.1038/339140a0

Davey JW, Hohenlohe PA, Etter PD, Boone JQ, Catchen JM, Blaxter ML (2011) Genome-wide genetic marker discovery and genotyping using next-generation sequencing. Nat Rev Genet 12:499–510. doi:10.1038/nrg3199

Ellison CE, Hall C, Kowbel D, Welcha J, Bremb RB, Glassa NL, Taylora JW (2011) Population genomics and local adaptation in wild isolates of a model microbial eukaryote. Proc Natl Acad Sci U S A 108:2831–2836. doi:10.1073/pnas.1014971108

Fiers W, Contreras R, Duerinck F, Haegeman G, Iserentant D, Merregaert J, Min Jou W, Molemans F, Raeymaekers A, Van den Berghe A, Volckaert G, Ysebaert M (1976) Complete nucleotide sequence of bacteriophage MS2 RNA: primary and secondary structure of the replicase gene. Nature 260:500–507. doi:10.1038/260500a0

Fleischmann RD, Adams MD, White O, Clayton RA, Kirkness EF, Kerlavage AR, Bult CJ, Tomb JF, Dougherty BA, Merrik JM, McKenney K, Sutton G, FitzHugh W, Fields C, Gocayne JD, Scott J, Shirley R, Liu LI, Glodek A, Kelley JM, Weidman JF, Phillips CA, Spriggs T, Hedblom E, Cotton MD, Utterback TR, Hanna MC, Nguyen DT, Saudek DM, Brandon RC, Fine LD, Fritchman JL, Fuhrmann JL, Geoghagen NSM, Gnehm CL, McDonald LA, Small KV, Fraser CM, Smith HO, Venter JC (1995) Whole-genome random sequencing and assembly of *Haemophilus influenzae* Rd. Science 269(5223):496–512. doi:10.1126/science.7542800

Galagan JE, Calvo SE, Borkovich KA, Selker EU, Read ND, Jaffe D, FitzHugh W, Ma LJ, Smirnov S, Purcell S, Rehman B, Elkins T, Engels R, Wang S, Nielsen CB, Butler J, Endrizzi M, Qui D, Ianakiev P, Bell-Pedersen D, Nelson MA, Werner-Washburne M,

Selitrennikoff CP, Kinsey JA, Braun EL, Zelter A, Schulte U, Kothe GO, Jedd G, Mewes W, Staben C, Marcotte E, Greenberg D, Roy A, Foley K, Naylor J, Stange-Thomann N, Barrett R, Gnerre S, Kamal M, Kamvysselis M, Mauceli E, Bielke C, Rudd S, Frishman D, Krystofova S, Rasmussen C, Metzenberg RL, Perkins DD, Kroken S, Cogoni C, Macino G, Catcheside D, Li W, Pratt RJ, Osmani SA, DeSouza CP, Glass L, Orbach MJ, Berglund JA, Voelker R, Yarden O, Plamann M, Seiler S, Dunlap J, Radford A, Aramayo R, Natvig DO, Alex LA, Mannhaupt G, Ebbole DJ, Freitag M, Paulsen I, Sachs MS, Lander ES, Nusbaum C, Birren B (2003) The genome sequence of the filamentous fungus *Neurospora crassa*. Nature 422 (6934):859–868. doi:10.1038/nature01554

Galagan JE, Henn MR, Ma LJ, Cuomo CA, Birren B (2005) Genomics of the fungal kingdom: insights into eukaryotic biology. Genome Res 15(12):1620–1631. doi:10.1101/gr.3767105

Goffeau A, Barrell BG, Bussey H, Davies RW, Dujon B, Feldmann H, Galibert F, Hoheisel JD, Jacq C, Johnston M, Louis EJ, Mewes HW, Murakami Y, Philippsen P, Tettelin H, Oliver SG (1996) Life with 6000 genes. Science 274(5287):546–567. doi:10.1126/science.274.5287.546

Hibbett DS, Gilbert LB, Donoghue MJ (2000) Evolutionary instability of ectomycorrhizal symbioses in basidiomycetes. Nature 407(6803):506–508. doi:10.1038/35035065

Holley RW, Apgar J, Everett GA, Madison JT, Marquisee M, Merrill SH, Penswick JR, Zamir A (1965) Structure of a ribonucleic acid. Science 147(3664):1462–1465. doi:10.1126/science.147.3664.1462

Kirk PM, Cannon PF, Minter DW, Stalpers JA (2008) Dictionary of the fungi. CABI, Wallingford

Kohler A, Kuo A, Nagy LG, Morin E, Barry KW, Buscot F, Canbäck B, Choi C, Cichocki N, Clum A, Colpaert J, Copeland A, Costa MD, Doré J, Floudas D, Gay G, Girlanda M, Henrissat B, Herrmann S, Hess J, Högberg N, Johansson T, Khouja HR, LaButti K, Lahrmann U, Levasseur A, Lindquist EA, Lipzen A, Marmeisse R, Martino E, Murat C, Ngan CY, Nehls U, Plett JM, Pringle A, Ohm RA, Perotto S, Peter M, Riley R, Rineau F, Ruytinx J, Salamov A, Shah F, Sun H, Tarkka M, Tritt A, Veneault-Fourrey C, Zuccaro A, Consortium M g i, Tunlid A, Grigoriev IV, Hibbett DS, Martin F (2015) Convergent losses of decay mechanisms and rapid turnover of symbiosis genes in mycorrhizal mutualists. Nat Genet 47:410–415. doi:10.1038/ng.3223

Laessøe T, Hansen K (2007) Truffle trouble: what happened to the Tuberales? Mycol Res 111 (9):1075–1099. doi:10.1016/j.mycres.2007.08.004

Le Tacon F, Zeller B, Plain C, Hossann C, Bréchet C, Robin C (2013) Carbon transfer from the host to *Tuber melanosporum* mycorrhizas and ascocarps followed using a 13 C pulse-labeling technique. PLos One 8(5):e64626. doi:10.1371/journal.pone.0064626

Margulies M, Egholm M, Altman WE, Attiya S, Bader JS, Bemben LA, Berka J, Braverman MS, Chen YJ, Chen Z, Dewell SB, Du L, Fierro JM, Gomes XV, Godwin BC, He W, Helgesen S, Ho CH, Irzyk GP, Jando SC, Alenquer ML, Jarvie TP, Jirage KB, Kim JB, Knight JR, Lanza JR, Leamon JH, Lefkowitz SM, Lei M, Li J, Lohman KL, Lu H, Makhijani VB, McDade KE, McKenna MP, Myers EW, Nickerson E, Nobile JR, Plant R, Puc BP, Ronan MT, Roth GT, Sarkis GJ, Simons JF, Simpson JW, Srinivasan M, Tartaro KR, Tomasz A, Vogt KA, Volkmer GA, Wang SH, Wang Y, Weiner MP, Yu P, Begley RF, Rothberg JM (2005) Genome sequencing in microfabricated high-density picolitre reactors. Nature 437(7057):376–380. doi:10.1038/nature03959

Martin F, Aerts A, Ahrén D, Brun A, Danchin EG, Duchaussoy F, Gibon J, Kohler A, Lindquist E, Pereda V, Salamov A, Shapiro HJ, Wuyts J, Blaudez D, Buée M, Brokstein P, Canbäck B, Cohen D, Courty PE, Coutinho PM, Delaruelle C, Detter JC, Deveau A, DiFazio S, Duplessis S, Fraissinet-Tachet L, Lucic E, Frey-Klett P, Fourrey C, Feussner I, Gay G, Grimwood J, Hoegger PJ, Jain P, Kilaru S, Labbé J, Lin YC, Legué V, Le Tacon F, Marmeisse R, Melayah D, Montanini B, Muratet M, Nehls U, Niculita-Hirzel H, Oudot-Le Secq MP, Peter M, Quesneville H, Rajashekar B, Reich M, Rouhier N, Schmutz J, Yin T, Chalot M, Henrissat B, Kües U, Lucas S, Van de Peer Y, Podila GK, Polle A, Pukkila PJ, Richardson PM, Rouzé P, Sanders IR, Stajich JE, Tunlid A, Tuskan G, Grigoriev IV (2008) The genome of Laccaria bicolor provides insights into mycorrhizal symbiosis. Nature 452 (7183):88–92. doi:10.1038/nature06556

Martin F, Kohler A, Murat C, Balestrini R, Coutinho PM, Jaillon O, Montanini B, Morin E, Noel B, Percudani R, Porcel B, Rubini A, Amicucci A, Amselem J, Anthouard V, Arcioni S, Artiguenave F, Aury JM, Ballario P, Bolchi A, Brenna A, Brun A, Buee M, Cantarel B, Chevalier G, Couloux A, Da Silva C, Denoeud F, Duplessis S, Ghignone S, Hilselberger B, Iotti M, Marcais B, Mello A, Miranda M, Pacioni G, Quesneville H, Riccioni C, Ruotolo R, Splivallo R, Stocchi V, Tisserant E, Viscomi AR, Zambonelli A, Zampieri E, Henrissat B, Lebrun MH, Paolocci F, Bonfante P, Ottonello S, Wincker P (2010) Perigord black truffle genome uncovers evolutionary origins and mechanisms of symbiosis. Nature 464 (7291):1033–1038. doi:10.1038/nature08867

Metzker ML (2010) Sequencing technologies—the next generation. Nat Rev Genet 11(1):31–46. doi:10.1038/nrg2626

Molinier V, Murat C, Morin E, Gollotte A, Wipf D, Martin F (2013) First identification of polymorphic microsatellite markers in the Burgundy Truffle, *Tuber aestivum* (Tuberaceae). Appl Plant Sci 1(2):apps 1200220. doi:10.3732/apps.1200220

Murat C, Riccioni C, Belfiori B, Cichocki N, Labbé J, Morin E, Tisserant E, Paolocci F, Rubini A, Martin F (2011) Distribution and localization of microsatellites in the Perigord black truffle genome and identification of new molecular markers. Fungal Genet Biol 48(6):592–601. doi:10.1016/j.fgb.2010.10.007

O'Donnell K, Cigelnik E, Weber NS, Trappe JM (1997) Phylogenetic relationships among ascomycetous truffles and the true and false morels inferred from 18S and 28S ribosomal DNA sequence analysis. Mycologia 89(1):48–65. doi:10.2307/3761172

Olivier JM, Savignac JC, Sourzat P (2012) Truffe et trufficulture. Fanlac, Périgueux

Payen T, Murat C, Bonito G (2014) Truffle phylogenomics: new insights into truffle evolution and truffle life cycle. In: Martin F (ed) The ecological genomics of fungi. Wiley, New York, pp 211–234. doi:10.1016/B978-0-12-397940-7.00007-0

Payen T, Murat C, Gigant A, Morin E, De Mita S, Martin F (2015) A survey of genome-wide single nucleotide polymorphisms through genome resequencing in the Périgord black truffle (*Tuber melanosporum* Vittad.). Mol Ecol Resour 15(5):1243–1255. doi:10.1111/1755-0998.12391

Percudani R, Trevisi A, Zambonelli A, Ottonello S (1999) Molecular phylogeny of truffles (Pezizales: Terfeziaceae, Tuberaceae) derived from nuclear rDNA sequence analysis. Mol Phylogenet Evol 13(1):169–180. doi:10.1006/mpev.1999.0638

Plett JM, Kemppainen M, Kale SD, Kholer A, Legué V, Brun A, Tyler BM, Pardo AG, Martin F (2011) A secreted effector protein of *Laccaria bicolor* is required for symbiosis development. Curr Biol 21(14):1197–1203. doi:10.1016/j.cub.2011.05.033

Plett JM, Daguerre Y, Wittulsky S, Vayssières A, Deveau A, Melton SJ, Kholer A, Morrell-Falvey JL, Brun A, Veneault-Fourrey C, Martin F (2014) Effector MiSSP7 of the mutualistic fungus *Laccaria bicolor* stabilizes the *Populus* JAZ6 protein and represses jasmonic acid (JA) responsive genes. Proc Natl Acad Sci USA 111(22):8299–8304. doi:10.1073/pnas.1322671111

Sanger F, Nicklen S, Coulson AR (1977) DNA sequencing with chain-terminating inhibitors. Proc Natl Acad Sci USA 74(12):5463–5467

Smith SE, Read DJ (2010) Mycorrhizal symbiosis. Academic, New York

Spatafora J (2011) 1000 fungal genomes to be sequenced. IMA Fungus 2(2):41–45

Spatafora JW, Sung GH, Johnson D, Hesse C, O'Rourke B, Serdani M, Spotts R, Lutzoni F, Hofstetter V, Miadlikowska J, Reeb V, Gueidan C, Fraker E, Lumbsch T, Lücking R, Schmitt I, Hosaka K, Aptroot A, Roux C, Miller AN, David M, Geiser DM, Hafellner J, Hestmark G, Arnold AE, Büdel B, Rauhut A, Hewitt D, Untereiner WA, Cole MS, Scheidegger C, Schultz M, Sipman H, Schoch CL (2006) A five-gene phylogeny of Pezizomycotina. Mycologia 98(6):1018–1028. doi:10.3852/mycologia.98.6.1018

Tedersoo L, May TW, Smith ME (2010) Ectomycorrhizal lifestyle in fungi: global diversity, distribution, and evolution of phylogenetic lineages. Mycorrhiza 20(4):217–263. doi:10.1007/s00572-009-0274-x

Thiers HD (1984) The secotioid syndrome. Mycologia 76(1):1–8

Traeger S, Altegoer F, Freitag M, Gabaldon T, Kempken F, Kumar A, Marcet-Houben M, Pöggeler S, Stajich JE, Nowrousian M (2013) The genome and development-dependent transcriptomes of Pyronema confluens: a window into fungal evolution. PLoS Genet 9(9): e1003820. doi:10.1371/journal.pgen.1003820

Trappe JM, Molina R, Luoma DL, Cázares E, Pilz D, Smith JE, Castellano MA, Miller SL, Trappe MJ (2009) Diversity, ecology, and conservation of truffle fungi in forests of the Pacific Northwest. USDA, Forest Service, Pacific Northwest Research Station

van der Heijden MGA, Martin FM, Selosse MA, Sanders IR (2015) Mycorrhizal ecology and evolution: the past, the present, and the future. New Phytol 205(4):1406–1423. doi:10.1111/nph.13288

Part II
The Abiotic Environment

Chapter 10
Influence of Climate on Natural Distribution of *Tuber* Species and Truffle Production

François Le Tacon

10.1 Introduction

Species belonging to the genus *Tuber* are associated through ectomycorrhizas (ECMs) with the roots of deciduous or coniferous trees or shrubs. They are completely dependent on their host and cannot be found in treeless areas. Nevertheless, they can also be associated and form ECMs with non-tree species such as Cistaceae (Giovanetti and Fontana 1982). The genus *Tuber* displays a large natural geographic distribution, but almost only in the Northern Hemisphere. Truffles occur naturally throughout all of Europe, including Scandinavia (Gotland Island in Sweden) (Weden et al. 2004) and some species are found in North Africa (Ceruti et al. 2003). The genus *Tuber* is widespread in Asia (India, China, Mongolia) and present in North America (Bonito et al. 2013). The actual natural distribution of the genus *Tuber* results from current climatic conditions and soil characteristics in the Northern Hemisphere but also from past long-distance migrations related to past climatic changes and the resulting co-migration of their hosts (Murat et al. 2004).

In this review, we will analyse first the past migrations of the genus *Tuber*. We will then describe the current natural distribution of the main commercialised species related to current climatic characteristics. We will follow with a discussion about the effects of annual climatic variations on *Tuber borchii* Vittad. and *Tuber melanosporum* Vittad. ascoma production, and we will conclude by proposing several considerations on the possible effect on truffle production of CO_2 increase and its consequence, predicted climate warming.

F. Le Tacon (✉)
INRA, UMR 1136, Université de Lorraine, Interactions Arbres-Microorganismes, Laboratoire d'excellence ARBRE, 54280 Champenoux Cedex, France
e-mail: le_tacon@nancy.inra.fr

A. Zambonelli et al. (eds.), *True Truffle (*Tuber *spp.) in the World*, Soil Biology 47,
DOI 10.1007/978-3-319-31436-5_10

10.2 Past Migrations of the Genus *Tuber*

The Tuberaceae are present in Southern and Northern Hemispheres suggesting that this family has a Pangean origin. But *Tuber* species have been recorded in natural sites almost only in the Northern Hemisphere suggesting that the genus *Tuber* would have appeared after the break-up of Pangea.

Jeandroz et al. (2008) suggested that the *Tuber* common ancestor originated from Europe or Eurasia. Genetic marker analysis and the use of molecular clocks tell us that differentiation of the *Tuber* genus would have taken place between 205 and 184 Mya (Jeandroz et al. 2008) or later, between 161.6 and 121.8 Mya (Bonito et al. 2013). A first diversification event occurring more than 150 Mya led to the differentiation of a first clade, the Aestivum clade, comprising the Aestivum, Magnatum and Macrosporum groups only found in Europe (Jeandroz et al. 2008). Later, four other clades would arise in Asia or Eurasia, the Excavatum clade 110 Mya, the Rufum clade 70 Mya, the Melanosporum clade between 85 and 25 Mya and, finally, the Puberulum clade between 65 and 53 Mya. The presence of the Rufum, Melanosporum and Puberulum clades in Europe, Asia and North America could be explained by migrations between Asia and Europe and migrations towards America from Asia through the North Atlantic Land Bridge prior to its break 45 Mya (Jeandroz et al. 2008).

In the middle of the Miocene, 16 Mya, a new connection between North America and Eurasia occurred with the appearance of the Bering Land Connection, allowing a new route of terrestrial migration for these three groups. The exchanges between Asia and North America stopped 3.5 Mya, when the land connection disappeared (Behrensmeyer et al. 1992). The Bering Land Bridge would appear again during the ice ages. But the absence of trees excluded migration of *Tuber* species via this route.

During the last glacial period, the European forest was restricted to the Mediterranean region, which could have induced a loss of genetic diversity and a bottleneck in remaining populations of *T. melanosporum* (Bertault et al. 1998). After the end of the last glacial period, the recolonisation of France by *T. melanosporum* took place from the two refugia of Spain and Italy following the recolonisation by oak (Murat et al. 2004). This scenario is probably also applicable to the other European *Tuber* species. Similarly, migrations might have occurred in Asia and North America during the last ice age.

From these past long-distance migrations and from those which took place more recently, it is clear that *Tuber* species have climatic requirements, which partly explain their present natural distribution.

10.3 Climate and Current Natural Distribution of the Main Five Commercialised *Tuber* Species

10.3.1 Tuber melanosporum

Tuber melanosporum is limited in Europe between a latitude of 40° and 48° North. The highest latitude where *T. melanosporum* has been recorded is that of Lorraine near Commercy in France (48.7° North). The countries where *T. melanosporum* is most frequently found in its natural habitat are France, Italy and Spain. *Tuber melanosporum* has also been recorded in Portugal, the Czech Republic, the Slovak Republic, Switzerland, Croatia, Serbia, Bulgaria, Romania, Greece and Turkey (Ceruti et al. 2003). Its presence in Germany and Poland is doubtful. *Tuber melanosporum* is typically a Mediterranean species. It is found in lowlands and extends from the sea level to highlands up to 1300 m AMSL. The Mediterranean climate is characterised by hot and dry summers with strong soil water deficit, with relatively humid and mild winter months. However, the variability of the Mediterranean climate is high and is characterised by a number of interactions at different scales (rainfall, altitude, topography, circulation systems).

The main factors limiting the natural distribution of *T. melanosporum* are the summer drought and the winter frost. *Tuber* ECMs appear to be very resistant to summer drought and winter frost, while the fruiting process is very sensitive. The young ascomata born at the end of the spring develop slowly during the summer and become mature from November to March of the following year. This long cycle, in contrast to the very short time necessary for the development of fruiting bodies of epigeous fungi, makes the survival of *Tuber* ascomata highly sensitive to annual climatic conditions. It is the reason why *T. melanosporum* is rare or not present in Mediterranean regions, where annual rainfall is less than 600 mm. Similarly, due to winter frost, *T. melanosporum* cannot be found at latitudes higher than 48° North. Nevertheless, Zampieri et al. (2011) demonstrated that cold temperatures modify *T. melanosporum* gene expression, mainly heat shock protein-coding genes and genes involved in cell wall and lipid metabolism. These genes could be involved in the adaptation of *T. melanosporum* to low temperatures. The *T. melanosporum* ECMs are more resistant to frost than the ascomata and are capable of vegetative propagation. While this might allow for *T. melanosporum* to be present in the form of mycorrhizas in Northern latitudes, ascoma formation may not be possible.

10.3.2 Tuber indicum

The Asian black truffle *Tuber indicum* Cooke and Massee was first described from a dried sample harvested in January 1892 in India at an altitude of about 2000 m (Cooke and Massee 1892). Much later, other similar species have been described: *Tuber sinense* K. Tao and B. Liu; *Tuber himalayense* B. C. Zhang and Minter;

Tuber pseudohimalayense G. Moreno, Manjón, J. Díez and García-Mont.; and *Tuber formosanum* H. T. Hu. It is now agreed that these five taxa are synonymous and form a single species *T. indicum* with different populations, or groups or cryptic species (Zhang et al. 2005; Wang et al. 2006; Chen et al. 2011; Kinoshita et al. 2011; Belfiori et al. 2013). *Tuber indicum* has a very large distribution area in Asia and can be found from 75° East (India) to 140° East (Japan) and from 23° to 40° North. In China, *T. indicum* is found mainly in the provinces of Yunnan and Sichuan between 25° and 30° North and at an altitude of between 1500 and 3000 m. In these two provinces, the tropical climate is moderated by the altitude. The average annual temperature is about 15 °C (Chuxiong 16 °C, Hui Dong 15 °C, Kunming 14.9 °C, Gongshan 12 °C). At these altitudes, the annual rainfall is 1000 mm or more (Hui Dong 1153 mm, Miyi 1148 mm, Huize 1102 mm, Kunming 1011 mm, Gongshan 968 mm, Chuxiong 863 mm). Moreover, rainfall is abundant during the three summer months (cumulated rainfall of June, July and August: Hui Dong 646 mm, Huize 620 mm, Kunming 587 mm; Gongshan 615 mm, Chuxiong 497 mm), eliminating water stress, which is favourable to the development of the young ascomata. The coldest month is January with an average temperature ranging between 6 and 8 °C. While occasionally the minimum temperature can drop to −5 °C, freezing days during truffle production are uncommon.

10.3.3 Tuber aestivum

Tuber aestivum Vittad. is spread throughout all Europe and can be found in North Africa (Morocco) between a latitude of 35° and 57° North (Ceruti et al. 2003). The highest latitude where *T. aestivum* has been recorded is that of the Gotland Island on the eastern coast of Sweden (Weden et al. 2004). Towards the west, the highest longitude is 8° West in Ireland and Portugal. To the east, our knowledge is fragmentary. *Tuber aestivum* has been found in Poland, in Russia (Ceruti et al. 2003) and in Azerbaijan (Fekete et al. 2014). We do not know if this species crossed the Ural. The presence of *T. aestivum* has been reported in China (Chen et al. 2005), but several phylogenetic analyses seem to prove that Chinese *T. aestivum* should be placed in a clade different from that of Europe (Zambonelli et al. 2012) and named *T. sinoaestivum* (Zhang et al. 2012). At all events, *T. aestivum* has adapted to a broad range of climatic conditions covering the entire natural area of *T. melanosporum* distribution characterised by a Mediterranean climate and including temperate rainy climates (Ireland, Scotland and England), cold maritime climates (Sweden) and very cold continental climates (Deutschland, Poland, Hungary, Russia). *Tuber aestivum* can be found until an altitude of 1400–1600 m. This astonishing broad natural distribution could be explained by the fact that, contrary to *T. melanosporum*, the *T. aestivum* cycle is continuous as was first reported by Geoffroy (1711). The young *T. aestivum* ascomata will produce continuously given favourable soil moisture and temperature conditions (Stobbe et al. 2013). The cycle is rated by two climatic parameters: the summer drought and the winter frost. In Mediterranean conditions, the cycle is often stopped short by the summer drought. In May or June or even later, white nonmature ascomata are harvested in southern Europe. However, if the summer is not too dry, mature ascomata can be harvested in autumn. In Atlantic and temperate conditions, mature ascomata can be found in autumn or at the same period in continental conditions before the winter frost. Nonmature ascomata, however, can be found in autumn or in winter depending on the summer rainfall or winter temperatures.

10.3.4 Tuber borchii

Tuber borchii also has a wide distribution in Europe. It can be found from 37° to 61° North from southern Finland to Sicily and from Ireland to Poland (Ceruti et al. 2003). To date, *T. borchii* has not been recorded in Russia. It cannot be found above an altitude of 1000 m.

10.3.5 Tuber magnatum

Contrary to *T. aestivum* and *T. borchii*, *Tuber magnatum* Pico exhibits a narrow distribution range. It can be found between 40° and 46° North from South East of France to Italia, Slovenia, Croatia and Serbia from low elevation to an altitude of about 700–900 m ASLM (Ceruti et al. 2003; Hall et al. 2007). *Tuber magnatum* ascomata seem very sensitive to winter frost and summer drought, which could explain this narrow distribution and the fact that *T. magnatum* is mainly found in riparian areas (Hall et al. 2007).

10.4 Climatic Variations and Truffle Ascoma Production

Truffles are often harvested from naturally occurring forests in Europe, Asia and North America. Nevertheless, very little data exists concerning the relationship between ascoma production and climatic conditions in natural forests. The only data available for natural forests are those of Salerni et al. (2014) on *T. borchii*. For truffle plantations we have more data, but they remain scarce and often not reliable.

10.4.1 Climatic Variations and Ascoma Production of T. borchii in Natural Sites

Salerni et al. (2014) conducted surveys for 4 years (1996–1999) on *T. borchii* ascoma production in Central Italy (Parco Regionale della Maremma, Tuscany) (lat. 42°39′ North, long 11°2′ East, altitude 5 m, mean annual rainfall 590 mm and mean annual temperatures 14.7 °C). The production of *T. borchii* ascoma was observed to positively correlate with rainfall occurring 3 months before harvest. But a negative effect was observed as a result of rainfall during ascoma harvesting, indicating that an excess of soil moisture in autumn could stop the development of ascoma or could induce their rotting. Salerni and colleagues also found that low temperatures in autumn had a positive effect on *T. borchii* ascoma production.

10.4.2 Climatic Variations and T. melanosporum Ascoma Production in Plantations

Plantations of trees for *T. melanosporum* production occurred throughout the nineteenth century in France and Italy and then in Spain. From that time forward, production of natural sites has progressively been replaced by that of plantations, mainly in France where it is almost exclusively the plantations that produce Périgord black truffles.

Büntgen et al. (2012a) found, for the period going from 1970 to 2006, a significant positive correlation of total black truffle production of Southern France (Périgord), Northeastern Spain (Aragón) and Northern and Central Italy (Piedmont and Umbria) with June–August rainfall and a negative correlation with temperature maxima. Spanish and French production displayed significant positive correlations with summer rainfall, while Italian production was poorly correlated with summer precipitations (Büntgen et al. 2012a).

In France, the peak of truffle production was reached at the beginning of the twentieth century with a record of approximately 1000 t in 1904. Today, annual truffle production does not exceed 80 t (source Fédération Française des Trufficulteurs).

Le Tacon et al. (2014) studied sales in French wholesale markets for the past 25 years. Sales were stable for the whole of France, with an average of 12.9 t per year. This stability, however, masks important annual variations ranging from 4 to 25 t per year (Fig. 10.1).

An analysis of the Richerenches and Carpentras wholesale markets revealed that the main factor explaining these variations was the cumulative late spring and summer hydric balance (HB in mm = precipitation, potential evapotranspiration), which accounted for 36.9 % in the variance of truffle sales in Carpentras and 45.7 % in Richerenches (Fig. 10.2).

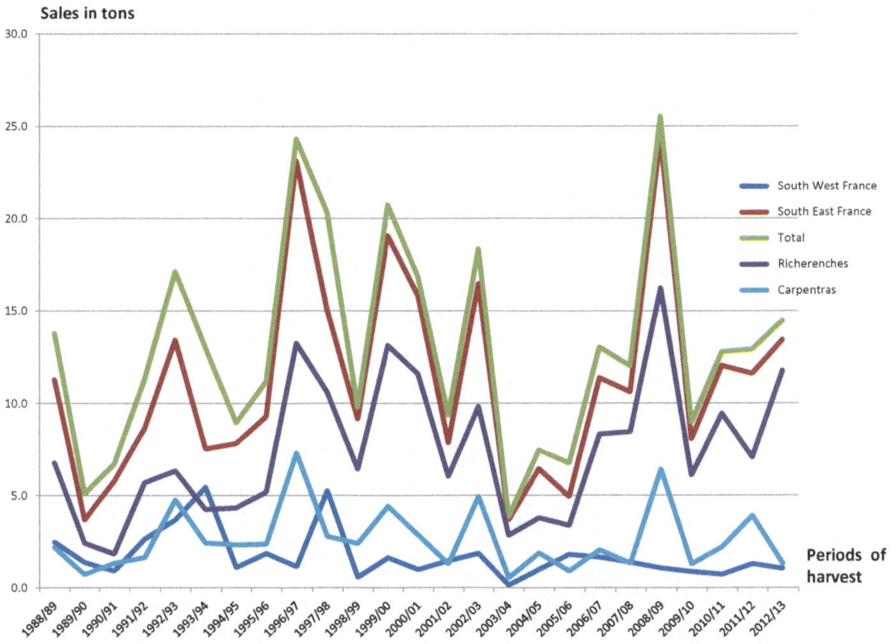

Fig. 10.1 Truffle sales in wholesale markets in South Western France (*dark blue*), South Eastern France (*red*), total France (*green*), Richerenches (*violet*) and Carpentras (*light blue*) from 1988–1989 to 2012–2013 (in tons per year, source "Réseau des Nouvelles des Marchés" of the French Ministry of Agriculture, adapted from Le Tacon et al. (2014))

The second factor explaining variations of sales in the Richerenches basin was the number of freezing days with minimum temperatures equal to or less than −5 °C, which accounted for 14 % of the variance in truffle sales.

Additional factors for both sites, such as cumulative rainfall for October and November, affected truffle sales but were less consequential than the summer HB. Rainfall in October and November allowed for the growth of young truffles which had survived the summer months. Good water availability in late winter could also serve to support the formation of new ECMs, thus allowing for sexual reproduction.

Models obtained from 1988–1989 to 2012–2013 for detrended sales were used to simulate the evolution of the sales from 1965–1966 to present day using the meteorological data available for the Richerenches and Carpentras basins. The simulation of potential truffle sales from 1965–1966 to 2012–2013 considering climatic conditions for that period showed stability in truffle sales these past 48 years despite a degradation of the summer HB, compensated in part (for the Richerenches site) by reducing the number of freezing days. These results show us that a decline in French truffle production over the last 48 years could be attributed more to changes in rural communities than to climatic changes (see Chap. 2). Nevertheless, the Mediterranean climate has warmed by about 1 °C over the past

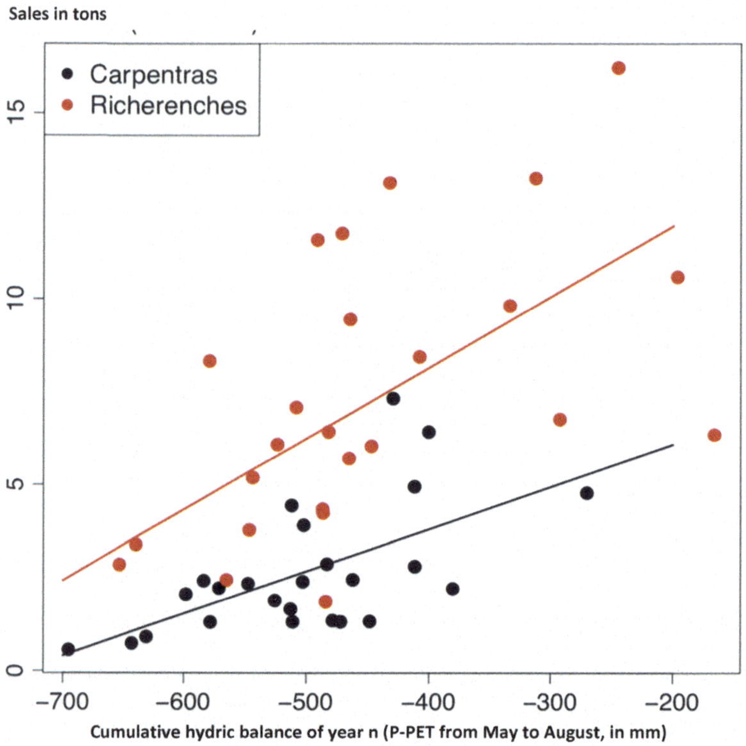

Fig. 10.2 The relationship between sales (in tons per year) from 1988–1989 to 2012–2013 and the cumulative hydric balance of year *n* (P-PET from May to August, in mm) in Carpentras (*black*) ($R^2 = 0.607$, p-value < 0.01) and Richerenches (*red*) ($R^2 = 0.676$, p-value < 0.01), adapted from Le Tacon et al. (2014)

100 years. While this warming trend might have played a role in the decreased production of black truffles over the past century in France, it does not solely explain the sharp fall in production.

10.5 Future Global Change and Truffle Production

Kauserud et al. (2012) and Büntgen et al. (2012b) showed that in Europe, the fructification period for mushrooms growing in natural conditions may or may not respond to important shifts in climate changes and that it depends on both the species and the locations. The fruiting period and the ascoma production of the *Tuber* species could be both affected by climate changes. Büntgen et al. (2011) suggested that a shift of *T. aestivum* biome has already occurred in line with extended growing season. On the contrary, Splivallo et al. (2012) suggested that climate change has had little effect on the present natural distribution of *T. aestivum*.

In order to discuss how climate change may affect natural truffle distribution or ascoma production, it is necessary to consider possible direct and indirect effects of rising concentrations of atmospheric CO_2 (Norby and Luo 2004).

10.5.1 Direct Effects of Elevated CO_2 Concentrations

In addition to the greenhouse effect induced by increased CO_2 concentrations, it is well known that the first consequence of this increase is more efficient photosynthesis and improved plant growth in the absence of other limiting factors such as water availability and nitrogen or phosphorus supply. Increase in atmospheric CO_2 also generally improves water use efficiency (Morison 1985). Plants growing in enriched atmospheric CO_2 allocate more photosynthates to roots. The result is an increase of carbon availability to the associated fungi, higher numbers of ECMs, more fungal biomass and an increase in fungal respiration (Ineichen et al. 1995; Rouhier and Read 1998; Nehls and Hampp 2000; Treseder and Allen 2000; Alberton et al. 2005; Frannson et al. 2007). Although no experiments currently exist which specifically focus on possible effects of increased atmospheric CO_2 levels on truffles, we can deduce with some certainty that future increases in atmospheric CO_2 should have beneficial effects on *Tuber* ECM status. This could also have positive effects on ascoma production, which is dependent on control of the host plant over carbon allocation.

10.5.2 Indirect Effects of Elevated CO_2 Concentrations on Warming and Summer Water Stress

For the twenty-first century, almost all scenarios in Europe predict a trend towards warming, ranging from 2 to 4 °C. Almost all scenarios also predict decreased rainfall in the Mediterranean region, mainly in summer months (Giorgi and Lionello 2008). Similarly, Büntgen et al. (2012a) predicted intensified potential summer evapotranspiration in this area to occur up to the end of the twenty-first century. The result could be increased summer water stress, which, as discussed at the second international meeting on truffle in the Vaucluse region (Rousset-Rouard 2008), could greatly affect black truffle production.

As we have shown above, *T. magnatum* ascomata are mainly found in riparian areas and most commonly in natural conditions. Soils in these zones are deep with high water availability, even in summer months. In these conditions, degradation of the summer hydric balance should have a more limited effect on *T. magnatum* ascoma production than on that of *T. melanosporum*.

10.5.3 *Possible Influences on* T. melanosporum *Production and Management Techniques for Decreasing Summer Drought Stress*

Distribution of several tree species is likely to change in the Mediterranean region, particularly for white oak (*Quercus pubescens* L.) and holm oak (*Quercus ilex* L.) to which *T. melanosporum* is associated (Petit et al. 2005). Migration of their hosts could affect the presence of *T. melanosporum* in its natural habitats. At present, the black truffle is mainly produced in orchards where its hosts are planted. Nevertheless, while these hosts are artificially maintained, climatic changes could have direct effects on truffle production. Truffles are fungi, and we know that mushroom production in natural conditions depends first on rainfall and soil moisture levels (O'Dell et al. 1999; Bonet et al. 2004; Krebs et al. 2008).

Management techniques currently used in these orchards (irrigation, mulching, soil tilling, host pruning) can partly compensate for summer water deficits. However, if degradation of the summer hydric balance continues, it will be necessary to adopt improved management methods (Chevalier and Wehrlen 2008; Genola 2008; Gregori 2008). Far too often in France, black truffle orchards are installed on soils with limited water holding capacity (100 mm or less). To at least partly compensate for deficits in summer hydric balance, *T. melanosporum* orchards will need to be installed on soils with a high water holding capacity (200 mm or more).

The use of irrigation in black truffle orchards has begun to be more frequently applied on large scales in France, Spain and Australia. But irrigation is not sufficiently based on guidelines, although they have existed for a long time (Le Tacon et al. 1982). These authors showed that truffle orchard irrigation can be driven simply by measuring soil pF (force necessary to extract water); maintaining soil pF in the 10 first cm <3.5 during the summer drought greatly improved truffle production. Combining irrigation and mulching also contributed to improved control of soil water availability (Le Tacon et al. 1982).

It would also be necessary to implement all techniques that allow for improved summer hydric balance: diminishing the herbaceous vegetation cover by soil tilling and decreasing tree evapotranspiration rates by diminishing host stem density and by reducing tree crowns. While these are known techniques, they are not sufficiently implemented; when we do see them applied, they are used empirically without relevant guidelines. A challenge for the next decades to come will be to provide truffle growers with best practice management guidelines based on established water balance models of truffle orchards.

Another possible way to counteract global warming could be to move the black truffle plantations further north or to higher altitudes. In France, the truffle growing surface area of black truffle orchards could be increased in Poitou, Charente and South of Touraine. In New Zealand, Renowden holds that global warming could allow for expanded areas suitable for truffle production, which would have a positive effect on truffle production in a country characterised by mild temperatures and moderately high rainfall (Guérin-Laguette and Renowden 2008). In Australia, the situation is different: while global warming could improve black truffle production in Tasmania and Victoria, in other regions it may be necessary to rely more heavily on irrigation (Malajczuk 2008).

10.6 Conclusions

From the past long-distance migrations and from those, which took place more recently, it is obvious that *Tuber* species have climatic requirements, which partly explain their present natural distribution. *Tuber melanosporum* is limited in Europe between a latitude of 40° and 48° North. *Tuber melanosporum* is typically a Mediterranean species. The main factors limiting its natural distribution are the summer drought and the winter frost.

Tuber indicum has a very large distribution area in Asia and can been found from 75° (India) to 140° (Japan) East and from 23° to 40° North. In China, *T. indicum* is found mainly between 25 and 30° North and at an altitude of between 1500 and 3000 m. The tropical climate, moderated by the altitude, is characterised by an abundant rainfall during the three summer months.

Tuber aestivum is spread throughout all Europe and can be found in North Africa (Morocco) between a latitude of 35° and 57° North. Towards the west, the highest longitude is 8° West in Ireland and Portugal. To the east, our knowledge is fragmentary. *Tuber aestivum* has adapted to a broad range of climatic conditions: Mediterranean, temperate rainy climates, cold maritime climates and very cold continental climates. In Mediterranean conditions, the cycle is often stopped short by the summer drought. In Atlantic and temperate conditions, mature ascomata can be found in autumn or at the same period in continental conditions before the winter frost.

Tuber borchii also has a wide distribution in Europe. It can be found from 37° to 61° North from southern Finland to Sicily and from Ireland to Poland. It cannot be found above an altitude of 1000 m.

Contrary to *T. aestivum* and *T. borchii*, *T. magnatum* exhibits a narrow distribution range. It can be found between 40° and 46° North from South East of France to Italia, Slovenia, Croatia and Serbia from low elevation to an altitude of about 700–900 m.

Very little data exists concerning the relationship between ascoma production and climatic conditions in natural forests. The only data available for natural forests are on *T. borchii*. For truffle plantations we have more data, but they remain scarce and often not reliable. An analysis of the French Richerenches and Carpentras *T. melanosporum* wholesale markets revealed that the main factor explaining the annual variations of sales was the cumulative late spring and summer water balance. The second factor explaining annual variation of sales in the Richerenches basin was the number of freezing days with minimum temperatures equal to or less than −5 °C. Models obtained from 1988–1989 to 2012–2013 were used to simulate the evolution of the sales from 1965–1966 to present day. The simulation showed

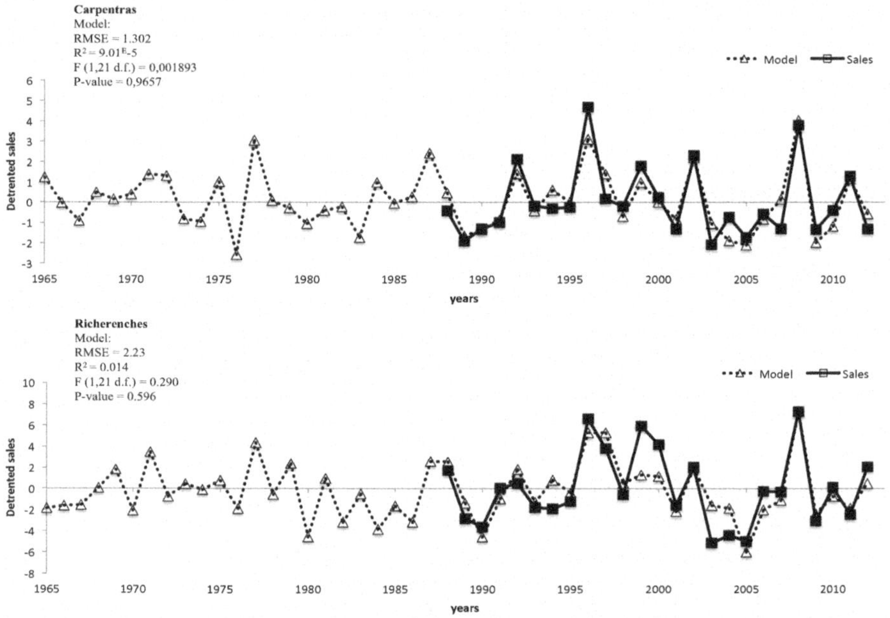

Fig. 10.3 Evolution of the detrended truffle sales in wholesale markets of Carpentras and Richerenches from 1988–1989 to 2012–2013 (*square continuous line*) and of the multiple linear regression model from 1965–1966 to 2012–2013 (*triangle dotted line*) in Carpentras and Richerenches. The statistics of the multiple linear regression of the predicted sales over time are indicated, adapted from Le Tacon et al. (2014)

that potential truffle sales would have been stable these past 48 years despite a degradation of the summer water balance, compensated in part (for the Richerenches site) by reducing the number of freezing days. The decline in French truffle production over the last 48 years could be attributed more to changes in rural communities than to climatic changes. Nevertheless, the Mediterranean climate has warmed by about 1 °C over the past 100 years. While this warming trend might have played a role in the decreased production of black truffles over the past century in France, it does not solely explain the sharp fall in production (Fig. 10.3).

For the twenty-first century, almost all scenarios in Europe predict a trend towards warming, ranging from 2 to 4 °C, and decreased rainfall in the Mediterranean region. The result could be increased summer water stress, which could greatly affect truffle production. In plantations, it would be necessary to implement all techniques that allow for improved summer hydric balance (irrigation, mulching, soil tilling, host pruning). A challenge for the next decades to come will be to provide truffle growers with best practice management guidelines based on established water balance models of truffle orchards. Another possible way to counteract global warming could be to move the truffle plantations further north or to higher altitudes.

Acknowledgements A part of this work was supported by a grant overseen by the French National Research Agency (ANR) as part of the "Investissements d'Avenir" programme (ANR-11-LABX-0002-01, Laboratory of Excellence ARBRE). We would like to thank the anonymous reviewers for their interest in our work and for their comments or suggestions, which allowed us to greatly improve the first version of this chapter. We are thankful to Aimee Orsini for having corrected the English of this manuscript.

References

Alberton O, Kuyper TW, Gorissen A (2005) Taking mycocentrism seriously: mycorrhizal fungal and plant responses to elevated CO_2. New Phytol 167(3):859–868. doi:10.1111/j.1469-8137.2005.01458.x

Behrensmeyer AK, Damuth JD, DiMichele WA, Potts R, Sues HD, Wing SL (1992) Terrestrial ecosystems through time. Evolutionary paleoecology of terrestrial plants and animals. University of Chicago Press, Chicago

Belfiori B, Riccioni C, Paolocci F, Rubini A (2013) Mating type locus of Chinese black truffles reveals heterothallism and the presence of cryptic species within the *T. indicum* species complex. PLoS One 8(12):e82353. doi:10.1371/journal.pone.0082353

Bertault G, Raymond M, Berthomieu A, Callot G, Fernandez D (1998) Trifling variation in truffles. Nature 394(6695):734. doi:10.1038/29428

Bonet JA, Fischer CR, Colinas C (2004) The relationship between forest age and aspect on the production of sporocarps of ectomycorrhizal fungi in *Pinus sylvestris* forests of the central Pyrenées. For Ecol Manage 203(1–3):157–175. doi:10.1016/j.foreco.2004.07.063

Bonito G, Smith ME, Nowak M, Healy RA, Guevara G, Cázares E, Kinoshita A, Nouhra ER, Domínguez LS, Tedersoo L, Murat C, Wang Y, Moreno BA, Pfister DA, Nara K, Zambonelli A, Trappe JM, Vilgalys R (2013) Historical biogeography and diversification of truffles in the *Tuberaceae* and their newly identified southern hemisphere sister lineage. PLoS One 8(1):e52765. doi:10.1371/journal.pone.0052765

Büntgen U, Tregel W, Egli S, Stobbe U, Sproll L, Stenseth NC (2011) Truffles and climate change. Front Ecol Environ 9(3):150–151. doi:10.1890/11.WB.004

Büntgen U, Egli S, Camarero JJ, Fischer EM, Stobbe U, Kauserud H, Tegel W, Sproll L, Stenseth NC (2012a) Drought-induced decline in Mediterranean truffle harvest. Nat Clim Change 2 (12):827–829. doi:10.1038/nclimate1733

Büntgen U, Kauserud H, Egli S (2012b) Linking climate variability to mushroom productivity and phenology. Front Ecol Environ 10(1):14–19. doi:10.1890/110064

Ceruti A, Fontana A, Nosenzo C (2003) Le specie europee del genere *Tuber:* Una revisione storica. Museo Regionale di Scienze Naturali, Regione Piemonte, Torino

Chen J, Liu PG, Wang Y (2005) Notes on *Tuber aestivum* Vittad (Tuberaceae, Ascomycota) from China. Acta Bot Yunnanica 27(4):385–389

Chen J, Guo SX, Liu PG (2011) Species recognition and cryptic species in the *Tuber indicum* complex. PLoS One 6(1):e14625. doi:10.1371/journal.pone.0014625

Chevalier G, Wehrlen L (2008) Quelques principes de lutte intégrée contre le réchauffement climatique en trufficulture. In: Rousset-Rouard Y, Savignac JC (eds) L'avenir de la truffe face au réchauffement climatique: la truffe européenne est-elle en danger ? Actes des 2e rencontres internationales de la truffe. Albin Michel, Paris, pp 48–62

Cooke MC, Massee G (1892) Himalayan truffles. Grevillea 20:67

Fekete AO, Bagi I, Conde IP, Iotti M, Zambonelli A (2014) First report on the truffles of Azerbaijan. In: Résumés du deuxième symposium sur les champignons hypogés du bassin mediterranéen (HYPOGES2) et cinquième congrès Tuber *aestivum/uncinatum* du groupe scientifique européen (TAUESG5), Université Mohammed V, Rabat, 9–13 Apr 2014, p 19

Fransson PMA, Anderson IC, Alexander IJ (2007) Does carbon partitioning in ectomycorrhizal pine seedlings under elevated CO_2 vary with fungal species? Plant Soil 291(1–2):323–333. doi:10.1007/s11104-007-9203-y

Genola L (2008) Peut-on envisager une trufficulture performante face aux nouvelles contraintes climatiques? In: Rousset-Rouard Y, Savignac JC (eds) L'avenir de la truffe face au réchauffement climatique: la truffe européenne est-elle en danger? Actes des 2e rencontres internationales de la truffe. Albin Michel, Paris, pp 86–97

Geoffroy CJ (1711) Observations sur la végétation des Truffes. In: Mémoires de l'Académie royale des Sciences, Memoires de mathématique et de physique, pp 29–44

Giorgi F, Lionello P (2008) Climate change projections for the Mediterranean region. Global Planet Change 63(2–3):90–104. doi:10.1016/j.gloplacha.2007.09.005

Giovanetti G, Fontana A (1982) Mycorrhizal synthesis between *Cistaceae* and *Tuberaceae*. New Phytol 92(4):533–537. doi:10.1111/j.1469-8137.1982.tb03412.x

Gregori GN (2008) Trufficulture en Italie: experiences, problèmes et évolutions possibles. In: Rousset-Rouard Y, Savignac JC (eds) L'avenir de la truffe face au réchauffement climatique: la truffe européenne est-elle en danger? Actes des 2e rencontres internationales de la truffe. Albin Michel, Paris, pp 65–76

Guérin-Laguette A, Renowden G (2008) Réchauffement climatique et trufficulture en Nouvelle-Zélande. In: Rousset-Rouard Y, Savignac JC (eds) L'avenir de la truffe face au réchauffement climatique: la truffe européenne est-elle en danger? Actes des 2e rencontres internationales de la truffe. Albin Michel, Paris, pp 113–116

Hall IR, Brown GT, Zambonelli A (2007) Taming the truffle. The history, Lore, and Science of the ultimate mushroom. Timber Press, Portland

Ineichen K, Wiemken V, Wiemken A (1995) Shoots, roots and ectomycorrhiza formation of pine seedlings at elevated atmospheric carbon dioxide. Plant Cell Environ 18(6):703–707. doi:10.1111/j.1365-3040.1995.tb00572.x

Jeandroz S, Murat C, Wang Y, Bonfante P, Le Tacon F (2008) Molecular phylogeny and historical biogeography of the genus *Tuber*, the 'true truffles'. J Biogeogr 35(5):815–829. doi:10.1111/j.1365-2699.2007.01851.x

Kauserud H, Heegaard E, Büntgen U, Halvorsen R, Egli S, Senn-Irlet B, Krisai-Greilhuber I, Dämon W, Sparks T, Nordén J, Høilanda K, Kirk P, Semenov M, Boddy L, Stenseth NC (2012) Warming-induced shift in European mushroom fruiting phenology. Proc Natl Acad Sci USA 109(36):14488–14493. doi:10.1073/pnas.1200789109

Kinoshita A, Sasaki H, Nara K (2011) Phylogeny and diversity of Japanese truffles (*Tuber* spp.) inferred from sequences of four nuclear loci. Mycologia 103(4):779–794. doi:10.3852/10-138

Krebs CJ, Carrier P, Boutin S, Boonstra R, Hofer E (2008) Mushroom crops in relation to weather in the southwestern Yukon. Botany 86(12):1497–1502. doi:10.1139/B08-094

Le Tacon F, Delmas J, Gleyze R, Bouchard D (1982) Influence du régime hydrique du sol et la fertilisation sur la fructification de la truffe noire du Périgord (*Tuber melanosporum* Vitt.) dans le Sud-Est de la France. Acta Oecol-Oecol Appl 3(4):291–306

Le Tacon F, Marçais B, Courvoisier M, Murat C, Montpied P, Becker M (2014) Climatic variations explain annual fluctuations in French 'Périgord black truffle' wholesale markets but do not explain the decrease in 'black truffle' production over the last 48 years. Mycorrhiza 24(Suppl 1):S115–S125. doi:10.1007/s00572-014-0568-5

Malajczuk N (2008) Réflexion sur le changement climatique et son impact sur la culture australienne. In: Rousset-Rouard Y, Savignac JC (eds) L'avenir de la truffe face au réchauffement climatique: la truffe européenne est-elle en danger? Actes des 2e rencontres internationales de la truffe. Albin Michel, Paris, pp 128–130

Morison JML (1985) Sensitivity of stomata and water use efficiency to high CO_2. Plant Cell Environ 8(6):467–474. doi:10.1111/j.1365-3040.1985.tb01682.x

Murat C, Diez J, Luis P, Delaruelle C, Dupré C, Chevalier G, Bonfante P, Martin F (2004) Polymorphism at the ribosomal DNA ITS and its relation to postglacial re-colonization routes of the Perigord truffle *Tuber melanosporum*. New Phytol 164(2):401–411. doi:10.1111/j.1469-8137.2004.01189.x

Nehls U, Hampp R (2000) Carbon allocation in ectomycorrhizas. Physiol Mol Plant Pathol 57 (3):95–100. doi:10.1006/pmpp.2000.0285

Norby RJ, Luo Y (2004) Evaluating ecosystem responses to rising atmospheric CO_2 and global warming in a multi-factor world. New Phytol 162(2):81–293. doi:10.1111/j.1469-8137.2004.01047.x

O'Dell TE, Ammirati JF, Schreiner EG (1999) Species richness and abundance of ectomycorrhizal basidiomycete sporocarps on a moisture gradient in the *Tsuga heterophylla* zone. Can J Bot 77 (12):1699–1711. doi:10.1139/b99-144

Petit RJ, Hampe A, Cheddadi R (2005) Climate changes and tree phylogeography in the Mediterranean. Taxon 54(4):877–885. doi:10.2307/25065474

Rouhier H, Read DJ (1998) Plant and fungal responses to elevated atmospheric carbon dioxide in mycorrhizal seedlings of *Pinus sylvestris*. Environ Exp Bot 40(3):237–246. doi:10.1016/S0098-8472(98)00039-2

Rousset-Rouard Y (2008) L'avenir de la truffe face au réchauffement climatique: la truffe européenne est-elle en danger? Actes des 2e rencontres internationales de la truffe. Albin Michel, Paris

Salerni E, Perini C, Gardin L (2014) Linking climate variables with Tuber borchii sporocarps production. Nat Resour 5(8):408–418. doi:10.4236/nr.2014.58038

Splivallo R, Rengenier R, Valdez N, Chevalier G, Molinier V, Wipf D, Karlovsky P (2012) Is climate change altering the geographic distribution of truffles? Front Ecol Environ 10 (9):461–462. doi:10.1890/12.WB.020

Stobbe U, Egli S, Tegel W, Peter M, Sproll L, Büntgen U (2013) Potential and limitations of Burgundy truffle cultivation. Appl Microbiol Biotechnol 97(12):5215–5224. doi:10.1007/s00253-013-4956-0

Treseder KK, Allen MF (2000) Mycorrhizal fungi have a potential role in soil carbon storage under elevated CO_2 and nitrogen deposition. New Phytol 147(1):189–200. doi:10.1046/j.1469-8137.2000.00690.x

Wang YJ, Tan ZM, Zhang DC, Murat C, Jeandroz S, Le Tacon F (2006) Phylogenetic and populational study of the *Tuber indicum* complex. Mycol Res 110(9):1034–1045. doi:10.1016/j.mycres.2006.06.013

Weden C, Chevalier G, Danell E (2004) *Tuber aestivum* (syn. *T. uncinatum*) biotopes and their history on Gotland, Sweden. Mycol Res 108(3):304–310. doi:10.1017/S0953756204009256

Zambonelli A, Iotti M, Piattoni F (2012) Chinese *Tuber aestivum sensu lato* in Europe. Open Mycol J 6:22–26. doi:1874-4370/12

Zampieri E, Balestrini R, Kohler A, Abbà S, Martin F, Bonfante P (2011) The Perigord black truffle responds to cold temperature with an extensive reprogramming of its transcriptional activity. Fungal Genet Biol 48(6):585–591. doi:10.1016/j.fgb.2010.09.007

Zhang L, Yang ZL, Song DS (2005) A phylogenetic study of commercial Chinese truffles and their allies: taxonomic implications. FEMS Microbiol Lett 245(1):85–92. doi:10.1016/j.femsle.2005.02.028

Zhang JP, Liu PG, Chen J (2012) *Tuber sinoaestivum* sp. nov., an edible truffle from southwestern China. Mycotaxon 122(10):73–82. doi:10.5248/122.73

Chapter 11
Soil Characteristics of *Tuber melanosporum* Habitat

Benoît Jaillard, Daniel Oliach, Pierre Sourzat, and Carlos Colinas

11.1 Introduction

Tuber melanosporum Vittad. belongs to the genus *Tuber* that only includes mycorrhizal fungi living and fruiting underground within the soil environment. All their life cycle, as mycorrhizas, mycelia and fruiting bodies, occurs within the soil environment. *Tuber melanosporum* may leave the soil only for a short time if ascomata are eaten and digested by an animal, before their spores return again to the soil. The soil is also the medium where host trees find the water and nutrients necessary for their growth and where they establish the mycorrhizal partnership with fungi. The soil organisation; its physical, chemical and organic properties; and its chemical and biological activities are of utmost importance for *T. melanosporum*. Yet most soil data collected in the field about truffle ecology are related just to truffle fruiting, and little is known about truffle "living" before

B. Jaillard (✉)
Institut National de la Recherche Agronomique, Eco&Sols UMR 1222, 2 place Pierre Viala, 34060 Montpellier cedex 2, France
e-mail: benoit.jaillard@montpellier.inra.fr

D. Oliach
Centre Tecnològic Forestal de Catalunya, Ctra. Sant Llorenç de Morunys, km 2, 25280 Solsona, Spain

P. Sourzat
Station d'Expérimentation sur la Truffe, Lycée professionnel agricole, Lacoste, 46090 Le Montat, France

C. Colinas
Centre Tecnològic Forestal de Catalunya, Ctra. Sant Llorenç de Morunys, km 2, 25280 Solsona, Spain

Department of Crop and Forest Sciences, University of Lleida, Av. Rovira Roure 191, 25198 Lleida, Spain

A. Zambonelli et al. (eds.), *True Truffle (*Tuber *spp.) in the World*, Soil Biology 47,
DOI 10.1007/978-3-319-31436-5_11

any fruiting takes place. Only a successful mycelium growing comfortably in a friendly soil environment will produce the treasured black truffles, the "black diamond" of Mediterranean gastronomy.

11.2 Soil Location, Organisation and Stoniness

11.2.1 Soil Location in Landscape

Tuber melanosporum grows and fruits best in well-structured, porous and aerated soils (Delmas and Poitou 1973, 1974; Poitou 1988, 1990). The soil profile must allow for water to flow freely through all the soil horizons and excess water to drain out of the topsoil layers where tree roots and fungi live. *Tuber melanosporum* thrives in soils with low concentrations of mineral nutrients, especially phosphorus of which the availability is generally under 18 mg kg^{-1} (Sáez 1991; Sáez and De Miguel 1995). Soils favourable to *T. melanosporum* are consequently most often on summits, ridges or slopes where free water can run off suctioning fresh air in its wake (Fig. 11.1). This ecological requirement leaves *T. melanosporum* to grow on marginal lands occupied by shrubs and forests or abandoned agricultural areas previously planted with vineyards, olive or almond trees, the least demanding crops in terms of water and fertilisers. In France and Italy, truffle hunting was traditionally associated with sheep or goat herding, and truffles were thus harvested in rangelands and forests.

11.2.2 Soil Organisation and Depth

Highlands and marginal agricultural lands often have stony and shallow soils, sometimes lying directly on bedrock (Fig. 11.2). Soil depth is a very important parameter for the growth of many plant species because it determines the soil's water reserve. Even though productive truffle beds can be found in skeletal soils hardly 10 cm deep (Reyna 2000), the optimal soil depth for truffle production is around 30–50 cm to ensure a good growth of both the tree and the fungus. Moreover, bedrock must be highly and finely fractured to ensure drainage of excess water from the soil profile and to allow roots to dig deep into the parent material to get the water necessary for tree growth (Grente et al. 1974; Callot and Jaillard 1996; Callot 1999; Weiller 2002). Many wild truffle beds are located on stony substrates, on hillside gravels and on rock debris that allow very efficient water drainage. Such geological formations are frequent on limestones of the Jurassic and Cretaceous period, but also in mountainous regions as the French Central Massif, the Pyrenees and the Apennines (Raglione et al. 2001; Jaillard et al. 2014).

Fig. 11.1 Location of truffle orchards in the landscape: (**a**) planting of *green* oaks, *white* oaks and hazelnut trees on limestone terraces; (**b**) planting of *green* oaks on a limestone plateau

Regardless of the prospective truffle orchard location, it is always necessary to dig soil profiles that allow the direct observation of all the soil profile and the bedrock condition since they can vary locally from site to site (Callot 1999; Weiller 2000, 2002). Moreover, soil profiles give the opportunity to observe in detail possible discontinuities between soil horizons that can induce local water accumulation inside. Soil profiles also allow to observe lithological discontinuities at the profile bottom that can block the growth of taproots deep into the bedrock that would help the trees and their partner fungi survive droughts through hydraulic lift (Querejeta et al. 2007). Callot (1999) studied the relationship between truffle production and soil and subsoil organisation based on a morphological study of soils. This relationship is weak or non-existing in sandy or sandy loam soils that are very favourable for *T. melanosporum*. It becomes general in sandy-clay-loam or

Fig. 11.2 Soil organisation and depth of truffle orchards: (**a**) cambisol on very fractured lithographical Jurassic limestone; (**b**) rendosol on Cretaceous limestone

sandy-clay soils that are less favourable to *T. melanosporum* and is determinant for truffle production in clay-loam or clay soils. For instance, Callot and Jaillard (1996) showed that the sites where ascoma were harvested were weakly related to the subsoil structure. The presence of a partially impermeable clay subsoil discourages fruiting of both epigeous and hypogeous fungi. Soils presenting a shallow calcareous crust discourage fruiting of truffles but encourage that of epigeous fungi. These authors showed that *T. melanosporum* fruits best in soils with a very porous horizon developed in contact with the calcareous bedrock. This horizon was characterised by an intensive biological activity. In another truffle orchard developed on a clay-loam soil, Callot (1999) showed that truffles were exclusively harvested along an ancient drainage system that corrected the naturally impermeable soil.

The organisation of the soil and its underlying parent material can only be studied by digging soil profiles down to 2 m deep or to the bedrock when possible. The analysis of the soil organisation must be completed by observations of the plot of interest and its surroundings and the results cautiously extrapolated. The digging of soil profiles is an essential prerequisite when planting a new truffle orchard. It is the only way to observe in detail soil organisation, although it only opens a few small windows to the real soil variability. Digging of soil profiles seems expensive and time-consuming, but this investment is minimal compared to the cost of a failed truffle orchard.

11.2.3 Soil Stoniness

Soil stoniness plays several roles on soil properties. In general, it increases the soil porosity and improves water drainage within the soil profile. Soil stoniness also increases aeration and gas movement in soil. Gravels (2–5 mm) and little stones (5–30 mm) are more efficient than larger stones. In the same way, stoniness decreases the mechanical cohesion of soil and thus facilitates root growth and the expansion of truffle ascomata. However, very rocky soils produce more irregularly shaped truffles that fetch lower prices in the market. Stones can themselves contribute to water retention when the parent rock is porous, and limestones contribute to maintain high soil alkalinity, regardless of the lime content inside the soil matrix.

In rocky soil (Fig. 11.3), the fine particles tend to settle among the larger stones, leaving the stones on the soil surface (Delmas and Poitou 1974; Poitou 1990; Reyna 2000). This effect gives the impression that the stones "rise up" to the soil surface. Rock fragments can consequently constitute a continuous mulch on the soil surface. In dry climates, this stony mulch decreases significantly water evaporation by partially reflecting the radiation of the sun. Moreover, rock mulches reduce soil surface compaction caused by raindrops and soil surface erosion caused by water run-off. Finally stoniness favours biological activities in the soil under the rock mulch by maintaining this microhabitat shady, cool and moist for long periods of time.

In Central Italy, Bencivenga and collaborators reported an average of 50 % stoniness, with a large range from 10 to 90 % (Bencivenga 1986; Bencivenga and Granetti 1988; Bencivenga et al. 1990). In France, Delmas et al. (1981) and Poitou (1990) reported a range from 0 to 75 % of large elements. In Spain, Reyna (2000) indicated that soil stoniness varies from 0 to 90 %. It can be so high that stones completely cover the soil surface. We have observed in France regularly productive truffle beds in massive limestone rock debris, where truffles are harvested under the blocks (Jaillard et al. 2014). Suz et al. (2008) studied the production onset and the productivity in truffle orchards. These authors showed that early producing trees had significantly lower surface rock cover than other trees. In contrary, the surface rock cover in a mature producing orchard was much higher under productive than under unproductive trees. These effects of rock cover on tree productivity may be

Fig. 11.3 Soil stoniness of truffle orchards: (**a**) harvesting of truffles in a vineyard planted with *white* oaks; (**b**) harvesting of wild truffles under limestone rocks

related to its direct effect on soil or to physical or chemical interactions between rocks and soil. Note that soil stoniness is rarely measured in the field because it would take moving large soil volumes: this parameter is often estimated and the surfacing of rock fragments induces a general overestimation of soil stoniness.

The truffle fungus does not need rock fragments to grow, but it grows frequently on stony soils because of their location on the landscape and the induced good aeration of soil. For this reason many authors have associated stony soils with truffle soils, because *T. melanosporum* fruits often and well in stony soils (Sourzat 1997; Oliach et al. 2005).

11.3 Soil Texture and Structure

The texture of a soil describes the size distribution of mineral particles that form the soil. Its measurement therefore requires destroying the soil structure beforehand. Soil texture informs on how the soil can be structured in view of its history, mineral and organic amendments that have been added and cultural practices. The soil texture contributes to shape, but does not determine, the soil structure (Duchaufour 1965). However, soil texture is stable and easy to measure in the laboratory. It is why soil texture is so often measured, as opposed to soil structure. It is also why many authors try to estimate field properties of soil as soil structure by using empirical relationships (pedotransfer functions) based on soil texture (Duchaufour 1965; Alonso Ponce et al. 2014).

The structure of a soil describes the organisation of mineral and organic particles that form the soil: it is a fundamental soil property because it determines how the soil liquid and gas phases work, and it also influences considerably its chemical, organic and biological dynamics. The soils in which the particles are clustered into aggregates have granular, crumb or blocky structures. However, the structure can change with time, climate and water regime, incorporation or loss of mineral and organic matter, biological activity or cultural practices. It results from any soil component (clays, metal oxides and organics) or process (liming, biological activity, drying/wetting cycles) that contributes to aggregate soil particles. This is why the soil structure must be primarily observed in the field. Soil structure plays a major role on *T. melanosporum* growth and fruiting that needs to be emphasised.

11.3.1 Soil Texture

Balanced textures, those where soil is formed by roughly equal proportions of clay, silt and sand (Delmas and Poitou 1974), are the most appropriate for the growth and fruiting of *T. melanosporum*. Indeed, balanced textures are well known to naturally generate loamy soils that are well-structured with crumb or subangular blocky structure, porous, aerated and generally without excess of water. Raglione et al. (2001) also identified sand and silt as positive and clay as negative soil properties to discriminate soils favourable to *T. melanosporum*. Excess of clay or silt limits water and gas exchange, which is unfavourable to *T. melanosporum*. Soils whose clay content is above 300–400 g kg^{-1} tend to develop large, compact and continuous clumps that are much less porous and permeable to water, both soil properties unfavourable to *T. melanosporum*. Note that the clay effect depends on the mineralogy of the clay: clays with low exchange capacity as kaolinite are very unfavourable to truffle by increasing soil compaction and decreasing water permeability. The lack of rock fragments also aggravates the effects of high clay content. Sandy soils pose a different problem. Fine sands tend to become solid massive clumps when dry, thus preventing any growth of underground ascomata in size.

However, coarse sands reduce the cohesion among soil particles effectively decreasing aggregate size, and sandy structures remain very porous to air and water. Soils with high contents of coarse sands do not retain enough water: this coarse texture should be compensated by a high content of organic matter.

11.3.2 Soil Structure

The most critical physical soil property, for the growth and fruiting of *T. melanosporum*, is the structure (Fig. 11.4). The best structure for *T. melanosporum* allows good soil aeration and water flow and easy growth of tree roots and fungal mycelium (Delmas and Poitou 1973, 1974; Poitou 1988, 1990). Soil aggregation should be crumb or granular with the aggregates of the size of a grain of wheat (Delmas 1973a, b; Poitou 1988, 1990). This specific structure is often called *coffee grounds* structure because of the lightness, looseness and consistency of the soil in the hand (Callot 1999). Lulli et al. (1999) showed that soil aggregation was granular, and aggregate size was 1–2 mm near the soil surface. Ricard (2003) noted that aggregates have a size of 0.25–2 mm in the top 15 cm of soil. Sourzat (1997, 2002, 2012) recommends a regular surface tillage to keep the soil soft and well aggregated.

This specific soil structure is associated to a great stability over time. Bragato and collaborators (Panini et al. 1993; Bragato 1997; Lulli et al. 1999; Castrignanò et al. 2000; Bragato et al. 2001) have measured the soil structure at the scale of a truffle bed. The authors showed that the spatial pattern of the probability of finding a productive *brûlé* was related to the structure of the soil, suggesting that *T. melanosporum* may prefer a soft and well-aerated soil environment to grow and fruit (Lulli et al. 1999; Castrignanò et al. 2000). They also showed that, inside the *brûlé*, soil aggregate size decreased, and the conditions were more oxidative than outside: this pattern was related to a 50 % decrease of total organic matter and microbial biomass (Bragato et al. 2001). The authors assumed that the disappearance of the grass cover in the *brûlé* and the increase in soil macroporosity might increase airflow in soil surface layers (Bragato 1997). This would be confirmed by the increase in DTPA-extractable manganese observed in productive *brûlés* (Lulli et al. 1999).

In the Pyrenees, the sites identified by truffle growers as the most productive have a sandy or pseudo-sandy structure and a crumb aggregation with an organic origin or a subangular blocky aggregation generated by a significant clay or oxyhydroxide content (Jaillard et al. 2013, 2014). Even in dry conditions, these soils have a low cohesion and can be dug easily by hand or with a small pocket-knife. Jaillard et al. (2014) showed analytically that the most productive soils were those whose aggregates remain stable during fast immersion in water. On the contrary, the least productive soils were those whose aggregation is blocky, aggregate size larger than 4 mm and that collapse in water. Indeed, soils unstable in water had the annoying property to harden during drying and to collapse during wetting,

Fig. 11.4 Soil aggregation in truffle orchards: (**a**) calcareous soil raised up by a growing truffle; (**b**) organic sandy soil on dolomitic limestone (reproduced from Jaillard et al. 2013)

while stable and well-aggregated soils stay granular and thus soft and uncohesive regardless of the moisture conditions. Soil structure is the single most relevant physical property of soil favourable to *T. melanosporum* development.

11.3.3 Origin and Role of Soil Structure

Soil structure results from several independent soil properties: soil exchange capacity; contents in clay, lime and oxides; soil organic matter; biological activity; plant cover; and soil tillage. This multifactorial dependence brings about a large spatial and temporal variability of soil structure. Indeed, if the texture and the contents of carbonate or metal oxides of soil are properties that generally vary only slightly in space and slowly over time, content of organic matter, plant cover and biological activity varies quickly in space and changes rapidly over time. Several authors (Bragato 1997; Bragato et al. 2001; García-Montero et al. 2009a) have also showed that the truffle fungus itself alters the soil structure: the *brûlé* area becomes ashy and powdery, following a decrease in organic matter content and an increase in fine and reactive limestone content. Recently, Bragato (2014) studied the distribution and compared the size of aggregates within and outside a *brûlé* of *T. melanosporum* in the Italian Apennines. He showed that the distribution of aggregates larger than 0.25 mm displayed a spatial pattern comparable to that of the *brûlé*, with sharp changes along the boundaries of the *brûlé* itself. This author suggested that the disappearance of grasses in the *brûlé* allows freeze-thaw cycles to decrease aggregate size in winter.

The large spatial and temporal variability of soil structure makes it a property that deserves to be regularly measured in time or at least estimated by hand in the field. Its measurement is not standardised. Several methods have been proposed (Emerson 1967; Oades and Waters 1991; Le Bissonnais 1996), and analytical laboratories generally do not offer this determination on their catalogue. However, the estimate of the structural stability of a soil can be easily made by any truffle farmer, because it is simple and requires very little equipment: a 5 mm screen and a glass of water. Aggregates are isolated by sieving dry soil and then immersed for a few minutes in water. If the aggregate remains intact, the soil structure is very stable in water. If the aggregates are broken but are still visible to the naked eye, the soil structure is quite stable in water. If the aggregates collapse, the soil structure is very unstable and soil surely unsuitable for the production of black truffle. Note that, as surprising as it may be, we still do not know today exactly why well-structured soils favour *T. melanosporum*.

Soil microfauna is very active in truffle-dominated soil ecosystems and contribute significantly to the development of soil structure. In particular, protozoa and small soil invertebrates (insects, mites, myriapods and annelids) are essential to ensure aeration of the soil, activation and dispersal of the ascospores and truffle development (Morcillo et al. 2015). Arthropods, together with earthworms, help to transform fresh organic matter and plant debris that drops on the soil surface into small faecal pellets. The excrements of all these animals form aggregates that improve soil structure and aeration. The organic matter that transits through the gut of these small animals, together with mineral particles, makes the nutrient elements more accessible to plants and mycorrhizal fungi (Olivier et al. 2002). Earthworms play an important role by incorporating exchangeable calcium, nitrate and available phosphorus and potassium, thus improving the chemical properties of the soil (Olivier et al. 2002). These annelids increase soil calcium carbonate by means of their calciferous glands (García-Montero et al. 2013), effectively fixing carbon in the soil from fresh organic matter or atmospheric CO_2 and raising soil pH (Canti 2009; García-Montero et al. 2013). These actions by earthworms have a similar impact to calcareous amendments, as they modify the availability of nutrients, alter ectomycorrhizal communities and improve the growth of roots (Monfort-Salvador et al. 2015).

11.4 Soil pH and Alkalinity

11.4.1 How to Measure Soil Alkalinity?

The alkalinity of a solution describes its ability to neutralise an acid. The inverse of alkalinity is the acidity. By extension, the alkalinity of a soil describes its ability to neutralise acids. This quantity thus determines how soils react to an acid intake. Every living organism has specific needs in terms of alkalinity/acidity of its environment. The pH of a soil measured in an aqueous solution (pH_{water}) measures the alkalinity of the soil solution at equilibrium. As reference, pH_{water} of soils is generally between 4.2 and 8.4. But soil pH_{water} can shift quickly and significantly with water or gas dynamics and biological activity (Hinsinger et al. 2002; Jaillard et al. 2003). To fill this gap, we often measure the pH_{KCl} (pH measured in an aqueous 1 M KCl solution) which takes more into account the alkalinity carried by the soil exchange complex: pH_{KCl} is always lower than pH_{water}.

Soil “active carbonate” is extracted with ammonium oxalate according to the Drouineau method (Drouineau 1942). The measurement of “active carbonate” takes into account the alkalinity of highly reactive carbonate minerals that mainly corresponds to the fine calcareous fraction of soil (Callot and Dupuis 1980). The Metson cation exchange capacity (CEC, generally in $cmol_+ \, kg^{-1}$ of soil) of the soil exchange complex is measured by extraction of cations, mainly calcium (Ca) and magnesium (Mg) in alkaline soils, with ammonium acetate (Metson 1956). In alkaline soils, this extractant can also dissolve an active fraction of carbonate minerals (Ciesielski and Sterckeman 1997). The ratio (Mg + Ca)/CEC (in $mol_+ \, mol_+^{-1}$) gives thus a good indicator of “reactive” alkalinity of a soil. The “reactive” alkalinity of a soil is much higher than the “equilibrium” alkalinity of soil solution: often, it much better describes the alkaline soil conditions that living organisms need over an extended period for growing and fruiting.

11.4.2 Soil pH

Tuber melanosporum grows and fruits best in alkaline soils (Delmas and Poitou 1973, 1974; Poitou 1988, 1990). The pH of productive truffle beds ranges from 7.0 to 8.9, with a median of 7.9 and a standard deviation near 0.4 units: that means that the pH of 95 % of truffle soils ranges between 7.5 and 8.3 (Bencivenga and Granetti 1988; Bencivenga et al. 1990; Raglione et al. 1992; Sáez and De Miguel 1995; Olivier et al. 1996; García-Montero et al. 2007b; Colinas et al. 2007). So, the soils favourable to *T. melanosporum* include neutral (7.0 < pH < 7.9), dolomitic (7.5 < pH < 8.3) and calcareous (8.0 < pH < 8.9) soils. Soil pH is most often used to characterise soil alkalinity because it is easy and inexpensive to measure. However, pH has the disadvantage of an asymptotic increase in the alkaline range of interest: in calcareous soils, pH is close to 8.3, that is, the pH_{water} value of a solution in equilibrium with lime. That is also the maximum soil pH adequate for *T. melanosporum* to grow. A higher soil pH indicates significant amounts of sodium or gypsum, both minerals detrimental to the black truffle fungus.

11.4.3 Exchange Complex

Alkaline soils are characterised by an exchange complex saturated by divalent Ca and Mg cations. The ratio between the sum of divalent cations and cation exchange capacity, i.e. (Mg + Ca)/CEC, has the advantage of a gradual variation from acid soils whose exchange complex is not or just barely saturated [(Mg + Ca)/CEC ≤ 1 $mol_+\ mol_+^{-1}$], up to highly calcareous soils that contain high amount of reactive limestone [(Mg + Ca)/CEC > 1 $mol_+\ mol_+^{-1}$] (Ciesielski and Sterckeman 1997). Dolomitic soils fit easily in this scale because their exchange complex is saturated and may contain a fraction of magnesium or calcium that is soluble and reactive. In addition (Mg + Ca)/CEC ratio avoids the nature of the soil minerals that support the alkalinity and takes into account the difference of cation exchange capacity between soils. Finally, exchangeable cations and cation exchange capacity are also commonly and inexpensively measured in routine soil analysis. Jaillard et al. (2013, 2014) have showed that (Mg + Ca)/CEC is likely the most pertinent indicator for alkalinity evaluation in truffle ecology: it ranges from 0.9 to 1.2 $mol_+\ mol_+^{-1}$ in neutral soils, from 1.4 to 1.9 $mol_+\ mol_+^{-1}$ in dolomitic soils and from 2.4 to more than 4 $mol_+\ mol_+^{-1}$ in calcareous or sandstone soils.

11.4.4 Soil Reactive Limestone

A soil exchange complex saturated by divalent cations generally results from the presence of calcium carbonate crystallised as calcite $CaCO_3$ or dolomite $CaMg(CO_3)_2$ in the soil mineral fraction. Indeed, this fraction is slowly dissolved releasing alkalinity as bicarbonate anion and associated calcium and magnesium cations. Several authors (Ourzik 1999; García-Montero et al. 2007b; Jaillard et al. 2014) have reported that the soils favourable to black truffle contained rather low amounts of active carbonate, on the order of only a few tens of grammes per kg of soil with an average of 30 g kg^{-1} of soil. Sourzat (2001) and Jaillard et al. (2008) found truffles grown on acidic soils (on granite or gneiss) whose upper horizons were carbonated by human practices (the lime of masonry walls, the limestone of road embankments, etc.) and had less than 20 g active carbonate kg^{-1} of soil. Grente et al. (1974), Callot and Jaillard (1996), Callot (1999) and Weiller (2000, 2002) have noted that the most productive truffle beds were on leached soils or at least on soils that had lost carbonate. These soils are brown to red, because of the mobilisation of iron or manganese (a process called "brunification", possibly followed by a "reddening" in Mediterranean climate) (FAO 2006). These data led some authors (Jaillard et al. 2007, 2008; García-Montero et al. 2007a, 2009a) to assume that the presence of active carbonate is critical for truffle production, even if it is in small amounts. Authors as Callot (1999), García-Montero et al. (2009a, b) and Demerson (2012) have even suggested that active carbonate could play a direct role in the nutrition of mycelium or fruiting of *T. melanosporum* ascomata. However, Sourzat (2008, 2012) reported *T. melanosporum* growing and fruiting in neutral soils with null carbonate content, and Jaillard et al. (2014) observed wild truffle beds on soils whose exchange complex was just close to saturation, with no carbonate in them. These observations clearly contradict the hypothesis of a possible role played by active carbonate on truffle nutrition. The data only are in good agreement with the mean pH value that indicates that *T. melanosporum* grows and fruits as well in neutral and dolomitic soils as in calcareous soils.

11.4.5 Soil Total Limestone

Soil content in total carbonate is, with pH in water, the most commonly measured soil property for determining truffle cropping potentiality. Indeed, black truffle is traditionally associated with the characteristic landscapes of Mediterranean limestone regions, with limestone hillsides in Southern France, Eastern Spain and Italy. However, soil content in total carbonate is not very informative. The lack of carbonate in a given soil does not imply that this soil will not produce *T. melanosporum*. The presence of carbonate in a given soil indicates only that soil exchange complex is very likely saturated by divalent cations and therefore that (Mg + Ca)/CEC ratio is higher than one. The capacity of a site to produce black truffles depends firstly on its location in the landscape and on the soil structure. Interestingly, Jaillard et al. (2014) analysed 220 truffle grounds in the Pyrenean regions colonised by *T. melanosporum* and reported that very calcareous soils may not be the best ones for truffle production, mostly when compared to dolomitic soils with an active carbonate content of only 4 g kg^{-1} on average. Very calcareous soils often result from the weathering of marls and marly limestones, rich in clay and loam, with an impermeable hardpan (petrocalcic horizon) in the lower soil horizons. This result reinforces the observations reported by García-Montero et al. (2007a, b, 2009a, b) and Jaillard et al. (2007) that suggested that the best truffle orchards frequently had low active carbonate, most often associated to low carbonate content. Valverde-Asenjo et al. (2009) also analysed 77 truffle beds colonised by *T. melanosporum* and *Tuber brumale* Vittad. in Aragón (Northeast Spain): they showed that the active carbonate content of *T. melanosporum* beds was lower in wild than in orchard soils, i.e. lower than 30 g kg^{-1} and higher than

100 g kg^{-1} in average, respectively. This suggests that truffle growers plant truffle trees in soils even more calcareous than wild truffle soils. This belief in the quality of soils with high carbonate content for truffle cultivation could cause a bias in the analysis of truffle soils.

11.4.6 Dolomitic Limestone and Calcareous Sandstone Soils

Special attention should be given to soils developed on dolomitic limestones and calcareous sandstones. Most limestones are composed of calcium carbonate $CaCO_3$ crystallised as calcite. Dolomite is a double carbonate of calcium and magnesium $CaMg(CO_3)_2$. Dolomitic limestones result from secondary processing of calcitic limestones. When a magnesium-rich solution percolates through calcitic limestones, a part of the calcium of the calcite is substituted by magnesium, and calcite gradually turns into dolomite. Calcite is much more soluble than dolomite. As a consequence in the field, the soil that results from the weathering of dolomitic limestones is made of dolomite grains quite free from calcite cement. The soil is mostly a sandy soil, alkaline, generally free from active carbonate and reacting very slowly to the addition of acid unlike a calcareous soil. Its pH is always less than 8.3, usually ranging from 7.5 to 7.9.

A calcareous sandstone is made of siliceous sand cemented by calcium carbonates. Calcareous sandstones weather by dissolving the carbonate cement and gradually releasing the sand. The resulting soil is also a sandy soil, but which can have high active carbonate content. In these calcareous sandy soils, the pH varies from 7.5 to 8.3 depending on the content of residual active carbonate. Both dolomitic limestones and calcareous sandstones produce alkaline sandy soils. With a high content in organic matter, the dolomitic limestone soils and the calcareous sandstone soils are among the most favourable soils for *T. melanosporum* (Jaillard et al. 2014).

11.5 Soil Organic Matter

11.5.1 Role of Organic Matter in Soil

Organic matter is a minor but essential component of soil because it considerably alters many soil properties, especially soil structure. Its content ranges generally from 30 to 150 g kg^{-1} in untilled and undisturbed soils and from 5 to 50 g kg^{-1} in cultivated soils. Organic matter incorporated in soil is usually plant debris. These fresh organic materials are first physically fragmented by the mesofauna, then chemically digested by the microfauna. The resulting organic matter is linked to minerals and forms a very stable organo-mineral complex because of the hydrophobicity and molecular structure of organic polymers (Buckman and Brady 1991).

Organic matter improves significantly the soil structure by increasing soil porosity, water-holding capacity and structure stability of soil to water. Organic matter also increases exchange capacity that, in turn, increases the soil capacity to retain mineral elements. Adding organic matter increases the soil CEC and, consequently, decreases the (Mg + Ca)/CEC ratio. As a consequence, organic matter is the main factor of soil structuring.

However, organic matter is labile: it is mineralised with time, mainly in the warm Mediterranean regions. Content of organic matter decreases with time, releasing carbon (C) as carbonic gas in the atmosphere and nitrogen (N) and phosphorus as available nutrients in soils. Plant debris have high values of C/N ratio, from 20 to 50 kg kg^{-1} and more. The mineralisation of plant debris is always associated with a decrease in C/N ratio, until reaching low values close to 8 or 10 kg kg^{-1} that correspond to relative C and N contents of microorganisms. The C/N ratio of soil organic matter thus provides useful information on biological activity in soil. Organic matter in a biologically very active soil generally has a low C/N ratio except if fresh organic matter has been recently added. But fresh organic matter in such a soil will be mineralised quickly returning the soil to a low C/N ratio in a short time. In alkaline soil, organic matter evolves towards calcic mull humus form and soil structure towards a crumb structure. Because of its lability, fresh organic matter must be regularly added to the soil as well composted plant residues, animal faeces or wood, in order to keep C/N ratio high and maintain its positive impact on other soil properties and functions (Callot 1999; Reyna and Colinas 2012).

11.5.2 Soil Organic Matter Favours T. melanosporum

The contents of organic matter in soils where the black truffle thrives range from 8 to 180 g kg^{-1}, with a mean near 50 g kg^{-1} and a reported maximum value above 370 g kg^{-1} (Delmas et al. 1981; Bencivenga and Granetti 1988). These are high values of organic matter content in soil. By comparing truffle productive and non-productive areas, Bragato (1997) showed that in truffle-producing areas, the content of total organic matter was lower, and the relative content of linked organic matter was higher than in non-productive areas. Callot and Salducci (1997, 1998) confirmed this result and suggested that high amounts of organic matter with high C/N ratio favour *T. brumale* rather than *T. melanosporum*. It is thus important to specify the quality of soil organic matter, especially the ratios of fresh organic matter to resulting stable organic matter. Poitou et al. (1983) and Delmas et al. (1981) reported that soils favourable to black truffle contained well-mineralised and stabilised organic matter, with a C/N around 10 kg kg^{-1} and ranging from 8 to 12 kg kg^{-1}. Callot (1999) and Reyna (2000) widen the range from 8 to 20 kg kg^{-1}, depending on the natural intake of fresh organic matter. Ricard (2003) reports that truffle beds contain generally low amounts of free organic matter (size > 20 μm), but can be rich in organic matter bound to mineral particles (size < 20 μm).

11.6 How to Choose and Improve the Soil for *T. melanosporum*?

11.6.1 Location and Physical Soil Properties Cannot Easily Be Changed

Tuber melanosporum is a thermophilic fungus of Mediterranean regions. Its edaphic requirements naturally confine it to calcareous or dolomitic mountains and calcareous sandstone plateaus of these regions. However, the development of truffle cultivation and planting of truffle orchards in the late twentieth century changed the distribution of this fungus. Former agricultural lands were often converted to truffle orchards, and many people that have inherited farms wish to convert them to truffle orchards. This new infatuation for truffle cultivation induces the plantation of many truffle orchards in areas that are not naturally favourable to truffle production. Traditionally preferred crop fields are often in flat plains convenient to use farming equipment on. They are flat because of the fine sediments that were deposited in previous eras. But those same sediments also make their clay content too high. The development of truffle cultivation raises the question of the choice of soil and its improvement through appropriate practices. We do not know how many of the 2500–3000 ha of truffle orchards planted every year in Europe are placed in inadequate sites. So far, even though the decline in production has stopped, we are still very far from the truffle crops of 100 years ago.

We have seen that the physical properties of the soil are essential. All plains, depressions and slope failures accumulate water permanently or temporarily on all or part of the soil profile: these areas are unfavourable to *T. melanosporum* and must be avoided. Moreover, accumulation of sediments in these areas generates rather deep and heavy soils with fine texture that contribute to hold water and humidity for a long time. Soils too clayey or too loamy should be avoided as they are too sensitive to excess of water, collapse and compaction (Sáez and De Miguel 1995; Olivier et al. 1996; Reyna 2000). Very sandy soils have poor water-holding capacity, but this problem can be corrected by adding clay and organic matter, and it is not a problem if the farm is located in an area of high precipitation or that can be irrigated. Generally, adding organic matter improves the physical properties of soil, but these improvements are limited to the soil surface. We must recall here the recommendation to previously observe the organisation of the soil and its underlying parent material by digging soil profiles down to two metres deep or to the bedrock when possible. The digging of soil profiles is not expensive by comparison with the investment of planting a truffle orchard that is expected to produce truffles over several decades. The site, its location in the landscape and the structure of its subsoil are paramount for the plantation of a black truffle orchard because these properties cannot be modified by any agricultural practices.

11.6.2 Other Soil Properties Can Be Improved

On the other hand, most other physical and chemical soil properties can be improved over time by adapted agricultural practices (Fig. 11.5). Soil structure can easily be modified (Le Bissonnais 1996). Soil tillage is the main and oldest practice used in truffle cultivation: it is recommended by most authors as a basic practice in truffle cultivation to keep the soil surface soft and well structured (Sourzat 1997, 2002, 2012). Adding coarse sands or rock fragments is also used to decrease the soil cohesion and improve the soil fragmentation in loamy soils. Liming is another common way to improve soil structure, but it is effective only in desaturated neutral soils. Finally the most commonly used practice is the addition of well-decomposed organic matter that can tremendously increase the structure stability of soils (Le Bissonnais 1996; Lulli et al. 1999). This culture practice is well known in horticulture, and it is only recently becoming common in truffle cultivation (Reyna and Colinas 2012). It should be encouraged to keep high the stability of soil structure in truffle beds.

Soil alkalinity can also be easily and successfully managed (Valverde-Asenjo et al. 2009) in the field: it increases with the addition of lime or calcium or magnesium carbonate and may decrease slightly with the addition of clay or carbon-rich organic matter. Moreover, the observation of truffle orchards in neutral soils implies that acidic soils with good physical properties can become productive truffle orchards by natural or artificial liming. The development of truffle culture in Australia, New Zealand, Chile and the USA (Sourzat 2012) has been largely based on this principle. For instance, Garland (1995, 1996) reported in North Carolina that soils with initial pH near 5.5 had been progressively limed over 10 years until their pH reached 8, and the site became a productive truffle orchard. Similar results are reported by Hall and Yun (2003) in New Zealand. We have observed in the Pyrenean mountains that some truffle growers are also liming their plantations to rise the truffle soil alkalinity.

11.7 Conclusions

Tuber melanosporum ultimately appears as a relatively ubiquitous fungus able to grow in a wide range of physical and chemical conditions of soils whose optimum is difficult to identify. This difficulty is enhanced by the meteorological (Alonso Ponce et al. 2010, 2014) and biological conditions that are favourable to the fungus and that we are only beginning to understand (see Chap. 10). The situation becomes even more confusing as our technology gives us the capacity to dramatically and

Fig. 11.5 Soil improvement in truffle orchards: (**a**) soil samples collected in a cultivated field (on the *left*) and a wild truffle woodland (on the *right*), both being a few meters apart (reproduced from Jaillard et al. 2014); (**b**) tillage of the soil surface by hand with a *bigot*

persistently change many soil properties. A soil favourable to *T. melanosporum* must be well-drained, well-structured and well-aggregated, alkaline and rather organic, which are all common soil properties, but that rarely occur together in the same site, which is the reason why soils naturally favourable to *T. melanosporum* are not very common.

References

Alonso Ponce R, Águeda B, Ágreda T, Modrego MP, Aldea J, Martínez-Peña F (2010) Un modelo de potencialidad climática para la trufa negra (*Tuber melanosporum*) en Teruel (España). For Syst 19(2):208–220

Alonso Ponce R, Ágreda T, Águeda B, Aldea J, Martínez-Peña F, Modrego MP (2014) Soil physical properties drive black truffle fructification in plantations. Mycorrhiza 24(Suppl 1): S55–S64. doi:10.1007/s00572-014-0558-7

Bencivenga M (1986) Le tartufaie in alcuni rimboschimenti dell'Alta Valle del Liri. Ann Fac Agrar Univ Stud Perugia 40:231–258

Bencivenga M, Granetti B (1988) Ricerca comparativa sulle esigenze ecologiche di *Tuber magnatum* Pico e *Tuber melanosporum* Vitt. dell'Italia centrale. Ann Fac Agrar Univ Stud Perugia 42:861–872

Bencivenga M, Calandra R, Granetti B (1990) Ricerche sui terreni e sulla flora delle tartufaie naturali di *Tuber melanosporum* Vitt. dell'Italia Centrale. In: Bencivenga M, Granetti B (eds) Proceedings of the 2nd international congress on truffle, Spoleto, 24–27 Nov 1988. Comunità Montana dei Monti Martani e del Serano, Spoleto, pp 337–374

Bragato G (1997) Modificazioni indotte dal tartufo nero pregiato (*Tuber melanosporum* Vitt.) sulla struttura e sulla sostanza organica del suolo. Monti Boschi 47:23–27

Bragato G (2014) Is truffle *brûlé* a case of perturbational niche construction? For Syst 23 (2):349–356. doi:10.5424/fs/2014232-04925

Bragato G, Lulli L, Castrignanò A, Bencivenga M (2001) Regionalization analysis of some soil factors realted to *Tuber melanosporum* production in an experimental truffle bed. In: Actes du Ve congrès international, Science et culture de la truffe et des autres champignons hypogés comestibles, Aix-en-Provence, 4–6 Mar 1999, Fédération Française des Trufficulteurs, Paris, pp 5.253–5.256

Buckman HY, Brady N (1991) Naturaleza y propiedades de los suelos. LIMUSA, Barcelona

Callot G (1999) La truffe, la terre, la vie. INRA, Paris

Callot G, Dupuis M (1980) Le calcaire actif des sols et sa signification. Sci Sol 1:17–26

Callot G, Jaillard B (1996) Incidences des caractéristiques structurales du sol sur l'entrée en production de *Tuber melanosporum* et d'autres champignons mycorhiziens. Agronomie 16 (7):405–419

Callot G, Salducci X (1997) Caractérisation microbiologique et organique de différents systèmes truffiers. Etude comparée de *Tuber brumale* et *Tuber melanosporum*. Rapport d'activité, Action Truffe, DADP. INRA, Paris

Callot G, Salducci X (1998) La truffe, une histoire de terre. Le Trufficulteur 22:11–14

Canti MG (2009) Experiments on the origin of ^{13}C in the calcium carbonate granules produced by the earthworm *Lumbricus terrestris*. Soil Biol Biochem 41(12):2588–2592. doi:10.1016/j.soilbio.2009.09.007

Castrignanò A, Goovaerts P, Lulli L, Bragato G (2000) A geostatistical approach to estimate probability of occurrence of *Tuber melanosporum* in relation to some soil properties. Geoderma 98(3):95–113. doi:10.1016/S0016-7061(00)00054-9

Ciesielski H, Sterckeman T (1997) A comparison between three methods for the determination of cation exchange capacity and exchangeable cations in soils. Agronomie 17(1):9–16

Colinas C, Capdevila JM, Oliach D, Fischer C, Bonet JA (2007) Mapa de aptitud para el cultivo de trufa negra (*Tuber melanosporum* Vitt.) en Cataluña. Centre Tecnològic Forestal de Catalunya, Solsona

Delmas J (1973a) La truffe: son écologie. Agriculture 363:82–86

Delmas J (1973b) Caractères écologiques de la trufficulture. Bull Tech d'Inf 283:721–731

Delmas J, Poitou N (1973) Contribution à la connaissance de l'écologie de *Tuber melanosporum*: la truffe du Périgord. C R Scéance Acad Agric Fr 59:1486–1494

Delmas J, Poitou N (1974) La truffe et ses exigences écologiques. Pépiniériste Hort Maraichers 144:33–39

Delmas J, Brian C, Delpech P, Soyer JP (1981) Application de l'analyse en composantes principales à une tentative de caractérisation physico-chimique des sols trufficoles français. Mushroom Sci 11:855–867

Demerson J (2012) CO_2, carbonate et calcaire actif dans le sol des truffières. Le Trufficulteur 81:13–14

Drouineau J (1942) Dosage rapide du calcaire actif des sols. Ann Agron 12:441–450

Duchaufour P (1965) Précis de Pédologie. Masson, Paris

Emerson WW (1967) A classification of soil aggregates based on their coherence in water. Aust J Soil Res 5(1):47–57

FAO (2006) World reference base for soil resources: a framework for international classification, correlation and communication. FAO, Rome

García-Montero LG, Casermeiro MA, Hernando I, Hernando J (2007a) Effect of active carbonate, exchangeable calcium, and stoniness of soil on *Tuber melanosporum* carpophore production. N Z J Crop Hort 35(1):139–146. doi:10.1080/01140670709510178

García-Montero LG, Manjon JL, Pascual C, García-Abril A (2007b) Ecological patterns of *Tuber melanosporum* and different *Quercus* mediterranean forest: quantitative production of truffles, burn sizes and soil studies. For Ecol Manage 242(2–3):288–296. doi:10.1016/j.foreco.2007.01.045

García-Montero LG, Quintana A, Valverde-Asenjo I, Díaz P (2009a) Calcareous amendments in truffle culture: a soil nutrition hypothesis. Soil Biol Biochem 41(6):1227–1232. doi:10.1016/j.soilbio.2009.03.003

García-Montero LG, Valverde-Asenjo I, Díaz P, Pascual C (2009b) Statistical patterns of carbonates and total organic carbon on soils of *Tuber rufum* and *T. melanosporum* (black truffle) *brûlé s*. Aust J Soil Res 47(2):206–212. doi:10.1071/SR08084

García-Montero LG, Valverde-Asenjo I, Grande-Ortíz MA, Menta C, Hernando I (2013) Impact of earthworm casts on soil pH and calcium carbonate in black truffle burns. Agrofor Syst 87 (4):815–826. doi:10.1007/s10457-013-9598-9

Garland F (1995) Cultivating the fabled black diamond (truffle). McIlvainea 1291:42–45

Garland F (1996) Truffle cultivation in North America. Garland Gourmet Mushrooms and Truffles, Hillsborough, NC

Grente J, Delmas J, Poitou N, Chevalier G (1974) Faits nouveaux sur la truffe. Mushroom Sci 9:815–846

Hall IR, Yun W (2003) Culture de la truffe noire du Périgord en Nouvelle-Zélande sur sols naturellement acides, amendés. In: Résumés de la Journée nationale de la trufficulture, Cuzance, 28 Mar 2003. CTIFL, Paris, pp 3–8

Hinsinger P, Plassard C, Tang C, Jaillard B (2002) Origins of root-mediated pH changes in the rhizosphere and their responses to environmental constraints: a review. Plant Soil 248 (1):43–59. doi:10.1023/A:1022371130939

Jaillard B, Plassard C, Hinsinger P (2003) Measurements of H^+ fluxes and concentrations in the rhizosphere. In: Rengel Z (ed) Handbook of soil acidity. Marcel Dekker, New York, pp 231–266

Jaillard B, Barry-Etienne D, Diette S, Moundy PJ, Plassard C, Salducci X (2007) Carbonates, matières organiques et pH des sols truffiers. In: World fungi 2007, Abstract of the 1st international conference on the conservation and sustainable use of wild fungi, Cordoba, 10–16 Dec 2007. Junta de Andalucia, Consejeria de Medio Ambiente, pp 75–76

Jaillard B, Barry-Etienne D, Diette S, Moundy PJ, Plassard C, Salducci X (2008) Relation entre carbonates, matière organique et pH des sols truffiers. Le Trufficulteur 64:8–10

Jaillard B, Barry-Etienne D, Colinas C, De Miguel AM, Genola L, Libre A, Neveu P, Oliach D, Saenz W, Sáez R, Salducci X, Souche G, Sourzat P, Villeneuve M (2013) Alcalinité et structure des sols déterminent la production truffière en régions pyrénéennes. Le Trufficulteur 85:16–21

Jaillard B, Barry-Etienne D, Colinas C, De Miguel AM, Genola L, Libre A, Neveu P, Oliach D, Saenz W, Sáez R, Salducci X, Souche G, Sourzat P, Villeneuve M (2014) Alkalinity and structure of soils determine the truffle production in the Pyrenean regions. For Syst 23 (2):364–377. doi:10.5424/fs/2014232-04933

Le Bissonnais Y (1996) Aggregate stability and assessment of soil crustability and erodibility: I. Theory and methodology. Eur J Soil Sci 47:425–437

Lulli L, Bragato G, Gardin L (1999) Occurrence of *Tuber melanosporum* in relation to soil surface layer properties and soil differentiation. Plant Soil 214(1):85–92. doi:10.1023/A:1004602519974

Metson AJ (1956) Methods of chemical analysis for soil survey samples. N Z Dep Sci Ind Res Soil Bur Bull 12:208

Monfort-Salvador I, García-Montero LG, Grande MA (2015) Impact of calcium associated to calcareous amendments on ectomycorrhizae in forests: a review. J Soil Sci Plant Nutr 15 (1):217–231. doi:10.4067/S0718-95162015005000018

Morcillo M, Sánchez M, Vilanova X (2015) Cultivar trufas, una realidad en expansión. Micologia Forestal & Aplicada, Barcelona

Oades JM, Waters AG (1991) Aggregate hierarchy in soils. Aust J Soil Res 29(6):815–828. doi:10.1071/SR9910815

Oliach D, Bonet JA, Fischer CR, Olivera A, Martínez de Aragón J, Suz LM, Colinas C (2005) Guia para el cultivo de trufa negra (*Tuber melanosporum* Vitt.). Centre Tecnològic Forestal de Catalunya, Solsona

Olivier JM, Savignac JC, Sourzat P (1996) Truffe et trufficulture. Fanlac, Périgueux

Olivier JM, Savignac JC, Sourzat P (2002) Truffe et trufficulture. Fanlac, Périgueux

Ourzik A (1999) Les sols argilo-calcaires à potentiel truffier de la Vienne. In: Actes du Ve congrès international, Science et culture de la truffe et des autres champignons hypogés comestibles, Aix-en-Provence, 4–6 Mar 1999, Fédération Française des Trufficulteurs, Paris, pp 5.273–5.275

Panini T, Lulli L, Bragato G, Pagliai M, Primavera F (1993) Suoli e siti del *Tuber melanosporum* Vitt. sulla Scaglia rossa di Volperino (PG). Monti Boschi 44:28–33

Poitou N (1988) Les sols truffiers. Choix du sol: prélèvement, analyse, correction, oligo-éléments. In: Journées nationales de la truffe, St Paul Trois-Châteaux, p 2190

Poitou N (1990) Les Sols Truffiers Français. In: Bencivenga M, Granetti B (eds) Proceedings of the second international congress on truffle, Spoleto, 24–27 Nov 1988. Comunità Montana dei Monti Martani e del Serano, Spoleto, pp 391–401

Poitou N, Villenave P, Baudet D, Delmas J (1983) Croissance *in vitro* du mycélium de *Tuber melanosporum* Vitt. et de certains compétiteurs en fonction du pH du milieu. C R Séance Acad Agric Fr 69:1363–1370

Querejeta JI, Egerton-Warburton LM, Allen MF (2007) Hydraulic lift may buffer rhizosphere hyphae against the negative effects of severe soil drying in a California Oak savanna. Soil Biol Biochem 39(2):409–417. doi:10.1016/j.soilbio.2006.08.008

Raglione M, Lorenzoni P, De Simone C, Monaco R, Angius A (1992) Osservazioni sulle caratteristiche pedologiche di alcuni siti di tartufo nero pregiato (*Tuber melanosporum*) in provincia di Rieti. Micol Veg Medit 7(1):211–224

Raglione M, Spadoni M, Cavelli S, Lorenzoni P, De Simone C (2001) Les sols des truffières naturelles de *Tuber melanosporum* Vitt. dans l'Apennin Central (Italy). In: Actes du Ve congrès international, Science et culture de la truffe et des autres champignons hypogés comestibles, Aix-en-Provence, 4–6 Mar 1999, Fédération Française des Trufficulteurs, Paris, pp 5.276–5.280

Reyna S (2000) Trufa, truficultura y selvicultura trufera. Mundi-Prensa, Madrid

Reyna S, Colinas C (2012) Truficultura. In: Truficultura, fundamentos y técnicas, 2nd Ed. Mundi-Prensa, Madrid

Ricard JM (2003) La truffe. Guide technique de trufficulture. CTIFL, Paris

Sáez R (1991) Trufa y truficultura en Navarra. Navarra Agraria 67:5–11

Sáez R, De Miguel AM (1995) Guía práctica de truficultura. ITG Agrícola SA y Universidad de Navarra, Pamplona

Sourzat P (1997) Guide pratique de trufficulture. Station d'expérimentation sur la truffe, Lycée professionnel agricole et viticole de Cahors, Le Montat
Sourzat P (2001) Les limites des critères agronomiques dans l'analyse de terre en trufficulture. In: Actes du Ve congrès international, Science et culture de la truffe et des autres champignons hypogés comestibles, Aix-en-Provence, 4–6 Mar 1999, Fédération Française des Trufficulteurs, Paris, pp 5.281–5.286
Sourzat P (2002) Guide pratique de trufficulture. Station d'expérimentation sur la truffe. Lycée professionnel agricole et viticole de Cahors, Le Montat
Sourzat P (2008) Les sols truffiers: synthèse des actions menées au cours des 11e et 12e Contrats de Plan Etat-Région. Fédération française des Trufficulteurs, Paris
Sourzat P (2012) Petit guide de trufficulture. Station d'expérimentation sur la truffe. Lycée professionnel agricole et viticole de Cahors, Le Montat
Suz LM, Martín MP, Oliach D, Fischer CR, Colinas C (2008) Mycelial abundance and other factors related to truffle productivity in *Tuber melanosporum-Quercus ilex* orchards. FEMS Microbiol Lett 285(1):72–78. doi:10.1111/j.1574-6968.2008.01213.x
Valverde-Asenjo I, García-Montero LG, Quintana A, Velázquez J (2009) Calcareous amendments to soils to eradicate *Tuber brumale* from *T. melanosporum* cultivations: a multivariate statistical approach. Mycorrhiza 19(3):159–165. doi:10.1007/s00572-008-0224-z
Weiller B (2000) Les sols truffiers. Référentiel pédologique. Promé-Terre, St-Léon-sur-Vézère
Weiller B (2002) Votre terrain est-il propice à la culture de truffes? Guide d'aide à la décision de plantation d'arbres mycorhizés. URTA-CET, Oniflhor

Chapter 12
Soil Characteristics for *Tuber magnatum*

Gilberto Bragato and Žaklina S. Marjanović

12.1 Introduction

The aim of this chapter is to provide useful information about the habitats of *Tuber magnatum* Pico to be further investigated in future environmental research. Until recently, most investigations on *T. magnatum* environments were done in Italy, and their results were reported in Italian journals that are not easily accessible.

Since recently, ascomata were the only observable manifestation of *T. magnatum* in the field. Accordingly, all investigations on its habitats were based on two assumptions: (1) *T. magnatum* mycelial phase requires the presence of symbiont trees, and (2) *T. magnatum* produces ascomata only in the right soil environment. In this chapter, we took the presence of symbionts for granted and focused our attention on the subset of soil environments that were suitable for ascoma production.

Soil is a natural body formed by mineral and organic constituents organized in structures that are (1) specific for the pedological medium and (2) in constant evolution (FAO 1998). Soil is layered in horizons that are differentiated by environmental processes acting on the underlying parent material. In areas where tectonics is active and soil contains finely dispersed carbonates, the most frequent sequence of horizons is A-AC/Bw-C, where A indicates the surface mineral horizon, AC and Bw subsurface horizons originated in the early phases of soil genesis and C the unconsolidated parent material. The characteristics of A and,

G. Bragato (✉)
Council for Agricultural research and Economics, Research Centre for the Soil-Plant System, Via Trieste 23, 34170 Gorizia, Italy
e-mail: gilberto.bragato@entecra.it

Ž.S. Marjanović
Institute for Multidisciplinary Research, University of Belgrade, Kneza Višeslava 1a, 11000 Beograd, Serbia

A. Zambonelli et al. (eds.), *True Truffle (*Tuber *spp.) in the World*, Soil Biology 47,
DOI 10.1007/978-3-319-31436-5_12

partially, of AC horizons are determined by the presence of organic matter that, coming from litter decomposition, interacts with mineral particles. These interactions are responsible for the arrangement of organic and mineral particles into aggregates. On the other hand, long-lasting physical and chemical processes originate Bw horizon. They result in the downward migration of primary carbonates and a more developed aggregation of soil particles.

In this chapter, we used the soil classification of the last edition of the World Reference Base for Soil Resources (IUSS Working Group WRB 2014). This classification allows to briefly summarize the main soil processes acting in the surveyed areas, and it will help the reader to identify areas that are potentially suitable for *T. magnatum* growth in national and regional soil maps.

12.2 The Beginnings

Since its first description by Pico in 1788 (Ceruti 1968), *T. magnatum* has been associated with the Langhe and Monferrato regions and the area around the city of Alba, in Italy. However, we had to wait until the pioneering work of Montacchini and Caramiello (1968) to get environmental data on areas suitable for *T. magnatum* growth. After an introduction to their geology—rugged hillslopes on Miocene and Pliocene marine sediments—the authors reported some chemical properties concerning the topsoil layer of 23 truffle grounds (Table 12.1) along with the characteristics of six soil profiles. The data of Table 12.1 delineate a soil environment characterized by the presence of calcium carbonates ($CaCO_3$), a high content of exchangeable Ca (Ca_{ex}), the presence of exchangeable Mg (Mg_{ex}) and a large range of variation of organic C (C_{org}).

Profiles were quite homogeneous, displaying an A–C horizon sequence, a brown to grey colour and two main texture classes: clay loam and sandy clay loam. Montacchini and Caramiello (1968) also recorded a granular structure in the surface horizon, with a total porosity ranging from 53 to 68 % and a subangular blocky or massive structure in depth. Unfortunately, even if they noticed the alluvial origin of

Table 12.1 Topsoil chemical characteristics of 23 truffle grounds of the Monferrato and Langhe areas (from Montacchini and Caramiello 1968)

	m	SD	Min	Max
C_{org} (%)	1.9	1.0	0.5	3.8
$CaCO_3$ (%)	25	17	1	68
Ca_{ex} (cmol(+) kg^{-1})	19.0	7.7	10.6	35.1
Mg_{ex} (cmol(+) kg^{-1})	2.10	0.85	0.73	3.71
NO_3^- (mg kg^{-1})	56.4	37.0	18.8	187.5
Total P (mg kg^{-1})	0.64	0.31	0.26	1.26

M mean, *SD* standard deviation, *Min* minimum value, *Max* maximum value some soils, they did not relate truffle grounds with the geomorphic features of the investigated areas.

These results were taken into consideration by Ceruti at the Congress on Truffles held in Spoleto in 1968. In the final part of his report, Ceruti listed the soil qualities he considered necessary for *T. magnatum* cultivation: near-neutral pH, good drainage, lack of waterlogging, calcareous and clay (but not too much) composition. Interestingly, he concluded by saying "I think that the use of mycorrhized seedlings it is not strictly necessary whereas it is essential that the soil is suitable to truffle". However, 8 years later, no further data on truffle soils were reported by Mannozzi-Torini (1976), who indicated the area of diffusion of *T. magnatum* known at that time (Northern and Central Italy) and underlined the prevailing frequency of truffle grounds in floodplains.

12.3 Systematic Collection of Chemical and Physical Data

Between 1982 and 1992, a lot of surveys in *T. magnatum* truffle grounds were done in Central Italy by the "Istituto Sperimentale per la Selvicoltura" of Arezzo (Mirabella 1983; Elisei and Zazzi 1985; Tocci 1985), the University of Perugia (Bencivenga and Granetti 1988) and the "Istituto Sperimentale per lo Studio e la Difesa del Suolo" of Florence (Lulli et al. 1991, 1992, 1993; Bragato et al. 1992b). Data about the A horizon, the soil layer where *T. magnatum* ascomata are usually found, are summarized in Table 12.2. All authors determined grain size distribution, C_{org}, pH and $CaCO_3$, and some papers also considered exchangeable cations and available P (P_2O_5).

Table 12.2 Physical and chemical data of *T. magnatum* soils in Central Italy

	ISS ($n = 322$)		*UNIPG* ($n = 68$)		*ISSDS* ($n = 138$)	
	m	SD	m	SD	m	SD
Sand (%)	55.4	15.7	31.6	15.4	33.8	18.5
Silt (%)	24.1	9.0	43.9	13.1	37.3	7.7
Clay (%)	20.5	9.4	24.5	7.4	18.0	8.3
pH	7.82	0.34	7.90	0.20	7.98	0.36
C_{org} (%)	1.96	0.94	1.51	0.64	2.12	1.20
$CaCO_3$ (%)	22.0	10.2	19.5	12.1	18.6	14.1
P_2O_5 (mg kg^{-1})	47.1	16.7	nd	nd	nd	nd
Ca_{ex} (cmol(+) kg^{-1})	16.24	7.72	nd	nd	nd	nd
Mg_{ex} (cmol(+) kg^{-1})	3.14	2.55	nd	nd	nd	nd
K_{ex} (cmol(+) kg^{-1})	0.42	0.75	nd	nd	nd	nd

ISS: Mirabella (1983), Elisei and Zazzi (1985); *UNIPG*: Bencivenga and Granetti (1988); *ISSDS*: Bragato et al. (1992b), Lulli et al. (1992, 1993)
n number of observations, *m* mean, *SD* standard deviation, *nd* not determined

A first insight on soil reactivity and exchange cations indicates a sub-alkaline pH ranging between 7.40 and 8.20, buffered by the presence of $CaCO_3$, which also explains the prevalence of Ca_{ex} and the presence of Mg_{ex}, representing 82 % and 16 % of exchangeable cations, respectively. Small-sized rock fragments were rarely found, suggesting the lack of fragments coarser than 2 mm as a compulsory requirement for *T. magnatum* growth.

We merged the data collected by Lulli and his colleagues (Lulli et al. 1991, 1992, 1993; Bragato et al. 1992b) with those reported by Elisei and Zazzi (1985), thus obtaining a set of 400 observations that helped us to improve the knowledge of the soil environment of *T. magnatum*. Figure 12.1 reports the ternary plot of grain size distribution over USDA texture classes (Soil Survey Division Staff 1993). About 73 % of the observations belong to the sandy loam (SL), sandy clay loam (SCL) and loam classes. The much less frequent silt loam (SiL) and silty clay loam (SiCL) observations refer to the Crete Senesi area in Tuscany, whereas clay samples mainly belong to the Schlier geological formation.

Figure 12.2 shows the histograms of some representative variables. These values are slightly skewed, the most frequent values indicating the average soil

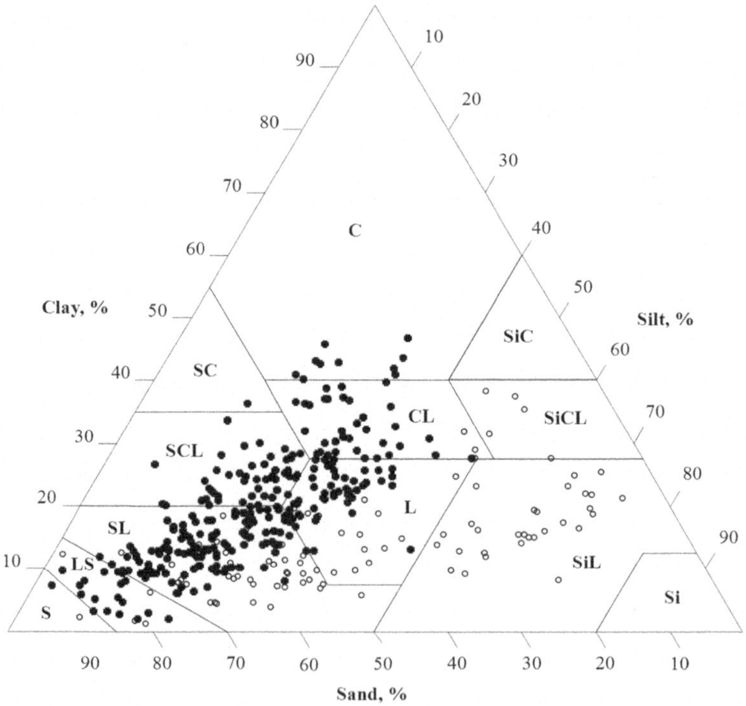

Fig. 12.1 Grain size distribution in the surface soil layer of 400 locations in Central Italy plotted over USDA texture triangle (Soil Survey Division Staff 1993). *Full dots*: data from Elisei and Zazzi (1985), *empty dots*: data from Bragato et al. (1992a, b), Lulli et al. (1992, 1993)

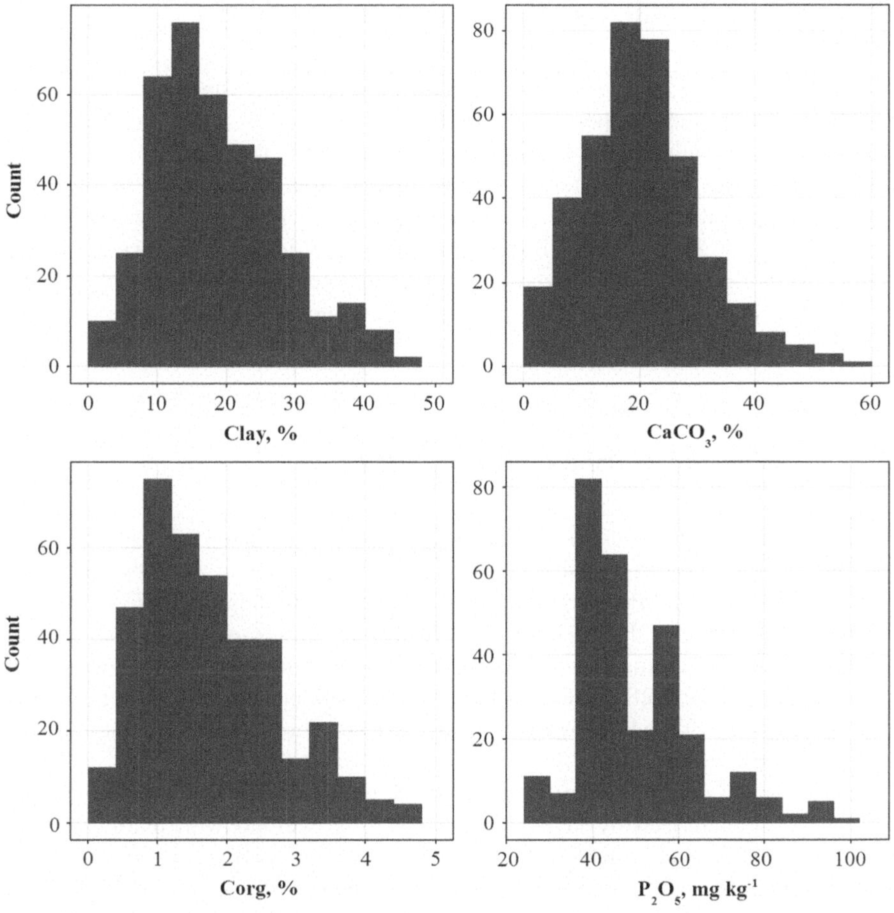

Fig. 12.2 Histograms of clay, $CaCO_3$, C_{org} and P_2O_5 in the surface soil layer of 400 *T. magnatum* truffle grounds in Central Italy. P_2O_5 data refers to 299 observations

characteristics of *T. magnatum* truffle grounds. The highest values, on the contrary, provide information on edaphic conditions found in specific regions. The largest values of clay are combined with high $CaCO_3$ percentages in truffle grounds on the Schlier formation, where small-sized carbonate particles dispersed in the clay fraction mitigate shrinking-swelling processes and the seasonal compaction of moist clay soils in autumn (Lulli et al. 1993). C_{org} values larger than 3 % are instead related to truffle grounds in mountain areas. They are the outcome of lower mineralization rates that, in the Apennines, are originated by a climate colder than that recorded in most of *T. magnatum* truffle grounds (Lulli et al. 1992). Unfortunately, a large range of variation also characterizes variables of Fig. 12.2, and as a consequence, none of them helps in delineating soils suitable for *T. magnatum*.

12.4 White Truffle Landforms and Soils in Italy

The data of Table 12.1 are common to many Italian soils, whereas *T. magnatum* fructifies in a limited number of locations. Geology itself was not very helpful in delineating areas suitable for *T. magnatum* colonization. The work done by Tocci (1985)—the largest ever in terms of investigated area and number of observations—reported the presence of *T. magnatum* in a broad range of geological formations, from Eocene and Miocene turbidites to Pliocene–Pleistocene marine sediments, which are characterized by more or less cemented sedimentary stratifications that, in the presence of active tectonics, provide widespread slope instability in hilly areas.

12.4.1 Pliocene Marine Clay and Silt Facies

Soils suitable for *T. magnatum* production were investigated by Lulli and his collaborators in Tuscany (Lulli et al. 1991, 1992; Panini et al. 1991; Bragato et al. 1992b). Lulli and Primavera outlined the ultimate findings of those researches in 2001. They focused the attention on soil-forming processes and factors not yet considered. Their approach was based on previous experiences of thematic soil mapping done by Rodolfi and Frascati (1979) and Lulli et al. (1980), who improved FAO procedures for land use assessment (FAO 1976) through the systematic use of geomorphology recordings.

In Tuscany, *T. magnatum* truffle grounds are mainly located on Pliocene marine sediments characterized by layered clays and silts overlaid with marine coastal sands. Combined with active tectonics, they originated two main types of landscapes connected to the presence of *T. magnatum* (Lulli et al. 1991; Bragato et al. 1992b). The landscape housing about 70 % of truffle grounds is that typical of the Crete Senesi area, where the outcrop of clay and silt sediments produced a sequence of rounded hills, dominated by cracking soils rich in expansible clays, and badlands with steep slopes where silt-sized sediments prevail.

Compactness and summer dryness make soils on slopes unsuitable for many tree species, while the size of mineral particles and geomorphic processes expose them to erosion. Eroded materials are captured by the dendritic drainage system of these hilly areas and redeposited by seasonal floods that chaotically release the coarsest particles along minor streams. In these frequently flooded areas, *T. magnatum* and its symbiont trees find a suitable environment to grow thanks to the presence of moist soils and loose sediments.

However, not all the floodplain is suitable for truffle production. Figure 12.3 reports the main land units present in narrow floodplains of the Crete Senesi area. Their relationships with *T. magnatum* can be summed up in terms of soil development by considering the sequence of horizons and soil structure organization

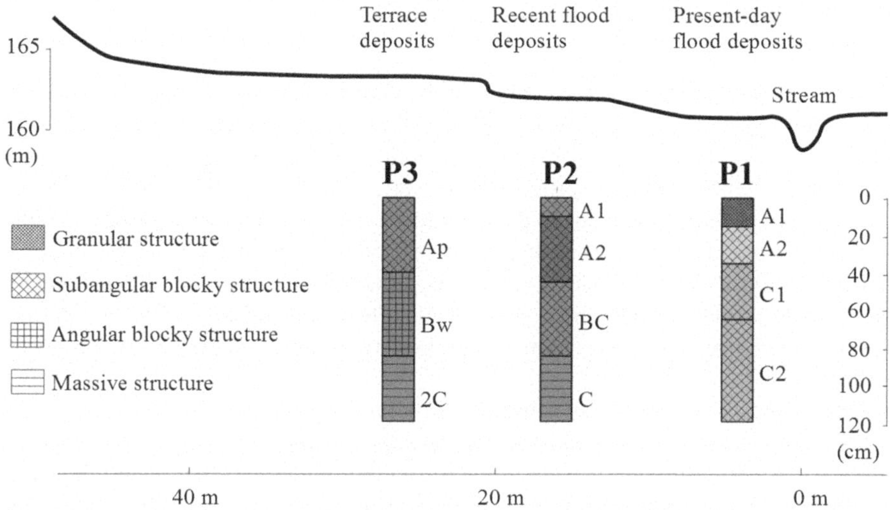

Fig. 12.3 Cross-section of a typical floodplain of the Crete Senesi area. The soil sequence from profile P1 to profile P3 is *Calcaric Fluvisol, Fluvic Calcaric Cambisol and Calcaric Cambisol*, respectively. *Tuber magnatum* is found in the *Calcaric Fluvisol* on present-day flood deposits

(Lulli et al. 1991). *Tuber magnatum* grows in the narrow strip of present-day flood deposits running parallel to the stream that originates moist soils with a loose, weakly organized granular to subangular blocky structure in the surface layer, while it is absent where floods do not occur anymore, like those on recent and terraced deposits. The time elapsed from the last deposition is enough to allow the precipitation and concentration of carbonates as nodules and the physical reorganization of clays from a poorly sorted sediment. Macroscopic results are a subangular to angular blocky structure, the occurrence of a Bw horizon and an early recurrence of cracks due to expansible clays inherited from the original marine sediments. As a consequence soils suitable for *T. magnatum* belong to *Calcaric Fluvisols*—i.e. very young alluvial soils rich in lime inherited from the parent material—whereas *Fluvic Calcaric Cambisols* and *Calcaric Cambisols* are found in recent and terrace deposits, respectively.

These results outline the fundamental role played by sedimentation dynamics on the suitability of soil for *T. magnatum*, whereas the size of the eroded particles influences the length of truffle grounds along streams. Compared to large-sized particles, small-sized ones remain suspended in water for a longer time, and as a result, their sedimentation takes place in longer portions of the floodplain. This outcome was used by Lulli et al. (1991) to distinguish two variants of *Calcaric Fluvisols* that correspond to the SL-L and the SiL-SiCL observations of Fig. 12.1. This idea was confirmed by Bragato et al. (1992b) who recorded the prevailing distribution of present-day deposits on the head of watercourses flowing from Pliocene sandy hills.

12.4.2 Pliocene Marine Coastal Sands

The second typical landscape is related to the cemented coastal sands layered with loose sandy-silty beds and compact silt beds that can be found west to the Chianti area in Tuscany (Panini et al. 1991; Bragato et al. 1992b). The undercutting of sand beds generate near-vertical, progressively retreating scarps. The slumped mass accumulated at their base triggers diffuse processes of soil creep and water erosion on lower slopes. The prevalence of one process on the other depends on the canopy density. When it is scarce, slope erosion is remarkable; otherwise, diffuse downward movements of unconsolidated materials affect the slope. Such instability can be further favoured by the surfacing of perched water tables flowing above compact silt beds.

Under forests, *T. magnatum* is usually found in slope channels where the energy of floods diminishes because of a sudden decrease in slope angle or the presence of obstacles to water flowing. However, the most common case is that of soft deposits originated by the accumulation of unconsolidated materials and soil aggregates moved downslope by mass movement (Panini et al. 1991). In Fig. 12.4, soils connected with *T. magnatum* are those of profiles P2 and P5, which show the same granular structure found in the P1 truffle soil of Fig. 12.3. These are classified as *Colluvic Calcaric Regosols,* because they lack any soil development and inherited carbonates from parent material. The prefix *colluvic* indicates their distinctive feature of being developed on materials deposited by mass movement. P2 displays a sequence of two equivalent soils, the lower one—the horizon sequence 3Ab-3Cb—buried under 50 cm of new materials fallen from the scarp. P5 shows upper-slope remnants of an A horizon mixed with AC1 horizon. In both cases, the accumulation of soil materials takes place in the presence of local changes of the slope angle: an abrupt slope decrease in P2, a concave slope segments in P5.

Soils without *T. magnatum* are instead representative of stable areas: summit, man-made terrace and linear mid-slope for P1, P3 and P4 profiles, respectively. P1 and P3 do not show any development and are classified as *Calcaric Regosols*. They differ for the thickness of the AC horizon, higher in P3 because of tillage practices that ended about 30 years earlier. When the slope is linear, like in P4, the soil has time enough to develop a B horizon, being thus classified as a *Calcaric Cambisol*. It is also worth noting that P4 is representative of the frequent layering of Pliocene sediments—i.e. horizons C, 2C and 3C—in which differentiation is mainly due to texture.

12.4.3 Eocene and Miocene Marine Formations

The information explicitly given by Tocci (1985) on landforms of Eocene and Miocene formations was very limited to relate landforms and soil processes observed in Tuscany with those of the Apennines. However, a careful analysis of

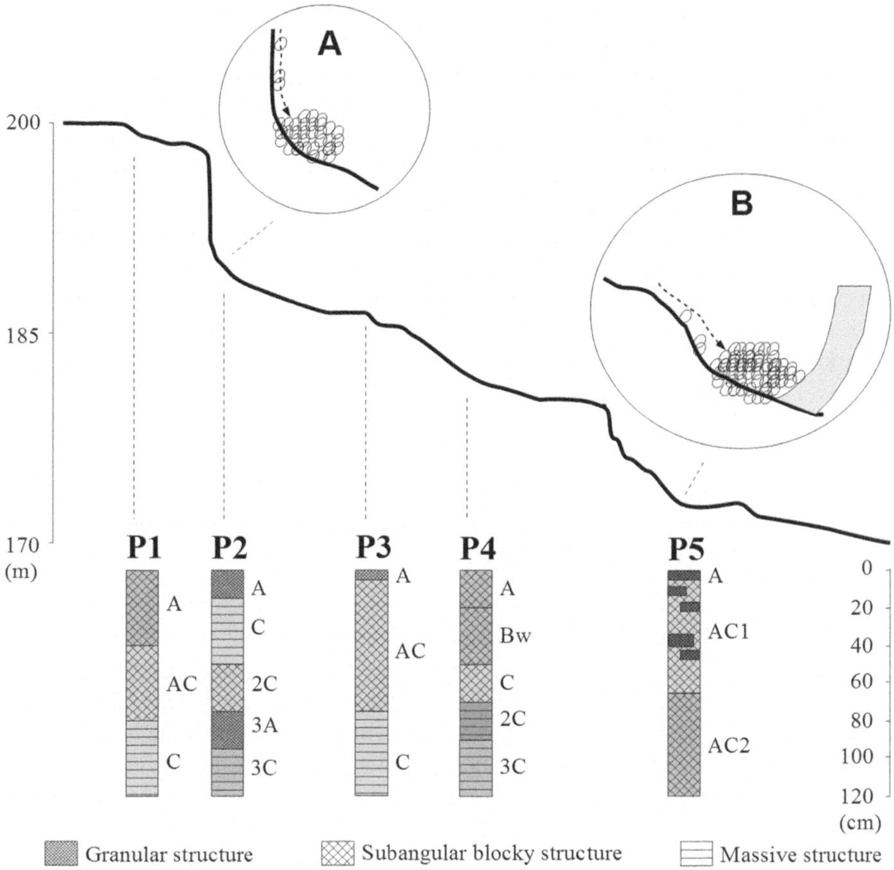

Fig. 12.4 Cross-section of a typical hillslope on Pliocene marine coastal sands in Tuscany. *Tuber magnatum* is found near profiles P2 and P5 (circles **A** and **B**). The soil sequence is represented by *Calcaric Regosols* (profiles P1 and P3), *Colluvic Calcaric Regosols* (P2 and P5) and *Calcaric Cambisols* (P4) (modified from Panini et al. 1991)

the punctual data reported in the annex of the paper indicates that more than 40 % of truffle grounds were located in flat or slightly sloping areas, about 20 % in slope channels, and the remaining 40 % in hillslopes sometimes explicitly described as uneven, i.e. possibly subjected to soil creep. These notations and those of Lulli et al. (1992) suggest that soil processes affecting *T. magnatum* in Central Apennines were almost the same of those observed in Tuscany. Chiucchiarelli et al. (2010) and Raglione et al. (2010) further confirmed these observations. The most part of soils of Abruzzo and Lazio regions surveyed by these authors belonged to *Calcaric* or *Eutric Cambisols* located in floodplains, colluvial areas, slope breaks or short slopes characterized by small-sized creep processes. All of them were characterized by a surface A horizon with granular or poorly developed subangular blocky aggregates, whereas none of them displayed vertic characteristics, i.e. cracks in surface and subsurface horizons due to the presence of expansible clay minerals.

Only the Miocene Schlier formation displayed specific characteristics that were investigated by Lulli et al. (1993) in the surroundings of Acqualagna (Marche region). Unlike hillslopes and floodplains of Tuscany, *T. magnatum* was widespread all over Schlier outcrops. This finding was explained with the specific properties of these marly calcareous sediments, in particular the high amount of clay and $CaCO_3$. Thanks to these characteristics, Schlier sediments originate rounded, gently sloping landforms and soft, clay loam to clay *Rendzic Phaeozems* and *Calcaric Cambisols* in forested and cultivated areas, respectively. In A horizon, these soils specifically combined a very open subangular blocky structure with a remarkable resilience to compaction that make them suitable for *T. magnatum* whenever the shape of landforms attracts water by gravity.

12.5 White Truffle Landscapes and Landforms in Istria and the Western Balkans

In fluvial landscapes of Central Italy, the *T. magnatum* areas were located in the upper portions of the floodplain. No data came from the medial part of floodplains—often characterized by a meandering shape—probably because these areas were interested by intensive cropping or underwent urbanization in the last decades. This gap was filled thanks to investigations done in truffle-producing areas on Eocene outcrops that characterize the Istria region in the northeastern part of the Adriatic Sea.

The first research concerned a portion of the river Mirna valley (Croatia), famous for *T. magnatum* production (Bragato et al. 2004). Marine sediments similar to those of Central Italy feed the Mirna floodplain, but truffle grounds cover most part of proximal and medial portions of the floodplain, the latter being also covered by an ancient mixed oak forest. The investigation focused on the comparison of three areas with different levels of truffle production: a productive one, an area where production is lacking in some years and an unproductive area located at the junction of the river with its main tributary.

Soils of the productive area were comparable to silty loam *Calcaric Fluvisols* present in Tuscany, and the wider distribution of present-day alluvial sediments was attributable to the higher energy of floods that sometimes submerged the entire floodplain. In accordance with Ceruti (1968), soils of the less productive and the unproductive area showed symptoms of more or less prolonged waterlogging periods. Below the A horizon, they displayed reductimorphic and oximorphic colour patches as signs of waterlogging. The percentage of these patches increased from the occasionally productive soil to the unproductive one, indicating that waterlogging does not favour truffle production. These soils can be classified as *Protostagnic Calcaric Fluvisols* and *Stagnic Calcaric Fluvisols*, respectively.

The data from the Mirna floodplain showed also a significant increase in clay from the *T. magnatum* productive area to the unproductive one. The local change in grain size distribution that determines soil conditions suitable for *T. magnatum* seemed related to fluvial landforms. This hypothesis was tested in Polje Čepić, a nearby reclaimed floodplain fed by the same Eocene sediments and wide enough to include the most representative landforms usually originated by fluvial dynamics (Bragato et al. 2010). The spatial distribution of clay content, in particular, helped to delineate the fluvial landforms obliterated by reclamation and farming. Clay distribution was useful in delineating the former distribution of natural levees and backswamps. Furthermore, it allowed to explain the correspondence of *T. magnatum* truffle grounds with natural levees as a result of flood turbulence during sedimentation. In backswamps, on the contrary, the slow sedimentation of fine particles and the resultant compactness achieved by soil are not favourable to *T. magnatum*. Figure 12.5 sketches the three main landforms of the area and their soil types confirming that *T. magnatum* grows in *Calcaric Fluvisols* and disappears in *Stagnic Calcaric Fluvisols*. As already pointed out in Italy, it is absent in *Calcaric Cambisols* of no more flooded areas located on the upper parts of the polje.

The relationship between *T. magnatum* and fluvial landforms can be a key to search for *T. magnatum* colonization in fluvial plains still affected by floods that deposit materials eroded from marine sedimentary formations. This is the case of Western Balkans, where *T. magnatum* is usually found in remnants of mixed oak and poplar forests that once covered the wide alluvial plains of the peri-Pannonian regions of Serbia (Milenković and Marjanović 2001; Marjanović et al. 2010a). These plains are connected to Pliocene-Pleistocene sedimentary formations of the

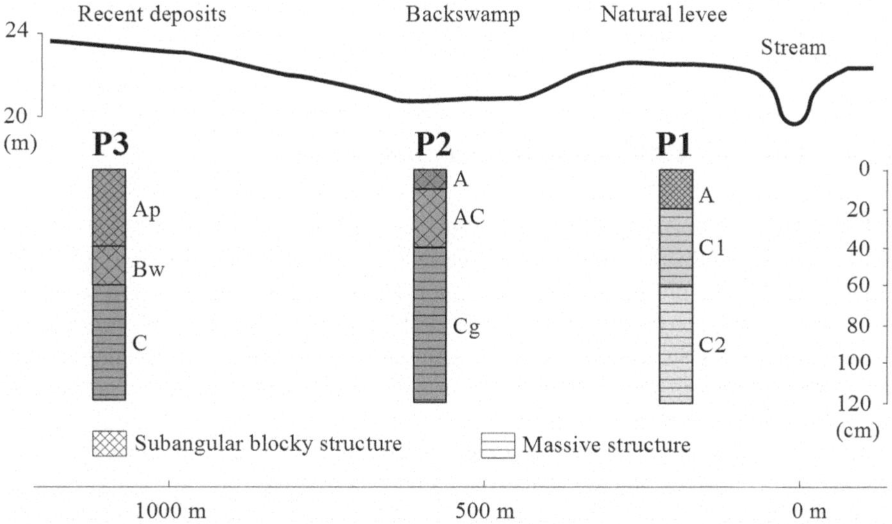

Fig. 12.5 Main landforms and related soil types in Polje Čepić : *Calcaric Fluvisols* (profile P1) in natural levees, *Stagnic Calcaric Fluvisols* (P2) in backswamps and *Fluvic Cambisols* (P3) in recent deposits. *Tuber magnatum* is found in natural levees

ancient Pannonian Sea. Marjanović et al. (2010b) compared these truffle grounds with those of Istria, chemical-physical characteristics of which were the same as those recorded in the river Mirna area.

The differences between the two areas concerned both climate and soil characteristics. Compared to the Mediterranean climate of Istria, Western Balkans has a continental climate strongly influenced by the cold Kosava wind that blows from Carpathian Mountains. The main characteristics of the region are a longer and hotter dry season, and a much colder winter with frequent snowfalls in the neighbouring Dinaric Alps. These differences also affect annual precipitations, which reach 1000 mm in Istria and range between 800 and 1300 mm in Italy but not exceed 600 mm in the fluvial plains of Serbia. The overall effect is a soil water regime where the decrease of precipitations is compensated by snow melting that recharges water tables and causes fluctuations of some metres in their height (Marjanović et al. 2015).

Truffle grounds of Serbia presented also remarkable differences with Italy and Croatia concerning soil characteristics. Serbia truffle grounds frequently lack $CaCO_3$, display pH values that range between 6.5 and 7.5 and have clay content sometimes larger than 50 %. Marjanović et al. (2010b) attributed these characteristics to the distance of *T. magnatum* areas from calcareous hills, a lower frequency of floods and a finer size of flooded particles. The result is that truffle grounds in Serbia are located on more developed soils where base saturation is kept high thanks to the renovation of Ca_{ex} and Mg_{ex} pool from water table vertical movements. These soils can be classified as *Fluvic Eutric Cambisols* that, despite their clay loam to clay texture, do not show vertic properties. The prefix *eutric* stands for high base saturation, a condition that we think is characteristic of truffle soils in Western Balkans. It means that while the saturation of exchangeable soil surfaces is required by *T. magnatum*, a high pH is likely not necessary for its fructification.

12.6 The Soil Environment of *T. magnatum*

All investigations conducted after 1990 indicated soil structure—the three-dimensional organization of solids and voids—as one of the most important properties related to the presence of *T. magnatum*. The solid component of soil structure is observed by recording shape, size and rupture resistance of aggregates. *T. magnatum* soils are characterized by granular to subangular blocky aggregates, the shape of which is determined by soil organic matter and grain size distribution. The granular structure is typical of light-textured *T. magnatum* soils. Soil organic matter binds coarse particles that otherwise tend to rearrange themselves, thus making the soil more compact. The subangular blocky shape prevails over granular structure when the clay content exceeds 10–15 %. In this case, *T. magnatum* soils differ from unproductive ones for a smaller size and a higher friability of their aggregates. Bragato et al. (2010) found that the size of dry-sieved aggregates in the unproductive area was twice as much those of *T. magnatum* productive soil. They

Table 12.3 Pore size classes determined with the pressure plate apparatus

	Pore size			
	>500 μm	50–500 μm	<50 μm	Total
Crete Senesi—sandy loam soil[a]				
Productive	3.4	13.6	36.9	53.9
Unproductive	1.8	7.7	34.5	43.9
Crete Senesi—silty loam soil[a]				
Productive	6.2	12.9	36.0	55.1
Unproductive	1.6	5.1	32.4	39.2
Slope on marine coastal sands[b]				
Productive	6.4	14.4	33.9	54.7
Unproductive	3.4	5.5	34.6	43.5

Pores larger than 50 μm (macropores) are involved in the movement of soil fluids, whereas the 0.5–50 μm size class stores soil water for plants (Greenland and Pereira 1977)
[a]Bragato et al. (1992a)
[b]Bragato et al. (1992b)

explained this result with a change in clay content. When the amount is low, like in truffle grounds, clay particles are diluted in the soil mass, and the distance between active surfaces limits their interactions. In the unproductive area, on the contrary, interparticle distances shorten, thus increasing adhesion between clay surfaces, size and rupture resistance of aggregates. This observation is further confirmed by some data of clay-rich truffle grounds of the Balkans, where the size of dry aggregates is significantly larger than that recorded in Istria (Marjanović, pers. comm.).

The importance of soil structure becomes clearer when shifting the focus from solids to voids. Bulk data on soil porosity obtained with the help of the pressure plate apparatus are reported in Table 12.3. Soils hosting *T. magnatum* show a higher porosity mainly attributable to pores larger than 50 μm (the macropores), which represent about 30 % and 20 % of total porosity in productive and unproductive areas, respectively. Since macropores are the main voids for the movement of fluids in soil, they are fundamental for aerobic species like *T. magnatum*, because they allow faster fluxes and higher oxygen tensions in soil. Table 12.3 also suggests minimum threshold values for soils suitable for *T. magnatum*: 53 % and 15 % of soil volume occupied by total porosity and macroporosity, respectively.

Porosity measurements are impractical for the delineation of soils suitable for *T. magnatum*. However, total porosity can be determined in much larger sets of locations by means of soil bulk density. Considering all the data, we obtained confidence limits of 1.04 and 1.12 g kg^{-1} for $n = 98$ and $p = 0.001$. Such values of bulk density correspond to a total porosity ranging between 59 and 62 %, values in line with those recorded by Montacchini and Caramiello (1968).

The bulk values of soil porosity do not fully explain the function of macropores, which also depends on their shape and spatial pattern (Pagliai and Vignozzi 2006). These characteristics were investigated with micromorphometric techniques based on the analysis of macrophotographs of soil thin sections (Bragato et al. 1992a; Lulli et al. 1993; Bragato et al. 2010). In Fig. 12.6, the productive location displays

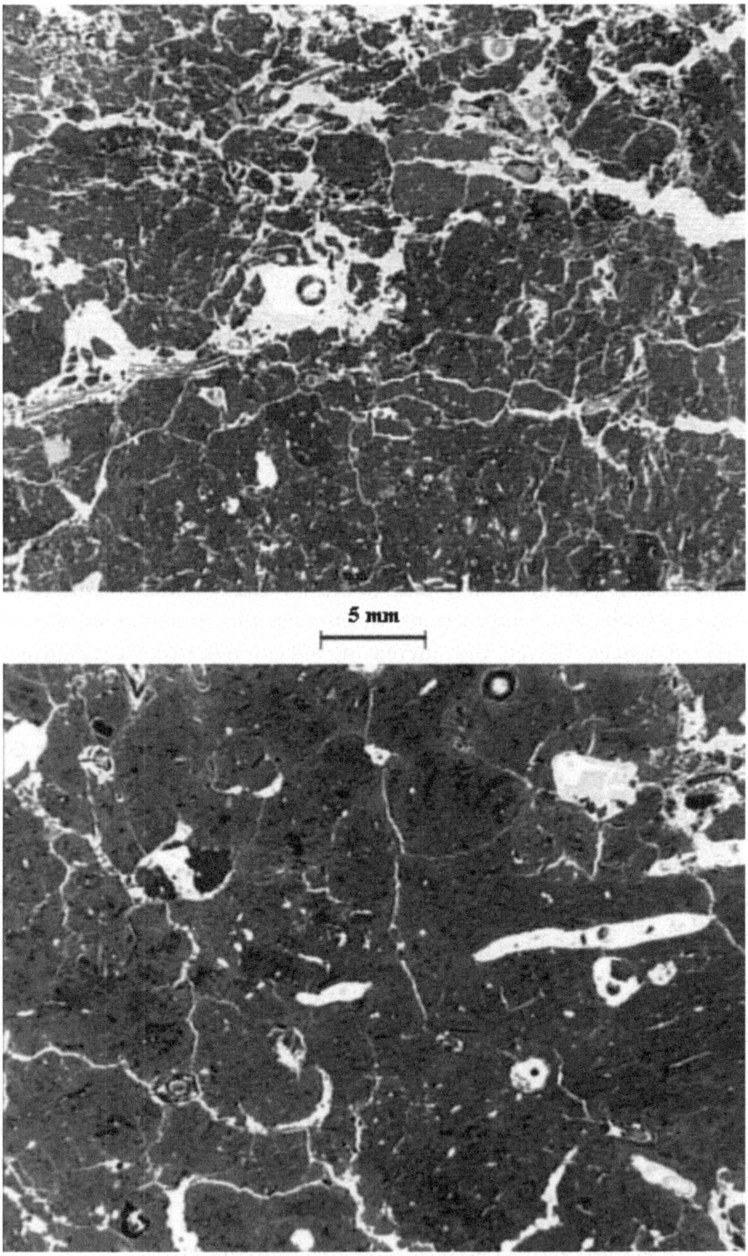

Fig. 12.6 Soil thin sections of a productive (*top*) and an unproductive location (*bottom*). Bright areas correspond to pores larger than 50 μm (Reproduced from Bragato et al. 2010)

Table 12.4 Total and elongated macropores measured in thin soil sections collected in soils producing or not *T. magnatum* ascomata

	Macroporosity (%)		Pore size 50–500 μm (%)		Pore size > 500 μm (%)	
	Total	Elongated	Total	Elongated	Total	Elongated
Crete Senesi—sandy loam soil[a]						
Productive	29.3	20.7	16.4	12.3	12.9	8.4
Unproductive	17.0	6.8	11.1	6.8	5.9	0.0
Crete Senesi—silt loam soil[a]						
Productive	31.9	21.1	16.9	10.9	15.0	10.2
Unproductive	13.2	1.8	6.6	1.8	6.6	0.0
Aqualagna[b]						
Productive	30.1	22	8.2	6.6	21.9	15.4
Unproductive	15.3	13.2	4.2	2.7	11.1	10.5
Polje Čepić[c]						
Productive	15.4	9.8	6.1	4.0	9.3	5.8
Unproductive	9.5	4.7	3.7	2.4	5.8	2.3

[a]Bragato et al. (1992a)
[b]Lulli et al. (1993)
[c]Bragato et al. (2010)

small-sized, loose aggregates and an evenly distributed macroporosity. The unproductive location is instead characterized by larger and more compact aggregates contoured by thin, scarcely connected macropores.

The visual analysis is supported by the data of Table 12.4, which reports the area percentage of total macroporosity and elongated pores. In Table 12.4, the macroporosity threshold of 15 % separates productive from unproductive areas, and many productive areas are well beyond that threshold. Micromorphometry showed that the 50–500 μm size class prevails in macropores and that about two-thirds of macropores in *T. magnatum* locations consist of elongated pores. It is this shape class that originates the high degree of pore connectivity and continuity in the productive location of Fig. 12.6.

Table 12.4 indicates two other interesting characteristics of *T. magnatum* environments. Compared to the floodplains of Tuscany, the lower values of macroporosity and elongated pores in Polje Čepić are likely related to the lower frequency of floods after its reclamation. Starting from a new flood and the related increase in macroporosity, mineral and organic particles reorganize themselves into aggregates that fill the soil volume more evenly. As a consequence, macroporosity decreases and the soil surface slightly lowers, resulting in a more compact soil. The longer the time elapsed from the last flood, the lower the macroporosity in soil (Bragato et al. 2010). The second characteristics to be underlined is the shift of elongated pores from the 50–500 μm to the >500 μm size class in Acqualagna. This pore shape is related to tillage and to the specific combination of high clay and $CaCO_3$ content in the Schlier formation: tillage produces very large pores that tend to collapse in time, but the stability of soil structure makes soils on Schlier's calcareous marls resilient to compaction.

The data on macroporosity delineate a very draining soil that, however, must retain enough water to sustain a dense tree canopy and to protect *T. magnatum* mycelium from strong summer radiation. These opposing requirements are met in specific morphological and microclimate conditions that explain both the high selectivity of *T. magnatum* for its soil habitat and the problems in delineating truffle-producing areas.

The morphology of the landscape defines the conditions that drain water from neighbourhood areas to *T. magnatum* locations, thus determining the specific soil moisture regime of truffle grounds. The most frequent presence of truffle locations in floodplains is explained by their importance as water attractors. Even in very dry summers, floodplains retain water in depth, thus allowing the host to access a permanent water source. The water stored in depth does not directly affect the soil moisture in surface strata colonized by *T. magnatum* but allows the growth of a dense tree cover that keeps surface strata cool during the summer. In the dry season, the phenomenon known as "nocturnal hydraulic lift" observed in oaks with deep taproots (Querejeta et al. 2003, 2009) might help the mycorrhizae of *T. magnatum* to survive as it drives water upwards to surface roots (Marjanović et al. 2015).

The volume of water retained increases with the size of the floodplain—and its watershed as well—and it decreases with the reduction of precipitations. In the former case, the reserve of water is large enough to sustain large fluvial forests that, like in Serbia, resist prolonged periods of drought. In the latter, its decrease causes the narrowing of the wooded strips along streams that, like in the narrow floodplains of the Crete Senesi area, takes place in the truffle grounds of Central and Southern Italy.

Truffle grounds on slopes in Central and Southern Italy are located on water attractors represented by concavities or areas of sudden decrease of the slope angle. The tree cover plays the same role observed in floodplains, by shading and cooling the soil surface. However, the lower amount of water retained in perched water tables allows the growth of symbiont tree species more resistant to dryness such as white oak, turkey oak and hop hornbeam. The lack of a permanent water source makes *T. magnatum* production on slopes strongly dependent on seasonal precipitation rates. This dependence on the precipitation regime explains the large seasonal fluctuations of truffle production in Italian hillslopes and the lack of truffle grounds on slopes of the Western Balkans reported by Marjanović et al. (2010b).

12.7 Conclusions

Tuber magnatum grows and produces ascomata in moist and drained environments of tectonically active areas. Soils are continuously rejuvenated by solid materials transported by floods in floodplains or by mass movement along slopes. The accumulation of particles and aggregates creates very porous, highly aerated, soft soils, with an interconnected macroporosity that allows this hypogeous fungus to "breathe" and its ascomata to grow in size without physical constraints.

Tuber magnatum also needs a microhabitat that must be neither too dry nor too moist thanks to water supplied by the surrounding environment. Such requirement is fulfilled in portions of the landscape that attract water and originate a specific soil moisture regime. The most important landscapes hosting *T. magnatum* are floodplains that, thanks to their capacity to permanently store water in depth, sustain the growth of a dense tree cover and, indirectly, the genesis of a cool environment in the soil surface. The colonization of *T. magnatum* on slopes is instead related to changes of the slope angle that locally originate areas of attraction for water and, again, a cool microhabitat.

Less limiting are soil chemical characteristics. The pH must be neutral to alkaline, but the presence of finely dispersed carbonates, present in the Mediterranean truffle grounds but lacking in those of Western Balkans, is not compulsory. Therefore, a strong prevalence of Ca and Mg cations in near-saturated to saturated exchangeable surfaces might be a basic requirement for *T. magnatum* colonization and fructification.

Soils suitable for *T. magnatum* colonization mostly belong to soil classes *Calcaric Fluvisols*, *Colluvic Calcaric Regosols*, *Calcaric Cambisols* and *Fluvic Eutric Cambisols*. Limited changes in adjacent soils influenced by the same soil-forming processes drastically decrease the probability of finding *T. magnatum* truffles, thus partially explaining their high market value.

Acknowledgements Many of the ideas presented here come from the suggestions of Luciano Lulli, whose knowledge in soil science was the solid base on which this chapter was written down. We also thank our fellow adventurers in the exploration of the soil environment of *T. magnatum*: Tiziano Panini, Lorenzo Gardin, Fabio Primavera and Aleksa Glišić.

References

Bencivenga M, Granetti B (1988) Ricerca comparativa sulle esigenze ecologiche di *Tuber magnatum* Pico e *Tuber melanosporum* Vitt. dell'Italia Centrale. Ann Fac Agrar Univ Perugia 42:861–872

Bragato G, Panini T, Pagliai M (1992a) Soil porosity and structural conditions in soils involved in white truffle production in the "Crete Senesi" area (Tuscany). Agric Medit 122:180–188

Bragato G, Lulli L, Panini T, Gardin L, Primavera F (1992b) I suoli delle tartufaie naturali della zona di San Miniato (Pisa). Monti Boschi 43:17–24

Bragato G, Sladonja B, Peršuric Đ (2004) The soil environment for *Tuber magnatum* growth in Motovun forest, Istria. Nat Croatica 13:171–186

Bragato G, Vignozzi N, Pellegrini S, Sladonja B (2010) Physical characteristics of the soil environment suitable for *Tuber magnatum* production in fluvial landscapes. Plant Soil 329 (1–2):51–63. doi:10.1007/s11104-009-0133-8

Ceruti A (1968) Biologia e possibilità di coltivazione dei tartufi. In: Atti del I° Congresso Internazionale sul Tartufo, Spoleto, 24–25 May 1968, pp 29–42

Chiucchiarelli I, Santucci S, Paolanti M (2010) I suoli delle tartufaie naturali in Abruzzo. In: Donnini D (ed) Proceedings of 3rd international congress on truffles, Spoleto, 25–28 Nov 2008. Comunità Montana dei Monti Martani, Serano e Subasio, pp 242–246

Elisei S, Zazzi A (1985) Caratteri fisico-chimici dei suoli tartufigeni del Centro Italia. Ann Ist Sper Selvicolt Arezzo 16:543–587

FAO (1976) A framework for land evaluation. FAO, Rome (Soil Bulletin 32)

FAO (1998) World Reference Base for Soil Resources. World Soil Resources Report n. 84. FAO, Rome

Greenland DJ, Pereira HC (1977) Soil damage by intensive arable cultivation: temporary or permanent? Philos T R Soc B 281(980):193–208. doi:10.1098/rstb.1977.0133

IUSS Working Group WRB (2014) World Reference Base for Soil Resources 2014. International soil classification system for naming soils and creating legends for soil maps. World Soil Resources Report n. 106. FAO, Rome

Lulli L, Primavera F (2001) *Tuber magnatum* Pico: environment of growth. In: Actes Vème congrès international science et culture de la truffe, Aix-en-Provence, 4–6 Mar 1999. Federation Française des Trufficulteurs, pp 269–272

Lulli L, Delogu F, Gregori E, Miclaus N (1980) Cartografia di base per la programmazione degli interventi in aree marginali. Area rappresentativa dell'Alta Valdera. I: Memorie illustrative della carta dei suoli. Ann Ist Sper Stud Difesa Suolo Firenze 11:33–96

Lulli L, Panini T, Bragato G, Gardin L, Primavera F (1991) I suoli delle tartufaie naturali delle Crete Senesi. Monti Boschi 42:31–39

Lulli L, Bragato G, Panini T, Gardin L, Primavera F (1992) I suoli delle tartufaie naturali della bassa valle del Santerno (Mugello – Toscana). Ital Forest Montana 47(5):251–267

Mannozzi-Torini L (1976) Manuale di tartuficoltura. Tartufi e tartuficoltura in Italia. Edagricole, Bologna

Lulli L, Pagliai M, Bragato G, Primavera F (1993) La combinazione dei caratteri che determinano il pedoambiente favorevole alla crescita del *Tuber magnatum* Pico nei suoli dei depositi marnosi dello Schlier in Acqualagna (Marche). Quad Sci Suolo 5:143–155

Marjanović Ž, Saljnikov E, Milenković M, Grebenc T (2010a) Ecological features of *Tuber magnatum* Pico in the conditions of West Balkans – soil characterization. In: Donnini D (ed) Proceedings of 3rd international congress on truffles, Spoleto, 25–28 Nov 2008. Comunità Montana dei Monti Martani, Serano e Subasio, pp 190–195

Marjanović Ž, Grebenc T, Marković M, Glišić A, Milenković M (2010b) Ecological specificities and molecular diversity of truffles (genus *Tuber*) originating from mid-west of the Balkan Peninsula. Sydowia 62:67–87

Marjanović Ž, Glišić A, Mutavdžić D, Saljnikov E, Bragato G (2015) Ecosystems supporting *Tuber magnatum* Pico production in Serbia experience specific soil environment seasonality that may facilitate truffle lifecycle completion. Appl Soil Ecol 95:179–190. doi:10.1016/j.apsoil.2015.05.007

Milenković M, Marjanović Ž (2001) Current results on *Tuber* spp. research in Yugoslavia. In: Actes Vème congrès international science et culture de la truffe, Aix-en-Provence, 4–6 March 1999. Federation Française des Trufficulteurs, pp 218–225

Mirabella A (1983) Indagine preliminare sui suoli di alcune tartufaie di *Tuber magnatum* Pico (tartufo bianco pregiato) nelle valli del Metauro, Foglia, Marecchia e Savio nelle Marche. Ann Ist Sper Selvicolt Arezzo 14:389–407

Montacchini F, Caramiello R (1968) Ecologia del "*Tuber magnatum*" Pico in Piemonte. Allionia 14:1–29

Pagliai M, Vignozzi N (2006) Image analysis and microscopic techniques to characterize soil pore system. In: Blahovec J, Kutilek M (eds) Physical methods in agriculture: approach to precision and quality. Kluwer Academic/Plenum, New York, pp 13–38

Panini T, Bragato G, Gardin L, Lulli L, Primavera F (1991) Suoli e siti tartufigeni di un versante tipico della zona di San Miniato, Toscana. Ital Forest Montana 46:373–393

Querejeta JI, Egerton-Warburton LM, Allen MF (2003) Direct nocturnal water transfer from oaks to their mycorrhizal symbionts during severe soil drying. Oecologia 134(1):55–64. doi:10.1007/s00442-002-1078-2

Querejeta JI, Egerton-Warburton LM, Allen MF (2009) Topographic position modulates the mycorrhizal response of oak trees to interannual rainfall variability. Ecology 90(3):649–662. doi:10.1890/07-1696.1

Raglione M, Owczarek M, Bonifazi A, Lorenzoni P (2010) Pedological aspects of natural truffle beds of *Tuber magnatum* Pico in the Rieti province (Central Apennines, Italy). In: Donnini D (ed) Proceedings of 3rd international congress on truffles, Spoleto, 25–28 Nov 2008. Comunità Montana dei Monti Martani, Serano e Subasio, pp 283–285

Rodolfi G, Frascati F (1979) Cartografia di base per la programmazione degli interventi in aree marginali. Area rappresentativa dell'Alta Valdera. I: Memorie illustrative della carta geomorfologica. Ann Ist Sper Stud Difesa Suolo Firenze 10:37–80

Soil Survey Division Staff (1993) Soil survey manual. Soil Conservation Service. US Department of Agriculture. Handbook 18

Tocci A (1985) Ecologia del *Tuber magnatum* Pico nell'Italia Centrale. Ann Ist Sper Selvicolt Arezzo 16:425–542

Chapter 13
Soil Characteristics for *Tuber aestivum* (Syn. *T. uncinatum*)

Christophe Robin, Noémie Goutal-Pousse, and François Le Tacon

13.1 Introduction

Tuber aestivum Vittad. has for several years been considered to be synonymous with *Tuber uncinatum* Chatin (Ceruti et al. 2003; Paolocci et al. 2004; Wedén 2004; Wedén et al. 2004a, b, 2005). A recent multigene phylogenetic study confirmed that *T. aestivum* and *T. uncinatum* are conspecific (Molinier et al. 2013b; see Chap. 3).

Among *Tuber* species of culinary interest, *T. aestivum*, the Burgundy truffle, displays the most extensive geographic range and is found in almost all European countries. Harvests of fruiting bodies have been recorded from Ireland to Azerbaijan and from North Africa to Sweden (see Chap. 3). Gotland Island at a latitude of 57–58 °N is considered to be the northernmost outpost of *T. aestivum* (Wedén and Danell 2001). It is a truffle with great economic value and mostly occurs in natural habitats and habitats that have been previously cultivated and recolonized. Its maturity period occurs in autumn, when soil temperatures remain above 0 °C, protecting fruiting bodies from frost damage.

The aim of this chapter is to define the soil characteristics necessary throughout the life cycle of the Burgundy truffle, including those essential for ascoma

C. Robin (✉)
INRA, UMR 1121 "Agronomie & Environnement" Nancy-Colmar, 2 avenue de la forêt de Haye, TSA 40602, 54518 Vandœuvre-les-Nancy Cedex, France
e-mail: Christophe.Robin@univ-lorraine.fr

N. Goutal-Pousse
Office National des Forêts, département Recherche Développement et Innovation, 11 rue de l'Ile-de-Corse, 54000 Nancy, France

F. Le Tacon
INRA, Université de Lorraine, UMR 1136, Interactions Arbres/Micro-organismes, Laboratoire d'Excellence ARBRE, 54280 Champenoux Cedex, France

A. Zambonelli et al. (eds.), *True Truffle (*Tuber *spp.) in the World*, Soil Biology 47,
DOI 10.1007/978-3-319-31436-5_13

formation. The beginning of this chapter describes bedrock characteristics, followed by the physical and chemical properties of soils suitable for the Burgundy truffle.

13.2 Methodology

Data presented in Table 13.1 and Figs. 13.2 to 13.10 have two origins. An initial soil data set was the result of direct contact with truffle growers, technicians, and scientists working on *T. aestivum*, both within and outside Europe. We requested their data from sites known to have produced *T. aestivum* ascoma at least once. A second data set was compiled from papers published in scientific journals and books or included in proceedings of international truffle conferences. We refined our search to consider only those papers which contained soil information clearly indicating that *T. aestivum* ascoma had been collected in these soils. Papers which posed any doubt were rejected.

We will present and discuss here the largest data set ever collected on soils favorable to *T. aestivum* in such a large geographic area; a total of 129 soils from 10 countries were included in this review (Fig. 13.1). This data set cannot be considered to be representative for all situations, yet it can help to form additional hypotheses which should be verified in the future.

Methods of soil analysis were carefully evaluated prior to integrating a data set in the study. We only considered those analyses which used the following standard soil testing methods:

- Soil granulometry with organic matter destruction but without carbonate dissolution prior to analysis with the following size threshold: clay between 0 and 2 μm, silt between 2 and 50 μm, sand between 50 and 2000 μm.
- pH in water (1:5 in volume, ISO 10390).
- Soil organic matter and total nitrogen content determined by dry combustion (ISO 10694 and ISO 13878, respectively).
- Total calcium carbonate content determined by volumetric method (ISO 10693).
- Available phosphorus according to Olsen (ISO 11263) or Joret-Hebert.
- Exchangeable cations (Ca, Mg, and K) and cationic exchange capacity (CEC) were determined either at soil pH (e.g., cobaltihexamine) or at pH 7 (e.g., Metson); both extraction types were considered separately as CEC values are highly dependent on soil pH (Ciesielski and Steckerman 1997).

As units differed from one method to another one, we recalculated several values prior to data processing in order to homogenize the data. We discarded data when soil analysis methods or units were not recorded in the file or the paper. R software (R core team 2013) was used to establish the soil texture triangle and to transform granulometry data with silt to sand thresholds of 63–50 μm (soil texture package: Moeys and Shangguan 2014) and to draw figures (ggplot2 package: Wickham 2009).

Table 13.1 List of the sites investigated and their characteristics, when available: mean annual precipitation (MAP), elevation, nature of the bedrock(s) and of soil (s), most putative host trees, natural site vs. orchard, and source of information

Site	Country	Region	MAP (mm)	Elevation (m AMSL)	Bedrock characteristics	Soil type(s)	Natural truffiere (Yes/No)	Host trees	Information source
Daix	France	Bourgogne	732	330–340	Upper Jurassic, Mid Oxfordian	Anthrosol (calcaric)	No	*Corylus avellana*	H Frochot (pers comm); Molinier et al. (2013a)
Boncourt-sur-Meuse	France	Lorraine	1056	250–300	Upper Jurassic, Oxfordian	Epileptic Cambisols (calcaric, novic), Rendzic Leptosol and Cambisol (hypereutric)/colluvial rendzina, rendzina, and brown calcisol	No	*C. avellana*	Communauté de Communes du Pays de Commercy (pers comm); 12 soil analyses
Rollainville	France	Lorraine	940	360	Jurassic, limestone plateau	Cambisol (calcic)/ brown calcisol	No	*C. avellana*	C Robin (pers comm)
Multi-sites (25)	France	Lorraine, Bourgogne, Franche-Comté, Auvergne	700–1060	700–1060	Cretaceous, upper and mid-Jurassic, Trias—Paleogene (Eocene and Oligocene)	Rendzic Leptosol and Cambisol (calcic)/ rendzina, brown rendzina, and brown calcisol	Yes	*Quercus* sp., *Corylus* sp., *Carpinus* sp., *Pinus* sp., *Tilia* sp., etc.	Le Tacon et al. (1997)
Otterswiller	France	Alsace	–	185–243	–		No	*C. avellana*	H Meyer (pers comm)
Munchhouse 1	France	Alsace	–	220	–	–	No	*C. avellana*, chêne, pin noir	B Vonflie (pers comm)
Munchhouse 2	France	Alsace	–	220	–	–	Yes	*C. avellana*	B Vonflie (pers comm)
Lavoye	France	Lorraine	–	200–270	–	–	No	–	The owner (pers comm)

(continued)

Table 13.1 (continued)

Site	Country	Region	MAP (mm)	Elevation (m AMSL)	Bedrock characteristics	Soil type(s)	Natural truffiere (Yes/No)	Host trees	Information source
Harmonville 1	France	Lorraine	940	313	–	–	No	*Corylus* av., *Quercus* sp., *Pinus nigra*, *Carpinus*, *Betula*, *Tilia*	S Philippe (pers comm)
Harmonville 2	France	Lorraine	940	313	–	–	No	–	S Philippe (pers comm)
Harmonville 3	France	Lorraine	940	313	–	–	No	–	S Philippe (pers comm)
Beine 1 (La Noue d'Aubigny)	France	Champagne-Ardennes	–	140–290	–	–	No	*C. avellana*, *Pinus nigra*, *Carpinus betulus*	JL Dubois (pers comm)
Beine 2 (Les Commelles)	France	Champagne-Ardennes	–	140–290	–	–	No	*C. avellana*, *Pinus nigra*, *Carpinus betulus*	JL Dubois (pers comm)
Humbauville 1	France	Champagne-Ardennes	–	130	–	–	No	*C. avellana*, *Pinus nigra*, oak	M Yverneau (pers comm)
Humbauville 2	France	Champagne-Ardennes	–	130	–	–	No	*C. avellana*, *Pinus nigra*, *Carpinus betulus*	M Yverneau (pers comm)
Blanzac-les-Matha	France	Poitou-Charentes	843	23–76	Jurassic	Calcaric Leptosol (clayic)	Yes	Tilia, *C. avellana*	A Tribot (pers comm)

La Foye-Monjault	France	Poitou-Charentes	814	59	–	Calcaric Leptosol (clayic)	No	*C. avellana*	JJ Sauvaget (pers comm)
Saint-Cybardeaux	France	Poitou-Charentes	750	24–114	–	Cambisol (calcaric, clayic)	Yes	Natural (*Tilia*)/implanted (fence *Corylus*, *Carpinus betulus*)	R Mesnier (pers comm)
Multi-sites (18)	Sweden	Gotland	528	40	Belonging to the Hemse group, stratified, predominantly crystalline, partly fine oolitic limestone, reef-shaped limestone, reef limestone, marly limestone and marlstone, sand limestone, and lime sandstone, and on Högklint limestone, stratified, more or less marly limestone and marlstone and reef limestone		Yes	*Quercus robur*, *C. avellana*	Wedén et al. (2004a, b)
Zapadné Slovensko	Slovakia	Zapadné Slovensko	–	–	–	Rendzic Leptosol, Epileptic Cambisols (calcaric), Luvisol/rendzinas, rendzina Cambisols, Luvisols	Yes	–	Miko et al. (2008); 35 soil analyses
Malé Karpaty	Slovakia	Malé Karpaty	–	–	–	Rendzic Leptosol, Cambisols, Luvisols	Yes	–	Miko et al. (2008)

(continued)

Table 13.1 (continued)

Site	Country	Region	MAP (mm)	Elevation (m AMSL)	Bedrock characteristics	Soil type(s)	Natural truffiere (Yes/No)	Host trees	Information source
Tribeč	Slovakia	Tribec	–	–	–	Rendzic Leptosol, Cambisols, Luvisols	Yes	–	Miko et al. (2008)
Vancouver Island. Nanaimo	Canada	British Columbia	967	32	Upper Cretaceous undivided sedimentary rocks	Canadian System of Soil Classification: orthic dystric brunisol, shallow lithic	No	–	S Berch (pers comm)
Pinczow M	Poland	Nida Basin	600	250	Mesozoic, marlstone	–	Yes	–	D Hilszczánska (pers comm); Hilszczánska et al. (2008, 2013)
Pinczow SA	Poland	Nida Basin	600	312	Mesozoic, marlstone	–	Yes	–	D Hilszczánska (pers comm); Hilszczánska et al. (2008, 2013)
Pinczow WR	Poland	Nida Basin	600	295	Mesozoic, marlstone	–	Yes	–	D Hilszczánska (pers comm); Hilszczánska et al. (2008, 2013)
Pinczow GR	Poland	Nida Basin	600	260	Mesozoic, gypsum	–	Yes	–	D Hilszczánska (pers comm); Hilszczánska et al. (2008, 2013)
Przedborz Upland	Poland	–	600	320	Jurassic limestone	–	Yes	–	D Hilszczánska (pers comm); Hilszczánska et al. (2008, 2013)

Chmielnik NW	Poland	Nida Basin	600	228	Marly limestone	–	Yes	–	D Hilszczánska (pers comm); Hilszczánska et al. (2008, 2013)
Wolynska Upland	Poland	–	550	200	Limestone	–	Yes	–	D Hilszczánska (pers comm); Hilszczánska et al. (2008, 2013)
Spoleto	Italy	Umbria	869	310	Fluvial deposits with coarse fragments from Cretaceous limestones	Eutric Cambisol (Aric)	Yes/No	–	
Six sites	Italy	Provincia di Parma		800–1200	Mesozoic (cretaceous) and Cenozoic (Paleocene, Eocene, Oligocene), marly limestone		Yes	Dominated by *Carpinus*	Gregori (2010)
Two sites	Italy	Provincia di Parma	–	–	–	–	–	–	Gregori (2010)
Provincia de Teramo	Italy	Provincia de Teramo	–	937	Cover detritus-eluvial of the Holocene origin	Mollic Cryosol	Not indicated	*Quercus pubescens*	Menta et al. (2014)
Provincia de Piacenza	Italy	Provincia de Piacenza	–	810	Lutetian limestones and marly limestones alternating marls and calcareous marls	Cambisol (calcaric)	Not indicated	*Q. pubescens*	Menta et al. (2014)
South Liege	Belgium	South Liège	–	240	–	–	Yes	–	The owner (pers comm)

(continued)

Table 13.1 (continued)

Site	Country	Region	MAP (mm)	Elevation (m AMSL)	Bedrock characteristics	Soil type(s)	Natural truffiere (Yes/No)	Host trees	Information source
Guadalajara	Spain	Guadalajara	797	1000	Jurassic and cretaceous limestone and dolomites	Rendzic Leptosol	–	*Quercus faginea*	Menta et al. (2014)
Velkà Chuchle	Czech Republic	Velkà Chuchle	–	–	Silurian limes (Paleozoic)	Rendzic Skeletic Leptosol	Yes	*Corylus*, *Tilia*, *Fraxinus*, etc.	Gryndler et al. (2013)
Kibbutz Bar'am	Israel	Upper Galilee	800	736	Cretaceous (Mesozoic)	Cambisols (dolomitic)	No	–	Y Sitrit (pers comm); Kagan-Zur et al. (2001)
Eger, Heves county	Hungary	Eger, Heves county	600	298	Limestone	Calcic Luvisol	No	–	Z Bratek, A Gogan; Gogan et al. (2012)
Hőgyész, Tolna county	Hungary	Hőgyész, Tolna county	630	210	Tertiary loess	Chernozem and Luvisol	No	–	Z Bratek, A Gogan; Gogan et al. (2012)
Jászság, 20 sampling places	Hungary	Jászság, 20 sampling places	515	<150	Middle Miocene volcanic ash covered with loess	Gleysol (63 %), Chernozem (stagnic) (23 %), Solonchak (4.5 %), Solonetz (9 %)	Yes	–	Z Bratek, A Gogan; Gogan et al. (2012)
South West Germany 1	Germany	South West Germany	890	659	Jurassic	Rendzic Leptosol	Yes	–	Stobbe et al. (2013)
South West Germany 2	Germany	South West Germany	857	772	Jurassic	Rendzic Leptosol	Yes	–	Stobbe et al. (2013)

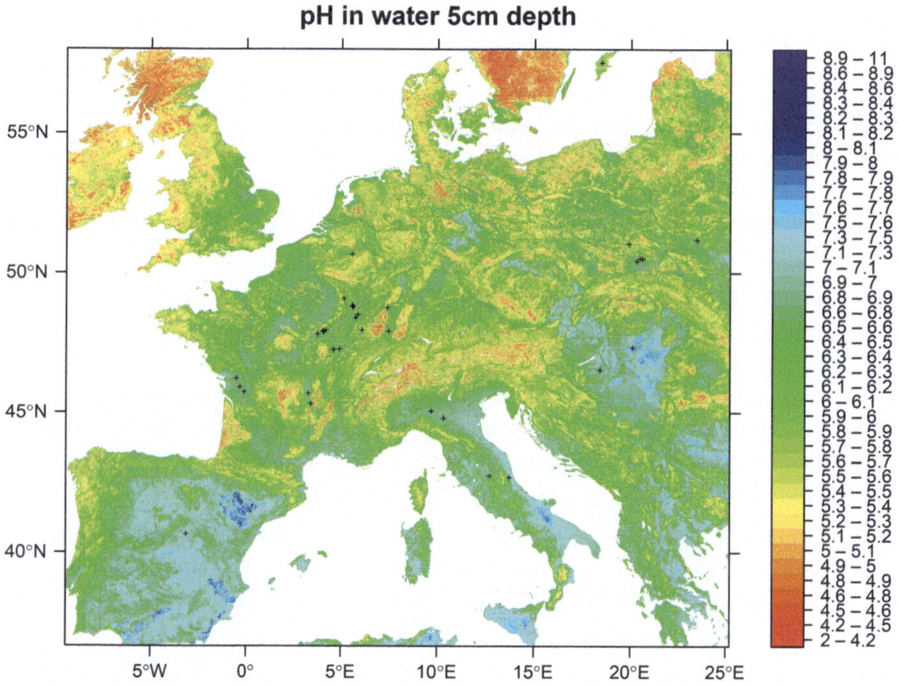

Fig. 13.1 Distribution of soil data set represented over the spatial predictions of soil pH in water at 5 cm depth according to ISRIC—World Soil Information (www.soilgrids.org). Each soil data set used in this study is represented by a cross. SoilGrids1km is a first approximation of soil property predictions at 1 km resolution, using automated global soil mapping. The absolute values of predictions available for download at www.soilgrids.org are presently of limited thematic and spatial accuracy and contain artifacts and missing pixels. Yet the relative values of soil pH in water can be useful to compare distribution of soils favorable to *T. aestivum* between regions

13.3 Nature of the Bedrock and Soil Types

Tuber aestivum fructifies naturally on a range of diverse geological substrates from the Paleozoic era (540 Mya) to present day (Cenozoic), including the Mesozoic (252–66 Mya) (Table 13.1). While bedrock which is sedimentary in origin is the most common trait, *T. aestivum* ascoma can occur on alkaline volcanic substrates and on quaternary formations (loess and glacial formations). Below are examples of fructification situations in countries for which information has been found.

France In northeastern France, the parent rock dates from the Jurassic, the Cretaceous, and, to a lesser extent, to the Trias periods. In central France, around Paris and Limagne (near Clermont-Ferrand), *T. aestivum* is found on limestone from the Tertiary (Eocene and Oligocene) period. In some cases, the parent material is a gravel formation dating from the glaciation period (Table 13.1; Le Tacon et al. 1997). In most documented cases, *T. aestivum* has been found on Rendzic Leptosol and Calcic to Calcaric Cambisol.

Czech Republic (Gryndler et al. 2013) A productive area near Prague is Rendzic Leptosol (Skeletic) formed on Silurian limestone.

Germany (South West, Baden-Württemberg, Stobbe et al. 2012, 2013) A broad range of calcareous bedrocks have been associated with truffle sites (*T. aestivum* was recorded on 116 of 121 sites where truffles have been found). The Jurassic rock formations of the Swabian Jura and the Rhine valley slopes, as well as Quaternary glacial deposits of the northern pre-Alps, are the most common bedrocks, followed by molasses, loess, and volcanic tuff.

Hungary (Bratek and Halász 2005; Gogan et al. 2012) *T. aestivum* is uniformly spread in the Carpathian basin. The bedrock has various origins: limestone, tertiary loess, middle Miocene volcanic ashes covered with loess, etc. The most famous *T. aestivum* habitat is located in the Jászság region. This area is situated in the middle of Hungary, between the Danube and Tisza rivers. This flatland area is basically covered by river alluviums leading to Chernozems, Fluvisols, Solonchaks, and Arenosols formation.

Italy The most represented situations in Tuscany are the marly limestones ("Alberese" and "Macigno di Londa" formations), the stratified limestones (Cretaceous and Jurassic), and the coarse sediments (sands mixed pebbles in sandy matrix) (Gardin 2005). Around Parma, burgundy truffle grows on land derived from sedimentary rock (limestone marl) Mesozoic (Cretaceous) and Cenozoic (Paleocene, Eocene, Oligocene), called Flysch (Belloli et al. 1999; Gregori 2010).

Poland Parent soil material suitable for *T. aestivum* in Poland is from Cretaceous marlstone, marly limestone, and gypsum (Hilszczánska et al. 2008) and from Miocene clays and sand. Soil cover consists primarily of Cambisols and Chernozems (Hilszczánska et al. 2013).

Slovakia (Miko et al. 2008) Ascoma are harvested in clayey soils rich in organic matter. Soils are Rendzic Leptosols, Cambisols, or Luvisols.

Spain (García-Montero et al. 2014; Menta et al. 2014) *T. aestivum* is mainly present in central Spain (province of Guadalajara) on Jurassic and Cretaceous limestones and dolomites. Soils are Lithic and Rendzic Leptosols.

Sweden (Wedén et al. 2004a) The bedrock belongs to five different groups (from the youngest to the oldest): (1) bedrock belonging to the Hemse group, partly fine oolitic limestone, reef-shaped limestone, reef limestone, marly limestone, and marlstone (nine sites); (2) Mulde marlstone, marlstone and marly limestone (one site); (3) Halla limestone, stratified, more or less marly limestone and reef limestone (one site); (4) Slite group, stratified, crystalline limestone and reef limestone, marlstone and marly limestone, sand limestone, and lime sandstone (six sites); and (5) Högklint limestone, stratified, more or less marly limestone and marlstone and reef limestone (one site) (Wedén et al. 2004a).

United Kingdom (Hall et al. 2007) Though rarely found, *T. aestivum* is mainly present in southern England, Scotland, and Wales, in shallow, lime-rich soils over Secondary and Tertiary limestones.

Israel *T. aestivum* does not appear to occur naturally in this country. Ascoma have been found in planted areas in Upper Galilee on bedrock from Cretaceous with dolomitic soils (Turgeman et al. 2012).

Canada (Berch pers comm) *T. aestivum* does not occur naturally in Canada and has only been found in one orchard planted following liming treatment on Vancouver Island on Mesozoic-upper Cretaceous undivided sedimentary rocks. According to Canadian System of Soil Classification, the soils are orthic dystric brunisol, shallow lithic, glaciomarine deposits (Berch and Bonito 2014).

13.4 Physical Properties

13.4.1 Soil Texture

Tuber aestivum can be found in soils with a large variety of textures as shown in the soil textural triangle (Fig. 13.2) established from 86 soil analyses of productive sites: clay, silty clay, silty clay loam, clay loam, loam, sandy loam, sandy clay loam, silt loam. The broad range in soil particle size distribution demonstrates that *T. aestivum* has a wider soil texture tolerance than the Périgord black truffle (see Chap. 11). Clearly, fruiting bodies can be harvested in soils that contain low to high sand percentages (2.8–79.8 %), low to high silt percentages (9.8–67.4 %), and low to medium clay percentages (5–55 %). Only soils with extreme textures have been excluded (sand, silt, and clay contents superior to 80 %, 70 %, and 60 %, respectively) (Fig. 13.2).

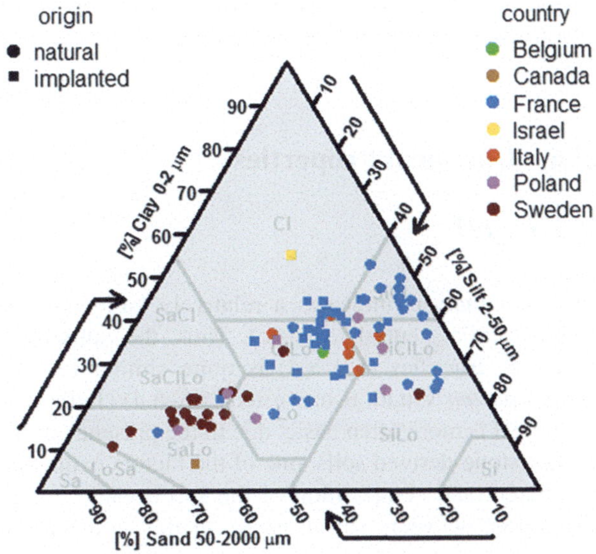

Fig. 13.2 Soil texture (USDA texture *triangle*) of the sites in which *T. aestivum* ascoma have been harvested (83 soils of 7 countries). Natural forest sites are represented by a *square* and implanted sites by a round point. Each site is represented by a colored point specific for each country

13.4.2 Soil Water-Holding Capacity and Drainage

Many soils of *T. aestivum* truffle orchards are characterized by a high water-holding capacity due to their silty clayey texture and/or their high organic matter content. High soil water availability is favorable to fruiting body production by lowering water stresses during drought periods. The topographic position at the base of slopes is particularly favorable, as observed in temperate climates. Under Mediterranean climate, ascoma production in summer months is often canceled due to insufficient water availability.

The bedrock is often cracked, and the presence of stones, associated with the granular and stable structure, ensures good drainage. Proper water infiltration prevents erosion (on sloped terrain) or waterlogging (on flat terrain), the latter being deleterious for mycorrhiza physiology. In Italy, the production sites are generally located on shallow soils, with depths of roughly 50–60 cm and consistently well drained (Gardin 2005).

13.4.3 Soil Structure

In sites where *T. aestivum* is harvested, the soil structure is generally good to excellent, being highly loose and stable (with a granular structure). Organic matter content together with the presence of $CaCO_3$ provides the soils with good structure, which can compensate for the clayey texture often found in eastern France. A loose and stable structure ensures high porosity throughout the year, which in turn increases soil water infiltration and aeration. Yet, soil structure and porosity also depend on cultivation practices. Soil compaction due to heavy traffic and/or intensive grazing in wet soils should be avoided.

13.5 Chemical Properties

13.5.1 pH

The meta-analysis indicates a relatively large range of soil pH variation (from slightly acidic, neutral to basic). Thus, the range of soil pH in water where *T. aestivum* fructifies (5.9–8.4) is quite similar to the pH of soils where *Tuber melanosporum* Vittad. is harvested (5.5–8.4) (Jaillard et al. 2014; see Chap. 11).

Soil pH is more often basic, due to the presence of limestone in all soil horizons. On limestone-derived soils, one of the factors explaining variations in pH levels is the content and composition of the soil organic matter. For example, a negative correlation between soil organic matter content and water pH (Spearman's rho = −0.72) can be observed in the implanted truffle grounds of France (solely limestone-derived soils) (Fig. 13.3).

The distribution of soil pH in water is similar for both natural sites and orchards (Fig. 13.4).

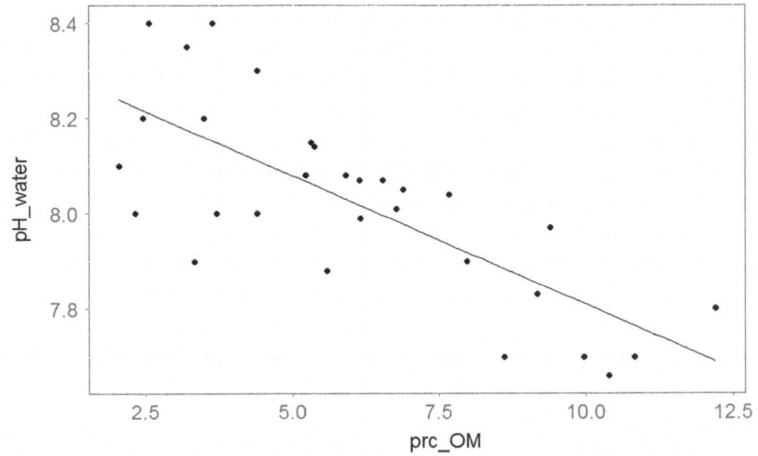

Fig. 13.3 Relationship between soil pH in water (pH_water) and organic matter content (in %, organic matter: prc_OM) for the French truffle orchard considered in this study (among 129 soils from 10 countries, we considered only limestone-derived soils that had been subject to the same historical uses prior to implanting *T. aestivum* and which provided a sufficient number of samples to assess the relationship between soil pH in water and organic matter content)

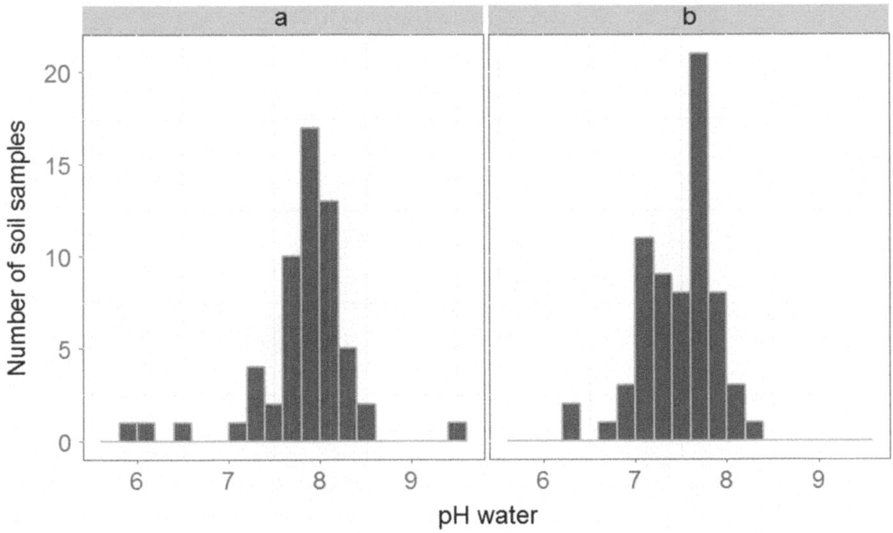

Fig. 13.4 Distribution of soil pH (in water) in sites where *T. aestivum* ascoma were harvested: orchards (**a**) and natural forest sites (**b**)

13.5.2 Carbonates

Soil carbonate content is highly variable, with values ranging between 0 and 69 %. Our data set shows differences of soil carbonate distribution between natural and implanted truffle orchards (Fig. 13.5). No direct correlation between Figs. 13.4 and 13.5 can be made as several soils were analyzed only for pH in water and not for carbonate content. *Tuber aestivum* orchards are implanted in soils that tend to be more carbonated than soils in which *T. aestivum* occurs naturally. The soil calcium carbonate content ranges from 0 to 55 % in natural forest soils where *T. aestivum* is harvested, but it is mainly harvested in soils with limited amounts of calcium carbonate (0.1–1.5 %) and most of the natural sites with a $CaCO_3$ content lower than 20 %. These results reinforce findings that unlike truffles with higher commercial value such as *Tuber magnatum* Pico and *T. melanosporum*, *T. aestivum* is able to live and form fruiting bodies in soils poor in carbonates. In productive orchards, the soil calcium carbonate content ranges from 3 to 80 %.

Although that most of the times *T. aestivum* orchards are implanted on carbonated soils, the contents of total or active calcium carbonate are not relevant descriptors as assumed by several authors (Callot 1999; García-Montero et al. 2007; Jaillard et al. 2008). Indeed, in natural habitats *T. aestivum* ascoma can be found up to pH 5.9 in the upper part of the soils (according to this data set which may be not representative of all situations).

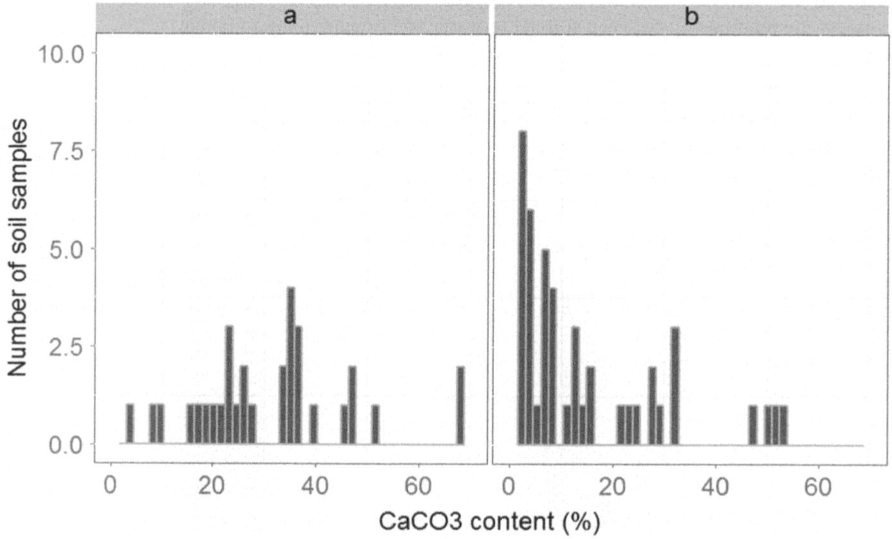

Fig. 13.5 Distribution of soil carbonates content in soils of orchards (**a**) and in soils of natural or secondary forest sites (**b**)

13.5.3 Organic Matter

The range of organic matter is considerable (0.7–21.2 %). In noncultivated carbonated soils, $CaCO_3$ coats the organic matter and prevents mineralization despite the high levels of biological activity. The organic matter combined with $CaCO_3$ is partly responsible for the exemplary structure of Burgundy truffle soils found in eastern France (Le Tacon et al. 1997). The data collected show that most of the truffle orchards display relatively low organic matter content, which is most likely related to the land's agricultural history (i.e., it is difficult to implant truffles in soils of a former forest), whereas natural truffle grounds display the highest organic matter content (Fig. 13.6).

The cationic exchange capacity (CEC) ranges from 5 to 30 cmol+ kg^{-1} in orchards and from 15 to 45 cmol+ kg^{-1} in natural or secondary forest sites where the content of organic matter is often higher than in truffle orchards. CEC of calcic and calcaric soils [Rendzic Leptosols, Epileptic Cambisols (calcaric), Cambisols (calcaric), Cambisols (hypereutric)] is higher than CEC of acidic soils although values vary considerably (Badeau et al. 1999). At neutral or basic pH levels, the CEC increases in proportion to the organic matter content (Badeau et al. 1999). The data collected show that, despite higher pH in soils with implanted truffles (mostly former agricultural soils), the proportion of high CEC soils is relatively low due to their lower organic matter content (Fig. 13.7).

Most of the time, the C/N ratio of the *T. aestivum* soils is quite low, at around 10–12, which is characteristic of fertile soils rich in stabilized organic matter. Only a few soils showed a C/N ratio higher than 20 (Fig. 13.8). A low C/N ratio confers to

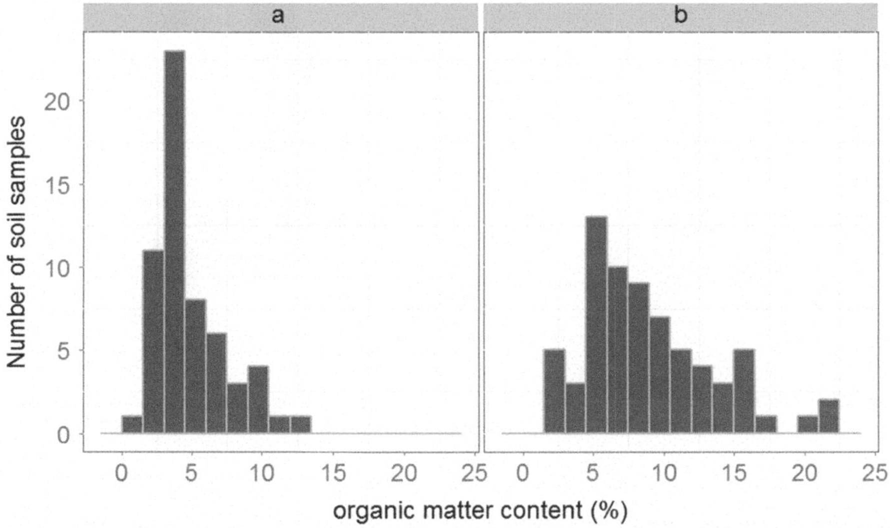

Fig. 13.6 Distribution of soils according to their organic matter content. Orchards (**a**) and natural forest sites (**b**)

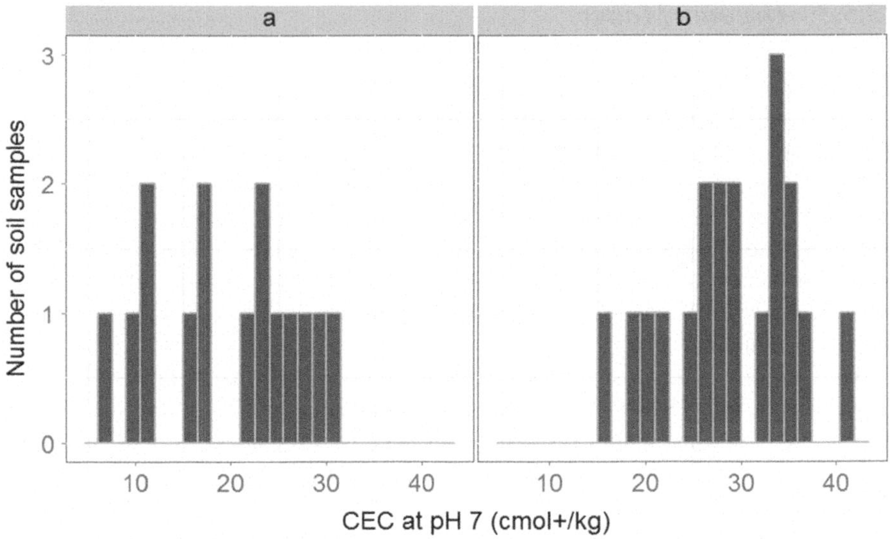

Fig. 13.7 Distribution of soils according to their CEC at pH 7. Orchards (**a**) and natural or secondary forest sites (**b**)

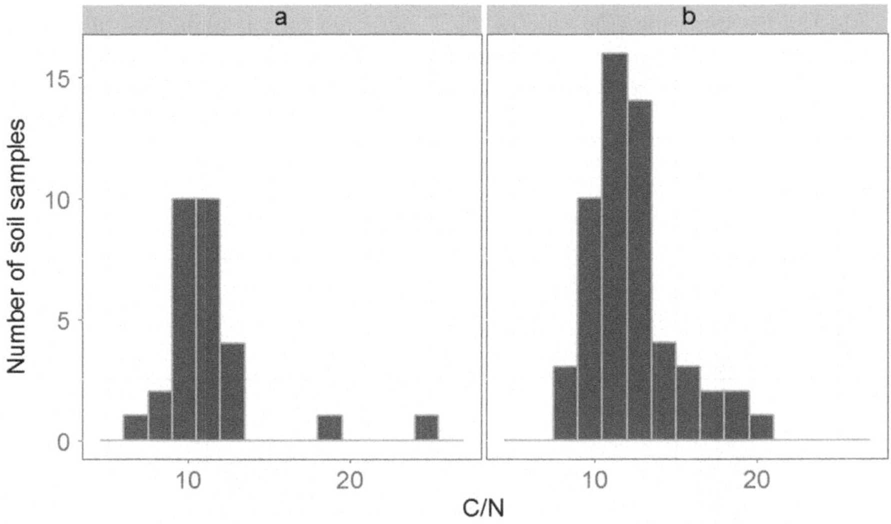

Fig. 13.8 Distribution of soils according to their C/N ratio. Orchards (**a**) and natural sites (**b**). *Tuber aestivum* ascoma have been harvested on each of the sites included in the graph

the microorganisms in soils the capacity to decompose the organic matter and, thus, to supply mineral nitrogen to the ecosystem. It is interesting to note that 90 % of agricultural soils have a C/N ratio less than 11 when compared to 8 % of forest soils. Nearly 50 % of agricultural soils have a C/N ratio ranging between 9 and

10 (Badeau et al. 1999). Thus, for this parameter, the collected data once again underlines that the main differences in soil properties between implanted and natural *T. aestivum* production sites relate directly to their original use (agricultural or forest land).

13.5.4 Nutrients

The available phosphorus content in soils (Joret-Hebert or Olsen methods) is comprised between 0.0035 and 0.51 g P kg^{-1}. Most of the soils show low concentrations of available P (between 0.0035 and 0.1 g P kg^{-1}) (Fig. 13.9) which is characteristic of the majority of limestone soils. Le Tacon et al. (1997) analyzed 25 sites in France which indicated a general range of available P values from 0.009 to 0.044 g P kg^{-1}.

Soils suitable for *T. aestivum* are saturated in exchangeable Ca (66–97 %), and the ratio of exchangeable K and Mg over the CEC is relatively low (1.3–15 % and 1.4–19 %, respectively). A K:Mg ratio over 2 has a negative effect on plants' uptake of Mg (Ericksson et al. 1997 cited in Hilszczánska et al. 2013) ; Mg availability problems may occur more often or may be more pronounced when soil available K content is twice as high as the Mg content. The exchangeable K to Mg ratio is never >2 in implanted *T. aestivum* orchards, whereas it is higher than 2 in a few naturally productive areas (Fig. 13.10). It should be noted that Mg deficiencies due to excess K content have to date not been documented for truffle production, and it

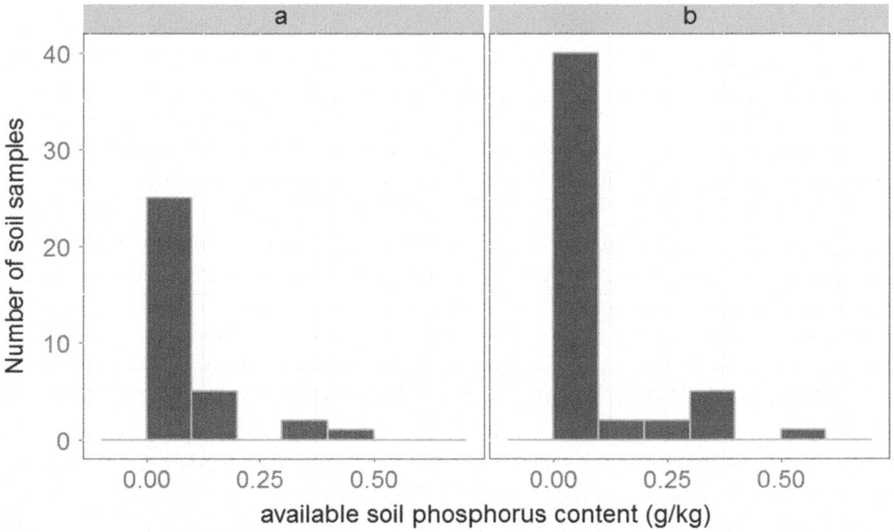

Fig. 13.9 Distribution of soils according to their phosphorus content. Orchards (**a**) and natural or secondary forest sites (**b**)

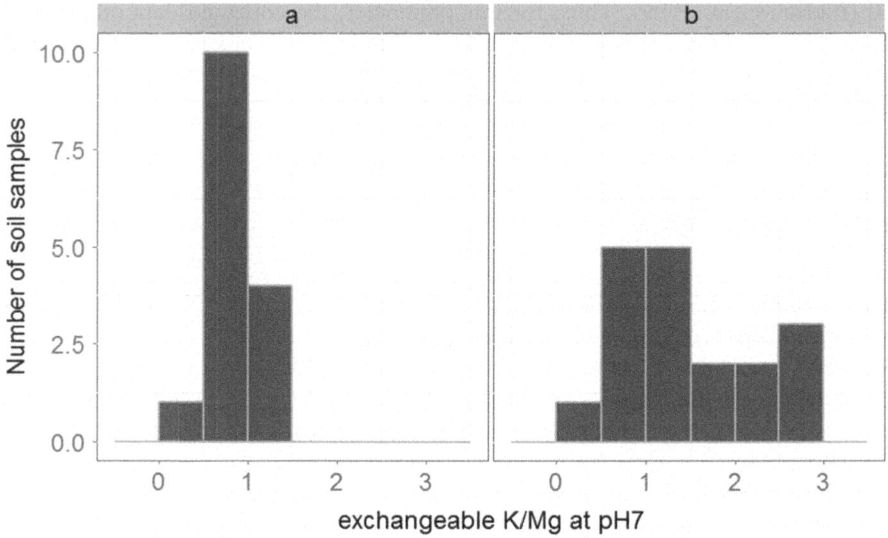

Fig. 13.10 Distribution of soils according to their K:Mg ratio measured at pH 7. Orchards (**a**) and natural sites (**b**). *Tuber aestivum* ascoma have been harvested on each of the sites included in the graph

remains difficult to draw conclusions for cases of higher than 2 K:Mg ratios in natural truffle orchards.

13.6 Conclusions

In natural or seminatural conditions, T. *aestivum* ascoma are found in soils exhibiting a wide range of pH levels, from slightly acidic to neutral and basic, when limestone is present. In plantations, the soil pH is more often basic. Many soils in which *T. aestivum* are found are characterized by a high water-holding capacity due to their silty clayey texture and/or their high organic matter content. Moreover, these sites are often well supplied with water due to their topographic positioning at the base of slopes. This high soil water availability is favorable to fruiting body production by decreasing water stress during drought periods. *Tuber aestivum* can be cultivated in almost all neutral to alkaline soils showing small to medium compactness (Bragato et al. 2009). The ecological and pedo-climatic requirements of the Burgundy truffle confer to this species an increased capacity to adapt to many environments. Anthropogenic areas such as road and rail embankments are emblematic and also favorable to the development of this truffle (Gardin 2005). The large diversity of host trees should also be emphasized (Table 13.1). This high flexibility explains why *T. aestivum* can be found spread across Europe. In many cases, in natural habitats or planted areas, *T. aestivum* is found together with other truffles species, including *T. melanosporum*.

The main challenges for future research related to *T. aestivum* soils will be to precisely identify optimal soil conditions for the initiation and growth of ascoma and to characterize conditions which facilitate the colonization and persistence of *T. aestivum* mycorrhizas in root systems. This knowledge should help to establish future best management practices for ascoma production and to optimize the current ones (irrigation, soil tillage, mulching, supplies of organic matter, micronutrients, etc.).

In order to extend the study to more sites, we invite the readers to send additional soil data sets of sites known to produce *T. aestivum*; please contact the corresponding author for guidelines.

Acknowledgments We would like to thank all colleagues and truffle growers listed below, including those who wish to remain anonymous and who provided soil data and/or published literature at local/regional/national levels: France, Jean-Sébastien Pousse (Codecom Pays de Commercy), Henri Frochot (Daix), Marc Yverneau (Humbauville), Bernard Vonflie (Munchhouse), Hubert Meyer (Otterswiller), Jean-Luc Dubois (Beine), Jean-Jacques Sauvaget (La Foye Montjault), Régis Mesnier (Saint-Cybardeaux), and Alain Tribot (Blanzac-les-Matha); Italy, Gilberto Bragato (CRA Gorizia) and Gianluigi Gregori (Sant'Angelo in Vado); Hungary, Andrea Gógán Csorbai (Szent István University, Gödöllő) and Zoltán Bratek (Eötvös Loránd University, Budapest); Poland, Dorota Hilszczanska (Forest Research Institute, Raszyn); Slovakia, Juraj Chlpik (Slovak University of Agriculture, Nitra); Spain, Daniel Oliach (Centre Tecnologic Forestal de Catalunya, Solsona); Israël, Yaron Sitrit (Ben-Gurion University of the Negev, Beer-Sheva); and Canada, Shannon Berch (British Columbia Ministry of Environment, Victoria).

References

Badeau V, Dambrine E, Walter C (1999) Propriétés des sols forestiers français résultant du premier inventaire systématique. Etude Gestion Sols 6(3):165–180

Belloli S, Bologna F, Gregori G, Zambonelli A (1999) Il tartufo nero di Fragno (*T. uncinatum* Chatin): Ecologia e coltivazione. In: Actes Vème congrès international science et culture de la truffe. Aix en Provence, France, 4–6 Mar 1999. Féderation Française des Trufficulteurs, pp 367–372

Berch SM, Bonito G (2014) Cultivation of Mediterranean species of *Tuber* (Tuberaceae) in British Columbia, Canada. Mycorrhiza 24(6):473–479. doi:10.1007/s00572-014-0562-y

Bragato G, Gardin L, Lulli L, Raglione M (2009) Edible truffles (*Tuber* spp.). In: Costantini EAC (ed) Manual of methods for soil and land evaluation. Science, Enfield, pp 254–266

Bratek Z, Halász K (2005) A *Tuber aestivum* Kárpát-medencei termőhelyei. In: Chevalier G, Frochot H, Bratek Z (eds) Az európai fekete szarvasgomba (Burgundi szarvasgomba – *Tuber uncinatum* Chatin). Első Magyar Szarvasgombász Egyesület, Budapest, pp 228–238

Callot G (1999) La truffe, la terre, la vie. INRA Editions, Paris

Ceruti A, Fontana A, Nosenzo C (2003) Le specie europee del genere *Tuber*. Una revisione storica. Museo regionale di scienze naturali, Torino

Ciesielski H, Sterckeman T (1997) Determination of cation exchange capacity and exchangeable cations in soils by means of cobalt hexamine trichloride. Effects of experimental conditions. Agronomie 17(1):1–7

García-Montero LG, Casermeiro MA, Hernando I, Hernando J (2007) Effect of active carbonate, exchangeable calcium, and stoniness of soil on *Tuber melanosporum* carpophores production. N Z J Crop Hortic 35(1):139–146. doi:10.1080/01140670709510178

García-Montero LG, Moreno D, Monleon VJ, Arredondo-Ruiz F (2014) Natural production of *Tuber aestivum* in central Spain: *Pinus* spp versus *Quercus* spp brûlés. Forest Syst 23 (2):394–399. doi:10.5424/fs/2014232-05112

Gardin L (2005) I tartufi minori in Toscana. Gli ambienti di crescita dei tartufi marzuolo e scorzone. Agenzia Regionale per lo Sviluppo e l'Innovazione nel settore Agricolo-forestale Regione Toscana, Quaderno ARSIA 1

Gogan AC, Nagy Z, Degi Z, Bagi I, Dimeny J (2012) Ecological characteristics of a Hungarian summer truffle (*Tuber aestivum* Vittad.) producing area. Acta Mycol 47(2):133–138. doi:10.5586/am.2012.015

Gregori G (2010) L'espèce *Tuber uncinatum* en Italie: écologie, commercialization et valorization. Österr Z Plizk 19:265–271

Gryndler M, Trilčová J, Hršelová H, Streiblová E, Gryndlerová H, Jansa J (2013) *Tuber aestivum* Vittad. mycelium quantified: advantages and limitations of a qPCR approach. Mycorriza 23 (5):341–348. doi:10.1007/s00572-012-0475-6

Hall IR, Brown GT, Zambonelli A (2007) Taming the truffle. The history, lore and science of the ultimate mushroom. Timber Press, Portland

Hilszczánska D, Sierota Z, Palenzona M (2008) New *Tuber* species found in Poland. Mycorrhiza 18(4):223–226. doi:10.1007/s00572-008-0175-4

Hilszczánska D, Rosa-Gruszecka A, Szmidla H (2013) Is Nida's basin habitat conducive to *Tuber aestivum* Vittad. cultivation? Wulfenia J 20(12):14–30

Jaillard B, Barry-Etienne D, Diette S, Moundy PJ, Plassard C, Salducci X, Callot G (2008) Relation entre carbonates, matière organique et pH des sols truffiers. Le Trufficulteur Français 64:8–10

Jaillard B, Barry-Etienne D, Colinas C, De Miguel AM, Genola L, Libre A, Neveu P, Oliach D, Saenz W, Saez M, Salducci X, Souche G, Sourzat P, Villeneuve M (2014) Alkalinity and structure of soils determine the truffle production in the Pyrenean regions. Forest Syst 23 (2):364–377. doi:10.5424/fs/2014232-04933

Kagan-Zur V, Freeman S, Luzzati Y, Roth-Bejerano N, Shabi E (2001) Survival of introduced *Tuber melanosporum* at two sites in Israel as measured by occurrence of mycorrhizas. Plant Soil 229(2):159–166. doi:10.1023/A:1004876504658

Le Tacon F, Chevalier G, Frochot H, Voillot JP (1997) Les sols du Bassin Parisien susceptibles de convenir aux truffières à *Tuber uncinatum*. In: Chevalier G, Frochot H (eds) La truffe de Bourgogne. Petrarque, Levallois Perret, pp 84–101

Menta C, García-Montero LG, Pinto S, Delia Conti F, Baroni G, Maresi M (2014) Does the natural "microcosm" created by *Tuber aestivum* affect soil microarthropods? A new hypothesis based on Collembolan in truffle culture. Appl Soil Ecol 84:31–37. doi:10.1016/j.apsoil.2014.06.012

Miko M, Gazo J, Priecel J, Vernarec J, Bratek Z (2008) Pôdne indikátori vhodnosti pestovatelskych podmienok pre hluzovku letnu na slovensku. Acta Fytotech Zootech 11 (3):70–72

Moeys J, Shangguan W (2014) Soil texture: functions for soil texture plot, classification and transformation. R package version 1.2.19. http://CRAN.R-project.org/package=soiltexture

Molinier V, Bouffaud ML, Castel T, Mounier A, Colombet A, Recorbet G, Frochot H, Wipf D (2013a) Monitoring the fate of a 30-year-old truffle orchard in Burgundy: from *Tuber melanosporum* to *Tuber aestivum*. Agroforest Syst 87(6):1439–1449. doi:10.1007/s10457-013-9649-2

Molinier V, van Tuinen D, Chevalier G, Gollotte A, Wipf D, Redecker D (2013b) A multigene phylogeny demonstrates that *Tuber aestivum* and *Tuber uncinatum* are conspecific. Org Divers Evol 13(4):503–512. doi:10.1007/s13127-013-0146-2

Paolocci F, Rubini A, Riccioni C, Topini F, Arcioni S (2004) *Tuber aestivum* and *Tuber uncinatum*: two morphotypes or two species? FEMS Microbiol Lett 235(1):109–115. doi:10.1111/j.1574-6968.2004.tb09574.x

R Core Team (2013) R: A language and environment for statistical computing. R Foundation for Statistical Computing, Vienna. http://www.R-project.org/

Stobbe U, Büntgen U, Sproll L, Tegel W, Egli S (2012) Spatial distribution and ecological variation of re-discovered German truffle habitats. Fungal Ecol 5(5):591–599. doi:10.1016/j.funeco.2012.02.001

Stobbe U, Stobbe A, Sproll L, Tegel W, Peter M, Büntgen U, Egli S (2013) New evidence for the symbiosis between *Tuber aestivum* and *Picea abies*. Mycorrhiza 23(8):669–673. doi:10.1007/s00572-013-0508-9

Turgeman T, Sitrit Y, Danai O, Luzzati Y, Bustan A, Roth-Bejerano N, Kagan-Zur V, Masaphy S (2012) Introduced *Tuber aestivum* replacing introduced *Tuber melanosporum*: a case study. Agroforest Syst 84(3):337–343. doi:10.1007/s10457-011-9478-0

Wedén C (2004) Black truffles of Sweden. Systematics, population studies, ecology and cultivation of *Tuber aestivum*, synonym *T. uncinatum*. PhD thesis. Faculty of Sciences and Technology, Uppsala

Wedén C, Danell E (2001) Scandinavian black truffles – distribution and habitats. In: Hall IR, Wang Y, Danell E, Zambonelli A (eds) Proceedings of the second international conference on edible mycorrhizal mushrooms. Christchurch, 3–6 July 2001, CD-rom

Wedén C, Chevalier G, Danell E (2004a) *Tuber aestivum* (syn. *uncinatum*) biotopes and their history on Gotland, Sweden. Mycol Res 108(3):304–310. doi:10.1017/S0953756204009256

Wedén C, Danell E, Camacho FJ, Backlund A (2004b) The population of the hypogeous fungus *Tuber aestivum* syn. *T. uncinatum* on the island of Gotland. Mycorrhiza 14(1):19–23. doi:10.1007/s00572-003-0271-4

Wedén C, Danell C, Tibell L (2005) Species recognition in the truffle genus *Tuber* – the synonyms *Tuber aestivum* and *Tuber uncinatum*. Environ Microbiol 7(10):1535–1546. doi:10.1111/j.1462-2920.2005.00837.x

Wickham H (2009) ggplot2: elegant graphics for data analysis. Springer, New York

Chapter 14
Soils and Vegetation in Natural Habitats of *Tuber indicum* in China

François Le Tacon, Yongjin Wang, and Noémie Goutal-Pousse

14.1 Introduction

Tuber indicum Cooke and Massee is an Asian black truffle species. *Tuber sinense* K. Tao and B. Liu, *Tuber himalayense* B. C. Zhang and Minter, *Tuber pseudohimalayense* G. Moreno, Manjón, J. Díez and García-Mont. and *Tuber formosanum* H. T. Hu and Y. Wang are additional names for the same species. These five taxa form a species complex, *T. indicum* consisting of different populations, groups or cryptic species (Zhang et al. 2005; Wang et al. 2006; Chen et al. 2011; Kinoshita et al. 2011; Belfiori et al. 2013). *Tuber indicum* has a broad geographic distribution in Asia and can be found from the 75° E meridian (India) to 140° E (Japan) and from 23° to 40° N. In China, *T. indicum* is found mainly in the Yunnan and Sichuan provinces and is less common in Tibet and Taiwan, and is found at N latitude of between 23° and 30° and at altitudes between 1000 and 3000 m AMSL, mainly at 1500–2500 m (Wang 1988, 1990).

The aim of this chapter is to define the vegetation and soil characteristics of the sites in which *T. indicum* is harvested in Yunnan, Sichuan and Tibet.

F. Le Tacon (✉)
INRA, Université de Lorraine, UMR 1136, Interactions Arbres/Micro-organismes, Laboratoire d'Excellence ARBRE, 54280 Champenoux Cedex, France
e-mail: le_tacon@nancy.inra.fr

Y. Wang
INRA, Université de Lorraine, UMR 1136, Interactions Arbres/Micro-organismes, Laboratoire d'Excellence ARBRE, 54280 Champenoux Cedex, France

SARL EU INTERNATIONAL, 132 rue Faubourg Saint Martin, 75010 Paris, France

N. Goutal-Pousse
département Recherche Développement et Innovation, Office National des Forêts, 11 rue de l'Ile-de-Corse, 54000 Nancy, France

A. Zambonelli et al. (eds.), *True Truffle (*Tuber *spp.) in the World*, Soil Biology 47,
DOI 10.1007/978-3-319-31436-5_14

14.2 Habitats, Vegetation Associated with *T. indicum* Complex and Host Trees in Sichuan and Yunnan

14.2.1 Habitats

The types of habitats in which *T. indicum* is found are complex environments. The subtropical evergreen oak forest is one of the original habitats of *T. indicum*. In these forests, evergreen oak species are associated with *Castanopsis* spp., *Cyclobalanopsis* spp., *Lithocarpus* spp. and numerous other species. A second natural habitat of *T. indicum* is constituted by the deciduous broad-leaved forests including, among other species, *Quercus* spp., *Populus yunnanensis* Dode, *Castanea mollissima* Blume, *Platycarya strobilacea* Siebold and Zuccarini and *Betula* spp. Most of these forests have been replaced by pine plantations. The pine plantations are now major habitats for *T. indicum*. *Pinus yunnanensis* Franch is the most important host species in the Pinaceae family for *T. indicum*. *Pinus yunnanensis* forms large areas of pure and mixed forests with *Keteleeria evelyniana* Mast. at elevations below 2500 m. *Pinus armandii* Franch is another important host species and comprises pure forests at higher elevations of 2300–2800 m or plantations (found at both higher and lower elevations). In Huidong county, *P. armandii* is mainly planted at lower elevations (1500–2300 m). In Yunnan, *P. armandii* forests grow naturally above 2300 m (Wang and Liu 2011).

To simplify, the habitats where *T. indicum* occurs in Sichuan ad Yunnan can be divided into two kinds of ecosystems: the hot-dry river valleys with mainly *K. evelyniana* forests and relatively cooler and wetter plateaus with *P. yunnanensis* or *P. armandii* forests or plantations (Fig. 14.1 and Fig. 14.2). Pine plantations are grazed by goats and ruminants. Shepherds working in these areas help to maintain a low density of young trees to favour grasslands. Open forests provide highly favourable conditions for truffle production. Nevertheless, overgrazing can lead to forest degradation, to the disappearance of the herbaceous vegetation and to soil erosion.

14.2.2 Associated Vegetation

To date, few phytosociological studies have been performed in the *T. indicum* natural habitats. In Huidong county, at an elevation of 2800 m, *Coriaria nepalensis* Wall., *Pyrus xerophila* Yu., *Pyracantha fortuneana* (Maxim.) Li., *Vaccinium fragile* Franch., *Dichotomanthus tristaniaecarpa* Kurz., *Rosa laevigata* Michx. and *Dendrobenthamia capitata* (Wall.) Hutch. (=*Cornus capitata* Wall.) have been identified as the indicative plants for *T. indicum* habitats (Zhang and Wang 2002). These species can also be found between 1700 and 2000 m associated with

Fig. 14.1 Open forest of *P. yunnanensis* (Sichuan, Huidong county, altitude 2500 m)

K. evelyniana and, surprisingly, with *Miscanthus floridulus* (Zhang Dacheng, pers comm).

14.2.3 Host Trees

Apart from *P. yunnanensis*, *P. armandii* and *K. evelyniana*, currently the most common hosts (Zhang and Wang 2002), *T. indicum* is also associated with species of the genus *Quercus*: *Quercus spinosa* David ex Franchet, *Quercus spinosa* David ex Franchet var. *monimotricha* Handel-Mazzetti (Ren 2003), *Quercus pannosa* Hand.-Mazz., *Quercus pseudosemicarpifolia* A. Camus, *Quercus rehderiana* Hand.-Mazz., *Quercus acutissima* Carr. (Zhang Dacheng, pers comm), *Quercus aliena* Blume (Zhang Dacheng, pers comm). They can also be associated with species of the genera *Betula* and *Picea* (Ren 2003). In small valleys with strong water supplies, *T. indicum* ascoma are found under *Alnus cremastogyne* Burk. (Zhang and Wang 2002).

Fig. 14.2 *Tuber indicum* site of *P. armandii* trees associated with *A. cremastogyne* (Sichuan, Huidong county, altitude 2800 m, small valley)

14.3 Nature of the Parent Rocks

In the Yunnan and Sichuan provinces, the majority of the parent rocks are either igneous crystal rocks or metamorphic rocks which occurred as a result of the transformation of sediments during the Hercynian orogenesis. During the Tertiary era, these rocks were strongly affected by the formation of the Himalaya mountains. The Yunnan and Sichuan provinces form a transitional zone between Tibet, the late Cretaceous Guilin karst in the Guangxi Zhuang Autonomous region and the Eastern coastal plain. The Longmen Mountains located in the centre of Sichuan represent the furthest northeastern limit of the Tibetan Plateau between the metamorphosed rocks of the Songpan-Ganze Fold Belt and the Proterozoic Sichuan basin (Dirks et al. 1994). The most common parent rocks are granites, diorites, peridotites, gabbros, high-grade metamorphic rocks, schales and sandstones. Carbonate rocks are relatively rare with the exception of the Guilin region. Substantial quantities of quaternary calcareous loess were deposited in Northern China. This loess is subjected to wind erosion, and today, regions across China including Sichuan, Yunnan and Tibet are subjected to dust fall which is often rich in calcium carbonate and which affects the chemical characteristics of the upper part of the soils (Derbyshire et al. 1998).

14.4 Soils

In 11 sites in Yunnan province, 10 in Sichuan and one in Tibet, truffle ascoma were located by examining cracks at the surface of the soil or through random digging. The collected ascoma were identified using morphological and molecular analysis which was performed by sequencing the internal transcribed spacer (ITS) regions.

Soil samples (0–10; 10–20; 20–30; 100–120 cm) (Table 14.1) were gathered under trees where *T. indicum* ascoma were found and were analysed as followed:

Table 14.1 List of soils of Sichuan, Yunnan and Tibet in which *T. indicum* ascoma have been harvested and which have been sampled for analysis

Place	Altitude (m)	Longitude	Latitude	Sampling depth (cm)
Huidong, Sichuan 1	2800	102°32′ E	26°39′ N	0–5, 10–40, 100–120
Huidong, Sichuan 2	3000	102°32′ E	26°39′ N	0–10, 10–20
Shalahe, Huidong, Sichuan 1	1700	102°32′ E	26°39′ N	0–10, 20–30
Shalahe, Huidong, Sichuan 2	1780	102°32′ E	26°39′ N	0–10, 20–30
Mindge, Huidong, Sichuan	2000	102°32′ E	26°39′ N	0–10, 20–30
Pingdi, Panzhihua, Sichuan 1	1717	101°51′ E	26°12′ N	0–10, 10–20
Pingdi, Panzhihua, Sichuan 2	1500	101°51′ E	26°12′ N	0–10
Pingdi, Panzhihua, Sichuan 3	1686	101°51′ E	26°12′ N	0–10, 10–20
Yakou, Miyi, Sichuan 1	1127	101°57′ E	26°47′ N	0–10, 10–20
Yakou, Miyi, Sichuan 2	1311	101°57′ E	26°47′ N	0–10, 10–20
Gongshan, Yunnan 1	1500	98°13′ E	27°90′ N	0–10, 20–30
Gongshan, Yunnan 2	1400	98°13′ E	27°90′ N	0–10, 20–30
Gongshan, Yunnan 3	1600	98°13′ E	27°90′ N	0–10, 20–30
Gongshan, Yunnan 4	1600	98°13′ E	27°90′ N	0–10
Gongshan, Yunnan 5	1500	98°13′ E	27°90′ N	0–10, 20–30
Gongshan, Yunnan 6	1500	98°13′ E	27°90′ N	0–10, 20–30
Gongshan, Yunnan 7	1400	98°13′ E	27°90′ N	0–10, 20–30
Gongshan, Yunnan 8	1500	98°13′ E	27°90′ N	0–10, 20–30
Gongshan, Yunnan 9	1500	98°13′ E	27°90′ N	0–10, 20–30
Gongshan, Yunnan 10	1500	98°13′ E	27°90′ N	0–10
Kunming, Yunnan	2000	102°41′ E	25°01′ N	0–10, 20–30
Bomi, Tibet	3000	97°16′ E	31°12′ N	0–10, 20–30

- Soil granulometry with organic matter destruction, while maintaining calcium carbonate, prior to analysis with the following size threshold: clay between 0 and 2 μm, silt between 2 and 50 μm and sand between 50 and 2000 μm.
- pH in water (1:5 in volume, ISO 10390).
- Soil organic matter and total nitrogen content determined by dry combustion (ISO 10694 and ISO 13878, respectively).
- Total calcium carbonate content determined by volumetric method (ISO 10693).
- Exchangeable cations (Ca, Mg and K) and cationic exchange capacity (CEC) at pH 7 (Metson method).
- Available phosphorus according to Duchaufour and Bonneau (1959) (double extraction with H_2SO_4 0.002 mol l^{-1} and NaOH 0.1 mol l^{-1}).

14.4.1 Soil Types

In the Sichuan and Yunnan provinces, the soils are influenced more by current and past climatic conditions than by the nature of the parent rocks. All of these soils are characterized by a deep paleo-alteration displaying long-lasting and intense weathering inherited from past tropical quaternary climates. They belong to the Ferralsols (WRB—World Reference Base for soil resources 2014) and are characterized by a deep ferralic horizon, often several metres deep (Fig. 14.3). The ferralic character is attenuated by conditions at altitude, inducing the formation of a cambic horizon. Above altitudes of 2800 m, the soils become Ferralic Cambisols. Higher than 3000 m, the ferralic characteristics disappear, and these soils can be considered as Cambisols. Despite their ferralic character and the acidity of many of the parent rocks, these soils generally display high base saturation, due to continuous exposure to airborne calcaric dust, as previously mentioned (Fig. 14.7).

14.4.2 Soil Physical Properties

The texture of the soils (Fig. 14.4) where *T. indicum* was harvested is relatively homogeneous: mainly loam, silty loam and clay loam. Only two soils exhibit a sandy loam texture. The percentage of clay never exceeds 40 % (Fig. 14.4).

Almost all the soils of *T. indicum* natural habitats are characterized by a high water-holding capacity due to the deep past soil weathering and to their loam texture. Moreover, they are often well supplied in water due to their topographic position located at the base of slopes or in small valleys. The majority of these soils are well drained despite their loam texture. Drainage is facilitated by these slopes, which are more often found high up in these mountainous regions. The structure of the upper horizons is generally granular and stable. It becomes blocky or prismatic in the ferralic horizons.

Fig. 14.3 A deep Ferralsol under *P. yunnanensis* (Sichuan Huidong county, altitude 2500 m)

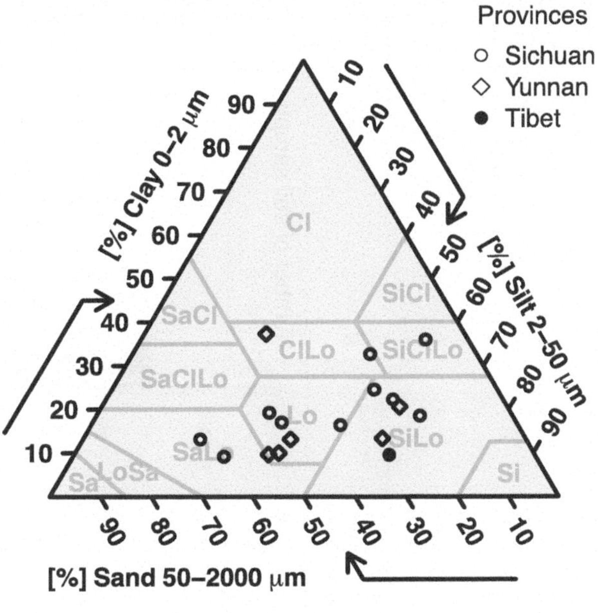

Fig. 14.4 Soil texture (USDA texture *triangle*) of the soils of Sichuan, Yunnan and Tibet in which *T. indicum* ascoma have been harvested

14.4.3 Soil Chemical Properties

14.4.3.1 Carbonates, pH, CEC, Saturation Rate and Exchangeable Cations (Figs. 14.5, 14.6 and 14.7)

Of all observed soils, only one was found to have developed on limestone. However, while the larger number of soils originally developed on acid rocks, six displayed small or negligible amounts of calcium carbonates in their upper horizons with a maximum of 0.28 %. For most cases, there is no carbonate under 20 or 30 cm. As previously mentioned, this carbonate occurrence originates from dust fall.

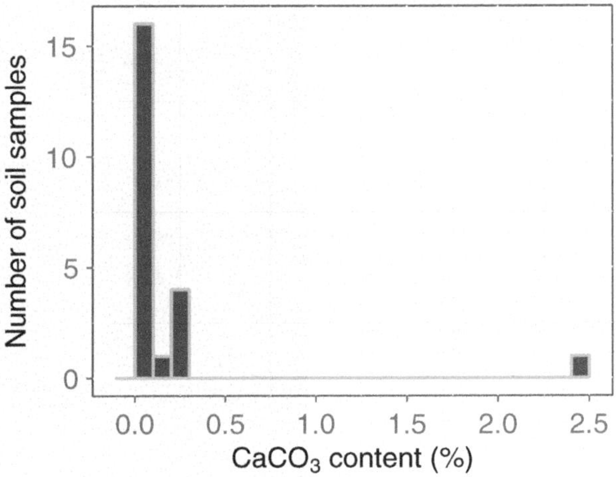

Fig. 14.5 Distribution of calcium carbonate content of upper horizons in soils of Sichuan, Yunnan and Tibet in which *T. indicum* ascoma have been harvested

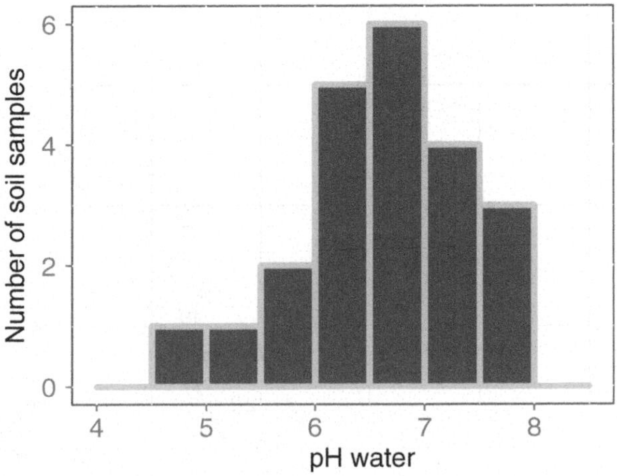

Fig. 14.6 Distribution of soil pH (in water) of upper horizons in sites of Sichuan, Yunnan and Tibet in which *T. indicum* ascoma have been harvested

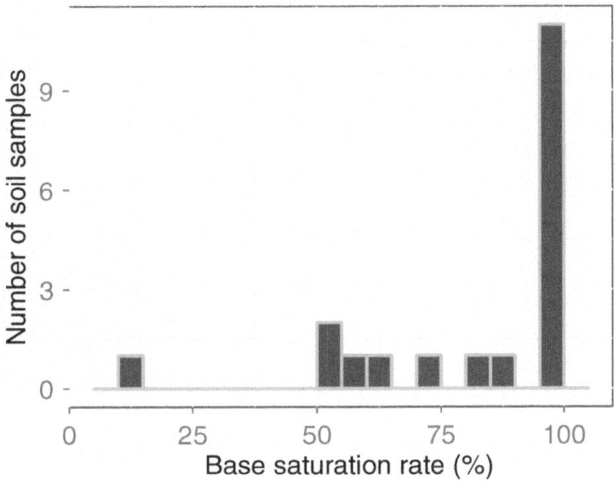

Fig. 14.7 Distribution of base saturation rates (ratio of the sum of exchangeable Ca, Mg and K over the CEC) of upper horizons of soils of Sichuan, Yunnan and Tibet in which *T. indicum* ascoma have been harvested

These dust fall events contribute to maintaining a high base saturation rate of the cationic exchange capacity in the upper horizons and, consequently, help to maintain their pH around the neutrality point. For example, a soil described in Huidong region at an elevation of 2800 m displays a pH of between 7.9 and 7.6 in the first 40 cm, while it is 5.2 for depths of between 100 and 120 cm. The same phenomenon is observed in several other soils such as those in the Miyi region, where one soil exhibits a pH of 7.37 in the first 10 cm and a pH of 5.52 between 10 and 20 cm.

For the 22 soils sampled, the pH of the upper horizon (0–10 cm) is in average of 6.68. Seven upper horizons displayed a pH higher than 7, six displayed a pH from 7 to 6.5, five from 6.5 to 6, three from 6 to 5 and only one displayed a pH lower than 5. These relatively acidic soils are found in the Gongshan region, which is likely less subjected to dust fall.

The average saturation rate is 83.8 % for the ten first cm. In the non-carbonated soils, this ranges from 53.4 to 100 %, with the exception of the acidic soil in Gongshan (14.3 %).

All of these soils displayed high levels of exchangeable cations, calcium potassium and magnesium, except for the more acidic soils found in Miyi.

14.4.3.2 Soil Organic Carbon and C/N Ratio (Figs. 14.8 and 14.9)

The percentage of soil organic carbon in the upper horizons ranges from 0.6 to 8.19 %. There is no correlation with altitude. It appears that the content of organic matter is determined by the rate of overgrazing.

The C/N ratio is relatively high ranging from 12.7 to 23.3 in the upper horizons. Nevertheless, these values, which rarely exceed 20, allow for sufficient

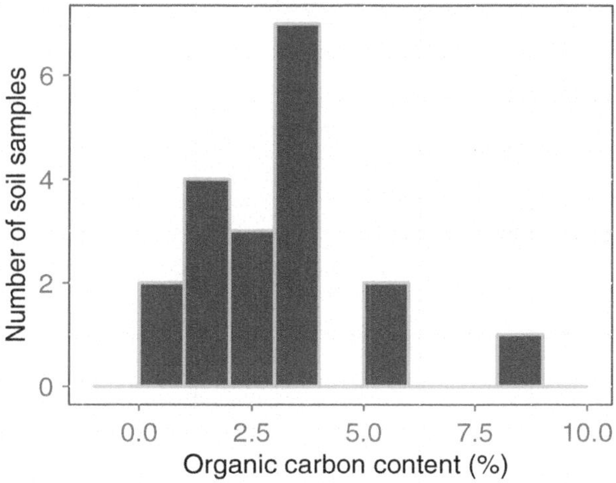

Fig. 14.8 Distribution of soil organic carbon contents in upper horizons of soils of Sichuan, Yunnan and Tibet in which *T. indicum* ascoma have been harvested

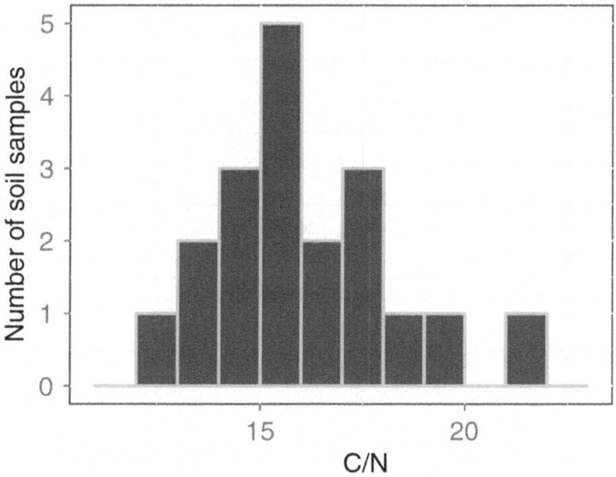

Fig. 14.9 Distribution of C/N ratio of upper horizons of soils of Sichuan, Yunnan and Tibet in which *T. indicum* ascoma have been harvested

mineralization of organic matter and provide a good supply of mineral nitrogen and other mineral elements.

14.4.3.3 Available Phosphorus (Fig. 14.10)

The available phosphorus content of the upper horizons expressed in P_20_5 (Duchaufour and Bonneau method) is comprised between 0.055 and 0.41 with an average of 0.18 g kg^{-1}, which is high. According to Duchaufour and Bonneau (1959), a value of 0.15 g kg^{-1} is considered as high with this method.

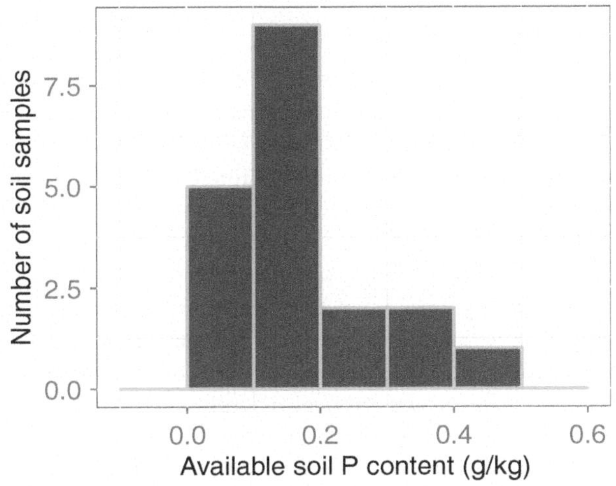

Fig. 14.10 Distribution of available phosphorus (Duchaufour and Bonneau method) in *upper* horizons of soils of the Sichuan and Yunnan provinces and in Tibet where *T. indicum* ascoma have been harvested

14.5 Conclusions

Ferralsols often display low water-holding capacity and low mineral fertility due to their sandy texture. They are frequently acidic and deficient in all essential nutrients, especially N, P, K, Ca and Mg. The Ferralsols of the Sichuan and Yunnan provinces, characterized by a deep paleo-alteration and a loamy texture, are the exception to this rule, mainly due to continuous exposure to airborne soil dust containing calcium carbonates and other mineral elements. This process contributes to maintaining a high base saturation rate of CEC in the upper horizons and, consequently, to maintaining a relatively neutral soil pH. The Ferralsols of Yunnan province are a little more acidic than those of Sichuan province which is probably due to lower levels of airborne soil dust. Contrary to common Ferralsols, the soils of Sichuan and Yunnan provinces are well provided with available phosphorus and other mineral elements. Ferralic characteristics of soil horizons are less pronounced at higher altitudes. Due to lower temperatures associated with altitude, soils found at elevations higher than 2800 m are Ferralic Cambisols, which become Cambisols above 3000 m. Rainfall is considerably high for the region, reaching as much as 1000 mm or more per year, and is particularly heavy in the summer months. Drought stress does not occur during the development of the young ascoma, and freezing very rarely occurs during truffle production. Consequently, the edaphic and climatic conditions in the Yunnan and Sichuan provinces and in the lower parts of the Tibetan plateau are highly favourable to the production of *T. indicum* ascoma.

Accidently introduced in Europe, *T. indicum* is generally considered to be an invasive species (Murat et al. 2008; see Chap. 2). However, despite the fact that *T. indicum* and *Tuber melanosporum* Vittad. are phylogenetically closely related, they differ in their ecological requirements. *Tuber indicum* grows between 1000 and 3000 m, while *T. melanosporum* is found below 1300 m. In the Sichuan and Yunnan provinces, tropical conditions are replaced by more temperate climates at higher elevations. *Tuber melanosporum* is mainly found in Mediterranean climates characterized by high summer drought stress, whereas *T. indicum* is less adapted to this type of stress. The soils of the natural habitats of the two species are also considerably different (Ferralsols, Ferralic Cambisols and Cambisols for *T. indicum*; Rendzic Leptosols and Calcaric Cambisols for *T. melanosporum*). Calcium carbonate is commonly found in *T. melanosporum* soils and rarely occurs in *T. indicum* soils. Nevertheless, differences in soil pH for the two species proved to be negligible, ranging from between 4.78 and 7.6 for *T. indicum* to between 5.5 and 8.2 for *T. melanosporum*. The texture of the soils where *T. indicum* are found is mainly loamy, while that of *T. melanosporum* soils varies. Differences in climatic and edaphic conditions of natural habitats of the Asiatic and European black truffles are significant. The adaptive capacities of *T. indicum* including its possible invasive behaviour in Europe are insufficiently understood and merit further investigations.

Acknowledgements We thank Zhu Ming TAN, Institute of Forest Ecology and Environment, Hunan Academy of Forestry, 410004 Changsha, Hunan, People's Republic of China and Da Cheng ZHANG, Committee of Science and Technology of Huidong County, Huidong, Sichuan, People's Republic of China, for their precious help. We would like to thank the anonymous reviewer for his comments or suggestions, which allowed us to greatly improve the first version of this chapter. We also thank A. Orsini for English language editing

References

Belfiori B, Riccioni C, Paolocci F, Rubini A (2013) Mating type locus of Chinese black truffles reveals heterothallism and the presence of cryptic species within the *T. indicum* species complex. PLoS One 8(12):e82353. doi:10.1371/journal.pone.0082353

Chen J, Guo SX, Liu PG (2011) Species recognition and cryptic species in the *Tuber indicum* complex. PLoS One 6(1):e14625. doi:10.1371/journal.pone.0014625

Derbyshire E, Meng X, Rob A, Kemp RA (1998) Provenance, transport and characteristics of modern aeolian dust in western Gansu Province, China, and interpretation of the Quaternary loess record. J Arid Environ 39(3):497–516. doi:10.1006/jare.1997.0369

Dirks PHGM, Wilson CJL, Chen S, Luo ZL, Liu S (1994) Tectonic evolution of the NE margin of the Tibetan Plateau; evidence from the central Longmen Mountains, Sichuan Province, China. J Southeast Asian Earth 9(1–2):181–192. doi:10.1016/0743-9547(94)90074-4

Duchaufour P, Bonneau M (1959) Une nouvelle methode de dosage du phosphore assimilable dans les sols forestiers. Bull Assoc Fr Etude Sols Sci Sol 4:193–198

Kinoshita A, Sasaki H, Nara K (2011) Phylogeny and diversity of Japanese truffles (*Tuber* spp.) inferred from sequences of four nuclear loci. Mycologia 103:779–794. doi:10.3852/10-138

Murat C, Zampieri E, Vizzini A, Bonfante P (2008) Is the Perigord black truffle threatened by an invasive species? We dreaded it and it has happened! New Phytol 178(4):699–702. doi:10.1111/j.1469-8137.2008.02449.x

Ren DJ (2003) Systematic study on *Tuber* in China. M Sc thesis. Fujian Agricultural and Forestry University

Wang Y (1988) Study on truffles and exploration of Chinese truffle resources. Mushroom 3:2–3

Wang Y (1990) First report of study on *Tuber* species from China. In: Bencivenga M, Granetti B (eds) Proceedings of 2nd international congress on truffles, Spoleto, 24–27 Nov 1988. Comunità dei Monti Martani, Serano e Subasio, Spoleto, pp 45–50

Wang Y, Liu PG (2011) Verification of Chinese names of truffles and their conservation in natural habitats. Plant Divers Resour 33(6):625–642. doi:10.3724/SP.1143.2011.11059

Wang YJ, Tan ZM, Zhang DC, Murat C, Jeandroz S, Le Tacon F (2006) Phylogenetic and populational study of the *Tuber indicum* complex. Mycol Res 110(9):1034–1045. doi:10.1016/j.mycres.2006.06.013

Zhang D, Wang Y (2002) Study on Chinese truffle and its ecology. Edible Fungi China 9(2):25–27

Zhang L, Yang ZL, Song DS (2005) A phylogenetic study of commercial Chinese truffles and their allies: axonomic implications. FEMS Microbiol Lett 245(1):85–92. doi:10.1016/j.femsle.2005.02.028

Part III
The Biotic Environment

Chapter 15
Tools to Trace Truffles in Soil

Javier Parladé, Herminia De la Varga, and Joan Pera

15.1 Introduction

Truffles are hypogeous ectomycorrhizal (ECM) fungi belonging to ascomycetes and producing fruiting bodies with high economic and social value (Boa 2004; Mello et al. 2006; Reyna and García-Barreda 2014). Bonito et al. (2010) reported at least 180 species of *Tuber* around the world, although only about a dozen species are considered to have commercial interest. The black truffle, *Tuber melanosporum* Vittad., is the most appreciated of the commercialized species (Reyna 2000; Reyna and García-Barreda 2014; Le Tacon et al. 2014; Murat 2015). This species is naturally distributed in Mediterranean calcareous soils in Spain, France, and Italy and has been successfully introduced in a range of countries by the establishment of orchards with inoculated seedlings (Yun and Hall 2004; Reyna and García-Barreda 2014; see Chap. 2). The culture of truffles provides economic returns in a sustainable way, considering both the conservation of natural areas and the social benefits derived from maintaining rural populations in marginal areas for agriculture (Reyna 2007). It is estimated that most of the commercialized European truffles already come from plantations (Olivier et al. 2012). Recent estimates establish between 80 and 90 % of the French production of *T. melanosporum* from truffle orchards

J. Parladé (✉) • J. Pera
IRTA, Sustainable Plant Protection, Centre de Cabrils, Ctra. de Cabrils km 2, 08348 Cabrils, Barcelona, Spain
e-mail: xavier.parlade@irta.cat

H. De la Varga
INRA, Laboratoire d'Excellence ARBRE, UMR 1136 Interactions Arbres-Microorganismes, 54280 Champenoux, France

Université de Lorraine, UMR1136 Interactions Arbres-Microorganismes, 54500 Vandoeuvre-lès-Nancy, France

A. Zambonelli et al. (eds.), *True Truffle (*Tuber *spp.) in the World*, Soil Biology 47,
DOI 10.1007/978-3-319-31436-5_15

(Reyna and García-Barreda 2014; Olivier et al. 2012), whereas this percentage lowers to around 50–60 % in Spain and Italy (Reyna and García-Barreda 2014).

In the 1960s and 1970s, the synthesis of *Tuber* mycorrhizas in controlled conditions started in Italy and France (Murat 2015). Different techniques were developed to improve the quality of the inoculated plants in collaboration with commercial nurseries (Reyna and García-Barreda 2014; Murat 2015). It is estimated that around 600,000 seedlings inoculated with truffles (mainly *T. melanosporum* and *Tuber aestivum* Vittad.) are produced yearly around the world. Truffle orchards occupy over 40,000 ha in France, Italy, and Spain, with a rate of plantation between 500 and 1000 ha per year (Reyna and García-Barreda 2014).

Ecological and agronomic requirements for the establishment of truffle orchards are well reported (Verlhac 1990; Chevalier and Pargney 2014; Ponce et al. 2014). Long-term studies on cultural factors influencing inoculated seedlings and ECMs as irrigation, fertilization, and weed control have been carried out by several authors (Bonet et al. 2006; Mamoun and Olivier 1997; Olivera et al. 2011; Ricard et al. 2003). However, to our knowledge only one study reported the effect of watering on fruiting body production (Le Tacon et al. 1982).

Given the long delay (around 10 years) between the establishment of the plantation and the formation of fruiting bodies (Shaw et al. 1996), short- and medium-term control of the survival and persistence of the fungal symbiont in plantations have to be evaluated.

Different methodological tools have been developed to trace truffles in soil, from fruiting body collections and mycorrhiza morphotyping to recent molecular techniques able to detect specific truffle DNA markers in the soil. The purpose of this chapter is to review these techniques and to discuss their implementation in a more rational and efficient truffle culture.

15.2 Tracing Truffles in the Soil

15.2.1 Fruiting Body Surveys

Truffles produce hypogeous fruiting bodies which are difficult to find without a trained dog. Moreover, the onset of fruiting starts between 5 and 10 years after seedling establishment in truffle orchards, depending on the tree species (Verlhac 1990), making the traceability of the fungal persistence a difficult task during the first stages of the plantation. Moreover, fruiting body surveys are limited to the fruiting season (from December to March for *T. melanosporum* in Europe, with a peak in mid-January) (Reyna and Colinas 2007).

Truffle productivity is highly dependent on annual variability of the climatic conditions. Le Tacon et al. (2014) showed strong annual fluctuations in black truffle production in France in the last 25 years. They identified factors as the cumulative hydric balance from May to August, the cumulative rainfall in November, and the number of freezing days with a minimum temperature equal to or less than −5 °C in the year "n" to explain most of the variation of truffle sales in three main markets in southeastern France. Estimates for *T. melanosporum* production in the last 23 years in France, Italy, and Spain were reported by Reyna and García-Barreda (2014) showing similar annual variability. Other long-term series of data showed also a high inter-annual variability of *Lactarius deliciosus* (L.) Gray and *Boletus edulis* Bull. fruiting in *Pinus sylvestris* L. forests (Martínez-Peña et al. 2012) and when considering ECM fungi altogether (Egli 2011).

15.2.2 The Brûlé

The zones with scanty vegetation around host plants colonized by some ECM fungi have been referred as brûlé (in French) or burn. This effect is based mainly on phytotoxic effects of metabolites emitted by some *Tuber* species, typically *T. melanosporum*, *T. aestivum*, and *Tuber indicum* Cooke & Massee, which affect the herbaceous cover and roots of host plants in this particular ecosystem (Streiblová et al. 2012). The brûlé is used for experts as an indicator to assess the presence of *Tuber* in the truffle orchards (García-Barreda et al. 2007). The intensity of the brûlé may be variable between the different *Tuber* species (Verlhac 1990). Also, there are sterile, unproductive brûlés and other fungal species different than truffles able to produce this effect. Some fungal species as the cosmopolitan genus *Astraeus* have a similar herbicidal-like action and can contribute to form a brûlé similar to that produced by the truffle mycelium (De Miguel et al. 2014).

García-Montero et al. (2007a) showed a correlation between *T. melanosporum* production and the size of 41 *Cistus laurifolius* L. burns, with burn size explaining 26 % of the variability in fruiting body production. In a study performed with 208 burns associated with *Quercus ilex* L. ssp. *ballota*, *Quercus faginea* Lam., *Corylus avellana* L., and *Tilia platyphyllos* Scop. in the center of the Iberian peninsula, it was found that burn size was a very significant factor and accounted for 38 % of the variance in fruiting body production (Garcia-Montero et al. 2007b).

A later study showed evidence that the area of *T. aestivum* brûlés differs among hosts (Garcia-Montero et al. 2014). In this study, the effect of both host and brûlé size on truffle production was significant, so truffle production depends on host and increases with the area of the brûlé.

15.2.3 Ectomycorrhizas

Gardes and Bruns (1996) surveyed the ECM community of a *Pinus muricata* D.Don forest by molecular techniques and found that the correspondence between above- (fruiting bodies) and belowground (mycorrhizas) was imprecise. More than half of the species forming mycorrhizas were poorly or not represented at all in the aboveground fruiting record. Among other possible causes, they indicated that differences exist among species in the phenology of mycorrhizas production. Thus, the information obtained from mycorrhizas can be dependent on the sampling date but may give more insights than fruiting body surveys only in the assessment of fungal persistence in soil. Other factors can influence the overlapping between below- and aboveground community: methodological differences in sampling (Smith et al. 2007) or nutritional strategies (Lalli et al. 2015).

Different approaches have been adopted for determining mycorrhizal persistence in both natural producing areas and truffle orchards (Kagan-Zur et al. 2001; Olivera et al. 2011; Reyna and De Miguel 2012). Sampling of truffle ECMs provides good information on the symbiotic activity of the fungus, but they may be a difficult task if the tree host roots are present in the deep layers of the soil, as in the case of *Quercus* species. Also, visual identification requires trained analysts, especially for discrimination among *Tuber* species. De Miguel et al. (2014) reviewed the ECM fungus diversity in both truffle orchards and natural forests. Among the 85 reviewed papers, 64 identify the ECMs of truffle plantations only according to morphological aspects (morphotypes), emphasizing the difficulties in identifying the ECMs to species level. Molecular techniques based on the amplification of the ITS rDNA region allowed for the accurate identification of ECMs belonging to species phylogenetically close to *Tuber* (Henrion et al. 1994; Mello et al. 1996; Paolocci et al. 1996, 1997, 1999). Zambonelli et al. (2000) were the first to combine morphological and molecular tools in truffle research. Studies combining both methodologies have increased significantly since then, with 18 articles published to date (De Miguel et al. 2014). The ECM fungi identified in truffle plantations worldwide reveal more than 100 species corresponding to 31 fungal genera. Studies by Chevalier (1994), Iotti et al. (2010), and Di Massimo et al. (2010) indicate that different *Tuber* species are the most important competitors in *T. melanosporum*, *Tuber magnatum* Pico, and *T. aestivum* stands.

Seasonal variability of *T. melanosporum* mycorrhizas is unknown. Studies carried out with other ECM fungi reveal differences in ECM abundance related to the sampling date. De la Varga et al. (2012) showed significant differences between the number of mycorrhizas of *B. edulis* sampled in September *vs* October and November. However, most of the studies involving ECMs of *T. melanosporum* have performed root sampling in spring (De Miguel et al. 2014).

Sánchez et al. (2014) compared two methods for ECM sampling in *T. melanosporum* plantations: direct root sampling and soil core collection. Both methodologies were equally effective in detecting this species, although soil core collection proved a better method to evaluate also ECM fungal diversity. According to these authors, detecting the presence of *T. melanosporum* ECMs in a homogeneous plantation is simple, as only a maximum of three trees and a single soil core per tree are needed. Similarly, one tree can be diagnosed with a maximum of three soil samples.

15.2.4 Extraradical Mycelium

The detection of extraradical mycelium of the target fungal species in the soil is a sound complement to fruiting body production and ECM sampling for tracking the fungal persistence along the whole biological cycle. Extraradical mycorrhizal mycelium has a key role in soil nutrient uptake and reciprocal transfer of carbon and nutrients between plants (Simard et al. 2002; Guidot et al. 2003; Landeweert et al. 2003a, b). However, because of the methodological difficulties for mycelium detection in soil, this is the most poorly understood phase of the symbiosis (Read 1992; Horton and Bruns 2001; Leake et al. 2004). The different methods to study structure, function, and turnover of extraradical mycorrhizal mycelium in soil such as direct measurements, biochemical and DNA-based markers, and root-free hyphal compartmentalization have been reviewed in Leake et al. (2004) and Wallander et al. (2013).

Selective primers for conventional PCR amplification have been designed for different *Tuber* species (Table 15.1). Moreover, specific amplification of soil DNA by conventional PCR has been applied to detect extraradical mycelium of *T. aestivum* (Gryndler et al. 2011), *T. melanosporum* (Suz et al. 2006), and *T. magnatum* (Zampieri et al. 2010).

Molecular methods have also been used for quantification of specific ECM fungi in soil to evaluate the ecological and functional impact of a given species in its natural environment. Guidot et al. (2002, 2003) quantified mycelium of *Hebeloma cylindrosporum* Romagn. in complex DNA mixtures extracted from forest soils by competitive PCR. Most of the recent fungal DNA quantification studies applied the real-time PCR technique (Heid et al. 1996; Schild 1996; Schena et al. 2004). This technique is based on the detection and quantification of a fluorescent signal generated by a fluorescently labeled sequence-specific probe or a intercalate dye (Higuchi et al. 1992; Lee et al. 1993; Livak et al. 1995). This signal increases in direct proportion to the amount of PCR product in a reaction. Thus, by recording the amount of fluorescence emission at each cycle, it is possible to monitor the PCR reaction during the exponential phase, where the first significant increase in the amount of PCR product correlates to the initial amount of target DNA template. The higher the starting copy number of the nucleic acid target, the sooner a significant increase in fluorescence is observed. Landeweert et al. (2003b) compared real-time PCR to conventional quantification techniques (total hyphal length, biochemical markers) for mycelial biomass estimates. They found that soil DNA extracts gave fluctuating results in the real-time quantitative PCR, whereas the standard samples with purified ITS plasmid inserts produced highly repetitive standard curves.

Schubert et al. (2003) used real-time PCR with specific primers and a TaqMan® probe for absolute quantification of extraradical hyphal biomass of *Piloderma croceum* Erikss. & Hjortst. in pure cultures and rhizotron samples. Raidl et al. (2005) calibrated the method to obtain a correlation between mycelial biomass of *P. croceum* and isolated rDNA ITS copies for absolute quantification. They

Table 15.1 Selective primers for conventional PCR amplification of different *Tuber* species

Tuber species	Sequence (5′-3′)	Primer name	Reference
T. borchii	TGTATGGGATGCCCTATCGGACT	TboI (fwd)	Amicucci et al. (1998)
"	CTATTACCACGGTCAACTTC	TboII (rev)	
"	GAAGTTGACCGTGGTAATAG	rTboII (fwd)	Amicucci et al. (2000)
"	TCCTCCGCTTATTGATATGC	ITS4 (rev)	
T. brumale	CAATGTCAGAGCCAATCTAATGC	ITSB (fwd)	Paolocci et al. (1999)
"	TGATATGCTTAAGTTCAGCGGG	ITS4LNG (rvs)	
"	TCGTCAGTGGTACACAATGT	SYLV1	Douet et al. (2004)
"	ATGACAGAACAAATAGAATGTAA	SYLV2	
T. dryophilum	ATCGGGCTCCCAAGCAAAACA	TdryI (fwd)	Amicucci et al. (1998)
"	TCTACTACCATGGTTCACTTT	TdryII (rvs)	
T. indicum	AACAACAGACTTTGTAAAGGGTTG	ITSCHCH (fwd)	Paolocci et al. (1999)
"	TGATATGCTTAAGTTCAGCGGG	ITS4LNG (rvs)	
"	ACCTGTGGGAGATCTCCAC	IndF1 (fwd)	Mabru et al. (2001)
"	GGCCATGTGTCAGATTTACTG	IndF2 (fwd)	
"	CATAGACTAGCAATTCACTCCTG	IndR (rvs)	
T. macrosporum	CGTCGCTCATCAAAGCAGTC	Tmacr For	Benucci et al. (2011)
"	CCGCCAGTACCACCAGGAG	TmacrRev	
T. maculatum	GACACAGGCTCCCGATAAAACAC	TmacI (fwd)	Amicucci et al. (1998)
"	CAGCAGCACTGATAGCCCCG	TmacII (rev)	
"	CGGGGCTATCAGTGCTGCTG	rTmacII (fwd)	Amicucci et al. (2000)
"	TCCTCCGCTTATTGATATGC	ITS4 (rvs)	
T. magnatum	GGATGCGTCTCCGAATCCTGAAT	TmagI (fwd)	Amicucci et al. (1998)
"	CGGGCCCTTTCTCAGACTGCTG	TmagII (rev)	
"	TCCTCCGCTTATTGATATGC	ITS4 (rev)	Amicucci et al. (2000)
"	CCTCCCAATTTGCAATACAC	tubmagnf (fwd)	Zampieri et al. (2010)
"	AAAGACGAAGTTATCTGGCCTGA	elytubr (rvs)	
T. melanosporum	TGGCCATGTGTCAGATTTAGTA	ITSML (fwd)	Paolocci et al. (1999)

(continued)

Table 15.1 (continued)

Tuber species	Sequence (5′-3′)	Primer name	Reference
"	TGATATGCTTAAGTTCAGCGGG	ITS4LNG (rvs)	
"	GTATTCCCGAACACAAACCT	ITS1TM	Suz et al. (2006)
"	AGACTTGTGACTGATCCAGG	ITS2TM	
"	CACCTGTGGGAGATCTCTAT	MELF	Douet et al. (2004)
"	TGTGACTGATCCAGGTCTTAT	MELR	
T. oligospermum	CTCCTGAGCTGAGGTGTC	Tolf (fwd)	Boutahir et al. (2013)
"	TCCTCCGCTTATTGATATGC	ITS4 (rev)	
T. puberulum	TCTGTTACCAGGGTCCACATT	TpuI (fwd)	Amicucci et al. (1998)
"	GGCTTCTGGGTTGAGGTGTTT	TpuII (rvs)	
T. rufum	TGCTTTCCCAGGTGGTTGG	Ru1f (fwd)	Iotti et al. (2007)
"	TTGCTTTCCCAGGGAATTGG	Ru2f (fwd)	
"	TCCTCCGCTTATTGATATGC	ITS4 (rvs)	
T. uncinatum	TGGGCCGCCGAAAA TTG	UncI	Mello et al. (2002)
"	CTGACGAGATGCCCCGGA	UncII	
"	AGAGCACCAAACCACAG	Tu1sekvF	Gryndler et al. (2011)
"	ACCACAGCGTCTACCAA	Tu1sekvR	

pointed out that the application of the technique in environmental samples strongly depends on the availability of fungal DNA extraction protocols for different types of soil.

Quantitative detection of *T. melanosporum* DNA in soil was carried out by Suz et al. (2008) and Zampieri et al. (2012) using real-time PCR with SYBR® Green dye for nucleic acid stain to compare the quantity of soil mycelium in productive versus nonproductive truffle orchards. More specific real-time PCR detection can be obtained using TaqMan® probes, which are theoretically able to discriminate a single nucleotide mismatch. Parladé et al. (2007) and Hortal et al. (2008, 2009) adapted the TaqMan® real-time PCR technique to quantify the mycelium of *L. deliciosus* in the soil. Further studies showed seasonal patterns in the distribution of extraradical mycelium of *B. edulis* in a forest soil (De la Varga et al. 2012, 2013) and the correlation between extraradical mycelium and mycorrhizas of *L. deliciosus* in pot experiments (Hortal et al. 2008). Similar correlations were found in *B. edulis* in natural Scots pine forests (De la Varga et al. 2012).

Parladé et al. 2013 designed specific primers and a TaqMan® probe for *T. melanosporum* and found significant differences between the mycelium found

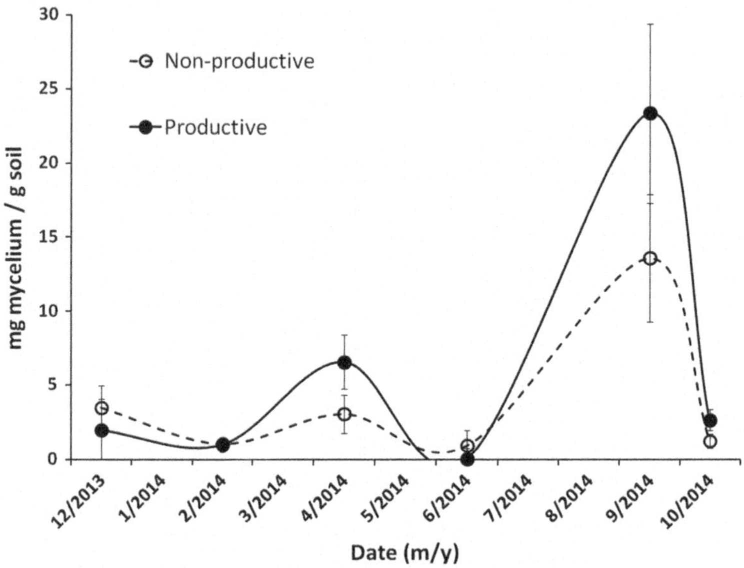

Fig. 15.1 Temporal dynamics of extraradical mycelium biomass of *T. melanosporum* under productive and nonproductive trees in a 12-year-old plantation. Data correspond to the mean of 4 productive trees and 8 nonproductive trees. Bars represent the standard errors

in truffle orchards subjected to different managements and in a natural truffle ground. Seasonal variability in extraradical fungal mycelium in truffle stands has been reported in Morcillo et al. (2015) showing the temporal dynamics of extraradical mycelium along 10 months (Fig. 15.1). These results indicate a pattern with a peak in spring and autumn and a lower average mycelium biomass detected under nonproductive trees as compared to productive ones. These results have been also obtained for *T. magnatum* (Iotti et al. 2014). The detection of fungal mycelium of this species is the only way to establish the presence of *T. magnatum* in soil in absence of fructifications since mycorrhizas are extremely rare or absent even in productive patches. Although these results are challenging to elucidate the relationship among temporal mycelium dynamics, plantation management, and fruiting body formation, more data are necessary to interpret the high variability found between sites and in different years.

Real-time PCR can be used for absolute quantification of DNA and for quantitation of gene products (gene expression) by isolating mRNA, converting it to DNA by reverse transcription and then quantifying the cDNA. The studies on application of the real-time methodology for either absolute DNA quantification or to study gene expression in ECM fungi are summarized in Table 15.2. Specific primers and probes have been reported for *T. aestivum*, *T. magnatum*, and *T. melanosporum*, mainly using TaqMan® chemistry, whereas all of the studies carried out with *Tuber borchii* Vittad. have been addressed to quantify gene expression using SYBR® Green chemistry. Gryndler et al. (2013) reported some limitations in using real-time

Table 15.2 Application of real-time PCR (qPCR) methodology (TaqMan® and SYBR®Green chemistries) for the quantification and gene expression (references in bold italics) of different ECM fungi

Fungus	qPCR chemistry	References
Amanita phalloides	SYBR®Green	Wolfe et al. (2010)
Boletus edulis	TaqMan®	De la Varga et al. (2012, 2013)
Hydnangium sp.	SYBR®Green	***Da Silva Coelho et al.*** *(*2010*)*
Lactarius deliciosus	TaqMan®	Parladé et al. (2007), Hortal et al. (2008, 2009)
Paxillus involutus	TaqMan®	Landeweert et al. (2003b)
Piloderma croceum	TaqMan®	Schubert et al. (2003), Raidl et al. (2005)
Pisolithus albus	SYBR®Green	***Majorel et al.*** *(*2012*)*
P. microcarpus	SYBR®Green	Drigo et al. (2012)
P. tinctorius	SYBR®Green	***Zhou et al.*** *(*2014*)*
Rhizopogon spp.	TaqMan®	Hortal et al. (2008)
Stipitate hydnoid fungi	SYBR®Green	Van der Linde et al. (2009)
Suillus bovinus	TaqMan®	Landeweert et al. (2003b)
Tuber borchii	SYBR®Green	***Guescini et al.*** *(*2003, 2007, 2009*)*, ***Guidi et al.*** *(*2006*)*, ***Abba et al.*** *(*2007*)*, ***Amicucci et al.*** *(*2010*)*, ***Zeppa et al.*** *(*2010*)*
T. aestivum	TaqMan®	Gryndler et al. (2013)
T. magnatum	TaqMan®	Iotti et al. (2012, 2014)
T. melanosporum	TaqMan® SYBR®Green	Suz et al. (2008), ***Zarivi et al.*** *(*2011, 2013*)*, ***Balestrini et al.*** *(*2012*)*, Zampieri et al. (2012), Parladé et al. (2013), Taschen et al. (2015)

PCR for soil mycelium quantification of *T. aestivum* based on the presence of steady extracellular fungal DNA in the soil. This DNA pool reaches a steady concentration, irrespective of the soil treatment, which should be known to determine the detection limit of the real-time PCR quantification. After autoclaving, the soil is recolonized by saprotrophic microorganisms and this conserved DNA pool can be eliminated.

15.2.5 Mating Types

The advances in the knowledge of the life cycle and reproductive biology of the two most economically important truffle species worldwide, *T. magnatum* and *T. melanosporum* (Rubini et al. 2005; Paolocci et al. 2006; Riccioni et al. 2008), as well as the recent sequencing of the *T. melanosporum* genome (Martin et al. 2010) have displayed new horizons in truffle research. From this information, it has been demonstrated that *T. melanosporum* is a heterothallic species and the genes involved in the sexual compatibility processes, with two mating-type (MAT) idiomorphs (MAT1-1 and MAT1-2), have been characterized (Martin et al. 2010; Rubini et al. 2011), allowing a greater understanding of the reproductive mechanisms of the black truffle. Also, this research is crucial to improving the management strategies of plantations aimed at increasing the controlled production of this edible fungus.

In truffle orchards, the production of fruiting bodies faces two crucial stages: the initiation of the sexual reproduction, which requires the meeting of the two compatible mycelium strains, and the growth of the fruiting bodies during several months. In order to better understand how sexual reproduction occurs, Murat et al. (2013) characterized the *T. melanosporum* small-scale genetic structure in two 20-year-old truffle plantations located in France and Italy. They highlighted that *T. melanosporum* truffle orchards are dynamic ecosystems which can contain several small black truffle genets in an area of 30 m^2. Moreover, an evidence of nonrandom distribution pattern of *T. melanosporum* was observed, resulting in a spatial separation between large soil patches exclusively colonized by fungal strains of the same mating types. In two recent reviews, Rubini et al. (2014) and Le Tacon et al. (2015) showed the practical implications of the competition existing between strains of opposite mating types in the root system of host plants, suggesting that a vegetative incompatibility occurs at ECM level. However, there are only a few works reporting the distribution of black truffle mating types in soil (Rubini et al. 2011; Zampieri et al. 2012; Murat et al. 2013). On those studies the authors detected one or both mating types by conventional PCR in soil samples, but they revealed that the detection is difficult by this technique and probably the results were not explaining the real distribution of both idiomorphs in soil. This could be due to the low amount of the target genes obtained from soil DNA extractions (MATs are single-copy genes), problem that might be solved by using more sensitive techniques than conventional PCR as the real-time PCR or the digital PCR. These techniques would allow to detect and quantify MAT1-1 and MAT1-2 idiomorphs in soil samples as well as to evaluate their temporal distribution changes, especially in spring when fertilization is supposed to occur. Recently, in 2015, a protocol was developed together between French research center (INRA) and ALCINA Fôrets, a start-up, to detect and quantify both *T. melanosporum* mating-type idiomorphs in soil samples using the real-time PCR.

15.3 Sampling Protocols

A considerable amount of variation is inherent in data collected from any soil sampling study and the utility of the resulting data is often dependent on the sampling design used. Variation of ECM fungal mycelium in the soil is likely to occur at a wide range of spatial scales. Pickles et al. (2010) used geostatistical tools to establish a core distance to avoid autocorrelations when sampling Scots pine ECMs. This requires an initial high investment in sampling units but can reap benefits later once optimum sampling distances are identified (Wallander et al. 2013). Most soil samples in studies on mycorrhizas and/or mycelium analysis of different truffle orchards and natural grounds have frequently been collected close to the base of the productive and nonproductive trees or in productive spots. Only one study considered small-scale sampling in a 1x1 m grid to evaluate the spatial genetic structure of *T. melanosporum* genets (Murat et al. 2013). Soil cores had variable diameter and ranged from 5 cm (Gryndler et al. 2013) to 30 cm depth (Iotti et al. 2014), although most of the studies used soil samples taken to a depth of 10–15 cm. Pooled samples were used in some studies to minimize soil variability (Suz et al. 2008; Antony-Babu et al. 2013; Parladé et al. 2013; Iotti et al. 2012, 2014). The limited number of samplings in many of the studies is due to restrictions by the owners of the truffle orchards and/or to the risk of breaking down the fungal web digging the soil of the productive truffle ground (Murat et al. 2005). DNA extractions were generally performed using commercial soil DNA isolation kits following the manufacturer's instructions, but, in some cases, the samples were previously submitted to an extraction phase using 2 % CTAB and then purified with a commercial kit (Iotti et al. 2012, 2014). Soil treatments before extraction included no treatment (from fresh soil) (Gryndler et al. 2013), freeze drying (Iotti et al. 2012, 2014), and low- or room-temperature drying (Parladé et al. 2013; Taschen et al. 2015).

Soil samples for *Tuber* spp. studies have been taken in a range of months from January to December. *Tuber magnatum* DNA was detected in most of the soil samples taken in a truffle ground in the Piedmont (Italy) collected in winter, while in spring the positive soils were fewer (Zampieri et al. 2010). On the other hand, Iotti et al. (2014) found *T. magnatum* mycelium all around the year in four sites along the Italian peninsula and it was more abundant in spring. Since the soil mycelium biomass might vary greatly along the year (Fig. 15.1), further studies on *Tuber* spp. mycelium dynamics are necessary to establish the relationships between spatial and temporal mycelium biomass allocation and fruiting body formation under different climatic-edaphic conditions.

15.4 Conclusions

Recent findings in disclosing the biological cycle of truffles based on molecular research have represented a great step forward toward rational truffle cultivation. The detection and quantification of truffle mycelium in the soil along the truffle development might be the key to understand the processes leading to fertilization and fruiting body formation. The real-time PCR methodology developed for soil mycelium traceability can also be used to control the quality of the plants produced by the nurseries and to detect the possible presence of other species in the inoculum used (especially *T. indicum* and *Tuber brumale* Vittad.). In 2014–2015 the research center INRA in Nancy (France) developed a quality protocol to control the inoculum of *T. melanosporum* and the quality of *T. magnatum* plants by real-time PCR in French nurseries.

On the other hand, the relationships between the different phases of the symbiosis, mycorrhizas, extraradical mycelium, and fruiting body formation, remain unsolved. High-sensitive techniques such as real-time PCR and the recent digital PCR could help to detect the single-copy MAT genes in the soil and might disclose the origin of the "male factor" in the biological cycle. Spatial segregation between mating types in adult trees is particularly interesting and seems to be linked with vegetative incompatibility. This indeed decreases the probability that compatible cells meet and therefore lowers the odds of mating. Therefore, this would be negative for truffle production, as indicated by Selosse et al. (2013). The perspectives for truffle cultivation are challenging to face this biological process which limits sexual encounters, by studying the spatial and temporal dynamics of the compatible genotypes. Hopefully, new technologies for tracing truffles in soil will contribute to unravel these uncertainties in the near future.

Acknowledgments Financial support for part of the studies presented by the authors was provided by the Spanish Ministry of Economy (MINECO), grant AGL2012-40035-C03-03, and the Generalitat de Catalunya, AGAUR, Innovative Pilot Project AAM/259/2013. The UMR1136 is supported by a grant overseen by the French National Research Agency (ANR) as part of the Investments for the Future program (ANR-11-LABX-0002-01, Lab of Excellence ARBRE).

References

Abba S, Balestrini R, Benedetto A, Rottensteiner H, De Lucas JR, Bonfante P (2007) The role of the glyoxylate cycle in the symbiotic fungus *Tuber borchii*:expression analysis and subcellular localization. Curr Genet 52(3):159–170. doi:10.1007/s00294-007-0149-7

Amicucci A, Zambonelli A, Giomaro G, Potenza L, Stocchi V (1998) Identification of ectomycorrhizal fungi of the genus *Tuber* by species-specific ITS primers. Mol Ecol 7 (3):273–277. doi:10.1046/j.1365-294X.1998.00357.x

Amicucci A, Guidi C, Zambonelli A, Potenza L, Stocchi V (2000) Multiplex PCR for the identification of white *Tuber* species. FEMS Microbiol Lett 189(2):265–269. doi:10.1111/j. 1574-6968.2000.tb09241.x

Amicucci A, Zambonelli A, Iotti M, Polidori E, Menotta M, Saltarelli R, Potenza L, Stocchi V (2010) Morphological and molecular modifications induced by different carbohydrate sources in *Tuber borchii*. J Mol Microbiol Biotechnol 18(2):120–128. doi:10.1159/000297915

Antony-Babu S, Murat C, Deveau A, Le Tacon F, Frey-Klett P, Uroz S (2013) An improved method compatible with metagenomic analyses to extract genomic DNA from soils in *Tuber melanosporum* orchards. J Appl Microbiol 115(1):163–170. doi:10.1111/jam.12205

Balestrini R, Sillo F, Kohler A, Schneider G, Faccio A, Tisserant E, Martin F, Bonfante P (2012) Genome-wide analysis of cell wall-related genes in *Tuber melanosporum*. Curr Genet 58 (3):165–177. doi:10.1007/s00294-012-0374-6

Benucci GMN, Raggi L, Di Massimo G, Baciarelli Falini L, Bencivenga M, Falcinelli M, Albertini E (2011) Species-specific primers for the identification of the ectomycorrhizal fungus *Tuber macrosporum* Vittad. Mol Ecol Resour 11(2):378–381. doi:10.1111/j.1755-0998.2010.02915.x

Boa E (2004) Wild edible fungi. A global overview of their use and importance to people, vol 17. Non wood forest products. FAO, Rome

Bonet JA, Fischer CR, Colinas C (2006) Cultivation of black truffle to promote reforestation and land-use stability. Agron Sustain Dev 26(1):69–76. doi:10.1051/agro:2005059

Bonito G, Gryganskyi AP, Trappe JM, Vilgalys R (2010) A global meta-analysis of *Tuber* ITS rDNA sequences: species diversity, host associations and long-distance dispersal. Mol Ecol 19 (22):4994–5008. doi:10.1111/j.1365-294X.2010.04855.x

Boutahir S, Iotti M, Piattoni F, Zambonelli A (2013) Morphological and molecular characterization of *Tuber oligospermum* mycorrhizas. Afr J Agric Res 8(29):4081–4087. doi:10.5897/AJAR2013.7354

Chevalier G (1994) Evolution des recherches sur les plants mycorhizes pour la truffe et perspectives de dévelopment. Giornale Botanico Italiano 128(1):7–18. doi:10.1080/11263509409437010

Chevalier G, Pargney JC (2014) Empirical or rational truffle cultivation? It is time to choose. For Syst 23(2):378–384. doi:10.5424/fs/2014232-04964

Da Silva CI, de Queiroz MV, Costa MD, Kasuya MC, de Araújo EF (2010) Identification of differentially expressed genes of the fungus *Hydnangium* sp. during the pre-symbiotic phase of the ectomycorrhizal association with *Eucalyptus grandis*. Mycorrhiza 20(8):531–540. doi:10.1007/s00572-010-0301-y

De la Varga H, Águeda B, Martínez-Peña F, Parladé J, Pera J (2012) Quantification of extraradical soil mycelium and ectomycorrhizas of *Boletus edulis* in a Scots pine forest with variable sporocarp productivity. Mycorrhiza 22(1):59–68. doi:10.1007/s00572-011-0382-2

De la Varga H, Águeda B, Ágreda T, Martínez-Peña F, Parladé J, Pera J (2013) Seasonal dynamics of *Boletus edulis* and *Lactarius deliciosus* extraradical mycelium in pine forests of central Spain. Mycorrhiza 23(5):391–402. doi:10.1007/s00572-013-0481-3

De Miguel AM, Águeda B, Sánchez S, Parladé J (2014) Ectomycorrhizal fungus diversity and community structure with natural and cultivated truffle hosts: applying lessons learned to future truffle culture. Mycorrhiza 24(Suppl 1):S5–S18. doi:10.1007/s00572-013-0554-3

Di Massimo G, Benucci GMN, Albertini E, Raggi L, Gigliotti G, Bencivenga M (2010) Ecologia di *Tuber magnatum* Pico nell'alta valle del Chiascio (Italia centrale). In: Donnini D (ed) Proceedings of 3rd international congress on truffles, Spoleto, Comunità Montana dei Monti Martani, Serano e Subasio, 25–28 Nov 2008, pp 196–204

Douet JP, Castroviejo M, Mabru D, Chevalier G, Dupré C, Bergougnoux F, Ricard JM, Médina B (2004) Rapid molecular typing of *Tuber melanosporum*, *T. brumale* and *T. indicum* from tree seedlings and canned truffles. Anal Bioanal Chem 379(4):668–673. doi:10.1007/s00216-004-2643-9

Drigo B, Anderson IC, Kannangara K, Cairney JWG, Johnson D (2012) Rapid incorporation of carbon from ectomycorrhizal mycelial necromass into soil fungal communities. Soil Biol Biochem 49:4–10. doi:10.1016/j.soilbio.2012.02.003

Egli S (2011) Mycorrhizal mushroom diversity and productivity—an indicator of forest health? Ann For Sci 68(1):81–88. doi:10.1007/s13595-010-0009-3

García-Barreda S, Reyna S, Pérez-Badía R, Rodríguez-Barreal JA (2007) Ecologia de la trufa y las áreas truferas. In: Reyna S (ed) Truficultura, fundamentos y técnicas. Mundi-Prensa, Madrid

García-Montero LG, Casermeiro MA, Manjón JL, Hernando I (2007a) Impact of active soil carbonate and burn size on the capacity of the rockrose *Cistus laurifolius* to produce *Tuber melanosporum* carpophores in truffle culture. Mycol Res 111(6):734–739. doi:10.1016/j.mycres.2007.03.017

Garcia-Montero LG, Manjón JL, Pascual C, Garcia-Abril A (2007b) Ecological patterns of *Tuber melanosporum* and different *Quercus* Mediterranean forests: quantitative production of truffles, burn sizes and soil studies. For Ecol Manage 242(2):288–296. doi:10.1016/j.foreco.2007.01.045

Garcia-Montero LG, Moreno D, Monleon VJ, Arredondo-Ruiz F (2014) Natural production of *Tuber aestivum* in central Spain: *Pinus* spp. versus *Quercus* spp. brûles. For Syst 23 (2):394–399. doi:10.5424/fs/2014232-05112

Gardes M, Bruns TD (1996) Community structure of ectomycorrhizal fungi in a *Pinus muricata* forest: above- and below-ground views. Can J Bot 74(10):1572–1583. doi:10.1139/b96-190

Gryndler M, Hršelová H, Soukupová L, Streiblová E, Valda S, Borovička J, Gryndlerová H, Gažo J, Miko M (2011) Detection of summer truffle (*Tuber aestivum* Vittad.) in ectomycorrhizae and in soil using specific primers. FEMS Microbiol Lett 318(1):84–91. doi:10.1111/j.1574-6968.2011.02243.x

Gryndler M, Trilčová J, Hršelová H, Streiblová E, Gryndlerová H, Jansa J (2013) *Tuber aestivum* Vittad. mycelium quantified: advantages and limitations of a qPCR approach. Mycorrhiza 23 (5):341–348. doi:10.1007/s00572-012-0475-6

Guescini M, Pierleoni R, Palma F, Zeppa S, Vallorani L, Potenza L, Sacconi C, Giomaro G, Stocchi V (2003) Characterization of the *Tuber borchii* nitrate reductase gene and its role in ectomycorrhizae. Mol Genet Genomics 269(6):807–816. doi:10.1007/s00438-003-0894-3

Guescini M, Zeppa S, Pierleoni R, Sisti D, Stocchi L, Stocchi V (2007) The expression profile of the *Tuber borchii* nitrite reductase suggests its positive contribution to host plant nitrogen nutrition. Curr Genet 51(1):31–41. doi:10.1007/s00294-006-0105-y

Guescini M, Stocchi L, Sisti D, Zeppa S, Polidori E, Ceccaroli P, Saltarelli R, Stocchi V (2009) Characterization and mRNA expression profile of the TbNre1 gene of the ectomycorrhizal fungus *Tuber borchii*. Curr Genet 55(1):59–68. doi:10.1007/s00294-008-0222-x

Guidi C, Zeppa S, Annibalini G, Pierleoni R, Guescini M, Buffalini M, Zambonelli A, Stocchi V (2006) The isoprenoid pathway in the ectomycorrhizal fungus *Tuber borchii* Vittad.: cloning and characterisation of the tbhmgr, tbfpps and tbsqs genes. Curr Genet 50(6):393–404. doi:10.1007/s00294-006-0097-7

Guidot A, Debaud JC, Marmeisse R (2002) Spatial distribution of the below-ground mycelia of an ectomycorrhizal fungus inferred from specific quantification of its DNA in soil samples. FEMS Microbiol Ecol 42(3):477–486. doi:10.1111/j.1574-6941.2002.tb01036.x

Guidot A, Debaud JC, Effosse A, Marmeisse R (2003) Belowground distribution and persistence of an ectomycorrhizal fungus. New Phytol 161(2):539–547. doi:10.1046/j.1469-8137.2003.00945.x

Heid CA, Stevens J, Livak KJ, Williams PM (1996) Real-time quantitative PCR. Genome Res 6 (10):986–994. doi:10.1101/gr.6.10.986

Henrion B, Chevalier G, Martin F (1994) Typing truffle species by PCR amplification of the ribosomal DNA spacers. Mycol Res 98(1):37–43. doi:10.1016/S0953-7562(09)80333-X

Higuchi R, Dollinger G, Walsh PS, Griffith R (1992) Simultaneous amplification and detection of specific DNA-sequences. Bio-Technol 10(4):413–417. doi:10.1038/nbt0492-413

Hortal S, Pera J, Parladé J (2008) Tracking mycorrhizas and extraradical mycelium of the edible fungus *Lactarius deliciosus* under field competition with *Rhizopogon* spp. Mycorrhiza 18 (2):69–77. doi:10.1007/s00572-007-0160-3

Hortal S, Pera J, Parladé J (2009) Field persistence of the edible ectomycorrhizal fungus *Lactarius deliciosus*: effects of inoculation strain, initial colonization level and site characteristics. Mycorrhiza 19(3):167–177. doi:10.1007/s00572-009-0228-3

Horton TR, Bruns TD (2001) The molecular revolution in ectomycorrhizal ecology: peeking into the black-box. Mol Ecol 10(8):1855–1871. doi:10.1046/j.0962-1083.2001.01333.x

Iotti M, Amicucci A, Bonito G, Bonuso E, Stocchi V, Zambonelli A (2007) Selection of a set of specific primers for the identification of *Tuber rufum*: a truffle species with high genetic variability. FEMS Microbiol Lett 277(2):223–231. doi:10.1111/j.1574-6968.2007.00963.x

Iotti M, Lancellotti E, Hall I, Zambonelli A (2010) The ectomycorrhizal community in natural *Tuber borchii* grounds. FEMS Microbiol Ecol 72(2):250–260. doi:10.1111/j.1574-6941.2010.00844.x

Iotti M, Leonardi M, Oddis M, Salerni E, Baraldi E, Zambonelli A (2012) Development and validation of a realtime PCR assay for detection and quantification of *Tuber magnatum* in soil. BMC Microbiol 12(1):93–101. doi:10.1186/1471-2180-12-93

Iotti M, Leonardi M, Lancellotti E, Salerni E, Oddis M, Leonardi P, Perini C, Pacioni G, Zambonelli A (2014) Spatio-temporal dynamic of *Tuber magnatum* mycelium in natural truffle grounds. PLoS ONE 9(12), e115921. doi:10.1371/journal.pone.0115921

Kagan-Zur V, Freeman S, Luzzati Y, Roth-Bejerano N, Shabi E (2001) Survival of introduced *Tuber melanosporum* at two sites in Israel as measured by occurrence of mycorrhizas. Plant Soil 229(2):159–166. doi:10.1023/A:1004876504658

Lalli G, Leonardi M, Oddis M, Pacioni G, Salerni E, Iotti M, Zambonelli A (2015) Expanding the understanding of a forest ectomycorrhizal community by combining root tips and fruiting bodies: a case study of *Tuber magnatum* stands. Turk J Bot 39(3):527–534. doi:10.3906/bot-1406-50

Landeweert R, Leeflang P, Kuyper TW, Hoffland E, Rosling A, Wernars K, Smit E (2003a) Molecular identification of ectomycorrhizal mycelium in soil horizons. Appl Environ Microbiol 69(1):327–333. doi:10.1128/AEM.69.1

Landeweert R, Veenman C, Kuyper TW, Fritze H, Wernars K, Smit E (2003b) Quantification of ectomycorrhizal mycelium in soil by real-time PCR compared to conventional quantification techniques. FEMS Microbiol Ecol 45(3):283–292. doi:10.1016/S0168-6496(03)00163-6

Leake J, Johnson D, Donnelly D, Muckle G, Boddy L, Read D (2004) Networks of power and influence: the role of mycorrhizal mycelium in controlling plant communities and agroecosystem functioning. Can J Bot 82(8):1016–1045. doi:10.1139/B04-060

Lee LG, Connell CR, Bloch W (1993) Allelic discrimination by nick-translation PCR with fluorogenic probes. Nucleic Acids Res 21(16):3761–3766. doi:10.1093/nar/21.16.3761

Le Tacon F, Delmas J, Gleyze R, Bouchard D (1982) Influence du régime hydrique du sol et de la fertilisation sur la fructification de la truffe noire du Perigord (*Tuber melanosporum* Vitt.) dans le Sud-Est de la France. Acta Oecol Oecol Appl 3:291–306

Le Tacon F, Marçais B, Courvoisier M, Murat C, Montpied P, Becker M (2014) Climatic variations explain annual fluctuations in French Périgord black truffle wholesale markets but do not explain the decrease in black truffle production over the last 48 years. Mycorrhiza 24 (Suppl 1):S115–S125. doi:10.1007/s00572-014-0568-5

Le Tacon F, Rubini A, Murat C, Riccioni C, Robin C, Belfiori B, Zeller B, De la Varga H, Akroume E, Deveau A, Martin F, Paolocci F (2015) Certainties and uncertainties about the life cycle of the Périgord black truffle (*Tuber melanosporum* Vittad.). Ann For Sci. doi:10.1007/s13595-015-0461-1

Livak KJ, Flood SJA, Marmaro J, Giusti W, Deetz K (1995) Oligonucleotides with fluorescent dyes at opposite ends provide a quenched probe system useful for detecting PCR product and nucleic acid hybridization. Genome Res 4(6):357–362

Mabru D, Dupré C, Douet JP, Leroy P, Ravel C, Ricard JM, Médina B, Castroviejo M, Chevalier G (2001) Rapid molecular typing method for the reliable detection of Asiatic black truffle (*Tuber indicum*) in commercialized products: fruiting bodies and mycorrhizal seedlings. Mycorrhiza 11(2):89–94. doi:10.1007/s005720100103

Majorel C, Hannibal L, Soupe ME, Carriconde F, Ducousso M, Lebrun M, Jourand P (2012) Tracking nickel-adaptive biomarkers in *Pisolithus albus* from New Caledonia using a transcriptomic approach. Mol Ecol 21(9):2208–2223. doi:10.1111/j.1365-294X.2012.05527.x

Mamoun M, Olivier JM (1997) Mycorrhizal inoculation of cloned hazels by *Tuber melanosporum*: effect of soil disinfestation and co-culture with *Festuca ovina*. Plant Soil 188(2):221–226. doi:10.1023/A:1004267405566

Martin F, Kohler A, Murat C, Balestrini R, Coutinho PM, Jaillon O, Montanini B, Morin E, Noel B, Percudani R, Porcel B, Rubini A, Amicucci A, Amselem J, Anthouard V, Arcioni S, Artiguenave F, Aury J-M, Ballario P, Bolchi A, Brenna A, Brun A, Buee M, Cantarel B, Chevalier G, Couloux A, da Silva C, Denoeud F, Duplessis S, Ghignone S, Hilselberger B, Iotti M, Marcais B, Mello A, Miranda M, Pacioni G, Quesneville H, Riccioni C, Ruotolo R,

Splivallo R, Stocchi V, Tisserant E, Viscomi AR, Zambonelli A, Zampieri E, Henrissat B, Lebrun M-H, Paolocci F, Bonfante P, Ottonello S, Wincker P (2010) Perigord black truffle genome uncovers evolutionary origins and mechanisms of symbiosis. Nature 464 (7291):1033–1038. doi:10.1038/nature08867

Martínez-Peña F, Ágreda T, Águeda B, Ortega-Martínez P, Fernández-Toirán LM (2012) Edible sporocarp production by age class in a Scots pine stand in Northern Spain. Mycorrhiza 22 (3):167–174. doi:10.1007/s00572-011-0389-8

Mello A, Nosenzo C, Meotto F, Bonfante P (1996) Rapid typing of truffle mycorrhizal roots by PCR amplification of the ribosomal DNA spacers. Mycorrhiza 6(5):417–421. doi:10.1007/s005720050141

Mello A, Cantisani A, Vizzini A, Bonfante P (2002) Genetic variability of *Tuber uncinatum* and its relatedness to other black truffles. Environ Microbiol 4(10):584–594. doi:10.1046/j.1462-2920.2002.00343.x

Mello A, Murat C, Bonfante P (2006) Truffles: much more than a prized and local fungal delicacy. FEMS Microbiol Lett 260(1):1–8. doi:10.1111/j.1574-6968.2006.00252.x

Morcillo M, Sanchez M, Vilanova X (2015) Truffle farming today, a comprehensive world guide. Micologia Forestal & Aplicada, Barcelona, 349 pp. ISBN 978-84-617-1307-3

Murat C (2015) Forty years of inoculating seedlings with truffle fungi: past and future perspectives. Mycorrhiza 25(1):77–81. doi:10.1007/s00572-014-0593-4

Murat C, Vizzini A, Bonfante P, Mello A (2005) Morphological and molecular typing of the below-ground fungal community in a natural *Tuber magnatum* truffle-ground. FEMS Microbiol Lett 245(2):307–313. doi:10.1016/j.femsle.2005.03.019

Murat C, Rubini A, Riccioni C, De La Varga H, Akroume E, Guaragno M, Le Tacon F, Robin C, Halkett F, Martin F, Paolocci F (2013) Fine scale spatial genetic structure of the black truffle (*Tuber melanosporum*) investigated with neutral microsatellites and functional mating type genes. New Phytol 199(1):176–187. doi:10.1111/nph.12264

Olivier JM, Savignac JC, Sourzat P (2012) Truffe et trufficulture. Editions Fanlac, Périgueux

Olivera A, Fischer CR, Bonet JA, Martínez de Aragón J, Oliach D, Colinas C (2011) Weed management and irrigation are key treatments in emerging black truffle (*Tuber melanosporum*). New For 42(2):227–239. doi:10.1007/s11056-011-9249-9

Paolocci F, Angelini P, Cristofari E, Granetti B, Arcioni S (1996) Identification of *Tuber* spp. and corresponding ectomycorrhizae through molecular markers. J Sci Food Agric 69(4):511–517. doi:10.1002/jsfa.2740690416

Paolocci F, Rubini A, Granetti B, Arcioni S (1997) Typing *Tuber melanosporum* and Chinese black truffle species by molecular markers. FEMS Microbiol Lett 153(2):255–260. doi:10.1111/j.1574-6968.1997.tb12582.x

Paolocci F, Rubini A, Granetti B, Arcioni S (1999) Rapid molecular approach for a reliable identification of *Tuber* spp. ectomycorrhizae. FEMS Microbiol Ecol 28(1):23–30. doi:10.1111/j.1574-6941.1999.tb00557.x

Paolocci F, Rubini A, Riccioni C, Arcioni S (2006) Reevaluation of the life cycle of *Tuber magnatum*. Appl Environ Microbiol 72(4):2390–2393. doi:10.1128%2FAEM.72.4.2390-2393.2006

Parladé J, Hortal S, Pera J, Galipienso L (2007) Quantitative detection of *Lactarius deliciosus* extraradical soil mycelium by real-time PCR and its application in the study of fungal persistence and interspecific competition. J Biotechnol 128(1):14–23. doi:10.1016/j.jbiotec.2006.09.010

Parladé J, De la Varga H, De Miguel AM, Sáez R, Pera J (2013) Quantification of extraradical mycelium of Tuber melanosporum in soils from truffle orchards in northern Spain. Mycorrhiza 23(2):99–106. doi:10.1007/s00572-012-0454-y

Pickles BJ, Genney DR, Potts JM, Lennon JJ, Anderson IC, Alexander IJ (2010) Spatial and temporal ecology of Scots pine ectomycorrhizas. New Phytol 186(3):755–768. doi:10.1111/j.1469-8137.2010.03204.x

Ponce RA, Ágreda T, Águeda B, Aldea J, Martínez-Peña F, Modrego MP (2014) Soil physical properties influence "black truffle" fructification in plantations. Mycorrhiza 24(Suppl 1):S55–S64. doi:10.1007/s00572-014-0558-7

Raidl S, Bonfigli R, Agerer R (2005) Calibration of quantitative real-time Taqman PCR by correlation with hyphal biomass and ITS copies in mycelia of *Piloderma croceum*. Plant Biol 7(6):713–717. doi:10.1055/s-2005-873003

Read DJ (1992) The mycorrhizal mycelium. In: Allen MF (ed) The mycorrhizal mycelium. Mycorrhizal functioning. Chapman and Hall, London, pp 102–133

Reyna S (2000) Trufa, truficultura y selvicultura trufera. Mundiprensa, Madrid

Reyna S (2007) Sostenibilidad de la truficultura: aspectos ecológicos, económicos y sociales. In: Reyna S (ed) Truficultura, fundamentos y técnicas. Mundi-Prensa, Madrid, pp 51–73

Reyna S, Colinas C (2007) Truficultura. In: Reyna S (ed) Truficultura, fundamentos y técnicas. Mundi-Prensa, Madrid

Reyna S, De Miguel AM (2012) Las micorrizas. Ciclo biológico de la trufa. In: Reyna S (ed) Truficultura. Fundamentos y Técnicas, 2nd edn. Mundi-Prensa, Madrid, pp 105–150

Reyna S, García-Barreda S (2014) Black truffle cultivation: a global reality. For Syst 23 (2):317–328. doi:10.5424/fs/2014232-04771

Ricard JM, Bergougnoux F, Callot G, Olivier JM, Pargney JC, Sourzat P (2003) La truffe. Guide technique de Trufficulture. Centre Technique Interprofessionel des Fruits et Légumes, CTIFL, Paris

Riccioni C, Belfiori B, Rubini A, Passeri V, Arcioni S, Paolocci F (2008) *Tuber melanosporum* outcrosses: analysis of the genetic diversity within and among its natural populations under this new scenario. New Phytol 180(2):466–478. doi:10.1111/j.1469-8137.2008.02560.x

Rubini A, Paolocci F, Riccioni C, Vendramin GG, Arcioni S (2005) Genetic and phylogeographic structures of the symbiotic fungus *Tuber magnatum*. Appl Environ Microbiol 71 (11):6584–6589. doi:10.1128/AEM.71.11.6584-6589.2005

Rubini A, Belfiori B, Riccion C, Tisserant E, Arcioni S, Martin F, Paolocci F (2011) Isolation and characterization of MAT genes in the symbiotic ascomycete *Tuber melanosporum*. New Phytol 189(3):710–722. doi:10.1111/j.1469-8137.2010.03492.x

Rubini A, Riccioni C, Belfiori B, Paolocci F (2014) Impact of the competition between mating types on the cultivation of *Tuber melanosporum*: *Romeo and Juliet* and the matter of space and time. Mycorrhiza 24(1):19–27. doi:10.1007/s00572-013-0551-6

Sánchez S, Ágreda T, Águeda B, Martín M, De Miguel AM, Barriuso J (2014) Persistence and detection of black truffle ectomycorrhizas in plantations: comparison between two field detection methods. Mycorrhiza 24(Suppl 1):S39–S46. doi:10.1007/s00572-014-0560-0

Schena L, Nigro F, Ippolito A, Gallitelli D (2004) Real-time quantitative PCR: a new technology to detect and study phytopathogenic and antagonistic fungi. Eur J Plant Pathol 110(9):893–908. doi:10.1007/s10658-004-4842-9

Schild TA (1996) Einführung in die real-time TaqMan^{TM} PCR technologie. PE Applied Biosystems GmbH, Weiterstadt

Schubert R, Raidl S, Funk R, Bahnweg G, Müller-Starck G, Agerer R (2003) Quantitative detection of agar-cultivated and rhizotron-grown *Piloderma croceum* Erikss. & Hjortst. by ITS1-based fluorescent PCR. Mycorrhiza 13(3):159–165. doi:10.1007/s00572-002-0212-7

Selosse MA, Taschen E, Giraud T (2013) Do black truffles avoid sexual harassment by linking mating type and vegetative incompatibility? New Phytol 199(1):10–13. doi:10.1111/nph.12329

Shaw PJA, Lankey K, Jourdan A (1996) Factors affecting yield of *Tuber melanosporum* in a *Quercus ilex* plantation in southern France. Mycol Res 100(10):1176–1178. doi:10.1016/S0953-7562(96)80177-8

Simard SW, Jones MD, Durall DM (2002) Carbon and nutrient fluxes within and between mycorrhizal plants. In: Sanders IR, van der Heijden M (eds) Mycorrhizal ecology. Springer, Berlin, pp 33–74

Smith ME, Douhan GW, Rizzo DM (2007) Ectomycorrhizal community structure in a xeric *Quercus* woodland based on rDNA sequence analysis of sporocarps and pooled roots. New Phytol 174(4):847–863. doi:10.1111/j.1469-8137.2007.02040.x

Streiblová E, Gryndlerová H, Gryndler M (2012) Truffle brûlé: an efficient fungal life strategy. FEMS Microbiol Ecol 80(1):1–8. doi:10.1111/j.1574-6941.2011.01283.x

Suz LM, Martín MP, Colinas C (2006) Detection of *Tuber melanosporum* DNA in soil. FEMS Microbiol Lett 254(2):251–257. doi:10.1111/j.1574-6968.2005.00029.x

Suz LM, Martin MP, Oliach D, Fischer CR, Colinas C (2008) Mycelial abundance and other factors related to truffle productivity in *Tuber melanosporum–Quercus ilex* orchards. FEMS Microbiol Lett 285(1):72–78. doi:10.1111/j.1574-6968.2008.01213.x

Taschen E, Sauve M, Taudiere A, Parladé J, Selosse MA, Richard F (2015) Whose truffle is this? Distribution patterns of ECM fungal diversity in *Tuber melanosporum* brûlés developed in multi-host Mediterranean plant communities. Environ Microbiol 17(8):2747–2761. doi:10.1111/1462-2920.12741

Van der Linde S, Alexander IJ, Anderson IC (2009) Spatial distribution of sporocarps of stipitate hydnoid fungi and their belowground mycelium. FEMS Microbiol Ecol 69(3):344–352. doi:10.1111/j.1574-6941.2009.00716.x

Verlhac A (1990) La truffe. Guide pratique. Centre Technique Interprofessionnel des Fruits et Légumes, CTIFL, Paris

Wallander H, Ekblad A, Godbold DL, Johnson D, Bahr A, Baldrian P, Björk RG, Kieliszewska-Rokicka B, Kjøller R, Kraigher H, Plassard C, Rudawska M (2013) Evaluation of methods to estimate production, biomass and turnover of ectomycorrhizal mycelium in forests soils—a review. Soil Biol Biochem 57:1034–1047. doi:10.1016/j.soilbio.2012.08.027

Wolfe BE, Richard F, Cross HB, Pringle A (2010) Distribution and abundance of the introduced ectomycorrhizal fungus *Amanita phalloides* in North America. New Phytol 185(3):803–816. doi:10.1111/j.1469-8137.2009.03097.x

Yun W, Hall IR (2004) Edible ectomycorrhizal mushrooms: challenges and achievements. Can J Bot 82(8):1063–1073. doi:10.1139/b04-051

Zambonelli A, Iotti M, Rossi I, Hall I (2000) Interactions between *Tuber borchii* and other ectomycorrhizal fungi in a field plantation. Mycol Res 104(6):698–702. doi:10.1017/S0953756299001811

Zampieri E, Murat C, Cagnasso M, Bonfante P, Mello A (2010) Soil analysis reveals the presence of an extended mycelial network in a *Tuber magnatum* truffle ground. FEMS Microbiol Ecol 71(1):43–49. doi:10.1111/j.1574-6941.2009.00783.x

Zampieri E, Rizzello R, Bonfante P, Mello A (2012) The detection of mating type genes of *Tuber melanosporum* in productive and non-productive soils. Appl Soil Ecol 57:9–15. doi:10.1016/j.apsoil.2012.02.013

Zarivi O, Bonfigli A, Colafarina S, Aimola PP, Ragnelli AM, Pacioni G, Miranda M (2011) Tyrosinase expression during black truffle development: from free living mycelium to ripe fruit body. Phytochemistry 72(18):2317–2324. doi:10.1016/j.phytochem.2011.08.025

Zarivi O, Bonfigli A, Colafarina S, Aimola PP, Ragnelli AM, Miranda M, Pacioni G (2013) Transcriptional, biochemical and histochemical investigation on laccase expression during *Tuber melanosporum* Vittad. development. Phytochemistry 87:23–29. doi:10.1016/j.phytochem.2012.11.019

Zeppa S, Marchionni C, Saltarelli R, Guidi C, Ceccaroli P, Pierleoni R, Zambonelli A, Stocchi V (2010) Sulfate metabolism in *Tuber borchii*: characterization of a putative sulfate transporter and the homocysteine synthase genes. Curr Genet 56(2):109–119. doi:10.1007/s00294-009-0284-4

Zhou AD, Wu XQ, Shen L, Xu XL, Huang L, Ye JR (2014) Profiling of differentially expressed genes in ectomycorrhizal fungus *Pisolithus tinctorius* responding to mycorrhiza helper *Brevibacillus reuszeri* MPt17. Biologia 69(4):435–442. doi:10.2478/s11756-014-0344-1

Chapter 16
True Truffle Host Diversity

Milan Gryndler

16.1 Introduction

Mycorrhizal symbiosis represents a type of coexistence of the host plant root with a specialized fungus. In general, the mycorrhizal fungus colonizes rhizodermis and primary root cortex of the finest roots, forming specific communication structures: mycorrhizas. In typical ectomycorrhizal (ECM) symbiosis, unlike the endomycorrhizal one, the fungus does not penetrate the host cell wall and stays in the intercellular spaces of the root cortex, forming a hyphal network called Hartig net. The root surface is covered by compact hyphal mantle. The mycorrhizas formed in ECM symbiosis are called ectomycorrhizas (ECMs) and are morphologically distinct from roots not colonized by the fungus (Smith and Read 2008).

In common view, the host plant in ECM symbiosis represents a partner responsible for energy supply, the saprotrophic nutrition being of marginal importance (Baldrian 2009). The solar energy is captured in the process of photosynthesis and is delivered to the mycorrhizal fungus in the form of energy-rich organic compounds, mostly carbohydrates, via ECMs. Ectomycorrhizal symbiosis involves complex physiological and ecological adaptations in both partners, which results in physiological and ecological specificity of the interaction (Smith and Read 2008). This means that, under field conditions, not all combinations of potential hosts with ECM fungi are possible. Instead, each fungal species can live in symbiosis with a specific range of hosts only. This is true also for true truffles as representatives of ECM fungi. Because the truffles are economically important fungal group, knowledge of the diversity of their host plant species is of a crucial importance.

M. Gryndler (✉)
Institute of Microbiology ASCR, v.v.i., Prague 4, Czech Republic

Faculty of Science, J. E. Purkinje University, Ústí nad Labem, Czech Republic
e-mail: gryndler@biomed.cas.cz

A. Zambonelli et al. (eds.), *True Truffle (*Tuber *spp.) in the World*, Soil Biology 47,
DOI 10.1007/978-3-319-31436-5_16

16.2 Concept of the Host in ECM Symbiosis with True Truffles

The concept of the ECM nutrition of true truffles (members of the genus *Tuber*, Pezizales, Ascomycota) and their close relation to the host has not been always fully accepted as valid (Barry et al. 1994; Callot 1999). *Tuber* fruiting bodies are formed on soil mycelium undoubtedly connected to a tree root via ECMs. In the course of their development, they become large carbon sink that may be localized far from active ECMs, apparently without sufficient physiological connection with them. It is not completely known whether they are always fed via direct transfer of carbohydrates from the host tree through the ECM and mycelial network or whether they become saprotrophic, independent on the symbiotic nutrition at some time after their formation. The existence of hypothetical saprotrophic nutrition of true truffles is suggested by the formation of extensive external mycelium clumps emerging from the ascoma peridium surface as well as by in vitro utilization of starch, cellobiose, and, probably in some extent, also cellulose as carbon sources (Mamoun and Olivier 1991). Saprotrophic capacity of *Tuber melanosporum* Vittad. pure cultures was also indicated by production of hydrolytic enzymes (unpublished results).

In the light of the newly gained information, however, the hypothesis of saprotrophic nature of the true truffle mycelia producing ascomata seems to be surpassed. It seems now that at least the mycelium of *T. melanosporum* at fructification stage gains its nutrition mainly from the host. The concept of colonized tree as exclusive energy supplier for ECM fungus was first experimentally supported by Zeller et al. (2008) who suggested, on the basis of the analysis of stable isotopes ^{15}N and ^{13}C, that nutrition of growing ascomata of four *Tuber* spp. is not saprotrophic. Similarly, the data obtained by Le Tacon et al. (2013, 2015) from a labeling experiment using the same stable isotopes indicates that almost all carbon allocated to the truffles (*T. melanosporum* ascomata) came from the tree-bearing ECMs.

Further, the sequencing of the *T. melanosporum* genome shows rather limited range of genes coding for hydrolytic enzymes involved in saprotrophic exploitation of complex forms of organic matter (Martin et al. 2010). This may be taken as other information supporting the idea of predominantly biotrophic nature of *T. melanosporum*. Evolution of ECM fungi has led to a loss of some enzymes involved in degradation of the plant cell wall. Most ECM fungi, including *T. melanosporum*, lack genes coding for cellobiohydrolases GH6 and GH7 but can maintain others, such as polysaccharide monooxygenases (Kohler et al. 2015). On the other hand, Martin et al. (2010) mentioned very high upregulation of endoglucanase and laccase genes in symbiotically living mycelium, in comparison to free-living mycelium of *T. melanosporum*. The same was not observed for another ECM fungus, *Laccaria bicolor* (Maire) P.D. Orton (Martin et al. 2008, 2010). Upregulation of the laccase gene transcription in ECM of *T. melanosporum* relative to free-living mycelium and non-colonized roots was independently confirmed by Zarivi et al. (2013) using quantitative PCR. As the laccase is involved in degradation of lignin components, its upregulation in fungus suggests the active utilization of the material of lignified host cell walls.

These findings indicate a potential of *T. melanosporum* ECM to hydrolyze particular macromolecules which can be met in proximity of the root and exploited for nutrition. At the same time, they reveal a difference between the physiology of *T. melanosporum* symbiosis and physiology of symbiosis of *L. bicolor* as representative of a "model" ECM fungus.

The above text might invoke the feeling that the mode of the fungal life is either (most likely) symbiotic/biotrophic or (less probably) saprotrophic. However, the reality might be much more complex. It seems that true truffles might use different life strategies involving, besides the mycorrhizal symbiosis, also commensalism, parasitism, or necrotrophy. If so, the *Tuber* mycelium utilizes soil organic matter as a source of nutrition auxiliary to the symbiotic one. The hypothesis of auxiliary saprotrophic nutrition of *Tuber* mycelium is supported, for example, by the existence of eccentric mycelial colonies (manifested as brûlé areas in some *Tuber* species) which may be very distant from host trees (Callot 1999, see Fig. 3–18B on p. 110; Sourzat 2002, see Fig. on p. 39). In these distant areas, the host root density may be very low which may result in symbiotic nutrition shortage and consecutive need for exploitation of other organic nutrients than those provided by ECMs.

The hyphae of *T. melanosporum* were observed to heavily colonize roots of *Thymus* sp. (Callot 1999, see Fig. on p. 109), whose tissues obviously undergo a kind of degradation, perhaps releasing mineral and organic nutrients that might be utilized by the mycelium. Formation of necroses probably induced by the same fungus has been observed in roots of plants traditionally not considered as hosts (Plattner and Hall 1995). These effects of *Tuber* mycelium on herbaceous plants may be due to the production of allelopathic compounds by the *Tuber* hyphae (Streiblová et al. 2012). A question appears whether these infected plants can be considered as symbiotic hosts of the invading fungus. It seems that, at least in the case of herbaceous plants that do not form ECM, the fungus may behave as a necrotroph exploiting mineral and/or organic nutrients released from moribund root tissues. This hypothesis has not been exhaustively tested till recent days but deserves attention because the fungus has the potential to induce the host cell death. This is documented by the formation of necroses induced by the interaction between the host and *Tuber* mycelia during the morphogenesis of ECM (Ragnelli et al. 2014).

The mycelium of *Tuber aestivum* Vittad. was detected in the microbial film attached to the surface of decaying layers of suberinized cells of host roots other than ECM (Gryndler et al. 2013) as well as in the roots of many herbaceous plant species present at the natural locality of *T. aestivum* (Gryndler et al. 2014) without signs of penetration of fungal hyphae into the tissue volume (Fig. 16.1).

This superficial contact may be either ineffective to induce plant defense reaction or the plant defense is induced but is not manifested in any visible change in root morphology. Again, it is unclear whether the plant, whose roots are superficially colonized, may be taken as a symbiotic host but it would be reasonable if a

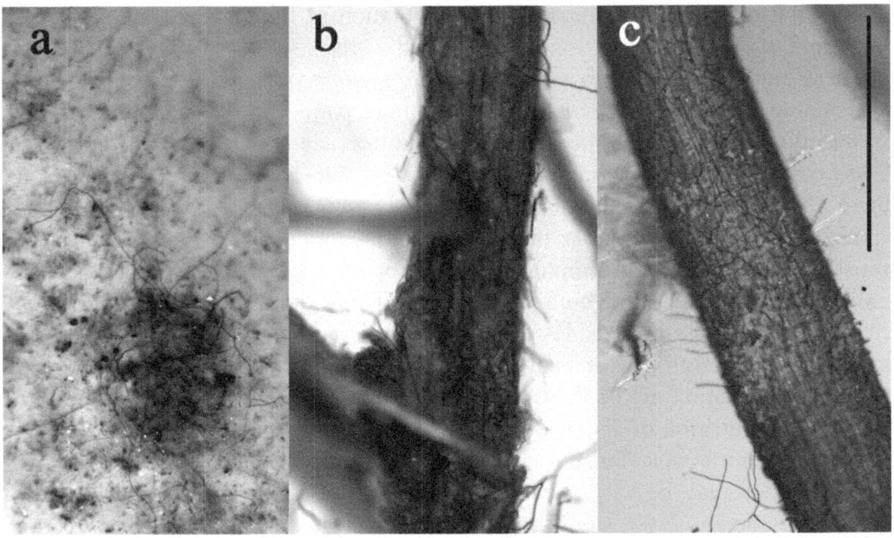

Fig. 16.1 Mycelium on the root surface of nonhost plants: *Silene vulgaris* (Moench) Garcke (**a**), *Carex muricata* L. (**b**), and *Achillea millefolium* L. (**c**) collected within the soil mycelial colony of *T. aestivum*. The root surface is covered by microbial film containing hyphae of various fungi. Highly dense mycelia of *T. aestivum* were detected there using quantitative real-time PCR (Gryndler et al. 2014). *Bar*: 1 mm

kind of signalization exists between such a "nonhost" root and the fungus. For example, bioactive compounds produced by the root tissues during secondary growth might serve as hypothetical signal mediators.

Similar, non-ECM association has been observed in *T. melanosporum*, which forms stromatic structures attached to the root bark surface (Pargney and Jalade 1995). These structures may represent an important, long-living contact surface between the mycelium and decaying root bark, distinct from ECMs. It is very interesting in this regard that the development of the *Tuber* soil mycelial colony seems to be stimulated by the presence of dead wood pieces, such as wooden fence pillar (Sourzat 2004), which may simulate the decaying root tissues. If this observation will be confirmed experimentally, it may indicate an association of the truffle mycelium with degradation of wooden root tissues.

It is also possible that the saprotrophic abilities vary among different *Tuber* species. In genomic data, different sets of genes coding for cazymes (carbohydrate active enzymes) involved in cellulose degradation were identified in different *Tuber* species (Claude Murat, pers comm; see also Chap. 9). This may be reflected in varying abundance of mycelia and ECMs in the soil. As the ECMs formed by *T. aestivum* (Pruett et al. 2009; Gryndler et al. 2013) or *T. melanosporum* (Rubini et al. 2011) are very abundant and can be easily extracted from the soil, the ECMs of other species, such as *Tuber magnatum* Pico, seem to be scarce or are even almost absent from soils in productive areas (Murat et al. 2005; Bertini et al. 2006; Leonardi et al. 2013; see Chap. 6). There may be no direct linkage among soil space occupied by *T. magnatum* mycelium and spatial distribution of mycorrhizas (Murat et al. 2005; Zampieri et al. 2010; Iotti et al. 2014).

The above text is dealing with only one aspect of mycorrhizal symbiosis: the nutrition. However, another important aspect should be taken into account when the host concept in mycorrhizal symbiosis is mentioned: the recognition of both partners. The partners must interchange the information on their identity, which results in specific modification of physiology at the level of cell, hyphae, or root tissues. The root may be colonized by either vegetative mycelium already established in soil or by the mycelium originating from germinating spores (ascospores, basidiospores). It was reported that the germination of spores of ECM fungi often requires the presence of a specific signal produced by a potential host root (Fries et al. 1987). Moreover, the germination may be enhanced in the presence of soil saprotrophic microflora (Birraux and Fries 1981). A cascade of events follows, which leads to the formation of ECMs and beginning of the nutrient exchange. This represents an important attribute of mycorrhizal symbiosis: the partners are tuned to each other, can accept a signal from each other, and produce an adequate response to the signal (Plett et al. 2011, 2014).

What we know about initial pre-colonization events occurring between *Tuber* spp. and their plant partners? The germination of *Tuber* ascospores under defined conditions is an extremely difficult task, which demands the prolonged storage at 5 °C (Grente et al. 1972), but is possible in the absence of any plant tissues, as it was proved also in our laboratory (Fig. 16.2). This fact indicates that *Tuber* ascospore germination is not induced by the root (does not depend on the presence of a potential host), though the germination may be enhanced by the root products (Guiochon 1959).

Similarly to the ascospore germination, the growth of vegetative mycelium is not dependent on the presence of host roots under axenic conditions. *Tuber* mycelia can be transferred to pure culture as it was demonstrated mainly for *T. aestivum* (e.g., by the author of this text and by Iotti et al. 2002), *T. melanosporum* (Fontana 1971; Iotti et al. 2002), *Tuber brumale* Vittad. (Chevalier 1972, 1973; Iotti et al. 2002), and *Tuber maculatum* Vittad., *Tuber macrosporum* Vittad., and *Tuber rufum* Pico (Iotti et al. 2002). Some strains can be maintained in pure culture infinitely, which is the proof of their saprotrophic capacity.

However, the above facts do not indicate the ability of free-living *Tuber* mycelia to compete with other soil microorganisms and a signal produced by the host may be necessary in this respect. The woody plant species traditionally accepted as hosts of true truffles may provide such a signal activating the growth of mycelial biomass which is nourished both saprotrophically and symbiotically. The role of other plants present at the locality, which are conventionally classified as "nonhost" species, is not known and is mostly neglected by scientific community.

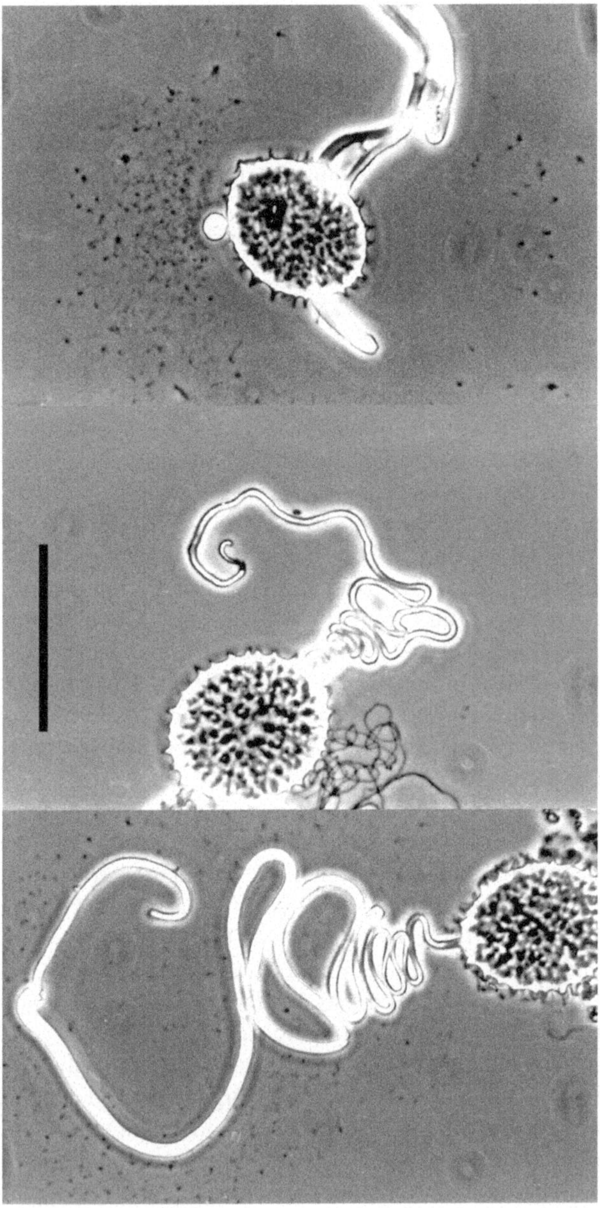

Fig. 16.2 Germination of ascospores, explanted from ripe ascoma of *T. melanosporum*, on water agar in the absence of the root. Ascospores were stored at 4 °C for 2 months and incubated at room temperature for 4 weeks. The uppermost photograph shows formation of multiple germ tubes per one ascospore. The germ tubes appear aseptate, in contradiction to figures presented in Granetti et al. (2005), p. 34, and in Delmas (1976), p. 15. The growth of these mycelia was limited and attempts to subculture them failed. Photographed by E. Streiblová and M. Gryndler, using Nomarski interference contrast optics. *Bar*: 50 μm

16.3 Hosts of True Truffles in Natural and Man-Made Environments

True truffles are inhabitants of various environments occupied by different known host plants (Table 16.1) as well as other plant species forming plant communities. These communities constitute a complex biological environment which, as a whole, support the existence and development of the *Tuber* mycelia. The communities associated with particular *Tuber* species were frequently studied, but the obtained data have been mostly published at national or even regional level and are thus hardly accessible to the majority of international scientific community.

One of the best accessible analysis of plant communities associated with *T. aestivum*, *Tuber mesentericum* Vittad., and *T. melanosporum* in Burgundy is presented by Chevalier and Frochot (1997, pp. 117–128) with conclusion that *T. aestivum* (syn. *Tuber uncinatum* Chatin) is typically – but not exclusively – found in the alliance *Carpinion betuli*, in particular in the association *Scillo-bifoliae-Carpinetum*, where *Carpinus betulus*, a European hornbeam, dominates as a host.

Voluminous list of plant species found in six *T. aestivum* localities in Poland is given by Hilszczanska et al. (2014). The authors registered the occurrence of five *Tuber* species on their localities but did not provide the details on possible host-fungus relationships.

Tuber spp. are also reported as associates or symbionts of orchids and some ericoid plants and are supposed to form typical orchideoid mycorrhizas (Selosse et al. 2004) and arbutoid mycorrhizas (Lancellotti et al. 2014; Taschen et al. 2015). This indicates that true truffles are able to produce at least three types of morphologically and functionally very different associations.

Among the orchid hosts, the presence of *T. aestivum* and *Tuber excavatum* Vittad. has been documented in roots of *Epipactis microphylla* (Ehrh.) Sw. (Selosse et al. 2004; Ouanphanivanh et al. 2008). Fungi similar to *T. rufum* and *Tuber maculatum* Vittad. have been detected in *Epipactis helleborine* (L.) Crantz using molecular identification (Ogura-Tsujita and Yukawa 2008). *Tuber maculatum* was confirmed in the root system of *E. helleborine* and *Cephalanthera damasonium* (Mill.) Druce by Ouanphanivanh et al. (2008).

It is obvious from the above reports and from the literature data sample summarized in Table 16.1 that true truffles are not strictly specialized on a particular host tree species and that the real range of hosts in the nature is relatively wide. However, the presence of *Tuber* species in the root system is not always accompanied by the production of fruiting bodies (Parádi and Baar 2006; Bonito et al. 2011).

Such a wide range of hosts of many *Tuber* species represents a factor which probably enhanced domestication of *T. melanosporum* and *T. aestivum*. Efficient hosts have been recruited during the decades of the field experimentation and this effort led to utilization of several woody species that could be easily managed in semiculture. *Quercus pubescens* Willd., *Quercus ilex* L., and *Corylus avellana* L. can be taken as examples of such suitable hosts. Even at the intraspecific level,

Table 16.1 Examples of ECM hosts reported for frequently studied *Tuber* species

Tuber sp.	Hosts
T. aestivum (syn. *uncinatum*)	*Abies alba*[DG], *Alnus cordata*[D], *Betula pendula*[G], *B. verrucosa*[BD], *Carpinus betulus*[BCDGH], *Carya illinoinensis*[A], *Castanea sativa*[BDG], *Cedrus atlantica* [BD], *C. deodara*[D], *Cistus* spp.[D], *Corylus avellana*[BCDEGHI], *C. colurna*[D], *Fagus sylvatica*[BCDGH], *Fumana procumbens*[D], *Ostrya carpinifolia* [BCDH], *Picea abies*[G], *P. excelsa*[D], *Pinus brutia*[D], *P. halepensis*[CD], *P. nigra*[BCD], *P. pinaster*[D], *P. pinea*[CD], *P. strobus*[D], *P. sylvestris*[BDG], *Populus* spp.[G], *P. nigra*[D], *Quercus boissieri*[I], *Q. calliprinos*[I], *Q. cerris*[BCDHI], *Q. ilex*[CDI], *Q. ithaburensis*[I], *Q. libani*[I], *Q. petraea*[BDGH], *Q. pubescens*[BCDHI], *Q. robur* [BCDGHI], *Tilia* spp.[BDGH], *Ulmus* spp.[G]
T. borchii	*Alnus cordata*[D], *Castanea sativa*[D], *Carya illinoinensis*[A], *Cedrus* spp.[C], *C. atlantica*[BD], *C. deodara*[D], *Corylus avellana*[DJ], *Fagus sylvatica*[CD], *Cistus* spp.[D], *C. incanus*[C], *Larix* sp.[D], *Ostrya carpinifolia*[D], *Picea excelsa*[D], *Pinus brutia*[D], *P. pinea*[CDJ], *P. nigra*[CD], *P. pinaster*[D], *P. strobus*[D], *P. sylvestris*[CD], *P. halepensis*[CD], *Populus* spp.[C], *P. alba*[D], *P. nigra*[D], *Quercus cerris*[CD], *Q. ilex*[D], *Q. petraea*[D], *Q. pubescens*[CD], *Q. robur*[DJ], *Salix alba*[D], *S. caprea*[D], *Tilia americana*[D], *T. cordata*[D], *Tilia x europaea*[D], *T. platyphyllos*[D]
T. brumale	*Abies alba*[G], *Corylus avellana*[ACEG], *Ostrya carpinifolia*[BC], *Fagus sylvatica* [CG], *Quercus cerris*[BC], *Q. ilex*[C], *Q. pubescens*[C]
T. excavatum	*Abies alba*[G], *Carpinus* spp.[C], *Corylus avellana*[G], *Fagus sylvatica*[G], *Picea abies*[G], *Pinus sylvestris*[G], *Populus* spp.[C], *Quercus* spp.[C], *Q. robur*[G], *Tilia* spp.[G]
T. fulgens	*Corylus avellana*[G], *Fagus sylvatica*[G], *Picea abies*[G], *Pinus sylvestris*[G], *Quercus robur*[G]
T. indicum	*Pinus armandii*[C], *P. yunnanensis*[C], *Quercus incana*[C]
T. macrosporum	*Betula verrucosa*[D], *Cedrus atlantica*[D], *C. deodara*[D], *Corylus avellana*[DG], *Ostrya carpinifolia*[CD], *Pinus pinaster*[D], *P. pinea*[D], *P. strobus*[D], *P. sylvestris* [D], *Populus* spp.[C], *P. alba*[D], *P. nigra*[D], *Quercus cerris*[CD], *Q. petraea*[D], *Q. pubescens*[CD], *Q. robur*[D], *Salix alba*[CD], *S. caprea*[CD], *Tilia x europaea*[D], *T. platyphyllos*[D]
T. maculatum	*Pinus strobus*[C]
T. magnatum	*Abies alba*[D], *Alnus cordata*[D], *Cedrus atlantica*[D], *C. deodara*[D], *Corylus avellana*[D], *Ostrya carpinifolia*[BD], *Pinus pinea*[D], *Populus* spp.[C], *P. alba*[D], *P. nigra*[D], *P. tremula*[D], *Quercus cerris*[D], *Q. ilex*[D], *Q. petraea*[D], *Q. pubescens*[BCD], *Q. robur*[CD], *Salix alba*[CD], *S. caprea*[CD], *Tilia* spp.[B], *T. cordata*[D], *Tilia x europaea*[D], *T. platyphyllos*[D]
T. melanosporum	*Abies alba*[D], *Alnus cordata*[D], *Betula* spp.[F], *Carpinus betulus*[DF], *C. orientalis* [C], *Castanea sativa*[BCF], *Cedrus* spp.[F], *C. atlantica*[D], *C. deodara*[D], *Cistus* spp. [D], *C. incanus*[CF], *Corylus avellana*[ABCDEFJ], *C. colurna*[BF], *C. heterophylla*[D], *Pinus nigra*[C], *P. pinaster*[D], *P. pinea*[CD], *P. strobus*[D], *P. sylvestris*[D], *Ostrya carpinifolia*[BCDF], *Populus* spp.[C], *P. alba*[D], *Quercus cerris*[BCD], *Q. coccifera* [D], *Q. faginea*[D], *Q. ilex*[CDJ], *Q. petraea*[CD], *Q. pubescens*[BCDF], *Q. robur*[BCDFJ], *Q. sessiliflora*[F], *Q. suber*[J], *Salix caprea*[D], *Tilia* spp.[CF], *T. cordata*[D], *Tilia x europaea*[D], *T. platyphyllos*[D]

(continued)

Table 16.1 (continued)

Tuber sp.	Hosts
T. mesentericum	*Betula pendula*[G], *Castanea sativa*[D], *Cedrus atlantica*[B], *Corylus avellana*[CDG], *C. colurna*[D], *Fagus sylvatica*[BCDG], *Ostrya carpinifolia*[BCD], *Pinus nigra*[CD], *P. pinea*[D], *P. sylvestris*[G], *Quercus pubescens*[CD], *Q. cerris*[CD], *Q. robur*[G], *Tilia cordata*[D], *Tilia x europaea*[D], *T. platyphyllos*[D]
T. oligospermum	*Abies* spp.[C], *Picea* spp.[C], *Pinus* spp.[C], *Pseudotsuga* spp.[C], *Quercus ilex*[C], *Q. pubescens*[C]
T. rufum	*Corylus avellana*[CG], *Fagus sylvatica*[CG], *Ostrya carpinifolia*[C], *Picea abies*[G], *Pinus nigra*[C], *P. halepensis*[C], *P. sylvestris*[G], *Quercus* spp.[C], *Q. robur*[G]

(A) Benucci et al. (2012); (B) Chevalier and Frochot (1997); (C) Granetti et al. (2005); (D) Hall et al. (2007); (E) Palenzona (1969); (F) Riousset et al. (2001); (G) Stobbe et al. (2012); (H) Stobbe et al. (2013); (I) Turgeman et al. (2012); (J) Zambonelli (2010)
Mycorrhizal association in underlined taxa has been proved experimentally see also Pacioni and Comandini (1999) and Hall et al. (2007)

the selection of efficient hosts was successful as documented for first time on *C. avellana* clone H219 with increased receptiveness toward *T. melanosporum* and enhanced resistance to colonization by competing ECM fungi (Mamoun and Olivier 1996).

The process of the host recruitment continues till present days as the new host species are tested and experimentally introduced to truffle grounds (Benucci et al. 2012; Turgeman et al. 2012). This effort is reasonable when the cultivation of truffles radiates to regions where traditionally used hosts do not perform well.

16.4 Colonization of Host Roots in vitro

Though the identity of the fungi present in ECM can be evaluated using efficient and reliable molecular methods (see Chap. 15), the experimental synthesis of ECM symbiosis should be performed under axenic conditions to prove the ECM status of the fungus. This will exclude the possibility that *Tuber* mycelium is present in the ECMs as an opportunist accompanying other ECM fungi (Murat et al. 2005). Generally, the mycorrhizal synthesis under fully controlled conditions is difficult and the failure to obtain ECMs with a particular combination of putative host with the tested fungus cannot be taken as a proof of impossibility of the association under natural conditions. Relatively few combinations of host plant/ECM fungus were hitherto tested for their ability to successfully form mycorrhizal symbiosis in vitro and the genus *Tuber* is not an exception of this rule.

First successful synthesis of *Tuber* axenic ECM symbiosis has been performed by Fontana (1967) and Fassi and Fontana (1967) with *Pinus strobus* L. and *T. maculatum*. Shortly after this initial success, Fontana and Palenzona (1969) continued these works by obtaining axenic mycorrhization of *P. strobus* and *Populus nigra* L. with *Tuber albidum* Pico (syn. *Tuber borchii* Vittad.). Chevalier (1973) used *T. brumale* as a fungal symbiont and *C. avellana*, *Pinus sylvestris* L., *Pinus halepensis* Mill., and *Pinus nigra* J. F. Arnold as hosts to fulfill the Koch's postulates necessary for ultimate proof of ECM status of *T. brumale* as ECM fungus. After that, axenic cultures have been reached for *T. melanosporum*, *T. mesentericum*, and *T. aestivum* with *C. avellana*, *P. sylvestris*, *P. halepensis*, *P. nigra*, *P. strobus*, *Q. pubescens*, *Quercus robur* L. (syn. *Quercus pedunculata* Ehrh.), *Quercus sessiliflora* Salisb, *Q. ilex*, *Quercus coccifera* L. *Fagus sylvatica* L., and *Tilia cordata* Mill. (Chevalier et al. 1973). These cultures were based on sterilized mineral substrate (vermiculite) or sterilized soil. Later, the synthesis of ECMs of *T. melanosporum* was obtained on agar medium (overlayered with sterilized soil) with *C. avellana* by Chevalier and Desmas (1977). In this case, very small fraction of ascospores present in the agar medium germinated. However, this approach confirmed that inoculation by ascospores in vitro may be successful. In vitro synthesis of ECMs of *T. borchii* with white poplar (Zambonelli et al. 1989) or with *Pinus radiata* D. Don (Duñabeitia et al. 1996) has been reached in axenic conditions in test tubes. This work confirmed that mycorrhization under axenic conditions can be met even with fungal isolates poorly growing in pure culture.

Structures morphologically resembling typical ECMs of *T. melanosporum* have been demonstrated by (Wenkart et al. 2001) on hairy (root-inducing T-DNA transformed) roots of *Cistus incanus* L. in liquid and solid cultures in vitro. The attempts to further cocultivate other clones of transformed roots of *C. incanus* with mycelia of *T. melanosporum* and *T. borchii* under various nutrition regimes resulted in obtaining structures corresponding to endomycorrhizas with hyphae penetrating root cells, or "pseudo-ectomycorrhizas" with hyphae forming loose hyphal mantle but without typical Hartig net (Pacioni et al. 2014). In this case, interestingly, the character of *T. melanosporum* colonization was determined by the root genotype: the *C. incanus* clone M2 tended to form preferentially the structures corresponding to endomycorrhizas, whereas the clone W51 was mostly willing to form "pseudo-ectomycorrhizas," the structures resembling the morphology of ECMs but lacking the typical fungal colonization. This indicates that behavior of the fungus may be dependent on its host genotype also under natural conditions. If this is the case, one fungal individuum (mycelial colony) may form a wide range of morphological structures with different plant genotypes.

16.5 Conclusions

In traditional view, true truffles represent a genus of ECM fungi which live in mycorrhizal symbiosis with a wide range of host woody plants, as it has been confirmed by axenic as well as greenhouse experiments. Besides this, however, the mycelium of true truffles may be associated with root tissues of "nonhost" woody and herbaceous plants. At the same time, some species of *Tuber* tend to form scarce ECMs. These facts suggest that *Tuber* mycelia may exploit some auxiliary sources of nutrition, other than these provided by the ECMs. Then, the dependence of free-living mycelia of some true truffles on their hosts may be relatively low, whereas the development of ascomata is probably indispensable from the symbiotic nutrition. The genus *Tuber* seems to possess relatively high biological multiformity, which may have enhanced the domestication of the economically most important species. This multiformity may induce the change of the host concept in *Tuber* symbiosis and enlarge the range of hosts by acceptation of other, hitherto neglected plants as symbiotic partners which do not form typical ECMs with *Tuber* mycelia but live in a functional interaction with them.

Acknowledgment The work was supported by the Czech Science Foundation within the frame of the project P504/10/0382.

References

Baldrian P (2009) Ectomycorrhizal fungi and their enzymes in soils: is there enough evidence for their role as facultative soil saprotrophs? Oecologia 161(4):657–660. doi:10.1007/s00442-009-1433-7

Barry D, Staunton S, Callot G (1994) Mode of the absorption of water and nutrients by ascocarps of *Tuber melanosporum* and *Tuber aestivum*. A radioactive tracer technique. Can J Bot 72 (3):317–322. doi:10.1139/b94-041

Benucci GMN, Bonito G, Falini LB, Bencivenga M (2012) Mycorrhization of Pecan trees (*Carya illinoinensis*) with commercial truffle species: *Tuber aestivum* Vittad. and *Tuber borchii* Vittad. Mycorrhiza 22(5):383–392. doi:10.1007/s00572-011-0413-z

Bertini L, Rossi I, Zambonelli A, Amicucci A, Sacchi A, Cecchini M, Gregori G, Stocchi V (2006) Molecular identification of *Tuber magnatum* ectomycorrhizae in the field. Microbiol Res 161 (1):59–64. doi:10.1016/j.micres.2005.06.003

Birraux D, Fries N (1981) Germination of *Thelephora terrestris* basidiospores. Can J Bot 59 (11):2062–2064. doi:10.1139/b81-267

Bonito G, Brenneman T, Vilgalys R (2011) Ectomycorrhizal fungal diversity in orchards of cultivated pecan (*Carya illinoinensis*; Juglandaceae). Mycorrhiza 21(7):601–612. doi:10.1007/s00572-011-0368-0

Callot G (1999) La truffe, la terre, la vie. INRA Editions, Paris

Chevalier G (1972) Obtention des cultures de mycélium de truffe à partir du carpophore et des mycorhizes. Académie d´Agriculture de France, extrait procès-verbal séance, 28 Juin 1972, pp 981–989

Chevalier G (1973) Synthèse axenique des mycorhizes de *Tuber brumale* Vitt. a partir de cultures pures du champignon. Ann Phytopathol 5:163–182

Chevalier G, Desmas C (1977) Synthèse des mycorhizes de *Tuber melanosporum* avec *Corylus avellana* sur gélose à partir de spores. Ann Phytopathol 9:531–540

Chevalier G, Frochot H (1997) La truffe de Bourgogne (*Tuber uncinatum* Chatin). Pétrarque, Levallois-Perret

Chevalier G, Grente J, Polacsek A (1973) Obtention de mycorhizes de différents *Tuber* par synthèse à partir de spores en conditions gnotoxéniques et à partir de cutures pures de mycélium en conditions axéniques et gnotoxéniques. Ann Phytopatol 5:107–108

Delmas J (1976) La truffe et sa culture. Editions SEI, INRA, Versailles, Etude No 60

Duñabeitia MK, Hormilla S, Salcedo I, Peña JI (1996) Ectomycorrhizae synthesised between *Pinus radiata* and eight fungi associated with *Pinus* spp. Mycologia 88(6):897–908. doi:10.2307/3761052

Fassi B, Fontana A (1967) Sintesi micorrizica tra *Pinus strobus* e *Tuber maculatum*. I. Micorrize e sviluppo dei semenzali nel secondo anno. Allionia 13:177–186

Fontana A (1967) Sintesi micorrizica tra *Pinus strobus* e *Tuber maculatum*. Giornale Botanico Italiano 101:298–299

Fontana A (1971) Il micelio di "*Tuber melanosporum*" Vitt. in coltura pura. Allionia 17:19–23

Fontana A, Palenzona M (1969) Sintesi di *Tuber albidum* in coltura pura con *Pinus strobus* e pioppo americano. Allionia 15:99–104

Fries N, Serck-Hanssen K, Häll-Dimberg L, Theander O (1987) Abietic acid an activator of basidiospore germination in ectomycorrhizal species of the genus *Suillus* (Boletaceae). Exp Mycol 11(4):360–363. doi:10.1016/0147-5975(87)90024-7

Granetti B, DeAngelis A, Materozzi G (2005) Umbria, terra di tartufi. Regione Umbria, Assessorato Regionale Agricoltura, Foreste, Caccia e Pesca

Grente J, Chevalier G, Pollacsek A (1972) La germination de l´ascospore de *Tuber melanosporum* et al. synthése sporale des mycorhizes. CR Acad Sci Paris Ser D 275:743–746

Gryndler M, Trilčová J, Hršelová H, Streiblová E, Gryndlerová H, Jansa J (2013) *Tuber aestivum* Vittad. mycelium quantified: advantages and limitations of a qPCR approach. Mycorrhiza 23 (5):341–348. doi:10.1007/s00572-012-0475-6

Gryndler M, Černá L, Bukovská P, Hršelová H, Jansa J (2014) *Tuber aestivum* association with non-host roots. Mycorrhiza 24(8):603–610. doi:10.1007/s00572-014-0580-9

Guiochon P (1959) Germination monospermes de la truffe, *Tuber melanosporum* Vitt. Mushroom Sci 4:294–297

Hall IR, Brown GT, Zambonelli A (2007) Taming the truffle. The history, lore, and science of the ultimate mushroom. Timber Press Inc, Portland

Hilszczanska D, Rosa-Gruszecka A, Szmidla H (2014) Characteristic of *Tuber* spp. localities in natural stands with emphasis on plant species composition. Acta Mycol 49(2):267–277. doi:10.5586/am.2014.024

Iotti M, Amicucci A, Stocchi V, Zambonelli A (2002) Morphological and molecular characterisation of mycelia of some *Tuber* species in pure culture. New Phytol 155(3):499–505. doi:10.1046/j.1469-8137.2002.00486.x

Iotti M, Leonardi M, Lancellotti E, Salerni E, Oddis M, Leonardi P, Perini C, Pacioni G, Zambonelli A (2014) Spatio-temporal dynamic of *Tuber magnatum* mycelium in natural truffle grounds. PLoS One 9(12), e115921. doi:10.1371/journal.pone.0115921

Kohler A, Kuo A, Lg N, Morin E, Kw B, Buscot F, Canbäck B, Choi C, Cichocki N, Clum A, Colpaert J, Copeland A, Costa MD, Doré J, Floudas D, Gay G, Girlanda M, Henrissat B, Herrmann S, Hess J, Högberg N, Johansson T, Khouja HR, LaButti K, Lahrmann U, Levasseur A, Lindquist EA, Lipzen A, Marmeisse R, Martino E, Murat C, Ngan CY, Nehls U, Plett JM, Pringle A, Ohm RA, Perotto S, Peter M, Riley R, Rineau F, Ruytinx J, Salamov A, Shah F, Sun H, Tarkka M, Tritt A, Veneault-Fourrey C, Zuccaro A, Mycorrhizal Genomics Initiative Consortium, Tunlid A, Grigoriev IV, Hibbett DS, Martin F (2015) Convergent losses of decay mechanisms and rapid turnover of symbiosis genes in mycorrhizal mutualists. Nat Genet 47(4):410–415. doi:10.1038/ng.3223

Lancellotti E, Iotti M, Zambonelli A, Franceschini A (2014) Characterization of *Tuber borchii* and *Arbutus unedo* mycorrhizas. Mycorrhiza 24(6):481–486. doi:10.1007/s00572-014-0564-9

Le Tacon F, Zeller B, Plain C, Hossann C, Bréchet C, Robin C (2013) Carbon transfer from the host to *Tuber melanosporum* mycorrhizas and ascocarps followed using a ^{13}C pulse-labeling technique. PLoS One 8(5), e64626. doi:10.1371/journal.pone.0064626

Le Tacon F, Zeller B, Plain C, Hossann C, Bréchet C, Martin F, Kohler A, Villerd J, Robin C (2015) Study of nitrogen and carbon transfer from soil organic matter to *Tuber melanosporum* mycorrhizas and ascocarps using ^{15}N and ^{13}C soil labelling and whole-genome oligoarrays. Plant Soil 395(1):351–373. doi:10.1007/s11104-015-2557-7

Leonardi M, Iotti M, Oddis M, Lalli G, Pacioni G, Leonardi P, Maccherini S, Perini C, Salerni E, Zambonelli A (2013) Assessment of ectomycorrhizal fungal communities in the natural habitats of *Tuber magnatum* (Ascomycota, Pezizales). Mycorrhiza 23(5):349–358. doi:10.1007/s00572-012-0474-7

Mamoun M, Olivier JM (1991) Influence du substrat carbone et de la forme d'azote mineral sur la croissance de *T. melanosporum* Vittad. en culture pure. Application à la production de biomasse mycelienne. Agronomie 11(6):521–527

Mamoun M, Olivier JM (1996) Receptivity of cloned hazels to artificial ectomycorrhizal infection by *Tuber melanosporum* and symbiotic competitors. Mycorrhiza 6(1):15–19. doi:10.1007/s005720050100

Martin F, Aerts A, Ahrén D, Brun A, Danchin EG, Duchaussoy F, Gibon J, Kohler A, Lindquist E, Pereda V, Salamov A, Shapiro HJ, Wuyts J, Blaudez D, Buée M, Brokstein P, Canbäck B, Cohen D, Courty PE, Coutinho PM, Delaruelle C, Detter JC, Deveau A, DiFazio S, Duplessis S, Fraissinet-Tachet L, Lucic E, Frey-Klett P, Fourrey C, Feussner I, Gay G, Grimwood J, Hoegger PJ, Jain P, Kilaru S, Labbé J, Lin YC, Legué V, Le Tacon F, Marmeisse R, Melayah D, Montanini B, Muratet M, Nehls U, Niculita-Hirzel H, Oudot-Le Secq MP, Peter M, Quesneville H, Rajashekar B, Reich M, Rouhier N, Schmutz J, Yin T, Chalot M, Henrissat B, Kües U, Lucas S, Van de Peer Y, Podila GK, Polle A, Pukkila PJ, Richardson PM, Rouzé P, Sanders IR, Stajich JE, Tunlid A, Tuskan G, Grigoriev IV (2008) The genome of *Laccaria bicolor* provides insights into mycorrhizal symbiosis. Nature 452 (7183):88–92. doi:10.1038/nature06556

Martin F, Kohler A, Murat C, Balestrini R, Coutinho PM, Jaillon O, Montanini B, Morin E, Noel B, Percudani R, Porcel B, Rubini A, Amicucci A, Amselem J, Anthouard V, Arcioni S, Artiguenave F, Aury JM, Ballario P, Bolchi A, Brenna A, Brun A, Buee M, Cantarel B, Chevalier G, Couloux A, Da Silva C, Denoeud F, Duplessis S, Ghignone S, Hilselberger B, Iotti M, Marcais B, Mello A, Miranda M, Pacioni G, Quesneville H, Riccioni C, Ruotolo R, Splivallo R, Stocchi V, Tisserant E, Viscomi AR, Zambonelli A, Zampieri E, Henrissat B, Lebrun MH, Paolocci F, Bonfante P, Ottonello S, Wincker P (2010) Perigord black truffle genome uncovers evolutionary origins and mechanisms of symbiosis. Nature 464 (7291):1033–1038. doi:10.1038/nature08867

Murat C, Vizzini A, Bonfante P, Mello A (2005) Morphological and molecular typing of the below-ground fungal community in a natural *Tuber magnatum* truffle-ground. FEMS Microbiol Lett 245(2):307–313. doi:10.1016/j.femsle.2005.03.019

Ogura-Tsujita Y, Yukawa T (2008) *Epipactis helleborine* shows strong mycorrhizal preference towards ectomycorrhizal fungi with contrasting geographic distributions in Japan. Mycorrhiza 18(6):331–338. doi:10.1007/s00572-008-0187-0

Ouanphanivanh N, Merényi Z, Orczán AK, Bratek Z, Szigeti Z, Illyés Z (2008) Could orchids indicate truffle habitats? Mycorrhizal association between orchids and truffles. Acta Biologica Szegediensis 52(1):229–232

Pacioni G, Comandini O (1999) *Tuber*. In: Cairney JWG, Chambers SM (eds) Ectomycorrhizal fungi: key genera in profile. Springer, Berlin, pp 163–186. doi:10.1007/978-3-662-06827-4_6

Pacioni G, Ragnelli AM, Aimola P, Leonardi M, Marinucci D, Roth-Bejerano N, Kagan-Zur V (2014) Endomycorrhiza and pseudo-ectomycorrhiza produced in vitro by two species of *Tuber* on transformed *Cistus incanus* roots. Int J Plant Biol Res 2(4):1021

Palenzona M (1969) Sintesi micorrizica tra *Tuber aestivum* Vitt, *Tuber brumale* Vitt, *Tuber melanosporum* Vitt. e semenzali di *Corylus avellana* L. Allionia 15:121–131

Parádi I, Baar J (2006) Mycorrhizal fungal diversity in willow forests of different age along the river Waal, The Netherlands. For Ecol Manage 237:366–372. doi:10.1016/j.foreco.2006.09.059

Pargney JC, Jalade M (1995) Cytological study of fungal structures forming stromas on roots of truffle mycorrhizal plantlets. Bulletin des Academie Societe Lorraines des Sciences 34 (1):27–44

Plattner I, Hall IR (1995) Parasitism of non-host plants by the mycorrhizal fungus *Tuber melanosporum*. Mycol Res 99(11):1367–1370. doi:10.1016/S0953-7562(09)81223-9

Plett JM, Kemppainen M, Kale SD, Kohler A, Legue V, Brun A, Tyler BM, Pardo AG, Martin F (2011) A secreted effector protein of *Laccaria bicolor* is required for symbiosis development. Curr Biol 21(14):1197–1203. doi:10.1016/j.cub.2011.05.033

Plett JM, Khachane A, Ouassou M, Sundberg B, Kohler A, Martin F (2014) Ethylene and jasmonic acid act as negative modulators during mutualistic symbiosis between *Laccaria bicolor* and *Populus* roots. New Phytol 202(1):270–286. doi:10.1111/nph.12655

Pruett GE, Bruhn JN, Mihail JD (2009) Greenhouse production of Burgundy truffle mycorrhizae on oak roots. New For 37(1):43–52. doi:10.1007/s11056-008-9108-5

Ragnelli AM, Aimola P, Maione M, Zarivi O, Leonardi M, Pacioni G (2014) The cell death phenomenon during *Tuber* ectomycorrhiza morphogenesis. Plant Biosyst 148(3):473–482. doi:10.1080/11263504.2013.788575

Riousset LG, Chevalier G, Bardet MC (2001) Truffles d'Europe et de Chine. INRA, Paris

Rubini A, Belfiori B, Riccioni C, Arcioni S, Martin F, Paolocci F (2011) *Tuber melanosporum*: mating type distribution in a natural plantation and dynamics of strains of different mating types on the roots of nursery-inoculated host plants. New Phytol 189(3):723–735. doi:10.1111/j.1469-8137.2010.03493.x

Selosse MA, Faccio A, Scappaticci G, Bonfante P (2004) Chlorophyllous and achlorophyllous specimens of *Epipactis microphylla* (Neottieae, Orchidaceae) are associated with ectomycorrhizal septomycetes, including truffles. Microb Ecol 47(4):416–426. doi:10.1007/s00248-003-2034-3

Smith SE, Read DJ (2008) Mycorrhizal symbiosis, 3rd edn. Academic, New York

Sourzat P (2002) Guide pratique de trufficulture. Quatrième edition. Lacoste, Le Montat

Sourzat P (2004) Questions d'écologie appliquées à la trufficulture. Lycée professionnel agricole de Cahors-Le Montat, Impression 43029, Le Montat

Stobbe U, Büntgen U, Sproll L, Tegel W, Egli S, Fink S (2012) Spatial distribution and ecological variation of re-discovered German truffle habitats. Fungal Ecol 5(5):591–599. doi:10.1016/j.funeco.2012.02.001

Stobbe U, Egli S, Tegel W, Peter M, Sproll L, Büntgen U (2013) Potential and limitations of Burgundy truffle cultivation. Appl Microbiol Biotechnol 97(12):5215–5224. doi:10.1007/s00253-013-4956-0

Streiblová E, Gryndlerová H, Gryndler M (2012) Truffle brûlé: an efficient fungal life strategy. FEMS Microbiol Ecol 80(1):1–8. doi:10.1111/j.1574-6941.2011.01283.x

Taschen E, Sauve M, Taudiere A, Parlade J, Selosse MA, Richard F (2015) Whose truffle is this? Distribution patterns of ectomycorrhizal fungal diversity in *Tuber melanosporum* brûlés developed in multi-host Mediterranean plant communities. Environ Microbiol 17 (8):2747–2761. doi:10.1111/1462-2920.12741

Turgeman T, Sitrit Y, Danai O, Luzzati Y, Bustan A, Roth-Bejerano N, Kagan-Zur V, Masaphy S (2012) Introduced *Tuber aestivum* replacing introduced *Tuber melanosporum*: a case study. Agrofor Syst 84(3):337–343. doi:10.1007/s10457-011-9478-0

Wenkart S, Roth-Bejerano N, Mills D, Kagan-Zur V (2001) Mycorrhizal association between *Tuber melanosporum* mycelia and transformed roots of *Cistus incanus*. Plant Cell Rep 20 (4):369–373. doi:10.1007/s002990100325

Zambonelli A (2010) Lecture de l'approche de la trufficulture dans l'hémisphère austral à la lumière de l'experience Italienne. In: Olivier JM, Vilatte C (eds) Les nouvelles techniques de culture de la truffe. MGD Imprimeurs à Sarlat, Sarlat, pp 68–77

Zambonelli A, Govi G, Previati A (1989) Micorrizazione in vitro di piantine micro-propagate di *Populus alba* con micelio di *Tuber albidum* in coltura pura. Micol Ital 18(3):105–111

Zampieri E, Murat C, Cagnasso M, Bonfante P, Mello A (2010) Soil reveals the presence of an extended mycelial network in a *Tuber magnatum* truffle-ground. FEMS Microbiol Ecol 71 (1):43–49. doi:10.1111/j.1574-6941.2009.00783.x

Zarivi O, Bonfigli A, Colafarina S, Aimola P, Ragnelli AM, Miranda M, Pacioni G (2013) Transcriptional, biochemical and histochemical investigation on laccase expression during *Tuber melanosporum* Vittad. development. Phytochemistry 87:23–29. doi:10.1016/j.phytochem.2012.11.019

Zeller B, Bréchet C, Maurice JP, Le Tacon F (2008) Saprotrophic versus symbiotic strategy during truffle ascocarp development under holm oak. A response based on ^{13}C and ^{15}N natural abundance. Ann For Sci 65(6):607. doi:10.1051/forest:2008037

Chapter 17
Truffle-Inhabiting Fungi

Giovanni Pacioni and Marco Leonardi

17.1 Introduction

True truffles (*Tuber* spp.) grow in a very complex and diverse biotic context, where they interact both as free mycelium and as mycorrhiza and finally also as ascoma. Plant communities determine the success of *Tuber* spp. in the competition to acquire photosynthates by establishing several types of mycorrhiza, which are not only represented by the ectomycorrhiza (ECM) but also by other types of mycorrhizal symbioses (Selosse et al. 2004; Gryndler et al. 2014; Pacioni et al. 2014). From the "stronghold" represented by the fine roots of the host plant, the mycelium radiates in the soil and establishes its web.

Very little is known about the relationships of *Tuber* with other free-living soil microorganisms both in terms of the functioning of food webs and as part of the community. In order to better understand these interactions, truffle-inhabited soils, natural areas, as well as orchards have been studied, first employing culture-dependent methods (Luppi-Mosca 1972; Luppi and Fontana 1977; Sbrana et al. 2002) and then by using molecular sequence data (Mello et al. 2013; Orgiazzi et al. 2013). In truffle habitats, the main target of these investigations was the "brûlé" (also called pianello or burned area) that characterizes the productive areas of some truffle species like *Tuber melanosporum* Vittad. and *Tuber aestivum* Vittad. (Napoli et al. 2010; Benucci et al. 2011; Streiblová et al. 2012; Mello et al. 2013). In recent years, however, careful attention has also been dedicated to the habitats of the most valuable species of white and whitish truffles *Tuber magnatum* Pico and *Tuber borchii* Vittad. (Mello et al. 2010; Iotti et al. 2010) and the black *Tuber macrosporum* Vittad. (Benucci et al. 2014).

G. Pacioni (✉) • M. Leonardi
Department of Life, Health and Environmental Sciences, University of L'Aquila, via Vetoio loc. Coppito, 67100 L'Aquila, Italy
e-mail: giovanni.pacioni@univaq.it

A. Zambonelli et al. (eds.), *True Truffle (*Tuber *spp.) in the World*, Soil Biology 47,
DOI 10.1007/978-3-319-31436-5_17

At the same time, mycorrhizal communities located within the production areas of truffles have also been extensively studied, both with traditional (morphotyping) and molecular (genotyping) methods. Within such communities, the understanding of the ecological interaction is still weak, although it should be highly significant because it involves a competitive struggle for the acquisition of photosynthates from host plants. For this reason, it has been the subject of many studies and two recent reviews (Zambonelli et al. 2012; De Miguel et al. 2014). Also, within this ecological niche, the interactions with all the other components of the microbiome, mainly bacteria, are intense, even though they have been barely studied in the wild and only touched on in laboratory investigations (Sbrana et al. 2000, Barbieri et al. 2000). Among the microbiological research, the most important aspect to consider is the microbioma of the fruiting bodies (ascomata) of *Tuber*, the prized truffles (Antony-Babu et al. 2014).

Truffles develop as a pellet of mycelium progressively growing into a spheroidal ascoma in which the outer cells differentiate into a more or less protective layer (peridium) (Pacioni et al. 1995; Zarivi et al. 2015). In black truffles, the peridium is robust and made of sclerified and melanized cells among which there are also pores, the outlet of sterile veins which represent authentic entryways (Pacioni 1990). In white truffles, the peridium appears continuous and consists of layers of, probably living, often fractured cells. Truffles are under attack from the fauna, mainly insects (Pacioni 1990) (see Chap. 19) which also create lesions and galleries inside the ascoma.

The presence of so many entry points implies that truffles cannot be sterile: bacteria (Archaea and Eubacteria) (see Chap. 18), fungi (yeasts and filamentous fungi) (Buzzini et al. 2005; Pacioni et al. 2007) and viruses (Stielow and Menzel 2010; see Chap. 20). A systematic investigation of the fungi which live inside truffle ascoma has never been carried out, even though their presence was unwittingly demonstrated when the first attempts to isolate *T. melanosporum* mycelium were made by both Boulanger and Matrouchot (1903).

Going through the old literature reported in Saccardo's Sylloge fungorum (1882–1913), it is possible to find a dozen reports on fungi occasionally found on moldy truffles or as ascospores within the gleba, such as *Melanospora* and *Battarrina*.

In the broad category of fungi interacting with other fungi (FF, fungicolous fungi; Hawksworth 1981), those belonging to the truffle mycobiome could be referred to as sporocarp-inhabiting fungi (SCIF) (Gams et al. 2004), but here the acronym TIF (truffle-inhabiting fungi) is preferred because we consider a more specific mycobiome which develops inside the *Tuber* ascomata.

17.2 Truffle-Inhabiting Fungi

Truffle-inhabiting fungi (TIF) is a neglected field of research. Only in recent years has it begun to be addressed in relation to diseases and the shelf life of truffles. TIF category includes both yeasts and filamentous fungi.

17.2.1 Yeasts

Marletto (1969) addressed the issue of the presence of yeasts in three *Tuber* (*T. magnatum*, *Tuber brumale* Vittad., and *T. borchii*), comparing the results of isolations in culture with those from productive niche soil. As a result, he was able to demonstrate the presence of 10 different species, nine of which were also present in the soil.

Later, Buzzini et al. (2005) isolated and molecularly identified the yeast community within the gleba of *T. melanosporum* and *T. magnatum*, in order to understand its ecological significance. They obtained 29 strains identified as *Candida saitoana* Nakase & M. Suzuki, *Cryptococcus* sp., *Debaryomyces hansenii* (Zopf) Lodder & Kreger-van Rij, *Rhodotorula mucilaginosa* (A. Jörg.) F.C. Harrison, and *Trichosporon moniliiforme* E. Guého & M.T. Sm. and essentially confirmed the findings of Zacchi et al. (2003), who found four yeasts in *T. aestivum*: *Cryptococcus albidus* (Saito) C.E. Skinner, *Cryptococcus humicola*, *D. hansenii*, and *R. mucilaginosa* (Table 17.1).

17.2.2 Filamentous Fungi

The checklist of the filamentous fungi reported to live in truffles is the product of different types of records: (a) occasional findings of forms sporulating inside the gleba or on fragments of truffles, (b) sporomata or conidiophores emerging from the peridium, and (c) the results of isolations in culture made from gleba of healthy and diseased truffles. To draft this checklist, some online databases including collections of living filamentous fungi (Table 17.2) were also accessed. The checklist does not include *Aspergillus calyptratus* Oudem. var. *italicum* Ferraris described from a peridium of a *T. melanosporum* preserved in alcohol and *Penicillium rubescens* Bainier sporulating on fragments of truffle (Saccardo 1882–1913), since they were from postharvest moldy material. Most of these species belong to the Ascomycota, but there are also some Zygomycota and Basidiomycota.

Table 17.2 shows species reported from the nineteenth century to the present day. The checklist is the result of direct observation of conidiophores or structures of sexual reproduction on or inside ascomata and in culture isolations from gleba. Some species, such as *Oospora placentiformis* (Corda) Sacc. & Voglino, *Oospora*

Table 17.1 Yeasts isolated from *Tuber* ascomata

Species	Classification	Ref.
Aureobasidium pullulans (de Bary & Löwenthal) G. Arnaud	Ascomycota Dothideales	(2)
Candida saitoana Nakase & M. Suzuki	Ascomycota Saccharomycetales	(1, 2)
Debaryomyces hansenii (Zopf) Lodder & Kreger-van Rij	Ascomycota Saccharomycetales	(1, 3)
Hansenula saturnus (Klöcker) Wick.	Ascomycota Saccharomycetales	(2)
Saccharomyces exiguus Reess ex E.C. Hansen	Ascomycota Saccharomycetales	(2)
Torulopsis inconspicua Lodder & Kreger-van Rij	Ascomycota Saccharomycetales	(2)
Cryptococcus albidus (Saito) C.E. Skinner	Basidiomycota Tremellales	(1, 3)
Cryptococcus diffluens (Zach) Lodder & Kreger-van Rij	Basidiomycota Tremellales	(2)
Cryptococcus humicola (Zach) Lodder & Kreger	Basidiomycota Tremellales	(3)
Cryptococcus laurentii (Kuff.) C.E. Skinner	Basidiomycota Tremellales	(2)
Cryptococcus sp.	Basidiomycota Tremellales	(1)
Rhodotorula mucilaginosa (A. Jörg.) F.C. Harrison	Basidiomycota Sporidiobolales	(1, 3)
Sporobolomyces roseus Kluyver & C.B. Niel	Basidiomycota Sporidiobolales	(2)
Trichosporon cutaneum (Beurm., Gougerot & Vaucher bis) M. Ota	Basidiomycota Tremellales	(2)
Trichosporon moniliiforme (Weigmann & A. Wolff) E. Guého & M.T. Sm.	Basidiomycota Tremellales	(1)
Trichosporon pullulans (Lindner) Diddens & Lodder	Basidiomycota Tremellales	(2)

(1) Buzzini et al. (2005); (2) Marletto (1969); (3) Zacchi et al. (2003)

tuberum (Corda) Sacc. & Voglino, *Battarrina inclusa* (Berk. & Broome) Clem. & Shear, *Melanospora zobelii* (Corda) Fuckel, and *Melanospora subterranea* L. Fan, C.L. Hou, P.F. Cannon, and Yu Li, which sporulate inside truffles, were observed in microscopic preparations of the gleba, sometimes spotted by colored spores of certain TIF, like *Melanospora*. Other species, instead, sporulate on the peridium surface, such as the anamorphs *Verticillium epimyces* Berk. & Broome, *Mucor nigrescens* Tiegh. (not Schumacher 1803), and *Trichothecium crotocinigenum* (Schol-Schwarz) Summerb., Seifert & Schroers and teleomorphs *Nectria inventa* Pethybr., *Hypomyces tubericola* (Schwein.) Sacc., *Claudopus byssisedus* (Pers.) Gillet, and perhaps *Cordyceps capitata* (Holmsk.) Link and *Boletus parasiticus* Bull. The taxon *Mucor nigrescens* Tieghem has not been well identified

Table 17.2 Checklist of TIF reported in or on *Tuber* ascomata

Species reported as	Current name or teleomorph	Classification	Ref.	
Absidia cylindrospora Hagem		Zygomycota, Mucorales	(9)	C
Arthrinium phaeospermum (Corda) M.B. Ellis		Ascomycota, Sordariomycetes	(11)	C
Mortierella bisporalis (Thaxt.) Björl.		Zygomycota, Mortierellales	CBS145.69	C
Mucor nigrescens Tiegh.		Zygomycota, Mucorales	(12)	S
Battarrina inclusa (Berk. & Broome) Clem. & Shear		Ascomycota, Hypocreales	(12)	S
Cylindrocarpon magnusianum Wollenw.	*C. obtusiusculum* (Sacc.) U. Braun	Ascomycota, Hypocreales	(9)	C
Clonostachys rosea (Link) Schroers, Samuels, Seifert & W. Gams	*Bionectria ochroleuca* (Schwein.) Schroers & Samuels	Ascomycota, Hypocreales	(7, 10)	C
Cordyceps capitata (Holmsk.) Link	*Elaphocordyceps capitata* (Holmsk.) G.H. Sung, J.M. Sung & Spatafora	Ascomycota, Hypocreales	(13)	S
Cylindrophora alba Bonord.		Ascomycota, incertae sedis	(12)	S
Fusarium sp.		Ascomycota, Hypocreales	(5)	C
Hypomyces tubericola (Schwein.) Sacc.		Ascomycota, Hypocreales	(12)	S
Lecanicillium psalliotae (Treschew) Zare & W. Gams		Ascomycota, incertae sedis	CBS154.70	C
Melanospora subterranea L. Fan, C.L. Hou, P.F. Cannon & Yu Li		Ascomycota, Hypocreales	(2)	S
Melanospora zobelii (Corda) Fuckel		Ascomycota, Hypocreales	(12)	S
Nectria inventa Pethybr.		Ascomycota, Hypocreales	(1)	S
Sporothrix tuberi A. Fontata & Fas.-Bonf.	*Nodulisporium tuberum* A. Fontana & Fas.-Bonf. ex de Hoog	Ascomycota, Xylariales	(3)	C
Tetracladium maxilliforme (Rostr.) Ingold	*Titaea maxilliformis* Rostr.	Ascomycota, Pleosporales	(9)	C
Talaromyces wortmannii (Klöcker) C.R. Benj	*Pyrenochaeta gardeniae* S. Chandra & Tandon	Ascomycota, Eurotiales	(9)	C

(continued)

Table 17.2 (continued)

Species reported as	Current name or teleomorph	Classification	Ref.	
Trichopezizella nidulus (J.C. Schmidt & Kunze) Raitv.		Ascomycota, Helotiales	(9)	C
**Trichothecium crotocinigenum* (Schol-Schwarz) Summerb., Seifert & Schroers	*Acremonium crotocinigenum* (Schol-Schwarz) W. Gams	Ascomycota, Hypocreales	(1) CBS129.64	S/C
Trichothecium roseum (Pers.) Link		Ascomycota, Hypocreales	(9)	C
Trichurus spiralis Hasselbr.		Ascomycota, Microascales	(7)	C
Verticillium epimyces Berk. & Broome		Ascomycota, incertae sedis	(4)	S
Verticillium lecanii (Zimm.) Viégas	*Lecanicillium muscarium* (Petch) Zare & W. Gams	Ascomycota, incertae sedis	CBS413.70A	C
Verticillium leptobactrum W. Gams		Ascomycota, incertae sedis	(9, 10)	C
Boletus parasiticus (Bull.)	*Pseudoboletus parasiticus* (Bull.) Šutara	Basidiomycota, Boletales	(13)	S
Claudopus byssisedus (Pers.) Gillet	*Entoloma byssisedum* (Pers.) Donk	Basidiomycota, Agaricales	(8)	S
Entoloma sp.		Basidiomycota, Agaricales	(6)	S
Oospora placentiformis (Corda) Sacc. & Voglino		Basidiomycota, Cantharellales	(12)	S
Peniophora cinerea (Pers.) Cooke		Basidiomycota, Russulales	(9)	C

(1) Eslick (2013), (2) Fan et al. (2012), (3) Fontana and Fasolo-Bonfante (1971), (4) Gams et al. (2004), (5) Gong et al. (2001), (6) Lancellotti et al. (2012), (7) Leonardi et al. (submitted), (8) Malençon (1942), (9) Pacioni et al. (2007), (10) Pavić et al. (2013), (11) Sabella et al. (2015), (12) Saccardo (1882–1913), (13) Szemere (1965), CBS (Centraalbureau voor Schimmelcultures Fungal Biodiversity Centre) living strains without published references. Names according to Index Fungorum and Mycobank

*Also reported as *A. crotocinigenum*

S: TIF sporulating in/on truffles; *C*: TIF cultured from truffles

taxonomically, but it could be *Absidia cylindrospora* Hagem. The external sporulation would seem to require oxygen since the old reports referred to damaged truffles, and these molds are frequent in postharvest truffles or on them when emerging from the soil. In fact, in his study on the etiology of a truffle rot of *T. melanosporum*, Eslick (2013) noted that sporulation occurred significantly on the exposed surface of truffles. The record of *C. byssisedus*, current name *Entoloma byssisedum* (Pers.) Donk, on *Tuber rufum* Pico (Malençon 1942) may actually be

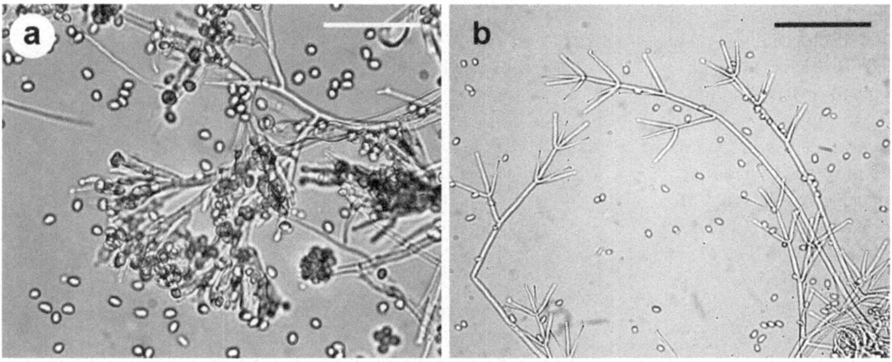

Fig. 17.1 Conidiophores of *B. ochroleuca* (**a**) and *V. leptobactrum* (**b**), two of the most frequent TIF isolated from *Tuber* spp. ascomata. *Bar* = 20 μm

attributed to *Entoloma parasiticum* (Quél.) Kreisel, which is a known parasite species of basidiomata. Lancellotti et al. (2012) found large patches of a whitish mycelial mat composed of parallel and clamped hyphae, molecularly identified as *Entoloma* sp., on the peridium of *T. borchii* ascomata. Sabella et al. (2015) recently reported the isolation of a filamentous fungus from *T. borchii* fruiting bodies, and it was identified as an *Arthrinium phaeospermum* (Corda) M.B. Ellis, an endophyte that forms various associations with healthy leaves, stems, and roots of plants. There is no macroscopic or molecular confirmation of the presence of *C. capitata* and *B. parasiticus*, two SCIF reported by Szemere (1965) on *Tuber* ascomata.

The list of TIF identified after isolation in culture from pieces of gleba, taken in an axenic manner, could have been expected to increase rapidly, thanks to metagenomics, but we have encountered major methodological difficulties (Leonardi et al. submitted; see below for more details). *Clonostachys rosea* (Link) Schroers, Samuels, Seifert & W. Gams [teleomorph *Bionectria ochroleuca* (Schwein.) Schroers & Samuels; Fig. 17.1a], *Verticillium leptobactrum* W. Gams (Fig. 17.1b), *Cylindrocarpon magnusianum* Wollenw., and *A. cylindrospora* seem to be species recurring in isolations from gleba.

17.3 Methods and Problems

The list reported in Tables 17.1 and 17.2 is indeed long enough to be considered exhaustive. The study of TIF presents great difficulties, and we are convinced that, in order to understand the actual composition of TIF communities, new strategies of approach should be found (Leonardi et al. submitted). The techniques of isolation, in fact, show a maximum of one or two species at a time, favoring the more aggressive or faster-growing species. The application of molecular methods, useful for identifying species in culture, does not currently provide a valuable aid for mycobiome resolution. Payen et al. (2014) argued that the presence of TIF in the

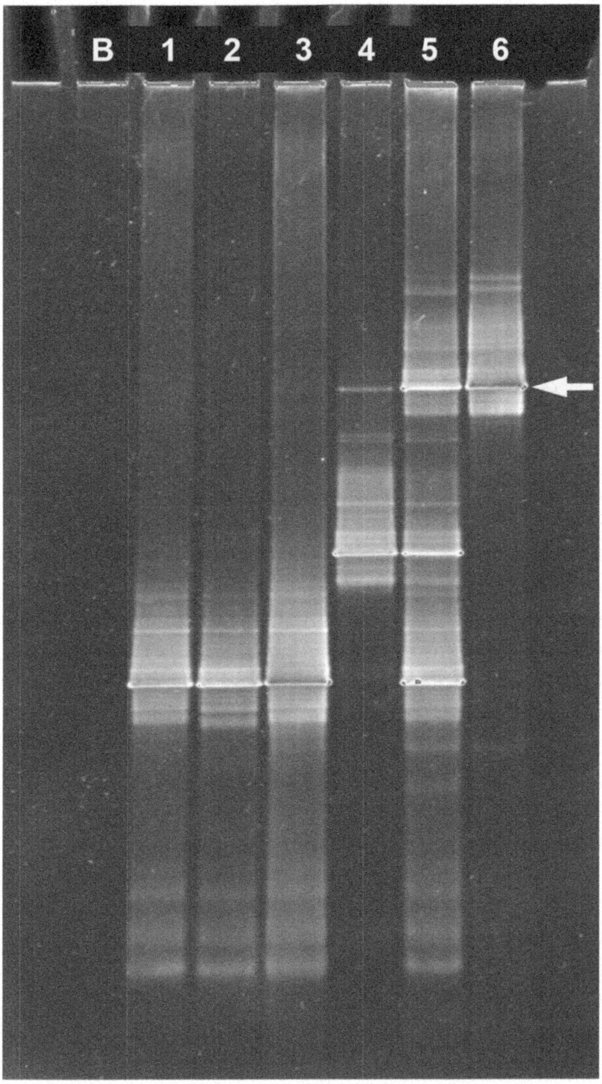

Fig. 17.2 The result of a DGGE electrophoresis of ITS: B- blank; (1–3) *T. melanosporum*; (4) decaying *T. mesentericum*; (5) mix of the previous four *Tuber* species ITS with that of *B. ochroleuca*; (6) ITS of *B. ochroleuca*. The *arrow* shows that *B. ochroleuca* ITS was detected in the mix and in a decaying *T. mesentericum*

gleba makes it difficult to study the genome of the host truffles by DNA sequencing from ascoma. Reversing the concept, to us it seems much more difficult to identify the guest TIF with mass sequencing because of the overwhelming presence of the genome of the host truffle. We have verified that currently it seems almost impossible to carry out a normal metagenomic survey to characterize the TIF community, because the truffle genome obscures that of its guest fungi (Leonardi et al. submitted).

The simultaneous presence of multiple TIF in the same truffle has now been established. In fact, using specific primers for three of the most common TIF species [*B. ochroleuca*, *V. leptobactrum*, and *Nectria mauritiicola* (Henn.) Seifert & Samuels], by means of qPCR the three species have been found to live together in almost all ascomata of *T. melanosporum* and *T. borchii* used for isolating mycelia. Instead DNA of *N. mauritiicola* has not been found in the total DNA extracted from *T. magnatum* gleba, while another two TIF have been detected. Since these TIF have often not been obtained by culture isolations from some of the previously analyzed ascomata, it is to be assumed that there is a complex TIF community inside truffles. After an accurate quantification of the DNA of *B. ochroleuca* in *T. melanosporum*, it was shown that the DNA copy number of the truffle is dramatically higher than that of a single TIF (10^6 vs 1). For this reason, survey systems such as denaturing gradient gel electrophoresis (DGGE) are not capable of separating the amplicons of the TIF from those of the host truffle (Leonardi et al. submitted). Only with a rotting sample of *Tuber mesentericum* Vittad. was a distinct band belonging to *B. ochroleuca* evident; on all other occasions, only the bands related to the ascomata of the tested *Tuber* were obtained (Fig. 17.2) (Leonardi et al. submitted). In this particular case, the amount of hyphae *B. ochroleuca* was obviously quite significant, being a species to which the ability to rot truffles is attributed (Eslick 2013; Pavić et al. 2013).

Though it is an extraordinary detection system, qPCR can only work with previously characterized TIF, and an application to mass sequencing with next-generation sequencers is likely to run into the same problems of "blackout" due to the huge quantity of the host truffle DNA. In fact, it is to be considered that in mass sequencing approach, the same primers have to be used for both TIF and host truffles.

Regarding the members of the TIF community, it is to be considered that its complexity could increase with the time of ascoma maturation due to subsequent and new fungal colonizations, as has been seen for bacteria (Antony-Babu et al. 2014). Indeed, Zacchi et al. (2003) isolated a single yeast species from immature *T. aestivum* ascomata, while they obtained three yeast species from mature truffles. This could be a possible explanation of the successes of *Tuber* mycelium isolation from immature truffles (Iotti et al. 2002; Iotti et al. 2012).

TIF are soilborne fungi, and *B. ochroleuca*, which seems to be the most frequent TIF, has been recorded as one of the seven-eight fungal species constantly present in Italian and French soils of different environments, also including production areas of *T. magnatum* and *T. melanosporum* (Orgiazzi et al. 2013). Based on this result, the authors suggested the existence of a stable core of fungi across the complex assemblage of soil fungi, which is either endowed with the capacity to sustain development in nutrient-poor soil conditions or with the ability to exploit organic resources (such as chitin) universally distributed in soils. This is probably why the anamorph of *B. ochroleuca*, which produces chitinase (Gan et al. 2007), is so frequently present in truffles, but it must be remembered that truffles seem to exert a greater selectivity against soil bacteria than towards fungi (Antony-Babu et al. 2014; Pacioni 1991).

17.4 Companions or Customers?

Like other microbiomes of living organisms, there is a continuum of relationships of guest microorganisms with the host also for the microbiome of truffles that, depending on its health condition, goes from neutralism to commensalism to minor parasitism or parasitism, with the decomposition agent on the dying or dead host. One type of relationship, in fact, might transition to another type if conditions change between antagonism and cooperation according to the state of equilibrium of both the biotic and chemical-physical conditions of the environment.

TIF appear very similar in behavior to fungal endophytes, and several species recorded as endophytes are also found as TIF, like *Tetracladium maxilliforme* (Rostr.) Ingold or *V. leptobactrum* (Pacioni et al. 2007). Most likely, there are unknown biochemical and molecular mechanisms that allow the penetration of TIF hyphae within the gleba even without symptoms. When co-cultured with *T. borchii* mycelium, almost all the TIF cultured by Pacioni et al. (2007) did not show any form of antagonism – with the only exception of an isolate of *V. leptobactrum* which showed hyphae with an incomplete entangling with those of *T. borchii*. This was in stark contrast to the marked incompatibility shown by *Agaricus macrosporus* Mont. (current name *A. crocodilinus* Murrill) towards all these TIF mycelia (Fig. 17.3), except for *V. leptobactrum*, which is known as a frequent FF isolated from several fruiting bodies of Basidiomycetes.

Trichothecium crotocinigenum [current name *Acremonium crotocinigenum* (Schol-Schwarz) W. Gams] seems to be the cause of an extensive rot epidemic affecting *T. melanosporum* cultivated in Australia. The epidemic persisted for some years and in some cases resulted in the loss of more than 50 % of the truffle harvest. Its pathogenicity was verified directly on truffles in the field, either by inoculating *T. crotocinigeum* conidia or gleba of rotten truffles (Eslick 2013). However,

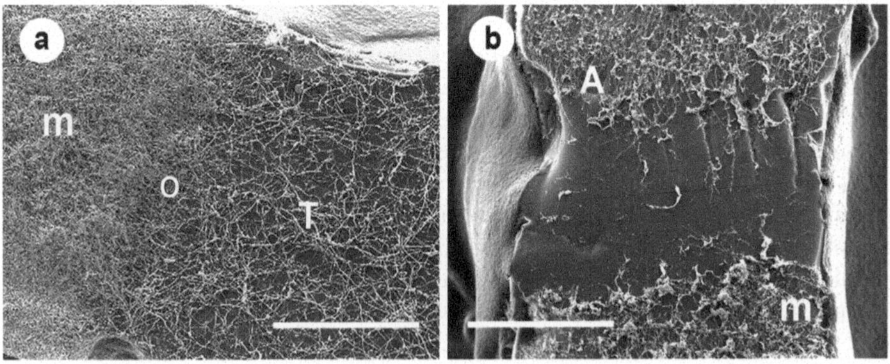

Fig. 17.3 SEM micrographs of mycelia of the test-species *T. borchii* and *A. macrosporus* in dual cultures with the mycelia of truffle guest fungi. (**a**) Interaction zone between the mycelia of *Tetracladium maxilliforme* from *T. melanosporum* (m) and *T. borchii* (T); (**b**) growth inhibition zone resulting from the cocultivation of *A. macrosporus* (**a**) and *Talaromyces wortmannii* (m) mycelia isolated from *T. rufum*. *Bar* = 1 mm

T. crotocinigenum was previously isolated from healthy ascocarps of *Tuber maculatum* Vittad. in the Netherlands (CBS number 566.84) (Eslick 2013). A similar species, *Trichothecium roseum* (Pers.) Link, was isolated from pink rotten Chinese truffles (Gong et al. 2001) and had previously been recorded on the desert truffles *Terfezia pinoyi* Maire and *Terfezia claveryi* Chatin after storage (Al-Delaimy and Ali 1970).

Nectria inventa, found sporulating on rotten *T. melanosporum* in Australia (Eslick 2013), is known as a destructive mycoparasite (Tsuneda and Skoropad 1980), but later investigations of Eslick (2013) did not seem to confirm its direct action on the disease itself.

The anamorph of *B. ochroleuca* (*C. rosea*), frequently isolated from healthy truffles and which was detected by qPCR in all samples of truffles examined by Leonardi et al. (submitted), was also isolated from rotten *T. melanosporum* ascomata in Australia (Eslick pers. com.) as well as, together with *V. leptobactrum*, from both healthy and rotting *T. magnatum* ascomata from Serbia (Pavić et al. 2013). This anamorph is considered a species with chitinolytic abilities, but Pavić et al. (2013) did not find evidence of these metabolic capabilities in its isolates from *T. magnatum*. However, we have seen that *B. ochroleuca* is highly aggressive towards other TIF mycelia when cultured in Petri dishes. For this reason, it is necessary to perform frequent subculturing of isolated TIF mycelia taking their superficial (aerial) hyphae in order to avoid transferring them together with the aggressive *B. ochroleuca*. Also other TIF has shown a chitinase capacity such as *A. cylindrospora* (Pacioni et al. 2007) (Fig. 17.4) but, probably, not all TIF colonize truffles to feed on chitin. Instead, it seems likely that many TIF could participate in the microhabitat represented by gleba truffle to acquire proteins and free amino acids as assumed by Buzzini et al. (2005). This aspect of the TIF commensalism seems to emerge from the test on the cellulase, chitinase, pectinase, and protease activities of truffle-inhabiting yeasts carried out by Zacchi et al. (2003).

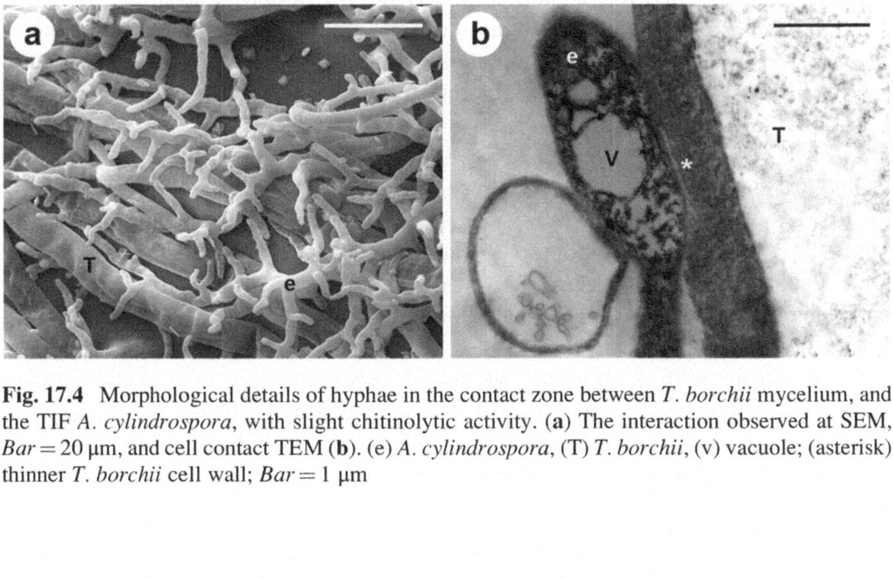

Fig. 17.4 Morphological details of hyphae in the contact zone between *T. borchii* mycelium, and the TIF *A. cylindrospora*, with slight chitinolytic activity. (**a**) The interaction observed at SEM, *Bar* = 20 μm, and cell contact TEM (**b**). (e) *A. cylindrospora*, (T) *T. borchii*, (v) vacuole; (asterisk) thinner *T. borchii* cell wall; *Bar* = 1 μm

The chitinolytic ability cannot, however, be excluded since some species of TIF are also known to be entomopathogen, like *Verticillium lecanii* (Zimm.) Viégas [current name *Lecanicillium muscarium* (Petch) Zare & W. Gams].

17.5 Effects of TIF on Truffle Biology

As for all "microbiomes," meant as the totality of the genetic and environmental interactions of all the microorganisms that live in an organism (Lederberg and McCray 2001), other effects in addition to any damage on the host ascoma can also be supposed for the truffle microbiome. As for bacteria, investigations are proceeding so that an active role can be envisaged in the formation and maturation of *T. melanosporum* ascoma (Antony-Babu et al. 2014) and also in the production of some components of volatile organic compounds (VOCs) (Splivallo et al. 2014). Previously, these bacterial interactions, positive as well as negative, were also shown in the establishment of mycorrhiza (Barbieri et al. 2005; Sbrana et al. 2002).

For TIF, to date, cultural experiences testing the possible in vivo interactions have shown that, most likely, there are effects on the VOC production and mycorrhizal symbiosis.

17.5.1 VOCs Production

Buzzini et al. (2005) tested the metabolic activity of 29 strains of yeasts isolated from *T. melanosporum* and *T. magnatum* ascomata, in culture media with L-methionine as the only nitrogen source, and detected nine volatile organic compounds, also characteristic of truffle aroma [2-methyl butanol, 3-methyl butanol, methanethiol, S-methyl thioacetate, dimethyl sulfide, dimethyl disulfide, dimethyl trisulfide, dihydro-2-methyl-3(2H)-thiophenone, and 3-(methylthio)-1-propanol]. Because of these results, they hypothesized a complementary role of truffle-associated yeasts in contributing to the final *Tuber* aroma through the synthesis of yeast-specific volatile constituents.

In one of our studies conducted several years ago (2008–2009, unpublished results), co-cultures of *T. borchii* mycelium (ATCC 96540) with various TIF identified by Pacioni et al. (2007) were examined by means of the solid-phase microextraction–gas chromatography–mass spectrometry (SPME-GC-MS) technique. On Potato Dextrose Agar (PDA), the co-cultured *T. borchii* and TIF showed the exclusive presence of several VOCs, not produced by any of the partners grown separately. Generally, they are molecules belonging to the terpenoids (Table 17.3).

In co-culture with *T. borchii*, while not producing any new compound of interaction, *Talaromyces wortmannii* C.R. Benj. continues to release a series of remarkable terpene compounds, such as menthone, caryophyllene, aromadendrene, β-cedrene, verticellol, and, likely, longifolene (unpublished data). A similar result

Table 17.3 Differential VOCs identified in the dual cultures of *T. borchii* (Tb) with some truffle-inhabiting mycelia

Co-cultured species	VOCs
Tb + *V. leptobactrum*	α-Canfolenal
Tb + *P. cinerea*	Camphor
Tb + *A. cylindrospora*	Camphor
Tb + *C. magnusianum*	β-Phellandrene 1,1-Dimethyldecylbenzene
Tb + *T. maxilliforme*	Caryophyllene oxide Docosane

was also obtained by Splivallo et al. (2014) applying the same approach (co-culture and SPME-GC-MS) to study the VOCs produced by truffle-inhabiting bacteria and demonstrating that some of the compounds present in the aroma of *T. borchii* resulted from bacterial activity.

17.5.2 Ectomycorrhiza

Several species of TIF are known to be pathogenic or endophytes of plants, thus potentially able to invade root tissues of the truffle host plants.

For this reason in 2008 (unpublished data), we wanted to investigate whether the mycelia of *A. cylindrospora*, *C. magnusianum*, *Peniophora cinerea* (Pers.) Cooke, *T. wortmannii*, *T. maxilliforme*, *Trichopezizella nidulus* (J.C. Schmidt & Kunze) Raitv., and *V. leptobactrum* isolated by Pacioni et al. (2007) affected the process of mycorrhization. For this purpose, an axenic system of mycorrhizal synthesis in Magenta vessels containing vermiculite, limestone sand, and peat was used. Micropropagated plantlets of *Populus alba* L. and the different fungal mycelia were tested, both individually and together with *T. borchii*. The process of mycorrhization was followed histologically using the terminal deoxynucleotidyl transferase dUTP Nick End Labeling (TUNEL) reaction to highlight any signs of programmed cell death. *Cylindrocarpon magnusianum* and *V. leptobactrum* were found not only in the intercellular spaces (apoplast) of root tips but also inside root hairs, whereas *A. cylindrospora* had an intracellular growth. Root cells did not seem to suffer colonization by these TIF, but no ECMs of *T. borchii* were found. From the results obtained, it seems that these species do not induce a defense response from the plant. In fact, at root level neither phenomena of programmed cell death nor excessive deposition of tannins occurs, these being the typical features generally found in the defense response implementation from the plant, in the case of attack by pathogens or mycorrhizal aggressive species like *Tuber* (Ragnelli et al. 2014).

This is a topic that deserves further research offering new insights. Vohník et al. (2007) isolated the ascomycete *Meliniomyces variabilis* Hambl. & Sigler from a hypogeous ascocarp of *Hydnotrya tulasnei* (Berk.) Berk. & Broome (Pezizales). By assaying it, they provided evidence that this fungus was able to produce both ECMs and ericoid mycorrhiza with host plant roots.

Another interesting effect on mycorrhization was attributed to *A. phaeospermum*. This fungus, tested with an in vitro symbiosis system between *Cistus creticus* L. and *T. borchii*, appears to be able to promote mycorrhiza formation in seedlings, inducing primary root shortening and an increase of secondary roots, similar to the effect of mycorrhization helper bacteria (MHB). Further bioassays suggested that volatiles released by the fungus alone are sufficient to alter root morphology in the early phase of interaction before the mycorrhiza formation. These results seem to indicate an influence of a truffle-hosted fungus on ECM symbiosis establishment, so as to claim the fungus as the first mycorrhization helper fungus (Sabella et al. 2015).

17.6 Conclusions

The road to a comprehensive assessment of TIF biodiversity and its practical implications on the entire truffle chain from crop to postharvest preservation is far from being solved. TIF isolation in pure culture is an unsatisfactory approach owing to the presence of multiple species which, likely, vary according to the geographical areas of truffle production and on the stage of ascoma maturation.

Since today there is every prospect of next-generation sequencing techniques leading to microbiological assessment on a global scale (Tedersoo et al. 2014), the solution of the minor problems posed by the truffle mycobiome could represent the challenge to be dealt with in the near future.

Acknowledgements We are grateful to Anna Maria Ragnelli and Pierpaolo Aimola for the microscopic images, Osvaldo Zarivi for the “real-time” qPCR analysis, Samantha Reale for the analysis of VOCs from co-cultures, and the former masters students Antonella Pompa and Michela Marcellitti, who could not complete their research on the effects of the TIF because of the devastating earthquake that struck the city of L’Aquila on April 6, 2009, respectively, for the formation of the ECM and the production of VOCs. Finally, we also have a debt of gratitude to Harry Eslick for useful information exchanges and Marta Lucia Pacioni for unpublished data on truffle DGGE.

References

Al-Delaimy KS, Ali SH (1970) Storage, spoiled and proximate food composition of Iraqi truffles. Beiträge zur Tropischen und Subtropischen Landwirtschaftlichen Tropenveterinärmedizin 8:77–81

Antony-Babu S, Deveau A, Van Nostrand JD, Zhou J, Le Tacon F, Robin C, Frey-Klett P, Uroz S (2014) Black truffle-associated bacterial communities during the development and maturation of *Tuber melanosporum* ascocarps and putative functional roles. Environ Microbiol 16:2831–2847. doi:10.1111/1462-2920.12294

Barbieri E, Potenza L, Rossi I, Sisti D, Giomaro G, Rossetti S, Beimfohr C, Stocchi V (2000) Phylogenetic characterization and *in situ* detection of a *Cytophaga-Flexibacter-Bacteroides* phylogroup bacterium in *Tuber borchii* Vittad. ectomycorrhizal mycelium. Appl Environ Microbiol 66:5035–5042. doi:10.1128/AEM.66.11.5035-5042.2000

Barbieri E, Bertini L, Rossi I, Ceccaroli P, Saltarelli R, Guidi C, Zambonelli A, Stocchi V (2005) New evidence for bacterial diversity in the ascoma of the ectomycorrhizal fungus *Tuber borchii* Vittad. FEMS Microbiol Lett 247:23–35. doi:10.1016/j.femsle.2005.04.027

Benucci GMN, Raggi L, Albertini E, Grebenc T, Bencivenga M, Falcinelli M, Di Massimo G (2011) Ectomycorrhizal communities in a productive *Tuber aestivum* Vittad. orchard: composition, host influence and species replacement. FEMS Microbiol Ecol 76:170–184. doi:10.1111/j.1574-6941.2010.01039.x

Benucci GMN, Raggi L, Albertini E, Csorbai AG, Donnini D (2014) Assessment of ectomycorrhizal biodiversity in *Tuber macrosporum* productive sites. Mycorrhiza 24:281–292. doi:10.1007/s00572-013-0538-3

Boulanger E, Matruchot L (1903) Le mycélium truffiers blancs. Oberthur, Rennes

Buzzini P, Gasparetti C, Turchetti B, Cramarossa MR, Vaughan-Martini A, Martini A, Pagnoni UM, Forti L (2005) Production of volatile organic compounds (VOCs) by yeasts isolated from the ascocarps of black (*Tuber melanosporum* Vitt.) and white (*Tuber magnatum* Pico) truffles. Arch Microbiol 184:187–193. doi:10.1007/s00203-005-0043-y

De Miguel AM, Águeda B, Sánchez S, Parladé J (2014) Ectomycorrhizal fungus diversity and community structure with natural and cultivated truffle hosts: applying lessons learned to future truffle culture. Mycorrhiza 24(Suppl 1):S5–S18. doi:10.1007/s00572-013-0554-3

Eslick H (2013) Identification and management of the agent causing rot in black truffles Part 2. Rural Industries Research and Development Corporation, No. 13/111. Canberra

Fan L, Hou CL, Cannon PF, Li Y (2012) A new species of *Melanospora* on truffles from China. Mycologia 104:1433–1442. doi:10.3852/11-338

Fontana A, Fasolo-Bonfante P (1971) Una nuova specie di *Sporothrix*. Allionia 17:5–13

Gams W, Diederich P, Põldmaa K (2004) Fungicolous fungi. In: Mueller GM, Bills GF, Foster MS (eds) Biodiversity of fungi: inventorying and monitoring. Elsevier, Amsterdam, pp 343–392

Gan Z, Yang J, Tao N, Yu Z, Zhang KQ (2007) Cloning and expression analysis of a chitinase gene Crchi1 from the mycoparasitic fungus *Clonostachys rosea* (syn. *Gliocladium roseum*). J Microbiol 45(5):422–430

Gong M, Zhong C, Chen Y, Wang F (2001) An investigation of diseases on the three species national truffles. Edible Fungi China 20:18–19

Gryndler M, Černá L, Bukovská P, Hršelová H, Jansa J (2014) *Tuber aestivum* association with non-host roots. Mycorrhiza 24:603–610. doi:10.1007/s00572-014-0580-9

Hawksworth DL (1981) A survey of the fungicolous conidial fungi. In: Cole GT, Kendrick K (eds) The biology of conidial fungi, vol 1. Academic, New York, pp 171–244

Iotti M, Amicucci A, Stocchi V, Zambonelli A (2002) Morphological and molecular characterization of mycelia of some *Tuber* species in pure culture. New Phytol 155:499–505. doi:10.1046/j.1469-8137.2002.00486

Iotti M, Lancellotti E, Hall I, Zambonelli A (2010) The ectomycorrhizal community in natural *Tuber borchii* grounds. FEMS Microbiol Ecol 72:250–260. doi:10.1111/j.1574-6941.2010.00844.x

Iotti M, Piattoni F, Zambonelli A (2012) Techniques for host plant inoculation with truffles and other edible ectomycorrhizal mushrooms. In: Zambonelli A, Bonito GM (eds) Edible ectomycorrhizal mushrooms, current knowledge and future prospects, vol 34, Soil Biology. Springer, Berlin, pp 145–161. doi:10.1007/978-3-642-33823-6_9

Lancellotti E, Franceschini A, Iotti M, Zambonelli A (2012) *Entoloma* sp. a parasite of *Tuber borchii* ascomata. J. Plant Pathol 94(Suppl 4):S4.65

Lederberg J, McCray AT (2001) 'Ome Sweet 'Omics - A genealogical treasury of words. Scientist 15:8

Leonardi M, Ascione S, Pacioni G, Cesare P, Pacioni ML, Miranda M, Zarivi O. (submitted) Towards an assessment of the truffle-inhabiting mycelia.

Luppi AM, Fontana A (1977) Studi sull'ecologia del *Tuber melanosporum*, IV. Analisi micologiche di terreni tartufiferi dell'Italia Centrale. Allionia 22:105–113

Luppi-Mosca AM (1972) La microflora della rizosfera nelle tartufaie. III. Analisi micologiche di terreni tartufiferi francesi. Allionia 18:33–40

Malençon GM (1942) Études de parasitisme mycopathologiques. Rev Mycol 7:27–52

Marletto F (1969) La microflora della rizosfera delle tartufaie, I. I blastomiceti dei tartufi e della rizosfera delle tartufaie. Allionia 15:155–171

Mello A, Miozzi L, Vizzini A, Napoli C, Kowalchuk G, Bonfante P (2010) Bacterial and fungal communities associated with *Tuber magnatum*-productive niches. Plant Biosyst 144:323–332. doi:10.1080/11263500903374724

Mello A, Ding GC, Piceno YM, Napoli C, Tom LM, De Santis TZ, Andersen GL, Smalla K, Bonfante P (2013) Truffle brûlés have an impact on the diversity of soil bacterial communities. PLoS One 8, e61945. doi:10.1371/journal.pone.0061945

Napoli C, Mello A, Borra A, Vizzini A, Sourzat P, Bonfante P (2010) *Tuber melanosporum*, when dominant, affects fungal dynamics in truffle grounds. New Phytol 185:237–247. doi:10.1111/j.1469-8137.2009.03053.x

Orgiazzi A, Bianciotto V, Bonfante P, Daghino S, Ghignone S, Lazzari A, Lumini E, Mello A, Napoli C, Perotto S, Vizzini A, Bagella S, Murat C, Girlanda M (2013) 454 Pyrosequencing analysis of fungal assemblages from geographically distant, disparate soils reveals spatial patterning and a core mycobiome. Diversity 5:73–98. doi:10.3390/d5010073

Pacioni G (1990) Scanning electron microscopy of *Tuber* sporocarps and associated bacteria. Mycol Res 94:1086–1089. doi:10.1016/S0953-7562(09)81338-5

Pacioni G (1991) Effects of *Tuber* metabolites on the rhizospheric environment. Mycol Res 95:1355–1358. doi:10.1016/S0953-7562(09)80384-5

Pacioni G, Ragnelli AM, Miranda M (1995) Truffles development and interactions with the biotic environment: molecular aspects. In: Stocchi V, Bonfante P, Nuti M (eds) Biotechnology of ectomycorrhizae. Plenum Press, New York, pp 213–227

Pacioni G, Leonardi M, Aimola P, Ragnelli AM, Rubini A, Paolocci F (2007) Isolation and characterization of some mycelia inhabiting *Tuber* ascomata. Mycol Res 111:1450–1460. doi:10.1016/j.mycres.2007.08.016

Pacioni G, Ragnelli AM, Aimola P, Leonardi M, Marinucci D, Roth-Bejerano N, Kagan-Zur V (2014) Endomycorrhiza and pseudo-ectomycorrhiza produced *in vitro* by two species of *Tuber* on transformed *Cistus incanus* roots. Int J Plant Biol Res 2:1021

Pavić A, Stanković S, Saljnikov E, Krüger D, Buscot F, Tarkka M, Marjanović Ž (2013) Actinobacteria may influence white truffle (*Tuber magnatum* Pico) nutrition, ascocarp degradation and interactions with other soil fungi. Fungal Ecol 6:527–538. doi:10.1016/j.funeco.2013.05.006

Payen T, Murat C, Bonito G (2014) Truffle phylogenomics: new insights into truffle evolution and truffle life cycle. In: Martin F (ed) Genomics of plant-related fungi, vol 70, Advances in Botanical Research. Academic, Elsevier, pp 211–234. doi:10.1016/B978-0-12-397940-7.00007-0

Ragnelli AM, Aimola P, Maione M, Zaviri O, Leonardi M, Pacioni G (2014) The cell death phenomenon during *Tuber* ectomycorrhiza morphogenesis. Plant Biosyst 148:473–482. doi:10.1080/11263504.2013.788575

Sabella E, Nutricati E, Aprile A, Miceli A, Sorce C, Lorenzi R, De Bellis L (2015) *Arthrinium phaeospermum* isolated from *Tuber borchii* ascomata: the first evidence for a 'Mycorrhization Helper Fungus'? Mycol Prog 14:59. doi:10.1007/s11557-015-1083-6

Saccardo PA (1882–1913) Sylloge fungorum omnium hucusque cognitorum, vol 25. Padova. doi:10.5962/bhl.title.5371

Sbrana C, Bagnoli G, Bedini S, Filippi C, Giovannetti M, Nuti MP (2000) Adhesion to hyphal matrix and antifungal activity of *Pseudomonas* strains isolated from *Tuber borchii* ascocarps. Can J Microbiol 46:259–268. doi:10.1139/w99-135

Sbrana C, Agnolucci M, Bedini S, Lepera A, Toffanin A, Giovannetti M, Nuti MP (2002) Diversity of culturable bacterial populations associated to *Tuber borchii* ectomycorrhizas and their activity on *T. borchii* mycelial growth. FEMS Microbiol Lett 211:195–201. doi:10.1111/j.1574-6968.2002.tb11224.x

Selosse MA, Faccio A, Scappaticci G, Bonfante P (2004) Chlorophyllous and achlorophyllous specimens of *Epipactis microphylla* (*Neottieae*, *Orchidaceae*) are associated with ectomycorrhizal septomycetes, including Truffles. Microbial Ecol 47:416–426. doi:10.1007/s00248-003-2034-3

Splivallo R, Deveau A, Valdez N, Kirchhoff N, Frey-Klett P, Karlovsky P (2014) Bacteria associated with truffle-fruiting bodies contribute to truffle aroma. Environ Microbiol 17:2647–2660. doi:10.1111/1462-2920.12521

Stielow B, Menzel W (2010) Complete nucleotide sequence of TaV1, a novel totivirus isolated from a black truffle ascocarp (*Tuber aestivum* Vittad.) Arch Virol 155:2075–2078. doi:10.1007/s00705-010-0824-8

Streiblová E, Gryndlerová H, Gryndler M (2012) Truffle brûlé: an efficient fungal life strategy. FEMS Microbiol Ecol 80:1–8. doi:10.1111/j.1574-6941.2011.01283.x

Szemere L (1965) Die unterirdischen Pilze des Karpatenbeckens. Akadémiai Kiadó, Budapest

Tedersoo L, Bahram M, Põlme S, Kõljalg U, Yorou NS, Wijesundera R, Villarreal Ruiz L, Vasco-Palacios AM, Thu PQ, Suija A, Smith ME, Sharp C, Saluveer E, Saitta S, Rosas M, Riit T, Ratkowsky D, Pritsch K, Põldmaa K, Piepenbring M, Phosri C, Peterson M, Parts K, Pärtel K, Otsing E, Nouhra E, Njouonkou AL, Nilsson RH, Morgado LN, Mayor J, May TW, Majuakim L, Lodge DJ, Lee SS, Larsson KH, Kohout P, Hosaka K, Hiiesalu I, Henkel TW, Harend H, Guo LD, Greslebin A, Grelet G, Geml J, Gates G, Dunstan W, Dunk C, Drenkhan R, Dearnaley J, De Kesel A, Dang T, Chen X, Buegger F, Brearley FQ, Bonito G, Anslan S, Abell S, Abarenkov K (2014) Global diversity and geography of soil fungi. Science 346:1078–1087. doi:10.1126/science.1256688

Tsuneda A, Skoropad WP (1980) Interactions between *Nectria inventa*, a destructive mycoparasite, and fourteen fungi associated with rapeseed. Trans Br Mycol Soc 74:501–507. doi:10.1016/S0007-1536(80)80049-0

Vohník M, Fendrych M, Kolařík M, Gryndler M, Hršelová H, Albrechtová J, Vosátka M (2007) The ascomycete *Meliniomyces variabilis* isolated from a sporocarp of *Hydnotrya tulasnei* (*Pezizales*) intracellularly colonises roots of ecto- and ericoid mycorrhizal host plants. Czech Mycol 59:215–226

Zacchi L, Vaughan-Martini A, Angelini P (2003) Yeast distribution in a truffle-field ecosystem. Ann Microbiol 53:275–282

Zambonelli A, Iotti M, Boutahir S, Lancellotti E, Perini C, Pacioni G (2012) Ectomycorrhizal fungal communities of edible ectomycorrhizal mushrooms. In: Zambonelli A, Bonito G (eds) Edible ectomycorrhizal mushrooms, vol 34, Soil biology. Springer, Berlin Heidelberg, pp 105–124. doi:10.1007/978-3-642-33823-6_7

Zarivi O, Cesare P, Ragnelli AM, Aimola P, Leonardi M, Bonfigli A, Colafarina S, Poma AM, Miranda M, Pacioni G (2015) Validation of reference genes for quantitative real-time PCR in Périgord black truffle (*Tuber melanosporum*) developmental stages. Phytochem 116:78–86. doi:10.1016/j.phytochem.2015.02.024

Chapter 18
Truffle-Associated Bacteria: Extrapolation from Diversity to Function

Elena Barbieri, Paola Ceccaroli, Deborah Agostini, Sabrina Donati Zeppa, Anna Maria Gioacchini, and Vilberto Stocchi

18.1 Introduction

In many ways truffles are still a mystery, especially if we consider their life cycle and the conditions stimulating ascoma development. Because of their economic and agricultural value, truffles are an extremely attractive crop for local cultivators.

Thanks to well-trained dogs or less frequently pigs, which can smell mature truffles through a layer of earth, truffles are usually harvested. A better understanding of environmental and soil conditions, as well as biotic and abiotic factors that may favor ectomycorrhizal (ECM) establishment and truffle maturation, is essential for improving truffle cultivation. Translational research in the field of truffle cultivation presents many challenges, but also holds great promise.

After the inoculated seedlings are planted, it takes several years for truffles to begin to produce fruiting bodies, but once they are established, their production can continue for decades. The truffle production quality, yield, and duration depend on many factors such as the type of soils, the species of host trees, cultural practices, etc. Some farms produce about 70 kg *per* acre each year, while others produce much less. Across Europe average yearly yields range between 10 and 16 kg *per* acre (Alessandra Zambonelli, pers comm).

A third partner in the symbiosis between the fungus and plant root is represented by natural bacterial communities, which seem to play pivotal role in the complex biological processes of exchange involving nutrients and signaling from the soil hyphae, ECMs, ascomata, and stromata. In the last few decades, various bacterial communities have been associated with different truffle species: *Tuber borchii* Vittad. (Sbrana et al. 2000; Barbieri et al. 2005; Splivallo et al. 2014), *Tuber*

E. Barbieri (✉) • P. Ceccaroli • D. Agostini • S.D. Zeppa • A.M. Gioacchini • V. Stocchi
Department of Biomolecular Sciences, University of Urbino Carlo Bo, via A. Saffi, 2, 61029 Urbino, PU, Italy
e-mail: elena.barbieri@uniurb.it

A. Zambonelli et al. (eds.), *True Truffle (*Tuber *spp.) in the World*, Soil Biology 47,
DOI 10.1007/978-3-319-31436-5_18

magnatum Pico (Barbieri et al. 2007), *Tuber aestivum* Vittad. (Gryndler et al. 2013), and *Tuber melanosporum* Vittad. (Antony-Babu et al. 2014). Specific bacterial species seem to be attracted by the peridium and gleba of truffles and appear to be selected from the soil communities during the early stage of truffle formation (Antony-Babu et al. 2014). A core microbiome of Bradyrhizobiaceae is a common denominator that has been consistently associated with all species studied to date (Barbieri et al. 2005, 2007; Antony-Babu et al. 2014). Despite the availability of new information regarding the microbial diversity of *Tuber*-associated bacteria in the different phases of the truffle ontogenetic cycle and in the surrounding environment, the role of truffle-associated bacteria still remains largely unexplored. Indeed, in this review, we sift through advanced knowledge regarding the structure, function, interaction, and dynamics of bacteria associated with truffles. In the process, we have tried to be critical in extrapolating data regarding truffle-associated microbial diversity to function. Molecular detection, identification, characterization, and quantification have driven an explosion of information about community memberships and vastly expanded the known diversity of microbial life associated with truffles. However, the new molecular approaches pose several great challenges regarding function and possible mechanistic linkages between microbial diversity and truffle ecosystem functioning. Recent high-throughput technologies for analyzing microbial community structures and functions will be critical for advancing our understanding of the mysterious truffle system.

18.2 The Diversity and Dynamics of Truffle-Associated Bacterial Communities

In each of the stages of its life cycle, spores, hyphae, ECMs, or ascomata, *Tuber* species harbor a diverse microbial community, which includes bacteria, yeasts, and filamentous fungi (Barbieri et al. 2005, 2007; Buzzini et al. 2005; Pacioni et al. 2007; Barbieri et al. 2012; Antony-Babu et al. 2014; see Chap. 17), but only the bacteria have been extensively analyzed.

The first natural bacterial communities to be studied were those associated with ascomata of the white truffle species such as *T. borchii* and *T. magnatum*: in the case of *T. borchii*, several studies examined the culturable fractions of the ascomata-associated bacterial communities suggesting various hypothetical roles for these bacteria in the fungal growth or nutrition during *T. borchii* ascoma development and maturation (Bedini et al. 1999; Gazzanelli et al. 1999; Citterio et al. 2001; Sbrana et al. 2000, 2002). Our group has identified several bacteria within *T. borchii* fruiting bodies by both cultivation and PCR-based 16S rRNA cloning library and sequence analyses. Most of the culturable isolates were affiliated with γ-Proteobacteria, mainly fluorescent pseudomonads; some isolates were members of the *Bacteroidetes* group and gram-positive bacteria, mostly Actinobacteria and

Bacillaceae. However, the majority of the clones from the library were α-Proteobacteria showing significant similarity values to members of the *Sinorhizobium/Ensifer* group, *Rhizobium*, and *Bradyrhizobium* spp., which had not been identified as *Tuber*-associated bacteria (Barbieri et al. 2005). The dual approach, i.e., cultivation and PCR-based 16S rRNA gene (rDNA) cloning library and sequence analyses, allowed us to obtain more information on the truffle bacterial community than approaches that rely on a single method such as cultivation or direct recovery of 16S rDNA sequences. Using culture-independent and culture-dependent 16S rRNA gene-based approaches, we were able to identify α-Proteobacteria as major constituents of a bacterial component also associated with *T. magnatum* ascomata, regardless of the degree of maturation (Barbieri et al. 2007). The most representative clones among the α-Proteobacteria are closely related to *Bradyrhizobium elkanii,* which is capable of fixing atmospheric nitrogen in symbiotic interactions with leguminous plants (Rumjanek et al. 1993). Subsequently, Barbieri and colleagues (Barbieri et al. 2010) demonstrated a nitrogenase activity within *T. magnatum* ascomata possibly attributed to Bradyrhizobia. Among black truffles, the partial sequence of the 16S rDNA was used for the isolation of the bacteria from ECMs of *T. aestivum*. In these samples several bacteria belonging to six orders were identified: Actinomycetales, Burkholderiales, Enterobacteriales, Pseudomonadales, Rhizobiales, and Xanthomonadales (Gryndler and Hršelová 2012). Gryndler et al. (2013) searched for associations between *T. aestivum* and prokaryotes using two approaches: (1) 454 sequencing of environmental samples, ECMs, and soil, collected within and outside of a linear transect where a single *T. aestivum* ascoma was found (truffle colony) and (2) microbial isolation from the ECMs within the truffle colony. Actinobacteria and Proteobacteria dominated prokaryotic communities in both ECMs and soil samples. Among bacteria, *Actinosynnema* and *Allokutzneria* showed the strongest correlation with *T. aestivum*. On the other hand, among the bacterial genera that negatively correlated with *T. aestivum* in ECMs, the authors found several taxa, including *Bradyrhizobium* spp., consistently present in mature ascomata of other *Tuber* spp. (Barbieri et al. 2007, 2010; Antony-Babu et al. 2014).

To date, very little is known regarding the natural communities that colonize the black truffle, *T. melanosporum*, and their potential impact on the life cycle of the fungus. Rivera et al. (2010) identified the microbial population present in the gleba and in the peridium of *T. aestivum* and *T. melanosporum* truffles belonging to *Pseudomonas* genus and the Enterobacteriaceae family. More recently, Antony-Babu et al. (2014) have studied the bacterial communities from *T. melanosporum* ascomata at three stages of maturation, the ectomycorrhizosphere, and the surrounding bulk soil using a combined approach based on temporal temperature gradient gel electrophoresis (TTGE) fingerprinting and 16S rDNA gene amplicon pyrosequencing. Bacteria from the ascomata were separated from those of the bulk soil, suggesting that the *T. melanosporum* ascomata create a niche that selects specific bacterial communities. As observed in other studies, the bulk soil was colonized by phyla such as Proteobacteria, Bacteroidetes, Verrucomicrobia, Actinobacteria, Firmicutes, and Acidobacteria (Gryndler et al. 2013; Mello et al. 2013), while the ascoma was dominated by α-Proteobacteria related to the *Bradyrhizobium* genus. Mello et al. (2013), using denaturing gradient gel electrophoresis (DGGE) and DNA microarrays of 16S rDNA fragments, compared the bacterial and archaeal communities inside and outside of the truffle brûlé (the burnt area around the *Tuber*-infected plants). Soil samples were collected from four *T. melanosporum*/*Quercus pubescens* truffle grounds located in France. A large proportion of operational taxonomic units (OTUs) belonging to Firmicutes (*Bacillus*), Actinobacteria (*Geodermatophilus*, *Rubrobacter*), Proteobacteria (*Massilia*), and a few Cyanobacteria had significantly higher signal intensities inside *vs* outside the brûlé, while relatively large proportions of OTUs belonging to Bacteroides (*Riemerella*, *Chryseobacterium*, *Flavobacterium*, *Pedobacter*) and Proteobacteria (*Pseudomonas*) had significantly higher signal intensities outside *vs* inside the brûlé.

All the data published to date provide insights into the interactions between truffles and bacteria, the potential role of these bacteria in truffle maturation, and how these organisms influence the productive and not productive niches. Thanks to recent advances in the analysis of genes in environmental samples, large-scale (next-generation) sequencing, RNA-centered meta-transcriptomic methods, and metabolomic techniques, we may be able to gain a better understanding of both the structure and function of the active microbial community.

18.3 Potential Roles of Microbiota in the Truffle Life Cycle

The life cycle of truffles starts with the germination of haploid spores and the growth of the haploid mycelium into the soil. The mycelium then establishes a symbiotic association with tree roots leading to the formation of ECMs. In the final stage, we have the formation of a hypogeous ascoma, which is linked to the ECMs through extramatrical hyphae until it matures completely (Le Tacon et al. 2013).

Hypogeous mycorrhizal fungi constantly interact with microorganisms present in the soil, but questions remain about when and how bacteria and mycorrhizal fungi interact and when and how bacteria become associated with truffles and the mycelium. Recent molecular approaches show that prokaryotes are associated not only with the extra radical hyphae of mycorrhizal fungi but also occur in their mycorrhizas and ascomata (see Sect. 18.2). There is little information available regarding endo- and epiphytic bacteria, which may colonize and/or invade fungal tissues throughout the truffle life cycle or for part of that cycle, hence becoming obligate bacterial partners.

18.3.1 Vegetative Phase: Mycelium-Associated Bacteria/Soil Hyphae

Unapparent and asymptomatic infections occurring during mycelia isolation from truffle fruiting bodies are probably linked to in vitro mycelia accompanying bacteria. During the mycelium isolation attempts, the timing and efficiency of the mycelial hypha to outgrowth from the truffle explant, the competition with the other ascoma-inhabiting microbes, and the capacity of these competitors to survive and grow in the saprophytic stage of in vitro growth, after the first subculture, are all important conditions to take into consideration, as also reported by Giomaro et al. (2005). Barbieri et al. (2000) described, as first direct observation, the presence of a not-yet-cultured *Cytophaga–Flexibacter–Bacteroides* (CFB) in *T. borchii* mycelia, ECMs, and ascomata, suggesting that certain bacteria occur within the symbiotic fungi during their entire life cycle. In this regard, an enhanced role of *Tuber*-associated bacteria, as mycelial helper bacteria in the progressive growth, adaptation, and nutritionally independent phase of the ascomata, cannot be excluded. Soil bacteria may have various effects on *Tuber* mycelium growth, ranging from inhibition to promotion. For example, one strain of *Staphylococcus aureus* has been shown to produce volatile organic compounds potentially involved in *T. borchii* mycelial growth inhibition (Barbieri et al. 2005). Moreover, Nuti and his collaborators (Sbrana et al. 2000) reported that *Pseudomonas* spp. isolates from *T. borchii* ascomata are able to produce phytoregulatory and biocontrol substances in pure culture, which affected *T. borchii* mycelial growth and morphogenesis. These bacteria, which exhibited antifungal and growth modulation activities, also showed a specific preferential adhesion to *Tuber* mycelium (Sbrana et al. 2000). Gryndler and Hršelová (2012) have studied the interactions of *T. aestivum* cultures with *Moraxella* sp., *Lysobacter* sp., *Phyllobacterium* sp., *Rhizobium* cf. *giardinii*, *Rhizobium* cf. *leguminosarum*, *Mesorhizobium* sp., *Xylophilus* sp. The authors reported that, after seven days of in vitro cocultivation, the interaction of truffle mycelium was observed only with the bacterial isolate *Rhizobium* cf. *leguminosarum* which caused an extensive hyphae vacuolization. No other cases of adverse or stimulatory effects were noted. The question of whether certain strains, such as *Rhizobia*-like strains/clones consistently found in the ascomata and ECMs of several *Tuber* spp. as well as the Pseudomonadaceae isolates described by Nuti and collaborators, are involved in the truffle life cycle is still unanswered (Sbrana et al. 2000).

Although recent molecular approaches provide insights into the structure of microbial communities in truffle grounds, potentially rich in *Tuber* hyphae, there is little available evidence on the roles of hypha-associated microbes in vivo ECM systems (Mello et al. 2010, 2013; Gryndler et al. 2013). These microorganisms could favor hyphal growth, branching and penetration of root tissues, and/or formation of truffle ECMs. Fungus–bacterium cocultures are easily set up and have often been used to screen for potential strains for mycorrhizal inoculation experiments as described by Frey-Klett et al. (2007). Thus, targeted co-inoculation of specific bacteria with *Tuber* spp. on commercial *Tuber* plantations could be used to increase positive selective pressure for truffles in natural habitats; hence, potentially increasing truffle yields.

18.3.2 Symbiontic Phase: Ectomycorrhiza-Associated Bacteria

The soil space directly affected by ECMs is called the mycorrhizosphere and is characterized by high biological activity (Chalot and Brun 1998). The presence of some bacteria may enhance the formation of ECM structures on roots and has been shown to change gene expression of the ECM fungal hyphae (Deveau et al. 2007). Some of these microbes, called "mycorrhiza helper bacteria" (MHB) (Garbaye 1994), are particularly important for their ecological and evolutionary implications, first of all for their capacity in stimulating mycorrhizal colonization of host roots. Sbrana et al. (2002) carried out the isolation and physiological and molecular characterization of culturable bacterial strains belonging to actinomycetes, pseudomonads, and aerobic spore-forming bacteria from *Quercus robur* L. var. *pedunculata to T. borchii* mycorrhizal root tips. The growth-stimulating activity shown by spore-forming bacteria represents a fundamental character of putative MHB. Furthermore, the presence of different culturable bacterial populations inside *T. borchii* ECMs suggests that bacterial incorporation into the ECM fungal mantle might represent an important event, and *Pseudomonas*, the dominant genus among isolates from *T. borchii* ascomata, establish a preferential interaction with *Tuber* mycelium (Sbrana et al. 2000). The effects of *Pseudomonas* sp. in the establishment of ECMs have been extensively studied in different ECM fungi (Rigamonte et al. 2010), but only a few studies have been conducted to demonstrate the influence of bacteria on *Tuber*–host plant interactions. Dominguez et al. (2012) and Dominguez-Nuñez et al. (2015) investigate combined effects of *Pseudomonas* and *T. melanosporum* on the quality of *Pinus* seedlings. The authors demonstrated that although the inoculation of *Pinus* seedlings with *P. fluorescens* (CECT 844) and *T. melanosporum* did not change the plant growth and N absorption more than the fungus alone, it doubled the level of root colonization by *T. melanosporum*. MHB could be used in the future as inoculants for artificially infected plants to improve the establishment and maintenance of a functional symbiosis, which would be particularly useful for plantation truffle production; also so far no commercial use of MHB is yet available. In *T. aestivum* ECMs, two members of the suborder *Pseudonocardineae* (*Kibdelosporangium* and *Lentzea*) have been identified (Gryndler et al. 2013). Interestingly, to date, there is no available information on interactions of the members of *Pseudonocardineae* with fungi or plants, but some other members of this group, e.g., *Pseudonocardia*, have been observed to associate with leaf-cutting ants, producing defensive antibiotic compounds protecting their fungal gardens (Schoenian et al. 2011). The ants carry the *Pseudonocardia* symbionts on their exoskeletons, which have some of the chemical properties of fungal hyphae, such as chitin content. Regarding *T. magnatum*, its ECMs are very rare in the soil, which is mostly occupied by other mycorrhizal fungi (Murat et al. 2005). Mello et al. (2010), using molecular analysis of microbial communities, found an association of *Moraxella osloensis* with the productive area of *T. magnatum*. However, the functional significance of such microbial associations remains largely unexplored, and since this work was carried out in one truffle field, it is still limited. In *T. melanosporum*, the ECM compartment has been shown to be significantly enriched in Actinobacteria sequences related to the *Streptomyces* and *Thermoleophilum* (Antony-Babu et al. 2014). These bacterial communities are distinct from ascoma-associated communities, and these differences may be attributed to the biochemical composition of gleba/peridium soils comparties. *Tuber melanosporum* ascomata accumulate carbohydrates, proteins, and lipids (Harki et al. 2006), while ECMs are responsible for the transfer of simple metabolites such as sugars, amino acids, organic acids, or fatty acids between the host and the fungus. Thus, the specific bacterial communities in the surrounding bulk soil are differentially recruited in the ascomata and in the ectomycorrhizosphere.

18.3.3 Fructification Phase: Ascoma-Associated Bacteria

The development of the ascomata begins with the primordium, with the disappearance of these hyphae, forms the peridium, and begins a protection for the inner part, the gleba. Over the next few months, the fruiting body is formed (Antony-Babu et al. 2014). During these months of development, the ascoma can select bacteria from the surrounding bulk soil, and they can participate in its development. Bacteria colonize both the peridium and gleba of the truffle and appear to be selected from the soil communities during the early stage of truffle formation (Antony-Babu et al. 2014).

Bacterial populations ranging from 10^7 to 10^8 cfu g^{-1} have been reported on the surface and within the truffle ascomata of *T. aestivum* (Gryndler et al. 2013), *T. borchii* (Sbrana et al. 2000; Citterio et al. 2001; Barbieri et al. 2005), *T. magnatum* (Barbieri et al. 2007), and *T. melanosporum* (Rivera et al. 2010). Numerous *Pseudomonas* strains have been isolated from the ascomata of *T. borchii* and *T. magnatum*, and some have been identified as potential helper bacteria of *T. borchii*. Some strains have been shown to restrict the growth of phytopathogenic fungi; to possess cellulolytic, chitinolytic, and proteolytic activities in vitro; and to release metabolites, both medium diffusible and volatiles. Pseudomonads isolated within *T. borchii* fruiting bodies have been found to be producers of biocontrol substances such as indole-3-acetic acid phytohormone at sufficient levels to have positive effects on the fungus (Bedini et al. 1999). Citterio et al. (2001) revealed the presence of a microbial population on *T. borchii* fruiting bodies regardless of the fruiting body maturation stage, through these analyses demonstrated a progressive bacterial modification during the maturation process. The constant presence of

Pseudomonads and spore-forming Bacillaceae, with proved chitinolytic and cellulolytic activities, suggests a possible role of these bacteria in the ascus opening and spore scattering. To build on these findings, Barbieri et al. (2001) carried out a bacterial genetic characterization; the chitinolytic pseudomonads isolated were *P. fluorescens*, *P. syringae*, and *P. aeruginosa* while chitinolytic spore-forming Bacillaceae belonging to *Bacillus subtilis*, *B. cereus*, and *Paenibacillus* sp. Regarding *T. magnatum*, the diversity of the bacterial communities did not vary with fruiting body maturation (Barbieri et al. 2007), but the presence of well-defined bacterial communities in completely immature ascomata is a new evidence.

Despite the presence of these bacteria, only Barbieri et al. (2010) have shown a nitrogenase activity within *T. magnatum* ascomata, possibly attributed to *Bradyrhizobia*. The acetylene reduction assay method was used to measure nitrogenase activity, and the values obtained with *T. magnatum* samples were comparable to those generally obtained in a test tube using legume root nodules. Furthermore, the expression of *nifH* genes detected from *Bradyrhizobia* suggests a potential role of these bacteria in ascoma N_2 fixation. Recently, Antony-Babu et al. (2014) have shown that the *T. melanosporum* ascoma-associated bacterial communities evolve during ascoma maturation. They observed that *Bacteroidetes* present in the early stage were replaced by *Firmicutes* in the fully mature stage, and these changes could be explained by the biochemical composition modifications of the truffle tissues during truffle maturation. These findings are in contrast with those reported for *T. magnatum*, but the two species have very different processes of development. Moreover, Pavic et al. (2013) have shown that Actinobacteria strains isolated from *T. magnatum* ascomata might be involved in improving truffle nutrition, degrading the ascoma, and establishing relationships with other soil fungi. In particular, recent evolutionary evidence on truffle nutrition strategies described by Kohler et al. (2015) and Martin et al. (2010) are reported in Chap. 9. Rivera et al. (2010) identified five genera of the Enterobacteriaceae family (*Enterobacter*, *Klebsiella*, *Rahnella*, *Raoultella*, and *Serratia*) in *T. aestivum* and *T. melanosporum* fruiting bodies. The microaerobic conditions where truffles grow are optimal for the development of members of this family. Moreover, truffles contain levels of glucose, and these levels rise as the fungus matures (Harki et al. 2006); hence, the Enterobacteriaceae could ferment this sugar, modifying the taste and aroma of the truffle. *Tuber melanosporum* fruiting body shows massive colonization by Bradyrhizobiaceae, but very weak colonization by *Pseudomonas*, which are absent from mature truffles (Antony-Babu et al. 2014). Are the bacteria helpers for truffle development or profiteers? In *T. melanosporum*, *nifH* genes have been detected in ascomata, although the gene was not related to Bradyrhizobiaceae but to unknown bacteria, suggesting the involvement of other N_2-fixing microbiota. Despite a similar taxonomic community structure, the bacterial community within the ascomata of *T. magnatum* and *T. melanosporum* would appear to differ at the functional level: bacteria have no role or a very limited role in the nitrogen nutrition of the *T. melanosporum* ascoma in contrast to what occurs in *T. magnatum* (Le Tacon et al. 2016). Furthermore, the harboring function of *Bacteroidetes* and Actinomycetes in cellulose and chitin degradation may contribute to the release of the ascospores. In this regard, Pavic et al. (2013) have shown that Actinobacteria strains isolated from *T. magnatum* ascomata might be involved in improving truffle nutrition, degrading the ascoma, and establishing relationships with other soil fungi.

18.4 The Bacterial Contribution of Truffle Fruiting Bodies Volatile Organic Compounds (VOCs)

It has been suggested that truffle-associated bacteria could be involved in the development and maturation of *T. borchii* and *T. melanosporum* and may even affect the aroma of these truffles (Barbieri et al. 2000; Splivallo et al. 2007; Antony-Babu et al. 2014; Splivallo et al. 2014; see Chap. 23). Recently, it has been demonstrated how a specific truffle-associated microbiome is selected from the fungus versus the ascomata in the black truffle *T. melanosporum* (Antony-Babu et al. 2014).

Several VOCs could play an important role in interactions with other organisms, such as the reported interaction of *T. borchii* with the *Tilia americana* L. during the pre-symbiontic ECM phase (Menotta et al. 2004). Some fungal VOCs can exhibit antibiotic effects (Kramer and Abraham 2012) and inhibit plant growth (Splivallo et al. 2007). ECM fungi emit complex, species-specific odor profiles, which distinguish them from fungi of other ecological niches (Müller et al. 2013). In this regard, VOCs produced by microbes (MVOCs) can affect plant fitness either positively or negatively and interfere with plant–animal interactions, thereby promoting growth and preventing the establishment of detrimental microbes (Barbieri et al. 2005; Junker and Tholl 2013).

Recently, the bacterial community of the white truffle *T. borchii* has been characterized, and the involvement of its microbiome in the production of sulfur-containing volatiles tested (Splivallo et al. 2014). *Tuber borchii* thiophene volatiles, only present in completely mature (71–100 % maturity) fruiting bodies (Zeppa et al. 2004), and increased upon storage at room temperature (Bellesia et al. 2001), are the result of the biotransformation of nonvolatile precursor (s) into volatile compounds by the microbiome inhabiting truffle fruiting bodies. The thiophene volatiles are only produced by bacteria isolated from *T. borchii* fruiting bodies, but not from mycelium, which is not able to produce the precursors of thiophenes in vitro (Splivallo et al. 2014), demonstrating that thiophene production is linked to the sexual stage of *T. borchii*. The upregulation in sulfur metabolism in *T. borchii* and *T. melanosporum* fruiting bodies (Zeppa et al. 2010; Martin et al. 2010; Splivallo and Maier 2011; Splivallo et al. 2011) provides indirect evidence that truffles are actively involved in the synthesis of sulfur volatiles, demonstrating an intimate interaction coexistence and cooperation among truffle and bacteria without the exclusion of yeasts.

18.5 From Metagenomics to Metabolomics

Emerging molecular approaches, which move us toward microbial ecology functional studies, may be applied to the complex truffle microbial system (Fig. 18.1). In general, a microbial ecosystem includes all the organisms (plants and animals) and microorganisms living in a particular niche. These actors interact in a dynamic way, reacting to several factors including environmental and chemical conditions.

Microorganisms interact in different ways to benefit each other, establishing synergistic relationships and exchanging metabolites. By synchronizing the behaviors of all members in the group, they act like a multicellular organism. Interspecies interactions are usually mutually beneficial, but they can also be interfering and antagonistic, and communication is likely to be one-way, two-way, multi-way in dynamic interactions (Mougi and Kondoh 2014). As described above, these microbe–host associations and their relationships have been only partially investigated in the truffle system. In the near future, this complex interaction could be analyzed using a systematic process that begins with an annotated genome and concludes with a predictive model of microbial physiology (Tang 2011).

In the last few years, several individual microbial genomes have been sequenced thanks to the high-throughput capability of sequencing technologies, allowing a large-scale gene assessment of community members (Qin et al. 2010) and shedding

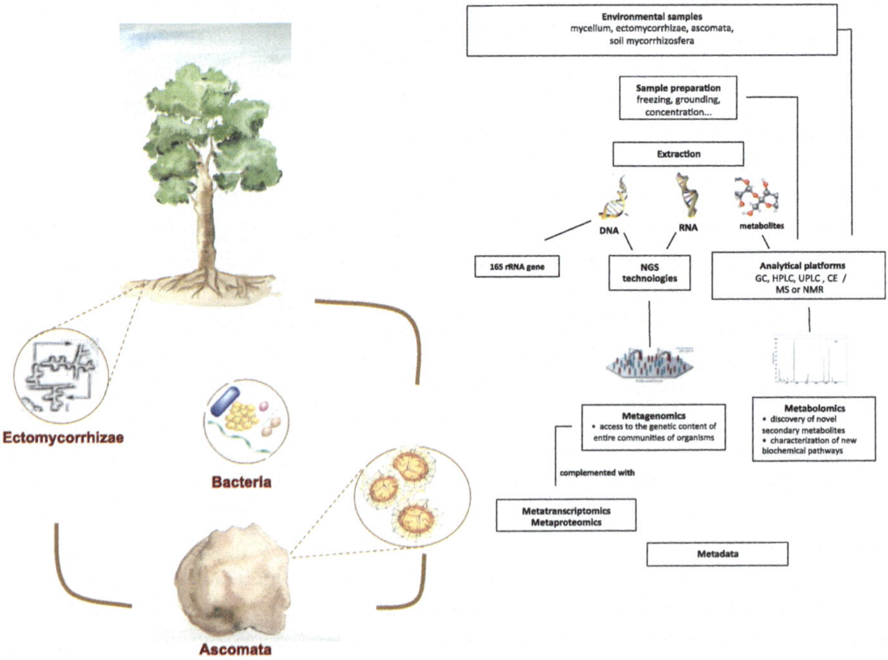

Fig. 18.1 Representation of metagenomic and/or metabolomic analysis of truffle-related microbiome

light on the biology of the systems. The genome of the black truffle *T. melanosporum*, published in 2010 by Martin and colleagues, allowed us to study and discover several truffle properties comprising 7496 protein-coding genes. However, the genes involved in the nutritional exchange within associated microbes, the functional cross talk, and interaction have not been investigated. As recently demonstrated in truffles using culture-independent molecular approaches, most microbes found in nature exist in complex, interdependent communities and cannot be readily isolated and grown in the laboratory. Indeed, new DNA sequencing technologies have also allowed the study of uncultivated members revealing previously unknown microbial diversity. Recent findings made by studying the microbial indigenous helper associated with truffles clearly show the benefits of using advanced metagenomics to unravel the complex network of biotic interactions in soil ecosystems (Barbieri et al. 2005, 2007; Gryndler et al. 2013; Antony-Babu et al. 2014; Mello et al. 2013). Moreover, high-throughput sequencing technology has been implemented by the incorporation of bar-coded primers or tags (Hamady et al. 2008).

Genes related to nitrogen were enriched in the *T. magnatum* ascomata as well as in *T. melanosporum* ascomata, which also expressed sulfur cycling genes (Barbieri et al. 2010; Antony-Babu et al. 2014). The pyrotagging technique, including targeted functional genes such as *nifH*, *amoA*, etc. (Mao et al. 2011), could provide new insights, even though this targeted approach does not allow us to obtain a complete picture of microbial genes, namely, a "metagenome." The rapid shotgun sequencing approach allows us to obtain a huge amount of data; however, a major challenge is the lack of efficient and effective strategies for assembling and annotating shotgun metagenomic data. Normalizing and partitioning the sequence reads may make it possible to develop more efficient assembly strategies (Jones et al. 2012). A common goal of metagenomic studies has been to link the structure of the microbial community with its function. Metagenomic data provide insights into the metabolic potential of the microbial community, but other approaches are needed to assess microbial activity in vivo, since many detected organisms could be quiescent or irrelevant, underlining the utility of metatranscriptomics.

Metaproteomics represents the most direct estimation of microbial activity, and several strategies for extraction and analysis of proteins from complex biological samples and soils have been set up. The development of shotgun proteomics, using two-dimensional liquid chromatography separation methods combined with tandem mass spectrometry analysis, has made it possible to identify thousands of proteins from soils (Keiblinger et al. 2012), although this approach is still little used compared to the metataxonomy approach due to its challenges. In this regard, in addition to the very limited proteome coverage of truffles, we are still in the preliminary stages of general biochemical or physiological assessments. Vita et al. (2013) applied proteomic analysis to characterize those differences occurring in samples and induced by the different environmental conditions present in the various Italian areas where *T. magnatum* can grow and differentiating between mature and immature truffles (Pierleoni et al. 2004). The use of the proteomic approach may help to study the functional characterization of novel proteins and improve our biological understanding of truffles (Islam et al. 2013); however, it is currently difficult to extrapolate only the bacterial protein profile within the fungal host tissue or ECM system.

As the metabolic complement of functional genomics, microbial metabolomics provides a more complete picture of cell-to-cell interaction. Metabolomics is the study of the global metabolite profiles of a cell under specific conditions; it not only sheds light on the enzymatic pathways and networks encoded within the genome, but also gives us an overview of the interplay of developmental processes and changing environments over the lifetime of an organism.

The main challenge in metabolomics is the identification and quantification of thousands of little, in most of the time unidentified, metabolites. In contrast to genome, transcriptome and proteome analyses, the products investigated by metabolomic studies are highly variable in their chemical structures and properties (Hollywood et al. 2006). They can be hydrophilic carbohydrates, volatile alcohols, ketones, amino and nonamino organic acids, hydrophobic lipids, and secondary metabolites such as antibiotics, pigments, non-ribosomal peptides, and cofactors (Villas-Boas et al. 2005). Hence, a simultaneous extraction and the determination of the entire set of metabolites at a given physiological state are extremely difficult (Garcia et al. 2008). Furthermore, their concentrations and compositions change rapidly in response to environmental stimuli (Villas-Boas et al. 2005). Separation and detection of metabolites are considered the crucial steps in metabolomic analyses. Separation techniques such as liquid chromatography (LC) in its high-performance (HPLC) or ultra-performance (UPLC) forms, gas chromatography (GC), capillary electrophoresis (CE), as well as the coupling of these instruments to detection techniques such as mass spectrometry (MS), nuclear magnetic resonance (NMR), and near-infrared spectrometry (NIR) have been used successfully. NMR and MS (Villas-Boas et al. 2005) are capable of handling a wide range of metabolites in a single measurement without preselection of specific analytes, allowing structural identification and molecule quantification (Nicholson and Lindon 2008). Solid-phase microextraction coupled to gas chromatography/mass spectrometry (SPME-GC-MS) has been employed to characterize the volatile profile of several truffle species (Díaz et al. 2002; Gioacchini et al. 2005), even during fruiting body maturation (Zeppa et al. 2004; Splivallo et al. 2014). As described above, metabolomic information can improve transcriptomics and proteomic data, since metabolomes are situated further downstream in functional analyses, reflecting microbial phenotypes.

18.6 Conclusions

All of these "omics" are useful and complementary as we strive to attain a predictive model of microbial physiology, and automated analysis should be coupled with manual techniques to fill gaps using known metabolic functions with genetic and biochemical data (Feist et al. 2009) alongside the metabolite profile. Phenotype experiments can validate predictive models obtained integrating the growing body of evidence provided by the various "omics" using mathematical and statistical methods. This is the basis of a systemic biological approach that aims to clarify metabolic networks and dynamic interactions between biological components. In this context the continuing application of these integrative and complementary approaches to truffle microbiome will be of great assistance in shedding light on truffle–bacteria interactions and their biological significance, also in relation to soil environmental characteristics. Such analyses will provide new insights into the truffle-associated microbial communities and their spatial and temporal dynamics. Together, these new techniques and the recent evidence will be useful in the ongoing collaboration and translational communication between scientists and truffle cultivators. It appears that we are about to enter a new era of predictive microbial ecology that will provide us with a much deeper understanding not only of the diversity of truffle-associated microbial communities, but also their function within the truffle ecosystem.

Acknowledgments We wish to thank Professor Timothy Bloom, Centro Linguistico di Ateneo of the University of Urbino Carlo Bo, for a critical reading of the manuscript.

References

Antony-Babu S, Deveau A, Van Nostrand JD, Zhou J, Le Tacon F, Robin C, Frey-Klett P, Uroz S (2014) Black truffle-associated bacterial communities during the development and maturation of *Tuber melanosporum* ascocarps and putative functional roles. Environ Microbiol 16 (9):2831–2847. doi:10.1111/1462-2920.12294

Barbieri E, Potenza L, Rossi I, Sisti D, Giomaro G, Rossetti S, Beimfohr C, Stocchi V (2000) Phylogenetic characterization and in situ detection of a *Cytophaga-Flexibacter-Bacteroides* phylogroup bacterium in *Tuber borchii* Vittad. ectomycorrhizal mycelium. Appl Environ Microbiol 66(11):5035–5042. doi:10.1128/AEM.66.11

Barbieri E, Potenza L, Stocchi V (2001) Molecular characterization of cellulosolytic-chitinolytic bacteria associated with fruit bodies of the ectomycorrhizal fungus *Tuber borchii* Vittad. Symbiosis 30(2):123–139

Barbieri E, Bertini L, Rossi I, Ceccaroli P, Saltarelli R, Guidi C, Zambonelli A, Stocchi V (2005) New evidence for bacterial diversity in the ascoma of the ectomycorrhizal fungus *Tuber borchii* Vittad. FEMS Microbiol Lett 247(1):23–35. doi:10.1016/j.femsle.2005.04.027

Barbieri E, Guidi C, Bertaux J, Frey-Klett P, Garbaye J, Ceccaroli P, Saltarelli R, Zambonelli A, Stocchi V (2007) Occurrence and diversity of bacterial communities in *Tuber magnatum* during truffle maturation. Environ Microbiol 9(9):2234–2246. doi:10.1111/j.1462-2920.2007.01338.x

Barbieri E, Ceccaroli P, Saltarelli R, Guidi C, Potenza L, Basaglia M, Fontana F, Baldan E, Casella S, Ryahi O, Zambonelli A, Stocchi V (2010) New evidence for nitrogen fixation within the Italian white truffle *Tuber magnatum*. Fungal Biol 114(11–12):936–942. doi:10.1016/j.funbio.2010.09.001

Barbieri E, Ceccaroli P, Palma F, Agostini D, Stocchi V (2012) Ectomycorrhizal helper bacteria: the third partner in the symbiosis. In: Zambonelli A, Bonito GM (eds) Edible ectomycorrhizal mushrooms, current knowledge and future prospect, vol 34, Soil biology. Springer, Berlin, pp 125–141. doi:10.1007/978-3-642-33823-6_8

Bedini S, Bagnoli G, Sbrana C, Leporini C, Tola E, Dunne C, Filippi C, D'Andrea F, O'Gara F, Nuti MP (1999) *Pseudomonads* isolated from within fruit bodies of *Tuber borchii* are capable of producing biological control or phytostimulatory compounds in pure culture. Symbiosis 26 (3):223–236

Bellesia F, Pinetti A, Tirillini B, Bianchi A (2001) Temperature-dependent evolution of volatile organic compounds in *Tuber borchii* from Italy. Flavour Fragrance J 16(1):1–6. doi:10.1002/1099-1026(200101/02)16:1<1::AID-FFJ936>3.0.CO;2-Y

Buzzini P, Gasparetti C, Turchetti B, Cramarossa MR, Vaughan-Martini A, Martini A, Pagnoni UM, Forti L (2005) Production of volatile organic compounds (VOCs) by yeasts isolated from the ascocarps of black (*Tuber melanosporum* Vitt.) and white (*Tuber magnatum* Pico) truffles. Arch Microbiol 184(3):187–193. doi:10.1007/s00203-005-0043-y

Chalot M, Brun A (1998) Physiology of organic nitrogen acquisition by ectomycorrhizal fungi and ectomycorrhizas. FEMS Microbiol Rev 22(1):21–44. doi:10.1111/j.1574-6976.1998.tb00359.x

Citterio B, Malatesta M, Battistelli S, Marcheggiani F, Baffone W, Saltarelli R, Stocchi V, Gazzanelli G (2001) Possible involvement of *Pseudomonas fluorescens* and *Bacillaceae* in structural modification of *Tuber borchii* fruit bodies. Can J Microbiol 47(3):264–268. doi:10.1139/w01-005

Deveau A, Palin B, Delaruelle C, Peter M, Kohler A, Pierrat JC, Sarniguet A, Garbaye J, Martin F, Frey-Klett P (2007) The mycorrhiza helper *Pseudomonas fluorescens* BBc6R8 has a specific priming effect on the growth, morphology and gene expression of the ectomycorrhizal fungus *Laccaria bicolor* S238N. New Phytol 175(4):743–755. doi:10.1111/j.1469-8137.2007.02148.x

Díaz P, Señoráns FJ, Reglero G, Ibañez E (2002) Truffle aroma analysis by headspace solid phase microextraction. J Agric Food Chem 50(22):6468–6472. doi:10.1016/j.chroma.2003.08.016

Dominguez JA, Martin A, Anriquez A, Albanesi A (2012) The combined effects of *Pseudomonas fluorescens* and *Tuber melanosporum* on the quality of *Pinus halepensis* seedlings. Mycorrhiza 22(6):429–436. doi:10.1007/s00572-011-0420-0

Dominguez-Nuñez JA, Medina M, Berrocal-Lobo M, Anriquez A, Albanesi A (2015) The combined effects of *Pseudomonas fluorescens* CECT 844 and the black truffle co-inoculation on *Pinus nigra* seedlings. iForest 8:624–630. doi:10.3832/ifor1334-007

Feist AM, Herrgard MJ, Thiele I, Reed JL, Palsson BØ (2009) Reconstruction of biochemical networks in microorganisms. Nat Rev Microbiol 7(2):129–143. doi:10.1038/nrmicro1949

Frey-Klett P, Garbaye J, Tarkka M (2007) The mycorrhiza helper bacteria revisited. New Phytol 176(1):22–36. doi:10.1111/j.1469-8137.2007.02191.x

Garbaye J (1994) Helper bacteria: a new dimension to the mycorrhizal symbiosis. New Phytol 128 (2):197–210. doi:10.1111/j.1469-8137.1994.tb04003.x

Garcia DE, Baidoo EE, Benke PI, Pingitore F, Tang YJ, Villa S, Keasling JD (2008) Separation and mass spectrometry in microbial metabolomics. Curr Opin Microbiol 11(3):233–239. doi:10.1016/j.mib.2008.04.002

Gazzanelli G, Malatesta M, Pianetti A, Baffone W, Stocchi V, Citterio B (1999) Bacteria associated to fruit bodies of the ecto-mycorrhizal fungus *Tuber borchii* Vittad. Symbiosis 26 (3):211–219. doi:10.10123/A:1020949531639

Gioacchini AM, Menotta M, Bertini L, Rossi I, Zeppa S, Zambonelli A, Piccoli G, Stocchi V (2005) Solid-phase microextraction gas chromatography/mass spectrometry: a new method for species identification of truffles. Rapid Commun Mass Spectrom 19(17):2365–2370. doi:10.1002/rcm.2031

Giomaro G, Sisti D, Zambonelli A (2005) Cultivation of edible ectomycorrhizal fungi by *in vitro* mycorrhizal synthesis. In: Declerck S, Stullu DG, Fortin JA (eds) *In vitro* culture of mycorrhizas, vol 4, Soil biology. Springer, Berlin, pp 253–267. doi:10.1007/3-540-27331-X_14

Gryndler M, Hršelová H (2012) Isolation of bacteria from ectomycorrhizae of *Tuber aestivum* Vittad. Acta Mycol 47(2):155–160. doi:10.5586/am.2012.018

Gryndler M, Soukupova L, Hršelová H, Gryndlerová H, Borovička J, Streiblová E, Jansa J (2013) A quest for indigenous truffle helper prokaryotes. Environ Microbiol Rep 5(3):346–352. doi:10.1111/1758-2229.12014

Hamady M, Walker JJ, Harris JK, Gold NJ, Knight R (2008) Error-correcting barcoded primers for pyrosequencing hundreds of samples in multiplex. Nat Methods 5(3):235–237. doi:10.1038/nmeth.1184

Harki E, Bouya D, Dargent R (2006) Maturation-associated alterations of the biochemical characteristics of the black truffle *Tuber melanosporum* Vitt. Food Chem 99:394–400. doi:10.1016/j.foodchem.2005.08.030

Hollywood K, Brison DR, Goodacre R (2006) Metabolomics: current technologies and future trends. Proteomics 6(17):4716–4723. doi:10.1002/pmic.200600106

Islam MT, Mohamedali A, Garg G, Khan JM, Gorse AD, Parsons J, Marshall P, Ranganathan S, Baker MS (2013) Unlocking the puzzling biology of the black Périgord truffle *Tuber melanosporum*. J Proteome Res 12(12):5349–5356. doi:10.1021/pr400650c

Jones DC, Ruzzo WL, Peng X, Katze MG (2012) Compression of next-generation sequencing reads aided by highly efficient *de novo* assembly. Nucleic Acids Res 40, e171. doi:10.1093/nar/gks754

Junker RR, Tholl D (2013) Volatile organic compound mediated interactions at the plant-microbe interface. J Chem Ecol 39(7):810–825. doi:10.1007/s10886-013-0325-9

Keiblinger KM, Wilhartitz IC, Schneider T, Roschitzki B, Schmid E, Eberl L, Riedel K, Zechmeister-Boltenstern S (2012) Soil metaproteomics: comparative evaluation of protein extraction protocols. Soil Biol Biochem 54:14–24. doi:10.1016/j.soilbio.2012.05.014

Kohler A, Kuo A, Lg N, Morin E, Kw B, Buscot F, Canbäck B, Choi C, Cichocki N, Clum A, Colpaert J, Copeland A, Costa MD, Doré J, Floudas D, Gay G, Girlanda M, Henrissat B, Herrmann S, Hess J, Högberg N, Johansson T, Khouja HR, LaButti K, Lahrmann U, Levasseur A, Lindquist EA, Lipzen A, Marmeisse R, Martino E, Murat C, Ngan CY, Nehls U, Plett JM, Pringle A, Ohm RA, Perotto S, Peter M, Riley R, Rineau F, Ruytinx J, Salamov A, Shah F, Sun H, Tarkka M, Tritt A, Veneault-Fourrey C, Zuccaro A, Mycorrhizal genomics initiative consortium, Tunlid A, Grigoriev IV, Hibbett DS, Martin F (2015) Convergent losses of decay mechanisms and rapid turnover of symbiosis genes in mycorrhizal mutualists. Nat Genet 47(4):410–415. doi:10.1038/ng.3223

Kramer R, Abraham WR (2012) Volatile sesquiterpenes from fungi: what are they good for? Phytochem Rev 11(1):15–37. doi:10.1007/s11101-011-9216-2

Le Tacon F, Zeller B, Plain C, Hossann C, Bréchet C, Robin C (2013) Carbon transfer from the host to *Tuber melanosporum* mycorrhizas and ascocarps followed using a $^{(13)}$C pulse-labeling technique. PLoS One 8(5), e64626. doi:10.1371/journal.pone.0064626

Le Tacon F, Rubini A, Murat C, Riccioni C, Robin C, Belfiori B, Zeller B, De la Varga H, Akroume E, Deveau A, Martin F, Paolocci F (2016) Certainties and uncertainties about the life cycle of the Périgord black truffle (*Tuber melanosporum* Vittad.) Ann Forest Sci: 73:105–117. doi:10.1007/s13595-015-0461-1

Mao Y, Yannarell AC, Mackie RI (2011) Changes in N-transforming archaea and bacteria in soil during the establishment of bioenergy crops. PLoS One 6(9), e24750. doi:10.1371/journal.pone.0024750

Martin F, Kohler A, Murat C, Balestrini R, Coutinho PM, Jaillon O, Montanini B, Morin E, Noel B, Percudani R, Porcel B, Rubini A, Amicucci A, Amselem J, Anthouard V, Arcioni S, Artiguenave F, Aury JM, Ballario P, Bolchi A, Brenna A, Brun A, Buee M, Cantarel B, Chevalier G, Couloux A, Da Silva C, Denoeud F, Duplessis S, Ghignone S, Hilselberger B, Iotti M, Marcais B, Mello A, Miranda M, Pacioni G, Quesneville H, Riccioni C, Ruotolo R, Splivallo R, Stocchi V, Tisserant E, Viscomi AR, Zambonelli A, Zampieri E, Henrissat B, Lebrun MH, Paolocci F, Bonfante P, Ottonello S, Wincker P (2010) Perigord black truffle genome uncovers evolutionary origins and mechanisms of symbiosis. Nature 464 (7291):1033–1038. doi:10.1038/nature08867

Mello A, Miozzi L, Vizzini A, Napoli C, Kowalchuk G, Bonfante P (2010) Bacterial and fungal communities associated with *Tuber magnatum*-productive niches. Plant Biosyst 144 (2):323–332. doi:10.1080/11263500903374724

Mello A, Ding GC, Piceno YM, Napoli C, Tom LM, De Santis TZ, Andersen GL, Smalla K, Bonfante P (2013) Truffle brûlés have an impact on the diversity of bacterial communities. Plos One 8(4), e61945. doi:10.1371/journal.pone.0061945

Menotta M, Gioacchini AM, Amicucci A, Buffalini M, Sisti D, Stocchi V (2004) Headspace solid-phase microextraction with gas chromatography and mass spectrometry in the investigation of volatile organic compounds in an ectomycorrhizae synthesis system. Rapid Commun Mass Spectrom 18(2):206–210. doi:10.1002/rcm.1314

Mougi A, Kondoh M (2014) Stability of competition-antagonism-mutualism hybrid community and the role of community network structure. J Theor Biol 360:54–58. doi:10.1016/j.jtbi.2014.06.030

Müller A, Faubert P, Hagen M, Castell WZ, Polle A, Schnitzler JP, Rosenkranz M (2013) Volatile profiles of fungi-chemotyping of species and ecological functions. Fungal Genet Biol 54:25–33. doi:10.1016/j.fgb.2013.02.005

Murat C, Vizzini A, Bonfante P, Mello A (2005) Morphological and molecular typing of the below-ground fungal community in a natural *Tuber magnatum* truffle-ground. FEMS Microbiol Lett 245(2):307–313. doi:10.1016/j.femsle.2005.03.019

Nicholson JK, Lindon JC (2008) System biology: metabonomics. Nature 455(7216):1054–1056. doi:10.1038/4551054a

Pacioni G, Leonardi M, Aimola P, Ragnelli AM, Rubini A, Paolocci F (2007) Isolation and characterization of some mycelia inhabiting *Tuber* ascomata. Mycol Res 111(12):1450–1460. doi:10.1016/j.mycres.2007.08.016

Pavic A, Stanković S, Saljnikove E, Krüger D, Buscot F, Tarkka M, Marjanović Ž (2013) Actinobacteria may influence white truffle (*Tuber magnatum* Pico) nutrition, ascocarp degradation and interactions with other soil fungi. Fungal Ecol 6(6):527–538. doi:10.1016/j.funeco.2013.05.006

Pierleoni R, Buffalini M, Vallorani L, Guidi C, Zeppa S, Sacconi C, Pucci P, Amoresano A, Casbarra A, Stocchi V (2004) *Tuber borchii* fruit body: 2-dimensional profile and protein identification. Phytochemistry 65(7):813–820. doi:10.1016/j.phytochem.2004.02.012

Qin J, Li R, Raes J, Arumugam M, Burgdorf KS, Manichanh C, Nielsen T, Pons N, Levenez F, Yamada T, Mende DR, Li J,[7] Xu J, Li S, Li D, Cao J, Wang B, Liang H, Zheng H, Xie Y, Tap J, Lepage P, Bertalan M, Batto JM, Hansen T, Le Paslier D, Linneberg D, Nielsen HB, Pelletier E, Renault P, Sicheritz-Ponten T, Turner K, Zhu H, Yu C, Li S, Jian M, Zhou Y, Li Y, Zhang X, Li S, Qin N, Yang H, Wang J, Brunak S, Doré J, Guarner F, Kristiansen K, Pedersen O, Parkhill J, Weissenbach J, MetaHIT Consortium, Bork P, Ehrlich SD, Wang J (2010). A human gut microbial gene catalogue established by metagenomic sequencing. Nature. 464(7285):59–65. doi:10.1038/nature08821

Rigamonte TA, Pylro VS, Duarte GF (2010) The role of mycorrhization helper bacteria in the establishment and action of ectomycorrhizae associations. Braz J Microbiol 41(4):832–840. doi:10.1590/S1517-83822010000400002

Rivera CS, Blanco D, Oria R, Venturini ME (2010) Diversity of culturable microorganisms and occurrence of *Listeria* monocytogenes and *Salmonella* spp. in *Tuber aestivum* and *Tuber melanosporum* ascocarps. Food Microbiol 27(2):286–293. doi:10.1016/j.fm.2009.11.001

Rumjanek NG, Dobert RC, van Berkum P, Triplett EW (1993) Common soybean inoculant strains in Brazil are members of *Bradyrhizobium elkanii*. Appl Environ Microbiol 59(12):4371–4373, doi:0099-2240/93/124371-0302.00/0

Sbrana C, Bagnoli G, Bedini S, Filippi C, Giovannetti M, Nuti MP (2000) Adhesion to hyphal matrix and antifungal activity of *Pseudomonas* strains isolated from *Tuber borchii* ascocarps. Can J Microbiol 46(3):259–268. doi:10.1139/w99-135

Sbrana C, Agnolucci M, Bedini S, Lepera A, Toffanin A, Giovannetti M, Nuti MP (2002) Diversity of culturable bacterial populations associated to *Tuber borchii* ectomycorrhizas and their activity on *T. borchii* mycelial growth. FEMS Microbiol Lett 211(2):195–201. doi:10.1111/j.1574-6968.2002.tb11224.x

Schoenian I, Spitelle M, Ghaste M, Wirth R, Herz H, Spiteller D (2011) Chemical basis of the synergism and antagonism in microbial communities in the nests of leaf-cutting ants. Proc Natl Acad Sci USA 108(5):1955–1960. doi:10.1073/pnas.1008441108

Splivallo R, Maier C (2011) Production of natural truffle flavours from truffle mycelium. US Patent Application No PCTIB2010052913

Splivallo R, Novero M, Bertea CM, Bossi S, Bonfante P (2007) Truffle volatiles inhibit growth and induce an oxidative burst in *Arabidopsis thaliana*. New Phytol 175(3):417–424. doi:10.1111/j.1469-8137.2007.02141.x

Splivallo R, Ottonello S, Mello A, Karlovsky P (2011) Truffle volatiles: from chemical ecology to aroma biosynthesis. New Phytol 189(3):688–699. doi:10.1111/j.1469-8137.2010.03523.x

Splivallo R, Deveau A, Valdez N, Kirchhoff N, Frey-Klett P, Karlovsky P (2014) Bacteria associated with truffle-fruiting bodies contribute to truffle aroma. Environ Microbiol 17 (8):2647–2660. doi:10.1111/1462-2920.12521

Tang J (2011) Microbial Metabolomics. Curr Genomics 12(6):391–403. doi:10.2174/138920211797248619

Villas-Boas SG, Mas S, Akesson M, Smedsgaard J, Nielsen J (2005) Mass spectrometry in metabolome analysis. Mass Spectrom Rev 24(5):613–646. doi:10.1002/mas.20032

Vita F, Lucarotti V, Alpi E, Balestrini R, Mello A, Bachi A, Alessio M, Alpi A (2013) Proteins from *Tuber magnatum* Pico fruiting bodies naturally grown in different areas of Italy. Proteome Sci 11:7. doi:10.1186/1477-5956-11-7

Zeppa S, Gioacchini AM, Guidi C, Guescini M, Pierleoni R, Zambonelli A, Stocchi V (2004) Determination of specific volatile organic compounds synthesised during *Tuber borchii* fruit body development by solid-phase microextraction and gas chromatography/mass spectrometry. Rapid Commun Mass Spectrom 18(2):199–205. doi:10.1002/rcm.1313

Zeppa S, Marchionni C, Saltarelli R, Guidi C, Ceccaroli P, Pierleoni R, Zambonelli A, Stocchi V (2010) Sulfate metabolism in *Tuber borchii*: characterization of a putative sulfate transporter and homocysteine synthase genes. Curr Genet 56(2):109–119. doi:10.1007/s00294-009-0284-4

Chapter 19
Biodiversity and Ecology of Soil Fauna in Relation to Truffle

Cristina Menta and Stefania Pinto

19.1 Introduction

In the last few decades, numerous studies have attempted to discover the "secrets" of truffles. Since the 3rd International Conference on Truffle in Spoleto (2008), researchers have made considerable progress in advancing understanding of the biology and ecology of *Tuber* species and in improving the sustainable productivity of these valuable fungi (Águeda et al. 2014). The entire genome of *Tuber melanosporum* Vttad. has been sequenced (Martin et al. 2010), and those of other *Tuber* species are in progress (see Chap. 9).

Some authors have studied the particular pedological, chemical and physical characteristics developed by some *Tuber* species in the brûlé, the area around the host plant where the germination and the growth of other plants are inhibited (Callot 1999; Ricard 2003; Granetti et al. 2005; Mello et al. 2013; García-Montero et al. 2014); others have explained the production and development mechanisms in this area (Sourzat 2004; Granetti et al. 2005; Streiblová et al. 2012). The substances produced by *Tuber* mycelium, which are well known and widely studied, adversely affect seed germination and the growth of young trees inside the brûlé (Plattner and Hall 1995). An interesting aspect that makes the truffles an engaging yet difficult case of study at the same time is their particular dispersal system. A truffle cannot release its spores, trapped as they are in their underground realm, and it needs an alternative dispersal system via animals. When an animal eats truffle, most of the flesh is digested, but the spores pass through unharmed and are defecated on the ground, where they can germinate if the conditions are right (Trappe and Claridge 2010).

C. Menta (✉) • S. Pinto
Department of Life Sciences, University of Parma, Strada Farini 90, 43121 Parma, Italy
e-mail: cristina.menta@unipr.it

A. Zambonelli et al. (eds.), *True Truffle (*Tuber *spp.) in the World*, Soil Biology 47,
DOI 10.1007/978-3-319-31436-5_19

We can assume that truffles emerged between 100 and 200 millions of years ago (Jeandroz et al. 2008; Bonito et al. 2013). As truffles retreated underground, mutations eventually led to the formation of aromatic compounds attractive to animals (Trappe and Claridge 2010). Many animals are known to be "truffle eaters": small mammals such as mice, squirrels and rabbits, big mammals like wild boars in the Northern Hemisphere (see Chaps. 21 and 22) and rat-kangaroos, armadillos and meerkats in the Southern Hemisphere. Molluscs are attracted to truffles, too, and insects may feed on truffles or lay eggs in them so that their larvae have a ready food source when they hatch (Trappe and Claridge 2010). Spore dispersal by animals can promote the fungal colonization of new habitats. Moreover, as highlighted by Bonito et al. (2013), spore deposition via animal mycophagy may be a more targeted dispersal mechanism than wind or water dispersal, because animals could enrich spores with nutrients (e.g. calcium, magnesium and potassium) and deposit their faecal pellets loaded with spores near the roots of suitable host trees. Indeed, the authors pointed out that truffle fruit bodies usually have durable, thick-walled spores that can withstand and possibly benefit from passing through the digestive tract of animals. These selected traits across a diversity of truffle lineages suggest that the transition from epigeous to hypogeous fruiting is driven by strong selection for traits that promote animal dispersal.

There is undoubtedly much more information still to be discovered. There are only a few studies on the interaction between truffles and animals (with the exception of mammals) or on the relationship between truffles and soil fauna. As we said above, truffles "live" in the soil and complete their whole life cycle within it. It would be really interesting to establish whether soil animals affect truffle, and if so, how, and conversely whether truffles affect soil fauna, by means of different soil chemical characteristics that allelopathic compounds make in the brûlé. In this chapter, we are going to discuss the various aspects of the relationship between truffles and soil fauna.

19.2 Role of Soil Fauna

The living component of the soil should be considered as the motor that drives soil functioning. Within it, invertebrates have been shown to influence almost every level of the decomposition cascade (Wolters 2000), and the ecosystem services provided by soil fauna are one of the most powerful arguments for the conservation of edaphic biodiversity. Soil fauna perform many different and very important functions within and for the soil. Invertebrates affect soil processes both directly and indirectly. Direct effects result from the incorporation and redistribution of various materials, while indirect effects result from soil invertebrates shaping the microbial community by both constructive (e.g. transport of fungal spores) and destructive means (e.g. selective reduction of viability; Shaw 1992). In addition, edaphic animals influence soil processes by altering distal factors controlling microbial performance through particulation, channelling and alteration of nutrient availability (Wolters 2000).

Although some soil animals are carnivorous, the most widespread ecosystematic activity of soil meso- and macrofauna is the "processing" and "mixing" of organic detritus in soil. Processing includes not only simple comminution of organic debris into smaller fragments but also various degrees of decomposition performed by enzymes and the gut microorganisms that feed on organic material (Killham 1994). Many organisms, such as isopods, myriapods, earthworms, springtails, a wide range of mites, larvae and adults of some insects, feed on vegetable and animals debris. This mechanical degradation facilitates the action of microorganisms and accelerates the degradation processes.

The results of food web analyses indicate that microfauna (organisms smaller than 100 μm) and, to a lesser extent, mesofauna (animals between 100 μm and 2 mm in size) have a particularly strong impact on C and N fluxes due to their rapid turnover rates and the consumption of microorganisms with low C/N ratios, while macrofauna (animals larger than 2 mm) make a less direct contribution to community metabolism because of longer generation times and the consumption of materials with high C/N ratios (Anderson 1995).

Another very important function of soil organisms is their burrowing activity, which causes direct and indirect chemical, physical and biological changes. Above all, earthworm action produces a remarkable effect not only on soil structure but also on its chemical composition, because the organic matter ingested is returned in a form easily usable by plants. They dig and eat soil, creating tunnels, channels and holes to live in. Horizontal and vertical burrowing by soil animals shifts organic material along the soil profile. Bioturbation significantly facilitates interactions with stabilizing inorganic fractions (Wolters 2000). Earthworms consume a large amount of mineral substances that are mixed with the organic matter and expelled in the form of casts (Fig. 19.1a), which are rich in nitrogen and other nutrients such as calcium, magnesium and potassium. They also contain a vast quantity of undigested bacteria, which proliferate easily in the soil, contributing to the humification and mineralization of organic matter. Casts are not expelled in the same environment. This could allow the widespread dispersal of hypogeous fungi spores far from their source of origin. This is another very important function of soil organisms, in particular for hypogeous fungi. Indeed, the hyphae of mycorrhizal fungi could make up a significant proportion of the total microbial biomass in some soils and can become one of the most important sources of food for fungus-grazing animals such as springtails (Menta et al. 2013). For example, through their faeces, springtails can spread still viable fungal spores to areas as far away as several metres from their point of origin. At the same time, fauna feeding on microorganisms can control specific plant pathogens. Root herbivory influences individual plant performance and higher-level processes both directly and indirectly (Brussaard 1998): directly by selectively feeding on seeds and seedlings, thereby preventing the establishment or abundance of certain plant species, and indirectly by interfering with the carbon and nutrient acquisition and the allocation of plants in interaction with above-ground herbivores. The action of some organisms, such as

Fig. 19.1 Earthworm cast (**a**) (photo by C. Menta); *Acerentomon italicum*, Protura (photo by L. Galli) (**b**); *Geogarypus minor*, Pseudoscorpionida (SEM, photo by M. Zinni); (**c**) *Folsomia candida*, Collembola (photo by C. Menta) (**d**)

protozoa, nematodes, rotifers, some springtails and mites, which feed on bacteria, actinomycetes and fungi (both hyphae and spores), is very important for density regulation and microorganism diffusion.

Soil fauna, especially molluscs and earthworms, also have an impact on soil, thanks to their cutaneous mucus secretions, which cement the soil, making it more stable. These secretions, together with the animal's faeces and body, influence the concentration of most of the nutrients in the soil, reducing the litter's C/N ratio and promoting breakdown. In addition, the ingestion of large amounts of mineral soil is particularly well documented by macro-soil engineers (Lavelle et al. 1998).

19.3 Relationships Between Soil Fauna and Truffles

An interesting environment present in the soil is the rhizosphere. The term rhizosphere, in its broadest sense, refers to the portion of soil surrounding roots in which the organisms are influenced by their presence; its extension varies greatly, but it is generally considered to be the cylinder of soil used by the root hairs and in which they emit exudates (Killham 1994). Protozoa and nematodes are the main consumers of microorganisms developing near living and dead roots and in hotspots in the soil matrix; rhizosphere/non-rhizosphere soil ratios are in the range 1.3–3.5 for protozoa numbers and 3–70 for nematode numbers (Brussaard 1998). The interaction between soil animals and plant roots can take a variety of forms that lead to benefits for or repress the growth of the plants and often involve interactions with the microbial populations of the soil. This relationship becomes more complex when it involves mycorrhizas. In this sense, the rhizosphere also becomes the soil portion where mycorrhizas and organisms interact. For example, at different stages of their lifecycle, truffles in particular release specific volatiles in order to interact with particular organisms (Splivallo et al. 2011). Secondary metabolites produced by fungi are one of the largest classes of natural products. Such compounds may not play a fundamental biochemical role in the normal growth or development of fungal cells, but they clearly play an ecological role. Secondary metabolism is important in many fungal processes, including stimulation or inhibition of organisms by allelochemicals and regulation of symbiotic and protective interactions with microbes and other interacting roots (Angelini et al. 2010). In truffles, these compounds are responsible for the formation of the brûlé, which is characterized by scarce plant cover due to the phytotoxic activity of the fungus (Splivallo 2008).

In general, the abundance and diversity of soil animals are greatest in soils with little or no disturbance, such as those of permanent grassland (Menta et al. 2011a) and natural woodland (Menta et al. 2013), where they play a vital role in nutrient cycling (Killham 1994). However, the particular conditions generated inside the brûlé by truffles' aromatic compounds could affect soil fauna in many different ways, both directly and indirectly. The brûlé is in fact characterized by lower levels of organic matter, sometimes higher pH values and lower retention capacity (Menta et al. 2014). Many of these factors could adversely affect organisms, which tend to avoid this environment, or could lead to a shift in the soil community, encouraging the presence of stress-resistant organisms. It has been demonstrated, for example, that the main factors driving Collembola distribution are vegetation, microflora, soil structure and soil moisture (Menta 2008).

Above all, the lack of vegetation cover could lead to soil moisture deficit. Soil moisture affects soil organisms because water is essential for life and for enzyme activities and metabolism and because soil moisture affects soil temperature and soil aeration. Furthermore, under wet conditions, oxygen does not diffuse through the soil as readily, so the levels available to organisms may become depleted, leading to anaerobic conditions (FAO 2015). In soils with sparse vegetation cover, such as the truffle brûlé, the high frequency of dry conditions on the surface causes nematodes to be more common at depths of 5–10 cm (Menta 2008). Diplura prefer soil with a high stable moisture content. Isopods, which avoid dry conditions, may also escape from the brûlé areas. Water stress also influences the size and activity of soil animal communities for two reasons, a reduction in water-filled pores and the thickness of soil water films (which controls soil aeration and the mobility of some animals) and a reduction in the free energy of the water, making water uptake increasingly difficult (Killham 1994). Furthermore, it has long been known that the production of earthworm casts, an excellent indicator of earthworm activity, is directly related to rainfall and soil moisture status (Killham 1994).

Burrowing species, such as *Lumbricus terrestris*, take refuge deep in damp soil, whereas species that live on the surface, such as *Allolobophora caliginosa* (*Nicodrilus caliginosus*) and *Eisenia rosea*, are declining in population (Menta 2008) due to soil moisture depletion.

Inside the truffle brûlé, the organic matter to be demolished comes in much smaller quantities compared to other ecosystems. This could lead to a different soil community structure, with a small presence of detritivores, and consequently predators, and possibly an increase in herbivores or organisms that feed on fungi.

Many works have documented the intimate relationship between soil fauna and organic matter. The amount and depth of organic matter on the forest floor, for instance, influences the distribution of springtails and acari Oribatida (Hasegawa 2001). Many soil invertebrates, such as symphylans, proturans, diplopods and earthworms, and many detritivores are more abundant in soils rich in organic matter. Similarly, pH could influence soil organisms because different species are active at different pH ranges. In a study conducted in ecologically analogous sites in oak-hickory forests, which are characterized by different amounts of acidic deposition, Kuperman (1996) found a positive correlation between pH and the abundance of different soil macroinvertebrates such as earthworms, gastropods, termites, carabids, staphylinids, cockroaches and dipteran larvae. Esher et al. (1993) reported huge effects of mild acid treatments on populations of earthworms. Several experiments have shown that many earthworms have particular limits in terms of tolerance to soil acidity and that they tend to avoid unfavourable pH levels. In laboratory experiments with *Eisenia foetida*, 100 % mortality was observed at $pH < 5$ or $pH > 9$ (Kaplan et al. 1980). Edwards and Lofty (1975) found that the numbers of Chilopoda and Symphyla were much greater between pH 5.0 and 6.0 than at higher or lower levels and none of these animals were present at a pH below 4.0. Wireworms and other insects were more numerous in plots with a pH between 4.0 and 7.0, i.e. very acid or alkaline soils did not support large numbers of these arthropods in their study (Kuperman 1996).

Kuperman (1996) performed a linear regression analysis to determine which parameter, pH or the amount of surface organic matter accumulation could affect soil macroinvertebrate community most. The regression analysis showed a significant relationship between soil pH and the total number of decomposers in study sites along the deposition gradient. Approximately 87 % of the variation in numbers of decomposers was explained by the soil pH. Furthermore, the relationship between the abundance of the macroinvertebrate decomposer community and the amount of surface organic matter accumulation along the deposition gradient was also highly significant. Approximately 71 % of the variation in the amount of surface organic matter accumulation was explained by the abundance of decomposers. The results of this study also show that the observed patterns of decreasing abundance of decomposers, increasing organic matter content and decreasing soil pH along the deposition gradient seem to be related.

The brûlé is also characterized by lower organic carbon content. It has been demonstrated, indeed, that organic carbon in the brûlé decreases by about 52 % (Bragato 1997). This parameter could significantly affect the soil fauna community as well. Kuperman (1996) found a significant negative correlation between this parameter in A_1 horizon and the numbers of Lumbricidae, Carabidae, Gastropoda and Staphylinidae. According to studies in a wide range of ecosystems, soil invertebrates mediate about 15 % of the C and 30 % of the N turnover (Anderson 1995).

19.4 Soil Microarthropod Biodiversity in the *Tuber* Brûlé

According to the phylogenetic hypothesis that hypogeous fungi, like truffles, evolved with an active system of spore dispersal, spore discharge must be by animal dispersal. Some animals have evolved a capacity for feeding on truffle-fruiting bodies, and in so doing, they have become active agents of spore dispersal (Pacioni et al. 1991). These animals have been dubbed "hydnophagous" (Pacioni et al. 1989), from the Greek *hydnon*, truffle, and *phagous*, eating, and this category includes species belonging to various taxonomic groups. Not just truffles but all mycorrhizal plants in the world are likely to have at least one (often many) species of insect herbivores that attack them (Koricheva et al. 2009). In the case of the truffle, the diet of certain animals, such as mammals (rodents, deer, boar), birds and slugs, includes these hypogeous fruiting bodies, which sometimes account for their entire diet: a case in point is provided by several species or genera of arthropods, mainly Coleoptera and Diptera. Various authors have described mechanisms explaining the impact of soil microarthropods on fungal communities. Hanlon and Anderson (1979), for instance, indicated that microarthropod feeding activities can exert a strong differential effect on fungal and bacterial populations, and some reviews on arbuscular mycorrhizae and soil fauna interactions suggest that Collembola have the potential to restrict mycorrhizal functioning in the field (Fitter and Sanders 1992; Fitter and Garbaye 1994). One of the most specialized fungivores that attack truffles is the beetle *Leiodes cinnamomea* (Coleoptera: Staphylinoidea, Leiodidae). It is univoltine, with a European distribution. It is known to feed on several species of truffle (*Tuber* spp.) and completes its lifecycle in or near the fungal fruiting body (Arzone 1970; Newton 1984). Hochberg et al. (2003) showed that adults of the beetle *L. cinnamomea*, though inflicting substantial damage to the fruiting bodies of the black truffle in the larval stage, are not attracted to ripe truffle odours. Some individuals, however, showed persistent attraction when tested repeatedly, showing that the odours can be perceived by the truffle beetles. Coleoptera are the most widely recorded, but there is also a large amount of data on Catopidae, Cryptophagidae, Leiodidae, Micropeplidae,

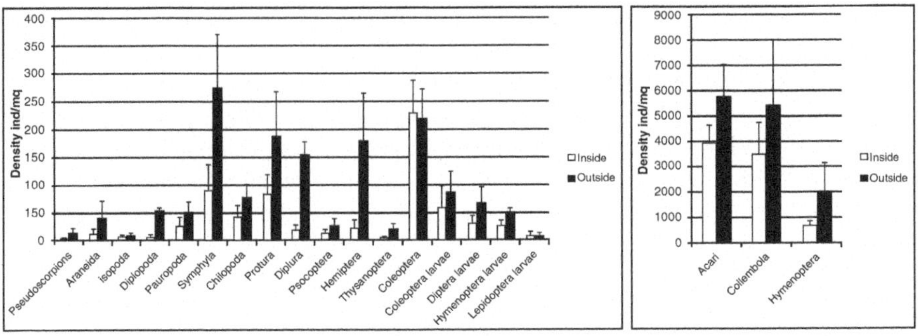

Fig. 19.2 Microarthropod community inside and outside the brûlé (based on data published by Menta et al. 2014)

Nitidulidae, Ptiliidae, Staphylinidae and Scarabaeidae. Some insects (Diptera and Coleoptera) have also been reported as fauna nutritionally related to the carpophores (Pacioni et al. 1989). The most notable insects are the "truffle flies", eight species of *Suillia* (Heleomyzidae) (Janvier 1963; Pacioni et al. 1989; Bratek et al. 1992; Callot 1999; García-Montero et al. 2004).

Besides the works mentioned above, there are only a few studies on the relationship between soil fauna and truffle biology (Queralt et al. 2014). Menta et al. (2014) tried to highlight differences among the microarthropod communities associated with the *Tuber aestivum* Vittad. brûlé and the area outside the brûlé. This research took into account three different areas, two in Italy (North and South) and one in central Spain (Guadalajara province). The authors did not observe a univocal trend, but almost all microarthropod groups were more abundant outside the brûlé (Fig. 19.2). These groups were Acari, Hymenoptera, Diplopoda, Symphyla, Pauropoda, Protura (Fig. 19.1b), Diplura, Hemiptera, Pseudoscorpionida (Fig. 19.1c) and Araneida. This trend suggests that these groups could find an unfavourable environment inside the brûlé, probably due to several factors, including the lower content of organic matter. This tendency was not followed by Coleoptera, or by Lepidoptera larvae, which probably find better conditions in the brûlé in terms of food resources and chemico-physical conditions. Menta et al. (2011b) and Tarasconi et al. (2011) reported interactions between microarthropods and *T. melanosporum* showing lower densities of mites and ants inside the brûlé. Within the collembolan group, Menta et al. (2014) showed that the families Hypogastruridae, Sminthurididae and Isotomidae were more abundant outside the brûlé, while within the Isotomidae family, *Folsomia* genus showed higher abundance inside the brûlé. One study (Hodge 2000) pointed out the presence of spores and extraradical mycelium of arbuscular mycorrhizal hyphae in the gut contents of *Folsomia candida* (Fig. 19.1d). Queralt et al. (2014) focus instead on mite communities associated with the *T. melanosporum* brûlé. They found that oribatid mites dominated the community in terms of abundance and species richness in wild and plantation soils. Most oribatid mites are mycophagous, and they could have a direct relationship with the black truffle cycle, interacting in the mycelium and spore dispersal. In some cases, they have been observed carrying spores attached to their bodies. The researchers also found a high-density population of the mites *Passalozetes ruderalis* Iberian endemism (Pérez-Íñigo 1993) and *Arthrodamaeus reticulatus* (Pérez-Íñigo 1997) in truffle areas and assumed that they could be potential truffle-related species. Soil fauna could have a positive impact on truffle development not only in terms of spore dispersal. Earthworms and ants, for example, as "ecosystem engineers" due to their important contribution to soil porosity, breakdown of organic matter and incorporation of organic matter into the soil, could have a positive effect on the truffle, altering the physical characteristics of the soil. The list of truffle-associated organisms most often includes, besides earthworms, nematodes and protozoa. Callot (1999), Ricard (2003) and Pargney et al. (2010) indicated that earthworms are often found around the truffle carpophores in the soil burns. These soils usually have low levels of organic matter, and earthworms therefore search near the carpophores for organic matter contained in the bacterial populations and droppings of microarthropods and macrofauna (García-Montero et al. 2013). These authors pointed out that earthworms benefit *T. melanosporum* development. Lulli et al. (1999) and Castrignano et al. (2000) indicated that this truffle requires considerable soil porosity, mainly originated by earthworms and ants. García-Montero et al. (2010) indicated that earthworm casts can have a great impact on the soil in the *T. melanosporum* brûlé. On the other hand, Callot (1999) found that Nematoda and Protozoa maintain the relationship between microbial activity and the truffle, while Acari and Collembola, besides regulating this relationship, also help to disseminate spores. Lastly, large animals, in addition to helping in the previously mentioned functions, also affect soil structure (Queralt et al. 2014).

19.5 Conclusions

Studies of the relationship between truffles and soil fauna are scanty, and they often concern particular species of truffle, such as *T. melanosporum* and *T. aestivum*. It would be important to analyse not only the effects of animals on truffles, in terms of dispersal of ingested spores, for example, but also the direct or indirect effects of truffles on animals. The truffle volatiles and the reduced plant cover present in the brûlé may both have a detrimental effect on some species. A further challenge will be to clarify whether *Tuber* sp. is able to create a natural "microcosm", the brûlé, as part of a larger macrocosm, the soil around it, not only in terms of chemical-physical characteristics and plant cover but also in terms of soil animal biodiversity and functionality. Lastly, the authors have tried to give a simple representation of

Fig. 19.3 A schematic representation of soil fauna community inside and outside the *T. aestivum* brûlé (based on the literature data cited in the text and on unpublished data collected by the authors). The dimensional proportions of the animals are not respected. Abbreviations: *S*, Symphylan; *P*, Pauropoda; *D*, Diplopoda; *N*, Nematodes; *Ps*, Pseudoscorpion; *E*, Earthworm; *A*, Acari; *C*, Collembola; *L*, *Leiodes cinnamomea*; *S*, *Suillia* spp. All the drawings were made by the authors

the soil fauna community inside and outside the *T. aestivum* brûlé, based on the literature data cited in the text and on unpublished data collected by the authors themselves (Fig. 19.3).

References

Águeda B, Zambonelli A, Molina R (2014) *Tuber* 2013: scientific advances in sustainable truffle culture. Mycorrhiza 24(1):1–4. doi:10.1007/s00572-014-0569-4

Anderson JM (1995) Soil organisms as engineers: microsite modulation of macroscale processes. In: Jones CG, Lawton JH (eds) Linking species and ecosystems. Chapman Hall, London, pp 94–106

Angelini P, Donnini D, Pagiotti R, Tirillini B, Granetti B, Venanzoni R (2010) Biological activities of methanolic extract from *Tuber aestivum*, *T. borchii*, and *T. brumale* f. *moschatum*. Österreichische Zeitschrift für Pilzkunde 19:281–290

Arzone A (1970) Reperti ecologici ed etologici di *Leiodes cinnamomea* Panzer vivente su *Tuber melanosporum* Vittodini (Coleoptera Staphylinoidea). Annali della Facoltà di Scienze Agrarie dell' Università degli Studi di Torino 5:317–357

Bonito G, Smith ME, Nowak M, Healy RA, Guevara G, Cázares E, Kinoshita A, Nouhra ER, Domínguez LS, Tedersoo L, Murat C, Wang Y, Moreno BA, Pfister DH, Nara K, Zambonelli A, Trappe JM, Vilgalys R (2013) Historical biogeography and diversification of

truffles in the tuberaceae and their newly identified southern hemisphere sister lineage. PLoS One 8(1), e52765. doi:10.1371/journal.pone.0052765
Bragato G (1997) Modificazioni indotte dal tartufo nero pregiato (*Tuber melanosporum* Vitt.) sulla struttura e sulla sostanza organica del suolo. Monti e Boschi 1:23–27
Bratek Z, Papp O, Merkl L, Ádám L, Takács V (1992) Insects living in truffles. Micologia e Vegetazione Mediterranea 1:103–107
Brussaard L (1998) Soil fauna, guilds, functional groups and ecosystem processes. Appl Soil Ecol 9(1):123–135. doi:10.1016/S0929-1393(98)00066-3
Callot G (1999) La truffe, la terre, la vie. INRA, Versailles
Castrignano A, Goovaerts P, Lulli L, Bragato G (2000) Ageostatistical approach to estimate probability of occurrence of *Tuber melanosporum* in relation to some soil properties. Geoderma 98(3):95–113. doi:10.1016/S0016-7061(00)00054-9
Edwards CA, Lofty JR (1975) The invertebrate fauna of the Park Grass Plots. I. Soil fauna. Report for 1974, Part 2, Rothamsted Experimental Station, pp 133–154
Esher RJ, Ursic SJ, Baker RL, Miller LC (1993) Responses of invertebrates to artificial acidification of the forest floor under southern pine. In: Proceedings of VIII American woodcock symposium, Biological Report 16. US Department of the Interior, Fish and Wildlife Service, Washington, DC, pp 75–83
FAO (2015) AGP—Physical factors affecting soil organisms. http://www.fao.org/agriculture/crops/thematic-sitemap/theme/spi/soil-biodiversity/soil-organisms/physical-factors-affecting-soil-organisms/en/
Fitter AH, Garbaye J (1994) Interactions between mycorrhizal fungi and other soil organisms. Plant Soil 159(1):123–132. doi:10.1007/BF00000101
Fitter AH, Sanders IR (1992) Interactions with the soil fauna. In: Allen MF, Allen MF (eds) Mycorrhizal functioning an integrative plant –fungal process. Chapman & Hall, London, pp 333–354
García-Montero LG, Pérez Andueza G, Díaz P, Manjón JL (2004) Contribución al conociemiento ecológico de los dípteros fitógafos (Diptera: Heleomyzidae) de las trufas en España. Boletín de Sanidad Vegetal Plagas 30:679–683
García-Montero LG, Valverde-Asenjo I, Díaz P, Pascual C, Menta C (2010) New data on impact of earthworms activity on black truffle soils. In: Donnini D (ed) Proceedings of 3rd international congress on truffles, Spoleto, 25–28 Nov 2008. Comunità Montana dei Monti Martani, Serano e Subasio, p 723
García-Montero LG, Valverde-Asenjo I, Grande-Ortíz MA, Menta C, Hernando I (2013) Impact of earthworms casts on soil pH and calcium carbonate in black truffle burns. Agroforest Syst 87 (4):815–826. doi:10.1007/s10457-013-9598-9
García-Montero LG, Kuyper T, Moreno D, Monleon V, Myrold D, Trappe J, Baham J, Valverde I, Parladé X, Álvarez A, Benito LF, Quintana JR, Parke J, Quintana A (2014) Could ectomycorrhizae perform "ecosystem engineering" on the host plant's rhizosphere? In: Bonet JA, Colinas C, Parladé X (eds) Proceedings of the encuentro en Cataluña del proyecto Micosylva. Centre Tecnològic Forestal de Catalunya, Lleida
Granetti B, De Angelis A, Materozzi G (2005) Umbria terra di tartufi. Regione Umbria, Terni
Hanlon RDG, Anderson JM (1979) The effects of Collembola grazing on microbial activity in decomposing leaf litter. Oecologia 38(1):93–99. doi:10.1007/BF00347827
Hasegawa M (2001) The relationship between the organic matter composition of a forest floor and the structure of a soil arthropod community. Eur J Soil Biol 37(4):281–284. doi:10.1016/S1164-5563(01)01099-8
Hochberg ME, Bertault G, Poitrineau K, Janssen A (2003) Olfactory orientation of the truffle beetle, *Leiodes cinnamomea*. Entomol Exp Appl 109(2):147–153. doi:10.1046/j.1570-7458.2003.00099.x
Hodge A (2000) Microbial ecology of the arbuscular mycorrhiza. FEMS Microbiol Ecol 32 (2):91–96. doi:10.1111/j.1574-6941.2000.tb00702.x

Janvier H (1963) La mouche de la truffle (*Heleomyza tuberiperda* Rondani). Bulletin de la Societé Entomoloaique de France 68:50–83

Jeandroz S, Murat C, Wang Y, Bonfante P, Le Tacon F (2008) Molecular phylogeny and historical biogeography of the genus *Tuber*, the 'true truffles'. J Biogeogr 35(5):815–829. doi:10.1111/j.1365-2699.2007.01851.x

Kaplan DL, Hartenstein R, Neuhauser EF, MNalecki MR (1980) Physicochemical requirements in the environment of the earthworm *Eisenia foetida*. Soil Biol Biochem 12(4):347–352. doi:10.1016/0038-0717(80)90008-5

Killham K (1994) Soil ecology. Cambridge University Press, Cambridge

Koricheva J, Gange AC, Jones T (2009) Effects of mycorrhizal fungi on insect herbivores: a meta-analysis. Ecology 90(8):2088–2097. doi:10.1890/08-1555.1

Kuperman RG (1996) Relationships between soil properties and community structure of soil macroinvertebrates in oak-hickory forests along an acidic deposition gradient. Appl Soil Ecol 4(2):125–137. doi:10.1016/0929-1393(96)00108-4

Lavelle P, Pashanasi B, Charpentier F, Gilot C, Rossi JP, Derpuard L, Andre J, Ponge JF, Bernier N (1998) Large-scale effects of earthworms on soil organic matter and nutrient dynamics. In: Edwards CA (ed) Earthworm ecology. St. Lucie Press, Boca Raton, pp 103–122

Lulli L, Bragato G, Gardin L (1999) Occurrence of *Tuber melanosporum* in relation to soil surface layer properties and soil differentiation. Plant Soil 214(1):85–92. doi:10.1023/A:1004602519974

Martin F, Kohler A, Murat C, Balestrini R, Coutinho PM, Jaillon O, Montanini B, Morin E, Noel B, Percudani R, Porcel B, Rubini A, Amicucci A, Amselem J, Anthouard V, Arcioni S, Artiguenave F, Aury JM, Ballario P, Bolchi A, Brenna A, Brun A, Buée M, Cantarel B, Chevalier G, Couloux A, Da Silva C, Denoeud F, Duplessis S, Ghignone S, Hilselberger B, Iotti M, Marçais B, Mello A, Miranda M, Pacioni G, Quesneville H, Riccioni C, Ruotolo R, Splivallo R, Stocchi V, Tisserrant E, Viscomi AR, Zambonelli A, Zampieri E, Henrissat B, Lebrun MH, Paolocci F, Bonfante P, Ottonello S, Wincker P (2010) Périgord black truffle genome uncovers evolutionary origins and mechanisms of symbiosis. Nature 464 (7291):1033–1038. doi:10.1038/nature08867

Mello A, Ding GC, Piceno YM, Napoli C, Tom LM, De Santis TD, Andersen GL, Smalla K, Bonfante P (2013) Truffle brûlé have an impact on the diversity of soil bacterial communities. PLoS One 8(4), e61945. doi:10.1371/journal.pone.0061945

Menta C (2008) Guida alla conoscenza della biologia e dell'ecologia del suolo—Funzionalità, degrado, indicatori. Alberto Perdisa Editore, Bologna

Menta C, Leoni A, Gardi C, Conti FD (2011a) Are grasslands important habitats for soil microarthropod conservation? Biodivers Conserv 20(5):1073–1087. doi:10.1007/s10531-011-0017-0

Menta C, Tarasconi K, García-Montero LG, Vicari M, Gregori G (2011b) Comunità edafica animale in suoli italiani e spagnoli interessati dalla presenza del tartufo. In: Gianguzza P, Chemello R (eds) Riassunti XXI Congresso della Società Italiana di Ecologia. Università degli Studi di Palermo, Palermo

Menta C, García-Montero LG, Conti FD, Pinto S (2013) The role and diversity of soil fauna in different woodlands. In: Manzanares L (ed) Woodlands: structure, species diversity and sustainable management. Nova, New York, pp 43–74

Menta C, García-Montero LG, Pinto S, Conti FD, Baroni G, Maresi M (2014) Does the natural "microcosm" created by *Tuber aestivum* affect soil microarthropods? A new hypothesis based on Collembola truffle culture. Appl Soil Ecol 84:31–37. doi:10.1016/j.apsoil.2014.06.012

Newton AF (1984) Mycophagy in Staphylinidae (Coleoptera). In: Wheeler Q, Blackwell M (eds) Fungus-insect relationships: perspective in ecology and evolution. Columbia University Press, New York, pp 302–353

Pacioni G, Bellina-Agostinone C, D'Antonio M (1989) Odour composition of the *Tuber melanosporum* complex. Mycol Res 94(2):201–204. doi:10.1016/S0953-7562(09)80614-X

Pacioni G, Bologna MA, Laurenzi M (1991) Insect attraction by *Tuber*: a chemical explanation. Mycol Res 95(12):1359–1363. doi:10.1016/S0953-7562(09)80385-7

Pargney JC, Boumaza O, Toutain F (2010) In situ micro-and ultrastructural study of ascocarp development in *Tuber mesentericum*. In: Donnini D (ed) Proceedings of 3rd international congress on truffles, Spoleto, 25–28 Nov 2008. Comunità Montana dei Monti Martani, Serano e Subasio, pp 355–364

Pérez-Íñigo C (1993) Acari, Oribatei, Poronota. In: Ramos MA (ed) Fauna Ibérica, vol 3. Museo de Ciencias Naturales, CSIC, Madrid

Pérez-Íñigo C (1997) Acari, Oribatei, Gymnonota I. In: Ramos MA (ed) Fauna Ibérica, vol 9. Museo de Ciencias Naturales, CSIC, Madrid

Plattner I, Hall IR (1995) Parasitism of non-host plants by the mycorrhizal fungus *Tuber melanosporum*. Mycol Res 99(11):42–45. doi:10.1016/S0953-7562(09)81223-9

Queralt M, Moraza ML, de Miguel AM (2014) Preliminary study of the mite community structure in different black truffle producing soils. For Syst 23(2):339–348. doi:10.5424/fs/2014232-04906

Ricard JM (2003) La truffe: guide technique de trufficulture. Centre Technique Interprofessionnel des Fruits et Légumes CTIFL, Paris

Shaw PJA (1992) Fungi, fungivores and fungal food webs. In: Carroll GC, Wicklow DT (eds) The fungal community—its organization and role in the ecosystem. Dekker, New York, pp 295–310

Sourzat P (2004) Questions d'icologie appliquies à la trufficulture. Lycée Professionnel Agricole de Cahors-Le Montat, Le Montat

Splivallo R (2008) Biological significance of truffle secondary metabolites. In: Karlovsky P (ed) Secondary metabolites in soil ecology, vol 14, Soil biology. Springer, Berlin, pp 141–165. doi:10.1007/978-3-540-74543-3_8

Splivallo R, Ottonello S, Mello A, Karlovsky P (2011) Truffle volatiles: from chemical ecology to aroma biosynthesis. New Phytol 189(3):688–699. doi:10.1111/j.1469-8137.2010.03523.x

Streiblová E, Gryndlerová H, Gryndler M (2012) Truffle brûlé: an efficient fungal life strategy. Microbiol Ecol 80(1):1–8. doi:10.1111/j.1574-6941.2011.01283.x

Tarasconi K, Menta C, García-Montero LG, Gregori G (2011) Effetti della presenza del tartufo sulla comunità edafica. In: Dazzi C (ed) Atti del workshop La percezione del suolo. Le Penseur, Palermo

Trappe JM, Claridge A (2010) The hidden life of truffles: not just for gourmands, truffles play essential roles in the health of ecosystems. Sci Am 302(4):78–84. doi:10.1038/scientificamerican0410-78

Wolters V (2000) Invertebrate control of soil organic matter stability. Biol Fert Soils 31(1):1–19. doi:10.1007/s003740050618

Chapter 20
Mycoviruses Infecting True Truffles

Claudio Ratti, Mirco Iotti, Alessandra Zambonelli, and Federica Terlizzi

20.1 Introduction

Viruses can be defined as small and simple obligate intracellular parasites. Virions, the complete, infectious virus particles, consist of a protein coat (capsid) enclosing the viral genome, that, in some cases, can be wrapped in an external coat derived by the host cell membrane. This set of nucleic acids encodes proteins necessary for replication, movement to adjacent plant cells (in the case of plant viruses), possibly proteins to interact with host defence and finally protein to help transmission among individuals of the host species.

Genome of known viruses lacks the genetic information necessary for the generation of metabolic energy (adenosine triphosphate, ATP) or for protein synthesis (ribosomes). Viruses are therefore dependent on the host cell for these functions. Moreover, even if viruses do not have cellular structure and do not growth through their own metabolism, they reproduce themselves and are able of adaptation to the environment by internally-originate changes, then they can be probably ascribed as living organisms. From a different point of view, viruses are living entities during the intracellular phase of their life cycle, according to the Virocell concept (Forterre 2011), and can be considered as a complex assemblage of metabolically inert chemicals outside the host cell.

Viruses can infect all living organisms including humans, animals, plants, insects, bacteria, archaea, algae, fungi and protozoa. The majority of infections

C. Ratti (✉) • A. Zambonelli • F. Terlizzi
Department of Agricultural Science, University of Bologna, viale Fanin 42, 40127 Bologna, Italy
e-mail: claudio.ratti@unibo.it

M. Iotti
Department of Life, Health and Environmental Sciences, University of L'Aquila, Via Vetoio, Coppito 1, 67100 L'Aquila, Italy

A. Zambonelli et al. (eds.), *True Truffle (*Tuber *spp.) in the World*, Soil Biology 47,
DOI 10.1007/978-3-319-31436-5_20

do not affect the health of the host and result in a latent replication of the virus, but in some cases, viruses interfere with normal cellular processes and become the causal agent of human, animal and plant diseases with enormous economic costs and impact on society. These diseases include smallpox, yellow fever, poliomyelitis, common flu, measles and AIDS in humans, avian influenza, foot-and-mouth disease of cattle, porcine reproductive and respiratory syndrome virus, infectious gastroenteritis and bronchitis in animals and diseases of cultivated plants such as potato, tomato, tobacco, stone fruits, coconut and citrus trees. For example, the porcine reproductive and respiratory syndrome virus causes a total cost of productivity losses in the US national breeding and growing pig estimated to US $ 664 million annually (Holtkamp et al. 2013).

20.2 Mycoviruses

More than 100 viral species infecting fungi are currently recognised by the International Committee on Taxonomy of Viruses (ICTV—Virology Division within the International Union of Microbiological Societies, IUMS), and many other viruses have been described and proposed as new species in the last few years. The taxonomy of mycoviruses is therefore in fast evolution, and huge amounts of information are awaited in the close future because of next-generation sequencing approaches in mycovirus discovery.

In the early history of mycovirology, the use of electron microscopy allowed to observe fungal tissues densely populated by variously shaped virus particles even if, in most cases, their presence is symptomless (Hollings 1962; Buck 1986).

Up to now, mycoviruses are classified within 12 families and 18 genera according to their genomic and morphological proprieties (Table 20.1). Most of the species possess icosahedral particles and double-stranded RNA (dsRNA) genome, but also filamentous or bacilliform particles have been reported with single-stranded positive sense RNA [ssRNA(+)] genome (Buck 1986; Ghabrial 1998). Moreover, some mycoviruses also lack true virions, and RNA can be included in pleomorphic vesicles or complexed in nucleoprotein aggregates (Nuss and Koltin 1990; Jacob-Wilk et al. 2006). Genome can be composed of just one DNA or RNA molecule (monopartite), two molecules forming two separate particles (bipartite), or divided into three or more molecules packaged separately (multipartite). Mycoviruses can be exchanged via anastomoses between hyphae (horizontal transmission) (Day and Anagnostakis 1973; Sonnenberg and van Griensven 1991; Ihrmark et al. 2002). Vertical transmission can also occur since infected spores can originate infected hyphae (Romaine et al. 1993; Ihrmark et al. 2002; Chu et al. 2003).

Mixed infections with two or more unrelated viruses are common (Ghabrial 1998); therefore, electron microscopy can fail in discriminating between morphologically similar but genetically distinct viruses (Romaine and Goodin 2002). Moreover, as mycoviruses are often asymptomatic and not encapsidated, the discrimination and identification of the virus of interest are mainly based on the

Table 20.1 Classification, proprieties and number and of viral species infecting fungi formally recognised by the International Committee on Taxonomy of Viruses (ICTV)

Genome	Family/subfamily	Genus	Number of species infecting fungi	Particle: Shape	Particle: Size (nm)	Genome: No. of segments	Genome: Size of segments (kbp or kb)
ssDNA	Unassigned		1	Icosahedral	22–22	1	2.1
dsRNA	*Chrysoviridae*	*Chrysovirus*	8	Icosahedral	35–40	4	3.6, 3.2, 3.0, 2.9
	Endornaviridae	*Endornavirus*	2	Pleomorphic RNA-containing vesicles	No virions; viral RNA contained in cytoplasmic vesicles	1	14.0–18.0
	Partitiviridae	*Alphapartitivirus*	6	Icosahedral	30–43	2	1.4–2.4 per segment
		Betapartitivirus	8				
		Gammapartitivirus	8				
		Unassigned	3				
	Reoviridae/ *Spinareovirinae*	*Mycoreovirus*	3	Turreted or non-turreted icosahedral capsid	60–85	9–12	19.0–32.0 total genome
	Totiviridae	*Totivirus*	7	Icosahedral, not enveloped	40	1	4.6–6.7
		Victorivirus	14				5.0
	Unassigned		5	Icosahedral, not enveloped	27	1	1.7–6.6
	Satellite dsRNAs		11	Helper virus		1–3	0.5–1.8
ssRNA-RT(+)	*Pseudoviridae*	*Hemivirus*	3	Icosahedral	60–80	1	5.0–9.0
		Pseudovirus	3				
	Metaviridae	*Metavirus*	5	Intracellular virus-like particles and/	Unknown	1	4.0–10.0

(continued)

Table 20.1 (continued)

Genome	Family/subfamily	Genus	Number of species infecting fungi	Particle		Genome	
				Shape	Size (nm)	No. of segments	Size of segments (kbp or kb)
				or enveloped extra-cellular virions			
ss(+) RNA	*Alphaflexiviridae*	*Botrexvirus*	1	Filamentous	10–15 × 470–800	1	6.0–9.0
		Sclerodarnavirus	1				
	Barnaviridae	*Barnavirus*	1	Bacilliform	18–20 × 48–53	1	4.0
	Gammaflexiviridae	*Mycoflexivirus*	1	Filamentous	13 × 720	1	6.8
	Hypoviridae	*Hypovirus*	4	Pleomorphic RNA-containing vesicles	No virions; viral RNA contained in cytoplasmic vesicles	1	9.0–13.0
	Narnaviridae	*Mitovirus*	5	No true virions Intracellular nucleo-protein complex	N/A	1	2.3–2.9
		Narnavirus	2				
	Unassigned		4	Icosahedral, not enveloped	27–32	1–3	0.9–5.5

Source: Virus Taxonomy release 2014—EC 46, Montreal, Canada, July 2014, Email ratification 2015 (MSL #29)—ICTV—Virology Division—IUMS isolation of its dsRNAs and on demonstrating its transmissibility. Electrophoretic separation of dsRNA results in a banding pattern typical, in number and size of segments, of a specific host-virus association. This method also allows detection of subviral agents that often accumulate in the presence of the main virus from which they are dependent for replication (satellite dsRNAs) and are thought to be generated from infectious virus genome by replicase error (defective RNAs) (Ghabrial 1998). However, a mycovirus similar to geminiviruses with a DNA genome and another with ssRNA(−) genome has been identified from the plant pathogenic fungus *Sclerotinia sclerotiorum* (Lib.) de Bary (Yu et al. 2010; Liu et al. 2014).

20.2.1 Viral Effects on Fungi

Latency or lack of symptoms found in most mycovirus-fungus relationships is probably the results of the long coevolution and coadaptation with their host (Ghabrial 1980; Buck 1986; Milgroom 1999). Viruses tend not to be extremely virulent as the death of the host results in complete virus elimination (Ghabrial 1998). In particular, viruses depending mainly on vertical transmission most likely evolved to keep their replication below the pathogenicity threshold. In contrast, viruses transmitted horizontally can be more deleterious to their hosts as they have more chances for survival (Milgroom 1999).

Mycovirus infection can be symptomless but also cause various effects on the hosts ranging from severe debilitation, hypovirulence, decreased growth rate, lack of sporulation, reduced germination of basidiospores and hypervirulence (Nuss and Koltin 1990; Ghabrial 1994; Moleleki et al. 2003; Ihrmark et al. 2004; Suzaki et al. 2005). In some cases, the presence of mycoviruses can result in a disadvantage for their fungal hosts. Plant pathogenic fungi *Cryphonectria parasitica* (Murrill) M.E. Barr, *Ophiostoma ulmi* (Buisman) Nannf. and *Helminthosporium victoriae* F. Meehan and H.C. Murphy become hypovirulent when infected by dsRNA viruses that have been suggested as a potential biological control agent for the fungal disease (Van Alfen et al. 1975; Ghabrial and Mernaugh 1983; Brasier 1991). The best studied of these is the chestnut blight fungus, *C. parasitica*, that became epidemic in Europe and in eastern North America with devastating effects until the natural appearance of hypovirulent strains causing superficial cankers only (Milgroom and Cortesi 2004). The mycovirus *Cryphonectria hypovirus 1-EP713* (CHV1-EP713) was shown to be associated with the hypovirulence phenotype (Shapira et al. 1991). In other cases, mycoviruses confer advantage to their fungi. Some yeast (*Saccharomyces cerevisiae* Meyen) or smut [*Ustilago maydis* (DC.) Corda] strains are named 'killer' and produce a viral toxin to which they are immune, but that is lethal to sensitive strains. In most cases, the toxin production is codified by dsRNA satellites (M-dsRNA, H-dsRNA, L-dsRNA), which are dependent on the helper dsRNA viruses *Saccharomyces cerevisiae virus L-A* and *Ustilago maydis virus H1* for replication and encapsidation (Hutchins and Bussey 1983; Kandel 1988; Koltin 1988).

20.3 Mycoviruses of Edible Mushrooms

Two diseases, one affecting *Agaricus bisporus* (J.E. Lange) Imbach and another affecting *Pleurotus ostreatus* (Jacq.) P. Kumm., are caused by the most important and best-studied mycoviruses of edible mushrooms. In addition to those, many others have been isolated mostly in asymptomatic fruiting bodies then often not clearly indicated as the direct cause of abnormalities.

The first mycovirus identified in diseased mushrooms was observed on *A. bisporus* (Hollings 1962), showing reduced production of basidiomata (Fig. 20.1a). Two main viral diseases affect *A. bisporus*: ‘La France’ disease and ‘mushroom virus X’ disease. The first one, also known as ‘X disease’, ‘watery stipe’, ‘brown disease’ or ‘dieback’, was first reported on the La France brothers’ farm in Pennsylvania, USA, in 1948 (Sinden and Hauser 1950) then reported in the UK, France, the Netherlands, Italy, Denmark, Australia and Turkey (Ghabrial 1994; Elibuyuk and Bostan 2010). During the 1970s, La France disease occurred at epidemic proportions in North America and Europe and was a major problem during commercial mushroom production (Romaine and Goodin 2002). It can affect all *A. bisporus* varieties but does not affect the alternative cultivated species *Agaricus bitorquis* (Quél.) Sacc. (Geels et al. 1988), which is often used as ‘virus breaker’ in order to interrupt the infective cycle of the virus. Infected mycelium does not permeate the casing layer of the commercial beds with delayed emergence of fruiting bodies and bare patches resulting in loss of yield. Symptoms of severe infections are associated with malformed fruiting bodies assuming a ‘drumstick’ phenotype, with elongated stems and small misshapen caps. Diseased fruiting bodies tend to mature prematurely and release infected spores (Romaine and Goodin 2002). Two types of isometric particles with diameters of 34–36 and 25 nm (Van Zaayen 1979) and bacilliform particles measuring 50×19 nm, named *Mushroom bacilliform virus* (MBV) (Revill et al. 1994), have been identified in La France diseased mushrooms. No obvious symptoms were observed in MBV singly infected mushrooms, while high correlation occurs between the disease and the presence of the six major dsRNA elements L1 (3.8 kbp), L2 (3.1 kbp), L3 (3.0 kbp), L4 (2.8 kbp), L5 (2.6 kbp) and M2 (1.3 kbp) (Harmsen et al. 1989; Romaine and Schlagnhaufer 1989). These dsRNA molecules are associated with the 34–36 nm virions, and this virus was initially named La France isometric virus (LIV) subsequently renamed as *Agaricus bisporus virus 1* (ABV1) (Goodin et al. 1992; Van der Lende et al. 1994). ABV1 is frequently found in co-infections with MBV, and a synergistic relationship between them is not excluded. Indeed, the titre of MBV can be 12-fold higher in mushrooms infected with ABV1 (Romaine et al. 1993). Viruses associated to La France can be transmitted horizontally from infected to healthy mycelium and vertically by basidiospores with a ratio between 65 and 75 % (Schisler et al. 1967; Sonnenberg and Van Griensven 1991; Romaine et al. 1993).

Another disease affecting *A. bisporus* and termed ‘mushroom virus X’ or ‘MVX’ disease was observed first in the UK then in the Netherlands, Belgium, Germany,

Fig. 20.1 Symptoms induced by mycoviruses in fruiting bodies of mushroom and truffle. (**a**) Production plots of healthy (*upper plot*) and diseased (*lower plot*) *A. bisporus* showing absence and/or delayed emergence of fruiting bodies. *Mushroom bacilliform virus*, La France isometric virus and a 9.4 kbp dsRNA element of mushroom virus X disease have been detected in symptomatic mushrooms. (**b**) Ascomata of *T. magnatum*: symptomless (*left*) and showing brown spots irregularly distributed on the peridium (*right*); a dsRNA fragment of about 17 kbp, isolated only from symptomatic fruiting bodies, shares significant amino acid identity with members of the genus *Endornavirus*

Ireland, Poland and Hungary (Gaze et al. 2000; Sonnenberg and Lavrijssen 2004; Rao et al. 2007; Pudelko 2010; Halász et al. 2014). Crops affected by MVX disease develop bare patches due to arrested development of pins or delay in fruiting body formation. Other symptoms include premature veil opening, brown discoloration and malformed basidiomata. Analyses of MVX-infected mushroom samples identified 27 dsRNA elements ranging in size from 0.64 to 20.2 kbp (Grogan et al. 2003) but no virus particles specifically associated with the disease (Sonnenberg and Lavrijssen 2004; Rao et al. 2007). Due to the various sizes and numbers of dsRNAs together with the diverse range of symptoms observed, it has been hypothesised that MVX disease complex might comprise more than one virus. A 12.75 kbp dsRNA fragment has also been sequenced and described as an endornavirus, *Agaricus bisporus* endornavirus 1 (AbEV1) (Maffettone 2007). Moreover, it is hypothesised that the 2.0 and 1.8 kbp molecules represent a single virus in the *Partitivirus* genus, and it is proposed that this should be named brown cap mushroom virus (BCMV; Green 2010). MVX dsRNAs can be transmitted by spores and mycelial fragments (Gaze et al. 2000; Grogan et al. 2003; Adie et al. 2004). In spite of the dsRNA sequences and epidemiological information revealed so far about the MVX disease, the specific viral aetiology remains enigmatic.

In *P. ostreatus*, spherical and bacilliform viruses were isolated from malformed basidiomata, and a correlation between the presence of dsRNA mycoviruses and slow-growing mycelium was hypothesised (Go et al. 1992; Van der Lende et al. 1995; Park and Kim 1996). Moreover, in Korea an epidemic disease of oyster mushroom, named 'dieback disease' which led to reduced yields in commercial farms, has been associated to an isometric virus with a 5.784 kb ssRNA genome, oyster mushroom spherical virus (OMSV) (Yu et al. 2003). The name of oyster mushroom isometric virus (OMIV) has been proposed for a novel dsRNA mycovirus identified in Korean *P. ostreatus* mushroom showing short and undeveloped stipes, flat and deformed caps and retarded abnormal mycelial growth on a solid medium (Ro et al. 2006). From fruiting bodies of *Pleurotus eryngii* (DC.) Quél., showing similar symptoms to those associated to OMIV, a virus with a ssRNA genome of 7.8 kbp and 31 nm diameter spherical particles has been isolated and was named *P. eryngii* spherical virus (PeSV) (Ro et al. 2007). Another (POSV, probably Pleurotus ostreatus spherical virus) has been associated to a serious degeneration of industrial cultivation of mushroom in China (Qiu et al. 2010). Successful sanitation of OMSV-infected mycelium of *P. ostreatus* has been obtained by in vitro growth onto a limited nutrient medium containing rifamycin and evidenced the negative effect of the virus on fruiting body production (Kwon et al. 2012). The partitivirus *Pleurotus ostreatus virus 1* (PoV1) and new viral strain from Korea (*Pleurotus* strain Shin-Nong—PoV-SN) infect *P. ostreatus* but are not associated with disease symptoms (Lim et al. 2005; Kim et al. 2008).

Furthermore, after the first virus-like particles were reported in *Lentinula edodes* (Berk.) Pegler, the second most cultivated edible mushroom in the world after *A. bisporus* (Inoue 1970; Boa 2004), several reports of mycovirus in this mushroom are available (Rytter et al. 1991; Ohta et al. 2008; Magae 2012; Kim et al. 2013). Infection of the viral agent tentatively designated as *Lentinula edodes* mycovirus (LeV) results in deleterious effect on mycelial growth in liquid culture (Kim et al. 2015). Moreover, icosahedral particles encapsidating a 12 kb ssRNA genome, for which the name *Lentinula edodes* spherical virus (LeSV) was suggested, have been recently associated with symptomatic fruiting bodies of *L. edodes*, showing underdeveloped, curled or cracked morphologies (Won et al. 2013).

Furthermore, two dsRNA elements (1.915 and 1.730 kbp), associated with 50 nm diameter virus-like particles, were detected in *Flammulina velutipes* (Curtis) Singer showing spontaneous brown colour change (discoloration) of fruiting bodies. The virus, tentatively named as Flammulina velutipes browning virus (FvBV), resulted most closely related to *Chondrostereum purpureum cryptic virus* of the *Partitivirus* genus (Magae and Hayashi 1999; Magae and Sunagawa 2010).

20.4 Mycoviruses of Endomycorrhizal and Ectomycorrhizal Fungi

Unlike plant or mushroom pathogenic species, mycoviruses infecting arbuscular mycorrhizal (AM) or ectomycorrhizal (ECM) fungi have been poorly investigated. Only in recent years, due to the progressive intensification in the use of next-generation sequencing platforms, a few reports regarding mycoviruses in AM fungi have been published. In particular, several dsRNA segments that differed in size from 2.5 to more than 10 kbp have been identified in the mycelium of *Glomus* sp. Among them, a 4.5 kbp segment, named *Glomus* sp. strain RF1 virus-like medium dsRNA (GRF1V-M), showing similarity with members of the genus *Giardiavirus*, was found to be responsible for changes in the symbiotic phenotypes such as spore productivity and the plant growth-promoting effect (Ikeda et al. 2012). More recently, a mitovirus has been reported in a fungus member of the phylum Glomeromycota. The name of *Rhizophagus clarus* mitovirus 1 strain RF1 (RcMV1-RF1) has been tentatively assigned to the 2.895 kb genome isolated from *R. clarus* (Kitahara et al. 2014).

Between the end of the fifties and beginning of the eighties, few reports described virus-like particles in the noncultivated basidiomata of *Boletus edulis* Bull., *Laccaria* sp. and *Cantharellus* sp. associated to symptoms such as morel-like or reduced pileus (Huttinga et al. 1975; Heinze 2012). The first small dsRNA of 2.2 kbp reported from ECM basidiomycetes has been probably isolated from the mycelium of *Hebeloma circinans* (Quél.) Sacc. (Cortinariaceae) (Bai et al. 1997). Later, evidence of viruses in ECM fungi was revealed from expressed sequence tag (ESTs) analysis of mycorrhizas and mycelium of *Pisolithus microcarpus* (Cooke & Massee) G. Cunn., but only little information is available. In particular, a single transcript was identified in *P. microcarpus* as deriving from a double-stranded RNA mycovirus, infecting the root rot fungus *Helicobasidium mompa* Nobuj. Tanaka, named *Helicobasidium mompa partitivirus V70* and classified as a member of the genus *Alphapartitivirus* within the *Partitiviridae* family (Osaki et al. 2002). The transcript seemed to be highly expressed in both mycelial and mycorrhizal tissues of *P. microcarpus* (Peter et al. 2003).

20.4.1 Mycoviruses of True Truffles

Only three different viruses have been reported to infect *Tuber aestivum* Vittad. and another to infect *Tuber excavatum* Vittad.; described infections are apparently symptomless as no alterations on morphology of the colonised fruiting bodies or mycelia were detected. However, recently, a mycovirus was found in *Tuber magnatum* Pico, which may be associated with the decay of its ascomata. Additionally, several sequences of putative replication initiation protein (Rep) related to geminivirus Rep (plant viruses with small circular single-stranded DNA genome) have been identified in the genome of *Tuber melanosporum* Vittad. (Périgord truffle) by systematic search for sequences related to circular single-stranded DNA (ssDNA) viruses in publicly available eukaryotic genome databases. The evolutionary meaning of these viral gene transfer into the eukaryotic genome of Périgord truffle is still an open question (Liu et al. 2011).

Moreover, the sequencing of *T. melanosporum* genome allowed to perform several studies on repeated sequences within its genome with particular attention to the transposable elements (TE) with RNA intermediate. The latter, named retrotransposons, are genetic elements that can self-amplify in a genome, are ubiquitous components of the DNA of many eukaryotic organisms and are considered viruses in classification and nomenclature. Six different clades (*Tmt1* to *Tmt6*) of Gypsy retrotransposons (family *Metaviridae*) have been identified in Périgord truffle corresponding to more than 10 % of its genome (Riccioni et al. 2008; Payen 2015). Gypsy retrotransposon clades identified are not expressed, except for one *Tmt1* element, but they contributed greatly to shape the *T. melanosporum* genome and also showed high genetic diversity rate among geographic accessions (Payen 2015). Around 58 % of the Périgord truffle genome consists of TE, half of which are retrotransposons; therefore, the impact of their activity needs to be addressed in true truffles (Martin et al. 2010).

20.4.1.1 Mycoviruses of *T. aestivum*

The first virus reported on *T. aestivum* which also represents the first evidence for the presence of mycoviruses in ECM fungi belonging to the phylum Ascomycota was isolated from an ascoma collected in a Hungarian-mixed beech forest. The virus, named *Tuber aestivum virus 1* (TaV1), is 4587 bp in length and possesses two open reading frames (ORFs) encoding for the coat protein (CP) and RNA-dependent RNA polymerase (RdRp) sharing high similarity with members of the genus *Totivirus* within the family *Totiviridae* (Stielow and Menzel 2010).

The same authors also reported, from mixed oak forests in different Hungarian regions, two distinct *T. aestivum* fruiting bodies infected one by a new endornavirus and the other by a new mitovirus species. The endornavirus, for which the name Tuber aestivum endornavirus (TaEV) was proposed, has a genome 9760 bp in length with a single ORF encoding for a 364.1 kDa polyprotein and lacks a capsid protein (Stielow et al. 2011a).

Similarly, a dsRNA fragment of 3480 bp was isolated from an ascoma of black truffle. Genomic characteristics suggest that it belongs to the genus *Mitovirus*, and it was tentatively named *Tuber aestivum* mitovirus (TaMV). The single ORF present in the genome encodes for a 71.3 kDa protein that shares identity ranging from 11.3 to 26.1 % with other members of the genus (Stielow et al. 2011b).

20.4.1.2 Mycoviruses of *T. excavatum*

To date, the only virus reported from the species *T. excavatum* has been identified in an ascoma collected in a mixed beech forest in Germany. The dsRNA genome is 3305 bp long and sequence analysis classified it within the genus *Mitovirus*, and for this reason, the name Tuber excavatum mitovirus (TeMV) was proposed. The single ORF present in the genome encodes for a protein with calculated molecular mass of 88.27 kDa sharing a sequence identity ranging from 11.0 to 23.6 % with the other members of the genus. In particular, an amino acid identity of 22 % can be calculated between ORF products of TeMV and TaMV (Stielow et al. 2012)

20.4.1.3 Mycoviruses of *T. magnatum*

To our knowledge, no viruses have been reported infecting *T. magnatum* with the exception of a recent identification of a new viral agent isolated from symptomatic fruiting bodies (unpublished results). Ascomata showing suberised consistency and brown spots irregularly distributed on the peridium have been collected in Italy and subjected to dsRNA extraction (Fig. 20.1b). Electrophoretic separation evidenced a single fragment of about 17 kbp not detected in symptomless samples. Random amplification of dsRNA fragment resulted in a sequence of 1327 kbp that showed significant nucleotide identity with mycoviruses. In particular, the sequence obtained shares identity ranging from 30 to 40 % with recognised members of the genus *Endornavirus* suggesting the existence of a new viral species within this genus that infect *T. magnatum*. Highest identity (54 %) has been calculated between the sequence obtained and the sequence of TaEV previously reported on *T. aestivum* (Stielow et al. 2011a). Additional studies are being carried out in order to completely characterise the identified virus for which the name Tuber magnatum endornavirus (TmEV) is proposed.

20.5 Conclusions

Micovirology of true truffle is an extremely young discipline, and only in recent years, it was demonstrated that these fungi are also hosts of several viral agents. As for other mycoviruses, the knowledge about how they interfere with the fungal (truffle) metabolism is still poorly understood. In particular, very little is known regarding the impact of mycoviruses on the evolution, adaptation and diversity of fungi and if they can spread in nature by ways different from spores or anastomosis. The increasing number of viral infection reported in fungi suggests that also viral agents able to infect truffles are more common than initially thought. Moreover, as for most mycoviruses, most of these infections are probably latent, unable to induce any phenotypic changes and therefore to raise the attention of the scientific community.

Many examples regarding devastating viral diseases of mushrooms suggest that we should not underestimate the potential risk of viral infection in truffles. La France and MVX diseases are known, in fact, to be responsible of extensive crop losses in the *A. bisporus* production, and different viral infections have been demonstrated to be related with symptomatic fruiting bodies of the most important cultivated mushrooms.

Moreover, four latent viruses have been described to infect fungi of the genus *Tuber*, and recently symptoms on ascomata of *T. magnatum* have been associated to a viral agent under characterisation. We cannot exclude that still unknown viral infections may play a role in some alterations that would reduce the economic value of ascomata. It is well known that low-quality fruiting bodies are often used for plant inoculations increasing the risk of spreading viral agents, eventually present in the inoculum, in new plants and in new areas. Altogether this aspects call for the need to improve knowledge regarding the viral population of true truffle worldwide, in particular considering the high economic impact of this crop.

References

Adie B, Choi I, Soares A, Holcroft S, Eastwood DC, Mead A, Grogan HM, Kerrigan RW, Challen M, Mills PR (2004) MVX disease and double-stranded RNA elements in *Agaricus bisporus*. In: Romaine CP, Keil CB, Rinker DL and Royse DJ (eds), Proceedings of the XVIth international congress on the science and cultivation of edible and medicinal fungi. Pennsylvania State University, pp 411–420

Bai X, Debaud JC, Schrunder J, Meinhardt F (1997) The ectomycorrhizal basidiomycete *Hebeloma circinans* harbors a linear plasmid encoding a DNA- and RNA polymerase. J Gen Appl Microbiol 43(5):273–279. doi:10.2323/jgam.43.273

Boa E (2004) Wild edible fungi a global overview of their use and importance to people. FAO, Rome

Brasier CM (1991) *Ophiostoma novo-ulmi* sp. nov., causative agent of current Dutch elm disease pandemics. Mycopathologia 115(3):151–161. doi:10.1007/BF00462219

Buck KW (1986) Fungal virology: an overview. In: Buck KW (ed) Fungal virology. CRC Press, Boca Raton, FL, pp 1–85

Chu YM, Lim WS, Yea SJ, Cho JD, Lee YW (2003) Complexity of dsRNA mycovirus isolated from *Fusarium graminearum*. Virus Genes 28(1):135–143. doi:10.1023/B:VIRU.0000012270.67302.35

Day PR, Anagnostakis SL (1973) The killer system *Ustilago maydis*: heterokaryon transfer and loss of determinants. Phytopathology 63:1017–1018. doi:10.1094/Phyto-63-1017

Elibuyuk IO, Bostan H (2010) Detection of a virus disease on white button mushroom (*Agaricus bisporus*) in Ankara, Turkey. Int J Agric Biol 12(4):597–600

Forterre P (2011) Manipulation of cellular syntheses and the nature of viruses: the virocell concept. C R Chim 14(4):392–399. doi:10.1016/j.crci.2010.06.007

Gaze RH, Calvo-Bado L, Challen MP, Adie BA, Romaine CP (2000) A new virus disease of *Agaricus bisporus*? Mushroom Sci 15(2):701–705

Geels FP, van de Geijin J, Rutjens AJ (1988) Pests and diseases. In: van Griensven LJLD (ed) The cultivation of mushrooms. Darlington Mushroom Laboratories Ltd, Rustington-Sussex, pp 361–422

Ghabrial SA (1980) Effects of fungal viruses on their hosts. Annu Rev Phytopathol 117:441–461. doi:10.1146/annurev.py.18.090180.002301

Ghabrial SA (1994) New developments in fungal virology. Adv Virus Res 43:303–388. doi:10.1016/S0065-3527(08)60052-4

Ghabrial SA (1998) Origin, adaptation and evolutionary pathways of fungal viruses. Virus Genes 16(1):119–131. doi:10.1023/A:1007966229595

Ghabrial SA, Mernaugh RL (1983) Biology and transmission of *Helminthosporium victoriae* mycoviruses. In: Compass RW, Bishop DHL (eds) Double stranded RNA viruses. Elsevier, New York, pp 441–449

Go SJ, Park YH, Shin GC, Wessels JGH (1992) Molecular analysis of double-stranded RNA in abnormal growing oyster mushroom, *Pleurotus florida* and *Pleurotus ostreatus* due to virus infection. Korean J Mycol 20(3):234–239

Goodin MM, Schlagnhaufer B, Romaine CP (1992) Encapsidation of the La France disease-specific double stranded RNAs in 36 nm isometric virus like particles. Phytopathology 82(3):285–290. doi:10.1094/Phyto-82-285

Green JM (2010) Investigating gene expression and brown colour development in the edible mushroom *Agaricus bisporus* during Mushroom Virus X infection. PhD thesis, Warwick University

Grogan HM, Adie B, Gaze RH, Challen MP, Mills PR (2003) Double stranded RNA elements associated with the MVX disease of *Agaricus bisporus*. Mycol Res 107(2):147–154. doi:10.1017/S0953756203007202

Halász K, Geosel A, Szarvas J, Viragh N, Hajdù CS, Lukacs N (2014) Occurrence of double-stranded RNA species in champignon and their relation to Mushroom Virus X symptoms. Acta Aliment 43(4):592–603. doi:10.1556/AAlim.2013.0009

Harmsen MC, van Griensven LJLD, Wessels JGH (1989) Molecular analysis of *Agaricus bisporus* double-stranded RNA. J Gen Virol 70(6):1613–1616. doi:10.1099/0022-1317-70-6-1613

Heinze C (2012) A novel mycovirus from *Clitocybe odora*. Arch Virol 157(9):1831–1834. doi:10.1007/s00705-012-1373-0

Hollings M (1962) Viruses associated with a die-back disease of cultivated mushroom. Nature 196(4858):962–965. doi:10.1038/196962a0

Holtkamp DJ, Kliebenstein JB, Neumann EJ, Zimmerman JJ, Rotto HF, Yoder TK, Wang C, Yeske PE, Mowrer CL, Haley CA (2013) Assessment of the economic impact of porcine reproductive and respiratory syndrome virus on United States pork producers. J Swine Health Prod 21(2):72–84. doi:10.2460/javma.2005.227.385

Hutchins K, Bussey H (1983) Cell wall receptor for yeast killer toxin: involvement of (1–6)-ß-D-glucan. J Bacteriol 154(1):161–169

Huttinga H, Wichers HJ, Dielemann-van Zaayen A (1975) Filamentous and polyhedral virus-like particles in *Boletus edulis*. Neth J Plant Pathol 81(3):102–106. doi:10.1007/BF01999860

Ihrmark K, Johannesson H, Stenstrom E, Stenlid J (2002) Transmission of double-stranded RNA in *Heterobasidion annosum*. Fungal Genet Biol 36(2):147–154. doi:10.1016/S1087-1845(02)00011-7

Ihrmark K, Stenström E, Stenlid J (2004) Double-stranded RNA transmission through basidiospores of *Heterobasidion annosum*. Mycol Res 108(2):149–153. doi:10.1017/S0953756203008839

Ikeda Y, Shimura H, Kitahara R, Masuta C, Ezawa T (2012) A novel virus-like double-stranded RNA in an obligate biotroph arbuscular mycorrhizal fungus: a hidden player in mycorrhizal symbiosis. Mol Plant Microbe Interact 25(7):1005–1012. doi:10.1094/MPMI-11-11-0288

Inoue T (1970) Virus-like particles from *Lentinus edodes*. Mushrooms 2:18–22

Jacob-Wilk D, Turina M, Van Alfen NK (2006) Mycovirus cryphonectria hypovirus 1 elements cofractionate with trans-Golgi network membranes of the fungal host *Cryphonectria parasitica*. J Virol 80(13):6588–6596. doi:10.1128/JVI.02519-05

Kandel JS (1988) Killer systems and pathogenic yeasts. In: Koltin Y, Leibowitz MJ (eds) Viruses of fungi and simple eukaryotes. Marcel Dekker, New York, pp 243–263

Kim YJ, Kim JY, Kim JH, Yoon SM, Yoo YB, Yie SW (2008) The identification of a novel *Pleurotus ostreatus* dsRNA virus and determination of the distribution of viruses in mushroom spores. JMicrobiol 46(1):95–99. doi:10.1007/s12275-007-0171-y

Kim JM, Yun SH, Park SM, Ko HG, Kim DH (2013) Occurrence of dsRNA mycovirus (LeV-FMRI0339) in the edible mushroom *Lentinula edodes* and meiotic stability of LeV-FMRI0339 among monokaryotic progeny. Plant Pathol J 29(4):460–464. doi:10.5423/PPJ.NT.03.2013.0037

Kim JM, Song HY, Choi HJ, Yun SH, So KK, Ko HK, Kim DH (2015) Changes in the mycovirus (LeV) titer and viral effect on the vegetative growth of the edible mushroom *Lentinula edodes*. Virus Res 197:8–12. doi:10.1016/j.virusres.2014.11.016

Kitahara R, Ikeda Y, Shimura H, Masuta C, Ezawa T (2014) A unique mitovirus from Glomeromycota, the phylum of arbuscular mycorrhizal fungi. Arch Virol 159(8):2157–2160. doi:10.1007/s00705-014-1999-1

Koltin Y (1988) The killer system of *Ustilago maydis*: secreted polypeptides encoded by viruses. In: Koltin Y, Leibowitz MJ (eds) Viruses of fungi and simple eukaryotes. Marcel Dekker, New York, pp 353–370

Kwon YC, Jeong DW, Gim SI, Ro HS, Lee HS (2012) Curing viruses in *Pleurotus ostreatus* by growth on a limited nutrient medium containing cAMP and rifamycin. J Virol Methods 185(1):156–159. doi:10.1016/j.jviromet.2012.06.002

Lim WS, Jeong JH, Jeong RD, Yoo YB, Yie SW, Kim KH (2005) Complete nucleotide sequence and genome organization of a dsRNA partitivirus infecting *Pleurotus ostreatus*. Virus Res 108(1–2):111–119. doi:10.1016/j.virusres.2004.08.017

Liu H, Fu Y, Li B, Yu X, Xie J, Cheng J, Ghabrial SA, Li G, Yi X, Jiang D (2011) Widespread horizontal gene transfer from circular single-stranded DNA viruses to eukaryotic genomes. BMC Evol Biol 11:276. doi:10.1186/1471-2148-11-276

Liu L, Xie J, Cheng J, Fu Y, Li G, Yi X, Jianga D (2014) Fungal negative-stranded RNA virus that is related to bornaviruses and nyaviruses. Proc Natl Acad Sci USA 111(33):12205–12210. doi:10.1073/pnas.1401786111

Maffettone E (2007) Characterization of a novel virus associated with the MVX disease of *Agaricus bisporus*. PhD Thesis. Cranfield health in collaboration with Warwick HRI (Wellesbourne)

Magae Y (2012) Molecular characterization of a novel mycovirus in the cultivated mushroom, *Lentinula edodes*. Virol J 9:60. doi:10.1186/1743-422X-9-60

Magae Y, Hayashi N (1999) Double-stranded RNA and virus particles in the edible basidiomycete *Flammulina velutipes* (Enokitake). FEMS Microbiol Lett 180:331–335. doi:10.1111/j.1574-6968.1999.tb08814.x

Magae Y, Sunagawa M (2010) Characterization of a mycovirus associated with the brown discoloration of edible mushroom, *Flammulina velutipes*. Virol J 7:342. doi:10.1186/1743-422X-7-342

Martin F, Kohler A, Murat C, Balestrini R, Coutinho PM, Jaillon O, Montanini B, Morin E, Noel B, Percudani R, Porcel B, Rubini A, Amicucci A, Amselem J, Anthouard V, Arcioni S, Artiguenave F, Aury JM, Ballario P, Bolchi A, Brenna A, Brun A, Buee M, Cantarel B, Chevalier G, Couloux A, Da Silva C, Denoeud F, Duplessis S, Ghignone S, Hilselberger B, Iotti M, Mello M, Miranda M, Pacioni G, Quesneville H, Riccioni C, Ruotolo R, Splivallo R, Stocchi V, Tisserant E, Viscomi AR, Zambonelli A, Zampieri E, Henrissat B, Lebrun MH, Paolocci F, Bonfante P, Ottonello S, Wincker P et al (2010) Perigord black truffle genome uncovers evolutionary origins and mechanisms of symbiosis. Nature 4(7291):1033–1038. doi:10.1038/nature08867

Milgroom MG (1999) Viral in fungal populations. In: Worrall JJ (ed) Structure and dynamics of fungal populations. Kluwer, Dordrecht, pp 283–306

Milgroom MG, Cortesi P (2004) Biological control of chestnut blight with hypovirulence: a critical analysis. Annu Rev Phytopathol 42:311–338. doi:10.1146/annurev.phyto.42.040803.140325

Moleleki N, Van Heerden SW, Wingfield MJ, Wingfield BD, Preisig O (2003) Transfection of *Diaporthe perjuncta* with Diaporthe RNA virus. Appl Environ Microbiol 69(7):3952–3956. doi:10.1128/AEM.69.7.3952-3956.2003

Nuss DL, Koltin Y (1990) Significance of dsRNA genetic elements in plant pathogenic fungi. Annu Rev Phytopathol 28:37–58. doi:10.1146/annurev.py.28.090190.000345

Ohta C, Taguchi T, Takahashi S, Ohtsuka K, Eda K, Ayusawa S, Magae Y (2008) Detection of double stranded RNA elements in cultivated *Lentinula edodes*. Mushroom Sci Biotechnol 16:155–158

Osaki H, Nomura K, Iwanami T, Kanematsu S, Okabe I, Matsumoto N, Sasaki A, Ohtsu Y (2002) Detection of a double-stranded RNA virus from a strain of the violet root rot fungus *Helicobasidium mompa* Tanaka. Virus Genes 25(2):139–145. doi:10.1023/A:1020105701017

Park JS, Kim YH (1996) Infectious RNA viruses in the edible mushroom *Pleurotus* spp. J Microbiol 34(1):61–67

Payen T (2015) Contribution à l'étude de l'évolution des genomes de champignons ectomycorhiziens du genre Tuber (Pézizomycètes) par génomique comparative. PhD Thesis, Lorraine University

Peter M, Courty PE, Kohler A, Delaruelle C, Martin D, Tagu D, Frey-Klett P, Duplessis S, Chalot M, Podila G, Martin F (2003) Analysis of expressed sequence tags from the ectomycorrhizal basidiomycetes *Laccaria bicolor* and *Pisolithus microcarpus*. New Phytol 159(1):117–129. doi:10.1046/j.0028-646x.2003.00796.x

Pudelko K (2010) Mushroom Virus V (MVX): a novel disease of mushrooms in Poland? J Plant Prot Res 50(3):366–371. doi:10.2478/v10045-010-0062-9

Qiu L, Li Y, Liu Y, Gao Y, Qi Y, Shen J (2010) Particle and naked RNA mycoviruses in industrially cultivated mushroom *Pleurotus ostreatus* in China. Fungal Biol 114:507–513. doi:10.1016/j.funbio.2010.04.001

Rao JR, Nelson DWA, McClean S (2007) The enigma of double-stranded RNA (dsRNA) associated with Mushroom Virus X (MVX). Curr Issues Mol Biol 9(2):103–122

Revill PA, Davidson AD, Wright PJ (1994) The nucleotide sequence and genome organization of mushroom bacilliform virus: a single-stranded RNA virus of *Agaricus bisporus* (Lange) Imbach. Virology 202(2):904–911. doi:10.1006/viro.1994.1412

Riccioni C, Rubini A, Belfiori B, Passeri V, Paolocci F, Arcioni S (2008) Tmt1: the first LTR-retrotransposon from a *Tuber* spp. Curr Genet 53(1):23–34. doi:10.1007/s00294-007-0155-9

Ro HS, Lee NJ, Lee CW, Lee HS (2006) Isolation of a novel mycovirus OMIV in *Pleurotus ostreatus* and its detection using a triple antibody sandwich-ELISA. J Virol Methods 138 (1–2):24–29. doi:10.1016/j.jviromet.2006.07.016

Ro HS, Kang EJ, Yu JS, Lee TS, Lee CW, Lee HS (2007) Isolation and characterization of a novel mycovirus, PeSV, in *Pleurotus eryngii* and the development of a diagnostic system for it. Biotechnol Lett 29(1):29–135. doi:10.1007/s10529-006-9206-4

Romaine CP, Goodin M (2002) Unravelling the viral complex associated with La France disease of the cultivated mushroom, *Agaricus bisporus*. In: Tavantzis SM (ed) dsRNA genetic elements: concepts and applications in agriculture, forestry, and medicine. CRC Press, Boca Raton, FL, pp 237–257

Romaine CP, Schlagnhaufer B (1989) Prevalence of dsRNAs in healthy and La France disease-affected basidiocarps of *Agaricus bisporus*. Mycologia 81(5):822–825. doi:10.2307/3759892

Romaine CP, Ulhrich P, Schlagnhaufer B (1993) Transmission of La France isometric virus during basidiosporogenesis in *Agaricus bisporus*. Mycologia 85(2):175–179. doi:10.2307/3760453

Rytter JL, Royse DJ, Romaine CP (1991) Incidence and diversity of double stranded RNA in *Lentinula edodes*. Mycologia 83(4):506–510. doi:10.2307/3760453

Schisler LC, Sinden JW, Sigel EM (1967) Etiology, symptomatology and epidemiology of a virus disease of cultivated mushrooms. Phytopathology 57(5):519–526

Shapira R, Choi G, Nuss DL (1991) Virus-like genetic organization and expression strategy for a double-stranded RNA genetic element associated with biological control of chestnut blight. EMBO J 10(4):731–739

Sinden JW, Hauser E (1950) Report on two new mushroom diseases. Mushroom Sci 1:96–100

Sonnenberg AS, Lavrijssen B (2004) Browning of mushroom and the presence of viral double-stranded RNA in Dutch mushrooms. In: Romaine CP, Keil CB, Rinker DL, Royse DJ (eds) Science and cultivation of edible and medicinal fungi. Mushroom science, vol 16. International Society for Mushroom Science, Penn State, USA, pp 541–546

Sonnenberg AS, Van Griensven LJLD (1991) Evidence for transmission of La France disease in *Agaricus bisporus* by dsRNA. In: van Griensven LJLD (ed) Genetics and breeding of agaricus. Pudoc, Wageningen, pp 109–113

Stielow B, Menzel W (2010) Complete nucleotide sequence of TaV1, a novel totivirus isolated from a black truffle ascocarp (*Tuber aestivum* Vittad.). Arch Virol 155(12):2075–2078. doi:10.1007/s00705-010-0824-8

Stielow JB, Klenk HP, Menzel W (2011a) Complete genome sequence of the first endornavirus from the ascocarp of the ectomycorrhizal fungus *Tuber aestivum* Vittad. Arch Virol 156 (2):343–345. doi:10.1007/s00705-010-0875-x

Stielow B, Klenk HP, Winter S, nzel W (2011b) A novel *Tuber aestivum* (Vittad.) mitovirus. Arch Virol 156(6):1107–1110. doi:10.1007/s00705-011-0998-8

Stielow B, Bratek Z, Klenk HP, Winter S, Menzel W (2012) A novel mitovirus from the hypogeous ectomycorrhizal fungus *Tuber excavatum*. Arch Virol 157(4):787–790. doi:10.1007/s00705-012-1228-8

Suzaki K, Ikeda KI, Sasaki A, Kanematsu S, Matsumoto N, Yoshida K (2005) Horizontal transmission and host-virulence attenuation of totivirus in violet root rot fungus *Helicobasidium mompa*. J Gen Plant Pathol 71(3):161–68. doi:10.1007/s10327-005-0181-8

Van Alfen NK, Jaynes RA, Anagnostakis SL, Day PR (1975) Chestnut blight: biological control by transmissible hypovirulence in *Endothia parasitica*. Science 189(4206):890–891. doi:10.1126/science.189.4206.890

Van der Lende TR, Harmsent MC, Wessels JGH (1994) Double stranded RNAs and proteins associated with the 34 nm virus particles of the cultivated mushroom *Agaricus bisporus*. J Gen Virol 75:2533–2536. doi:10.1099/0022-1317-75-9-2533

Van der Lende TR, Harmsen MC, Go SJ, Wessel JGH (1995) Double stranded RNA mycoviruses in mycelium of *Pleurotus ostreatus*. FEMS Microbiol Lett 125(1):51–56. doi:10.1111/j.1574-6968.1995.tb07334.x

Van Zaayen A (1979) Mushroom viruses. In: Lemke PA (ed) Viruses and plasmids in fungi. Marcel Dekker, New York, pp 239–324

Won HK, Park SJ, Kim DK, Shin MJ, Kim N, Lee SH, Kwon YC, Ko HK, Ro HS, Lee HS (2013) Isolation and characterization of a mycovirus in *Lentinula edodes*. J Microbiol 51(1):118–122. doi:10.1007/s12275-013-2351-2

Yu HJ, Lim D, Lee HS (2003) Characterization of a novel single-stranded RNA mycovirus in *Pleurotus ostreatus*. Virology 314(1):9–15. doi:10.1016/S0042-6822(03)00382-9

Yu X, Lia B, Fu Y, Jianga D, Ghabrial SA, Li G, Peng Y, Xie J, Cheng J, Huang J, Yi X (2010) A geminivirus-related DNA mycovirus that confers hypovirulence to a plant pathogenic fungus. Proc Natl Acad Sci USA 107(18):8387–8392. doi:10.1073/pnas.0913535107

Part IV
Spore Dispersal

Chapter 21
Truffles and Small Mammals

Alexander Urban

21.1 Introduction

Truffles completely depend on mycophagist animals for spore dispersal because they have lost the ability to forcefully eject their spores. Among mycophagist animals, small mammals are the most important (Trappe et al. 2009).

Small mammals feeding on true truffles (*Tuber* spp.) are as diverse as the habitats where truffles grow—from rangelands to thermophilic oak forests and from conifer mountain forest to alpine tundra (references in Table 21.1). They are different in terms of evolutionary history, habitat use, foraging behaviour, dietary preferences and digestive physiology. Most mammal species considered as obligate or preferential mycophagists are small; their body mass is typically about or (much) less than 1 kg (Johnson 1996; Claridge and Trappe 2005; Maser et al. 2008).

The identification of fungi consumed by mycophagists is facilitated by characteristic microscopic features of the spores of hypogeous fungi, which can be found abundantly in gut contents and fecal pellets of small mammal mycophagists. However, in microscopy-based studies, taxonomic resolution is typically limited at the level of genera or species groups (Table 21.1). By isolation of DNA from spores contained in fecal pellets (Schickmann et al. 2011) and by application of metabarcoding technologies, these limitations will be overcome, as soon as reliable reference sequences of hypogeous fungal species from diverse genera and regions will be available.

In this chapter the diversity of small mammals feeding on true truffles and their potential adaptations and specific capacities linked to mycophagy are discussed.

A. Urban (✉)
Department of Botany and Biodiversity Research, Faculty of Life Sciences, University of Vienna, Rennweg 14, 1030 Wien, Austria
e-mail: alexander.urban@univie.ac.at

A. Zambonelli et al. (eds.), *True Truffle (*Tuber *spp.) in the World*, Soil Biology 47,
DOI 10.1007/978-3-319-31436-5_21

Table 21.1 Small mammal species reported to feed on true truffles

Taxon	Common name	Distribution	Category	References
Insectivora, Soricidae, Soricinae				
Sorex araneus	Common shrew	W-Eurasia	Opportunistic	Schickmann et al. (2012)
Sorex minutus	Eurasian pigmy shrew	W-Eurasia	Opportunistic	Kataržytė and Kutorga (2011), Schickmann et al. (2012)
Lagomorpha, Leporidae, Leporinae				
Lepus californicus	Black-tailed jackrabbit	N-Am	Opportunistic?*	Maser et al. (1988)
Lepus townsendii	White-tailed jackrabbit	N-Am	Opportunistic?*	Maser et al. (1988)
Sylvilagus nuttallii	Mountain cottontail	N-Am	Opportunistic	Maser et al. (1988)
Lagomorpha, Ochotonidae, Ochotoninae				
Ochotona princeps	Pika or rock rabbit	N-Am	Opportunistic	Cázares and Trappe (1994)
Rodentia, Cricetidae, Arvicolinae				
Microtus agrestis	Field vole	Eurasia	Opportunistic	Schickmann et al. (2012)
Microtus californicus	California vole	N-Am	Preferential/ obligate	Frank et al. (2008)
Myodes californicus californicus	Western red-backed vole	N-Am	Obligate	Maser and Maser (1988)
Myodes californicus mazama	Mazama red-backed vole	N-Am	Preferential	Maser and Maser (1988)
Myodes californicus/ gapperi	Red-backed voles	N-Am	Preferential	Jacobs and Luoma (2008)
Myodes gapperi	Southern red-backed vole	N-Am	Preferential	Maser and Maser (1988), Pastor et al. (1996), Terwilliger and Pastor (1999)
Myodes glareolus	Bank vole	W-Eurasia	Preferential	Dróżdż (1966), Kataržytė and Kutorga (2011), Schickmann et al. (2012)

(continued)

Table 21.1 (continued)

Taxon	Common name	Distribution	Category	References
Rodentia, Cricetidae, Neotominae				
Neotoma fuscipes	Dusky-footed woodrat	N-Am	Opportunistic	Linsdale and Tevis (1951), Fogel and Trappe (1978)
Peromyscus leucopus	White-footed mouse	N-Am	Opportunistic	Maser and Maser (1987), Frank et al. (2008), Sidlar (2012)
Peromyscus maniculatus	Deer mouse	N-Am	Opportunistic	Maser and Maser (1987)
Peromyscus truei	Pinyon mouse	N-Am	Opportunistic	Maser and Maser (1987)
Reithrodontomys megalotis	Western harvest mouse	N-Am	Opportunistic?*	Frank et al. (2008)
Rodentia, Geomyidae, Geomyinae				
Thomomys bottae	Botta's pocket gopher	N-Am	Opportunistic?*	Taylor et al. (2009)
Rodentia, Gliridae, Glirinae				
Glis glis	Fat dormouse	Europe, W-Asia	Opportunistic	Schickmann et al. (2012)
Rodentia, Muridae, Murinae				
Apodemus flavicollis	Yellow-necked mouse	Europe, W-Asia	Opportunistic	Schickmann et al. (2012)
Rodentia, Sciuridae, Sciurinae				
Glaucomys sabrinus	Northern flying squirrel	N-Am	Preferential	Carey (1995), Rosentreter et al. (1997), Currah et al. (2000), Lehmkuhl et al. (2004), Gomez et al. (2005), Jacobs and Luoma (2008)
Sciurus aberti	Tassel-eared squirrel	N-Am	Preferential	Kotter and Farentinos (1984), Dodd et al. (2003)
Sciurus griseus	Western gray squirrel	N-Am	Preferential	Maser et al. (1988)

(continued)

Table 21.1 (continued)

Taxon	Common name	Distribution	Category	References
Sciurus vulgaris	Red squirrel	Europe	Preferential?*	unpubl. observation, Brive, SW France
Tamiasciurus hudsonicus	Red squirrel	N-Am	Preferential	Currah et al. (2000), Vernes et al. (2004), Sidlar (2012)
Rodentia, Sciuridae, Xerinae				
Callospermophilus lateralis	Golden-mantled ground squirrel	N-Am	Preferential	Maser et al. (1988)
Marmota caligata	Hoary marmot	N-Am	Opportunistic	Cázares and Trappe (1994)
Neotamias amoenus	Yellow-pine chipmunk	N-Am	Preferential	Sidlar (2012)
Neotamias siskiyou	Siskiyou chipmunk	N-Am	Preferential	Kotter and Farentinos (1984)
Neotamias townsendii	Townsend chipmunk	N-Am	Preferential	Maser et al. (1988), Jacobs and Luoma (2008)

Category of mycophagy (occasional, preferential, obligate) according to Claridge and Trappe (2005), Trappe et al. (2009) and Schickmann et al. (2012) or proposed here (*)
N-Am = North America

The nutritional qualities of true truffles are also summarized and confronted with digestive capacities and food preferences of small mammals.

21.2 Diversity of Small Mammals Feeding on True Truffles

Small mammal mycophagy of true truffles is present in various families of rodents; few reports are available for insectivores and lagomorphs. These orders are present across continents in the northern hemisphere. Small mammal species and genera differ considerably between the Palearctic and Nearctic regions (Wallace 1894).

The rodents comprise the majority of species reported to feed on truffles, including many preferential mycophagists. In the partly arboricolous voles of the genus *Myodes* (Cricetidae-Arvicolinae), mycophagy is of particular importance, in both North America (N-Am) and Europe. Three species (*M. californicus*, *M. gapperi*, *M. glareolus*) are considered as preferential mycophagists, *M. californicus* subsp. *californicus* even as obligately mycophagous (Maser and Maser 1988; Trappe et al. 2009). There is important variation in the degree of mycophagy in *Myodes* spp. across habitats, latitudes and altitudes, at the population level and at the subspecies level, in Europe and in N-Am (Hansson 1985; Maser and

Maser 1988), depending on the locally available choice of feed. *Myodes californicus* subsp. *californicus*, which inhabits conifer forests in the Coastal Ranges of Northern California and Western Oregon, was found more strictly mycophagous than subsp. *mazama*, which lives at higher elevations in the Cascade Range, feeding mostly on lichens during winter. *Myodes gapperi* subsp. *cascadensis* sampled in conifer forests in the Pacific Northwest was found to have ingested more and more diverse hypogeous fungi than other subspecies living in mixed and deciduous forests in the Central and Eastern USA (Maser and Maser 1988). *Myodes* spp. prefer woodlands with abundant ground cover to escape predation. The ground-dwelling and largely herbivorous voles of the genus *Microtus* were found to feed on truffles, too, in both N-Am and Europe; some species were classified as preferential or opportunistic mycophagists. At least in Central Europe they appear to be less mycophagous than *Myodes* spp. (Schickmann et al. 2012).

The Cricetidae-Neotominae diversified in the Nearctic, where some forest-dwelling species occupy ecological niches which are inhabited by wood mice (Muridae) in the old world. Species of the genera *Neotoma* and *Peromyscus* (Neotominae) were found to feed on true truffles, as was *Apodemus sylvaticus* (Muridae) in the Palaearctic (Schickmann et al. 2012).

Pocket gophers (*Geomyidae*) are burrowing occasional mycophagists living in North and Central America (Maser et al. 1988; Taylor et al. 2009). These medium small mammals (body mass ~ 120–250 g) are herbivorous and moderately long lived.

Murids, the most species-rich family of rodents and mammals, diversified in Eurasia, Africa and Australia. The yellow-necked mouse (*Apodemus flavicollis*) was found to feed on two different truffle species in Austria, Europe (from Puberulum group and Rufum group; Schickmann et al. 2012). The diversity of the fungal diet of *A. flavicollis* was only second to the bank vole, *M. glareolus*; however, the degree of mycophagy was significantly lower in the former (Schickmann et al. 2012). *Apodemus flavicollis* are excellent tree climbers; they use their habitat versatilely and exploit various highly nutritious food sources rather opportunistically. The species can be found in different types of forest and also in more open habitat. It was found abundant in early successional regeneration on a windthrow site (Schickmann et al. 2012), indicating that this species may be important for spore dispersal to disturbed forest sites, thereby potentially promoting forest regeneration by increasing available ectomycorrhizal (ECM) fungal diversity.

Most arvicoline rodents are herbivorous, while murine rodents require a richer diet including seeds and invertebrate prey. Anatomy, habitat use and feeding ecology of bank voles are intermediate between other arvicolids and the murine *Apodemus* spp. (Dróżdż 1966; Butet and Delettre 2011).

The arboricolous and relatively long-lived Glirids are restricted to the old world; mycophagy including feeding on *Tuber* sp. was demonstrated recently for the fat dormouse (*Glis glis*; Schickmann et al. 2012). Sciurids are particularly diverse in N-Am, and many are considered preferential mycophagists. Consumption of true truffles was reported in the Sciurinae, tree squirrels of the genus *Sciurus*, red pine squirrels (*Tamiasciurus hudsonicus*) and the northern flying squirrel (*Glaucomys sabrinus*) and in the Xerinae, chipmunks (*Neotamias* spp.), mantled ground squirrel (*Callospermophilus lateralis*) and hoary marmot (*Marmota caligata*) (Table 21.1). In Europe, reports on mycophagy in sciurids are limited to *Sciurus vulgaris*, which has been known for long to collect, cache and consume fungi (Kumerloeve 1968; Grönwall and Pehrson 1984; Lurz and South 1998; Bertolino et al. 2004). Consumption of *Tuber melanosporum* Vittad. by *S. vulgaris* was observed by a truffle grower in SW France (pers comm, Brive, France, 2013).

The fat dormouse and some sciurids are deep hibernators, e.g. marmots and certain ground squirrels. Most notorious mycophagists among the Sciurinae, such as *G. sabrinus* or *S. vulgaris*, do not hibernate. Food caching and hoarding, including truffles (Vernes and Poirier 2007) is a common survival strategy.

Shrews (*Sorex* spp.) are very small mammals feeding predominantly on invertebrate prey, which were only recently reported to consume *Tuber* spp. (Kataržytė and Kutorga 2011; Schickmann et al. 2012). Previously, soricids had been known to feed on *Endogone* spp. and Glomeromycota (Whitaker and Maser 1976; Maser et al. 1978). Ingestion of truffle spores by shrews was suspected to be an accidental side effect of preying on mycophagous invertebrates (Fogel and Trappe 1978; Schickmann et al. 2012). However, the quantity of truffle spores in fecal pellets was comparable in *Apodemus* spp. and *Sorex* spp. (Schickmann et al. 2012). High spore numbers of *Tuber* sp. and of some Russulaceae rather indicate direct consumption of fungal fruiting bodies, albeit possibly infested with insect larvae. Therefore, *Sorex* spp. were considered as selective occasional mycophagists (Schickmann et al. 2012). Inclusion of *Tuber* spp. in the diet of *Sorex* spp. is consistent with relatively high nutritional quality of true truffles (see below).

Maser et al. (1988) and Cázares and Trappe (1994) report the consumption of true truffles by Leporids and Ochotonids, respectively (Lagomorpha). Mycophagous small mammals in other orders (e.g. opossums, Didelphimorphia) might consume true truffles, too, but this has not been documented yet.

21.3 Potential Adaptations to Mycophagy and Life History Traits of Small Mammals

Mycophagy requires specific capacities, such as detection of truffles, recognition of edible versus poisonous fungi and effective digestion of fungal biomass. The diversity of mycophagous small mammals (Table 21.1) and the frequency of opportunistic and preferential mycophagy suggest that the adaptive value of mycophagy is commonly sufficient to compensate for the costs associated with adaptations to this diet. Small mammals might be preadapted for mycophagy, e.g. by a well-developed sense of smell, and specific capacities improving the benefits of mycophagy be acquired gradually. Obligate mycophagy, as reported for *M. californicus* subsp. *californicus*, a species with relatively fragile teeth not suitable for more abrasive food items of plant origin, appears to be very rare (Claridge and Trappe 2005). Preferential or occasional mycophagists avoid the risks of strict specialization and respond to the availability of diverse resources in a more versatile, flexible manner.

Life history traits of preferentially mycophagous small mammals are diverse. *G. sabrinus* has characteristic traits of K-selected species (sensu MacArthur and Wilson 1967): relative longevity, late reproduction, low annual fecundity, a long period of maternal investment and density-dependent population growth (Smith 2007). Similar life cycle characteristics are found among other tree squirrels (Sciurinae) and in the arboricolous *G. glis*. Longevity and associated traits of K-selection were attributed to reduced mortality by predation, as facilitated by arboreality (Shattuck and Williams 2010), and, more specifically, by gliding (Holmes and Austad 1994).

Most very small mycophagous mammals, such as red-backed voles (*Myodes* spp.), wood mice (*Apodemus* spp.) and forest-dwelling shrews (*Sorex* spp.), share characteristics of r-selected species: short lifespans of 1–2 seasons and prolific reproduction resulting in highly fluctuating populations, typically peaking in autumn (Gliwicz and Taylor 2002). Fast life cycles are considered a response to high extrinsic mortality rates, and furthermore they facilitate rapid adaptation of population sizes to food abundance (Bielby et al. 2007).

Body size, energetics and life history traits are generally correlated (Lovegrove 2000). Kleiber's law describes the relationship of body mass and metabolic energy demand, scaling with the three-fourth power of size. The metabolic cost of endothermy is increasingly high in temperate, boreal and subarctic regions, since mass-specific metabolic rates increase with decreasing body size and with increasing latitude. Small mammals in the Palearctic and Nearctic, which comprise most of the known distribution area of the genus *Tuber*, have particularly high size-corrected basal metabolic rates, compared to other zoogeographical zones (Lovegrove 2000). Allometric theory predicts that the digestive capacity of small mammals is constrained by body size and energy demands and that they need to select diets of high digestibility and nutrient quality (Johnson 1996), consistent with the observation that the lower limit of miniaturization of herbivorous mammals (e.g. *Microtus*) is at about 15 g (Gliwicz and Taylor 2002).

Extremely small mycophagous mammals such as *Sorex* spp. may be most informative concerning size-related physiological and ecological constraints acting in evolution. Shrews are highly specialized predators of invertebrates; body sizes are at the lower limit of mammal body size distribution (e.g. *S. minutus*, body mass ~ 4 g), and lifespans are short, typically less than 18 months. Mass-specific metabolic rates are extremely high; therefore, shrews require a high and continuous supply of highly nutritious food. Shrews consume up to 200 % of their body weight per day and can't afford extensive alimentary tracts or extended periods of gut passage. Without access to food, the animals starve rapidly (Hanski 1984). Shrews do not accumulate body fat reserves; nearly all the adipose tissue is brown adipose tissue, which serves for converting metabolic energy into thermal energy by non-shivering thermogenesis (Nieminen and Hyvärinen 2000). Hibernation is absent in shrews; instead, they continue foraging throughout winter, when energy costs of maintaining homeothermy are highest. A reduction of energy requirements is achieved by shrinking body size during winter (Hanski 1984). Short-time food caches assist in securing a continuous food supply.

Due to a solitary lifestyle, strict territoriality and short lifespan, *Sorex* spp. are unlikely to adapt their diet choice by learning; foraging behaviour most probably relies on innate and algorithmic cues. Poorly differentiated digestive systems resulting in limited capacity of digestion suggest that extremely small mammals are constrained to select diets of high digestibility and nutrient quality. These constraints are consistent with selective mycophagy on a subset of truffle species observed in *Sorex* spp. (Schickmann et al. 2012). The finding that the predominantly insectivorous shrews use truffles as a food source challenges the view that these are of low nutritional quality and raises the question whether shrews select truffle species of highest nutritional quality.

Preferential mycophagists are found along the entire r-K continuum in rodents, from *G. sabrinus*, an outstanding example of K-selection, to the mostly r-selected mammals of very small body size. Apparently, the mutualistic relationship between hypogeous fungi and small mammals is flexible and robust enough to be largely independent of the life cycle traits of main mycophagists. The evolution of life cycle characteristics appears to be lineage specific (phylogenetically conserved) and driven by top-down control (predation) rather than by bottom-up factors (food choice and supply; Prevedello et al. 2013).

Daily torpor and hibernation are widespread strategies to save energy in times of food shortage by temporarily abandoning euthermia. Hibernation results in a seasonal interruption of foraging, in contrast to daily torpor. Hibernation is rare or absent among very small mammals, such as *Myodes* spp., *Apodemus* spp. and *Sorex* spp., and among tree squirrels (Sciurinae), but common in ground squirrels (Xerinae), including preferential mycophagists such as *C. lateralis* (Ruf and Geiser 2015). *Glis glis*, an exemplary hibernator, was recently confirmed as mycophagous (Schickmann et al. 2012).

21.4 Truffles as a Diet

The nutritional quality of hypogeous fungi was discussed controversially (Claridge and Trappe 2005). Various lines of evidence and theoretical considerations support the dietary relevance of mycophagy in situ: (1) Fungi account for a major part of the food of certain small mammal species (e.g. 72 % of the yearly dietary volume in *Eutamias townsendii*; Fogel and Trappe 1978). (2) Positive relationships between the degree of mycophagy and diverse fitness parameters were observed, e.g. acquisition of hibernation fat in *Eutamias* chipmunks (Tevis 1952), body condition and fecundity of Tasmanian bettong (*Bettongia gaimardii*; Johnson 1994) and juvenile recruitment in *Sciurus alberti* (Dodd et al. 2003). (3) Truffle abundance and frequency was found to be the best predictor of population density of the highly mycophagous northern flying squirrel in Douglas-fir forests of various habitat characteristics (Gomez et al. 2005). (4) Isotopic evidence suggests that most organic C and nearly all N is derived from hypogeous fungi in the northern bettong (*Bettongia tropica*) (McIlwee and Johnson 1998). (5) The nutrient composition of fungi is complementary to a plant-based diet and can compensate for potential deficiencies of the latter concerning amino acid composition, vitamin D and trace minerals such as selenium (Claridge and Trappe 2005). (6) Fungi are present in habitats such as dense forests where other food source of high nutritional quality is scarce or highly seasonal. (7) The net nutritional value of truffles depends on foraging costs, which may be relatively low in terms of energy, time and predation risk when feeding on truffles (Fogel and Trappe 1978). (8) Seasonal and annual variation in fungal fruiting body production is relatively independent from variation of other major food items, e.g. mast; mycophagy can compensate for poor mast. Survival in times of hardiness and during population bottlenecks is of high selective value.

The risks and drawbacks of mycophagy need to be considered, too: many species of macrofungi are toxic, particularly when consumed raw; therefore, recognition of edible species is critical. Seasonal and annual variations in fungal fruiting are high, and nutrient availability is limited.

Experimental evidence from feeding experiments with captive small mammals showed negative (Cork and Kenagy 1989a) or marginally positive (Claridge and Cork 1994; Bozinovic and Muñoz-Pedreros 1995; Dubay et al. 2008) balances for N and energy, depending on the selection and diversity of species tested. Mixed diets of plant and fungal material were modelled to provide a more balanced nutrition than monospecific diets (Bozinovic and Muñoz-Pedreros 1995): truffle diversity is likely to be essential for the mycophagists' nutrition, consistent with the diversity of fungal species observed in many studies of mycophagy (e.g. Carey et al. 2002). Infestation of fungal fruiting bodies with insect larvae may provide a supplementary nutrient source (Bozinovic and Muñoz-Pedreros 1995). Even if the quantitative contribution of mycophagy to nutrition is limited (occasional mycophagy), fungi might serve as a dietary supplement which provides rare nutrients (see below).

21.5 Nutritional Value and Absence of Toxicity in True Truffles

Available data indicate a wide range of nutritional value within the diversity of epigeous and hypogeous fungi (Wallis et al. 2012) and within the genus *Tuber*. Orczán et al. (2012) found considerable differences in the mineral composition of diverse ascomycete and basidiomycete genera of hypogeous fungi. Notably, culinary species in the genera *Tuber* and *Mattirolomyces* had higher contents of P, K and Ca compared to other hypogeous fungi. Dry weight soluble protein content in ascomata of four culinary truffle species ranged between $8.7 \pm 0.83\,\%$ for *T. melanosporum* and $25.0 \pm 0.86\,\%$ for *Tuber magnatum* Pico. Values for *Tuber aestivum* Vittad. and *Tuber borchii* Vittad. were intermediate. Proteins were of high digestibility (Saltarelli et al. 2008), containing all essential amino acids including some (methionine, cystine, tryptophan and lysine) which are commonly limiting in many food sources of plant origin (Coli et al. 1990). Antioxidant activity of phenolic compounds contained in culinary truffles may offer additional fitness benefits (Villares et al. 2012).

True truffles are not toxic to humans, even when consumed raw. Edibility of fresh, untreated fungi is rather the exception than the rule. Among the epigeous relatives of truffles, *Helvella* and *Gyromitra* are known to cause various symptoms of intoxication when ingested raw or insufficiently processed. Gyromitrin was identified as causative agent of intoxications with *Gyromitra esculenta* (Pers.) Fr. (List and Luft 1968) and later found in many other ascomycetes (Andary et al. 1985), including *Helvella* spp. Gyromitrin is unstable and releases the toxic volatile monomethylhydrazine. It seems that gyromitrin is effective in preventing the consumption of these fungi by mammals, since traces of mammal mycophagy are usually absent even when *Gyromitra* or *Helvella* fruit abundantly (Alexander Urban, unpublished observation). Given the phylogenetic relationships of Discinaceae, Helvellaceae and Tuberaceae (e.g. Læssøe and Hansen 2007), it is possible that the epigeous ancestors of Tuberaceae contained gyromitrin and that the absence of gyromitrin in true truffles is an adaptation to mycophagy.

21.6 Digestion

Relatively simple gastrointestinal systems and short processing times suffice to digest high-quality forage. With decreasing digestibility and increasing fibre content, optimal digestion time and the importance of fermentation by gastrointestinal microbes increase. To satisfy their high mass-specific energy requirements, small mammals can increase throughput at the expense of optimal digestion efficiency. When analyzing fecal pellets of small rodents, starch granules staining with Lugol's iodine solution can be observed rather frequently, indicating that digestion is not perfect.

The preferential mycophagists *M. gapperi* and *G. sabrinus* reach about 70 % mean dry matter digestibility when feeding on diverse hypogeous fungi (Dubay et al. 2008). Nitrogen digestibility was 12.3 % and 24.9 % for squirrels and voles, respectively, confirming that most nitrogen was indigestible (Cork and Kenagy 1989a). Claridge et al. (1999) found that *M. californicus* and *G. sabrinus* were able to digest approximately 67 % and 66 %, respectively, of the dry matter from fruiting bodies of *Rhizopogon vinicolor* A.H. Sm. Apparent N digestibility was 35 % and 11 % for voles and squirrels, respectively. The greater assimilation of fungal N by voles than by the much larger squirrels indicates that the digestive systems of voles are very effective relative to their body size (Claridge et al. 1999; Dubay et al. 2008). The golden-mantled ground squirrel (*C. lateralis*) lost weight when fed with *Elaphomyces granulatus* Fr.; dry matter digestibility was approximately 60 % (Cork and Kenagy 1989a).

In addition to allometric constraints, important variation in the digestive capacities and nutritional requirements of small mammals of comparable size was found, linked to evolutionary history and specific adaptations of dietary physiology and anatomy. Differences in food choice and digestive capacity between arvicolid and murine rodents may serve as an example. The arvicolid bank voles, recently confirmed as preferential mycophagists (Schickmann et al. 2012), have a better capacity to digest low-energy food than the murid *Apodemus* spp. (Butet and Delettre 2011). Voles were found more effective in extracting N from fungal sources than squirrels, which may be explained by specific adaptations of the caecum. Maintenance nitrogen requirements in flying squirrels were lower than predicted by allometric equation. Northern flying squirrels did not prefer diets high in N, in contrast to red-backed voles. Saving nutrients might be an alternative strategy which allows to subsist during winter on poorer alternative food sources such as lichens (Dubay et al. 2008). Among mycophagous marsupials, a differentiated foregut is a common adaptation to improve the digestion of fungal forage (Claridge and Cork 1994).

Concentrate selection is a successful strategy of combining high throughput with high digestion efficiency. In concentrate selecting hindgut fermenters, increasing use of fibrous diets correlates with increasing relative size and complexity of the caecum and/or proximal colon which facilitate the selective retention of solutes and small particles (Cork 1994). *Microtine* rodents which have complex hindguts and separate digesta in the caecum are able to use food of low digestibility (down to 30 %). Intakes of dry matter by sciurids and murids, which apparently lack separation mechanisms, are consistently lower than those of microtines at digestibilities below ~ 70 %. Caecotrophy is typically combined with coprophagy (Sakaguchi 2003).

The gut systems of the insectivorous shrews differ importantly from the rodents: the post-gastric tract is relatively short (2–6 × head and body length), the caecum is reduced or absent, differentiation between small and large intestines is small and retention times are short, to cope with the extremely high mass-specific energy demands of shrews. These adaptations imply that high-quality food is needed for efficient digestion. Selectivity of mycophagy in shrews (Schickmann et al. 2012) may be a response to their specific requirements of higher-quality food. It is noteworthy that *Tuber* sp. was highly preferred by *Sorex minutus* (Schickmann et al. 2012), one of the smallest and probably most demanding species.

21.7 Food Choice: Selection of True Truffles?

There is variation in olfactory signals, nutrient content, seasonal availability, abundance and size among and within genera of hypogeous fungi. The question arises whether harvest and consumption of hypogeous fungi are neutral or selective (Johnson 1994). Theory would suggest that mycophagists preferentially forage for easily detectable, abundant and nutrient-rich fruiting bodies. Over evolutionary timescales, selection by mycophagist may have resulted in increased attractivity. From an anthropocentric perspective, culinary truffles are most attractive. This appreciation appears to be supported by data on the nutrient content of truffles (Orczán et al. 2012). Do small mammal mycophagists prefer the genus *Tuber* and, more specifically, the culinary species?

Glaucomys sabrinus was frequently reported to consume a diverse array of hypogeous fungi, but with a consistent preference for certain basidiomycete genera rich in P and K (*Gautieria*, *Rhizopogon*, *Gastroboletus*) (Zabel and Waters 1997; Lehmkuhl et al. 2004; Meyer et al. 2005; Dubay et al. 2008). True truffles appeared to be negatively selected by *G. sabrinus* (Gomez et al. 2005). Red-backed voles (*M. gapperi*) preferentially ingested *Hydnotrya variiformis* Gilkey, which is rich in N and lipids (Dubay et al. 2008). Schickmann et al. (2012) found that a species of Puberulum group was highly preferred by *S. minutus*.

21.8 Foraging Behaviour in Small Mammals and Dispersal of Truffle Spores

True truffles were detected in many studies of small mammal mycophagy (Table 21.1), but mostly at relatively low frequency (0.1–5 %), compared to the most abundant hypogeous basidiomycetes (e.g. *Rhizopogon* spp. in some coniferous forests). Seasonal, annual and habitat-specific variations of *Tuber* spore frequency are high and likely reflect variation in truffle availability. In addition, abundance of alternative diets (e. g. mast) influences foraging behaviour and food choice in the highly versatile small mammals. Low frequency implies that detection probability is highly dependent on sampling density, leaving an important role to chance and stochastic variation. In many studies, detection of *Tuber* spp. was site or season specific, in contrast to other, more frequent basidiomycete genera. Studies in ecosystems where true truffles are dominant, including plantations, are needed to obtain more data on mycophagy and true truffles.

Foraging behaviour of many small mammal species is highly flexible and adaptive. Home range size, microhabitat use and foraging behaviour influence the retrieval of truffles and the dispersal of truffle spores. Small mammals most likely rely on olfactory (Talou et al. 1990; Donaldson and Stoddart 1994) and tactile sensory perception when foraging for truffles. The visual sense is well developed in more arboreal species and may be reduced in strictly ground-dwelling species.

Evolution of a high degree of mycophagy can evolve in species with different habitat use; a ground-dwelling lifestyle of small mammals does not necessarily imply a higher degree of mycophagy, compared to more arboreal species. Many avid small mammal mycophagists are tree climbers (e.g. tree squirrels and flying squirrels, common dormouse, wood mice and, to some extent, bank voles) and most are nocturnal. Home range size, habitat use and complexity of foraging behaviour vary with body size, dietary specialization and phylogenetic affinity of small mammals. Small rodents are important prey of various predators, including diverse carnivores and birds of prey. Predation of mycophagists may result in accidental spore dispersal over longer distances, since the home ranges of predators are usually much larger than the home ranges of their prey.

Small mammals appear to be very efficient in locating hypogeous fungi of different species, and at least some species of hypogeous fungi are consumed quantitatively (Johnson 1994). Typically, small mammals consume more different species of hypogeous fungi than a specialized mycologist can find in a given area (Johnson 1996; Schickmann et al. 2012); therefore, the study of mycophagy holds promise in providing new data on local diversity and biogeography of hypogeous fungi. The diversity of fungal spores contained in fecal pellets may have multiple reasons: some small mammal species most likely eat all species they find and recognize as edible, to minimize foraging effort; in that case the diversity of fungi ingested would be representative of the diversity of fungal species encountered. Some species, such as *G. sabrinus*, select the most preferred species (Zabel and Waters 1997; Lehmkuhl et al. 2004; Meyer et al. 2005; Dubay et al. 2008). The diversity of spores encountered in fecal pellets may be higher than the diversity of fungi consumed at a given time, due to long retention times and mixing of fungal spores in the caecum. Cork and Kenagy (1989b) calculated mean retention times of *E. granulatus* spores in the gut systems of golden-mantled ground squirrels (*Spermophilus saturatus*; 31.8 ± 2.4 h) and deer mice (*Peromyscus maniculatus*; 12.0 ± 2.4 h). Ninety-five percent egestion required 52 ± 2 h in *S. saturatus*. Spore retention in the digestive tract of mycophagous rodents affects the temporal and spatial dynamics of spore dispersal (Danks 2012) and may result in an overestimation of the short-time diversity of truffle consumption. However, analyses of stomach contents, which can be expected to be more representative of most recent feeding activity than spore counts in fecal pellets, typically reveal diverse feeding habits, too (e.g. Maser and Maser 1988).

Many small mammals, including preferential mycophagists such as *Myodes* spp. or *G. sabrinus*, cache food, either as larder hoarders or as scatter hoarders, and fungi are no exception (Vernes and Poirier 2007). Shrews maintain short-time food caches, to satisfy their need of continuous food supply. Squirrels dry fungi on tree branches; sun exposure massively increases vitamin D content in dried fungi (Maser et al. 2008). Pilferage of caches and re-caching by conspecifics or by different animal species (e.g. birds) is not uncommon and may further extend and entangle the mycophagy network (Maser et al. 2008).

Recognition of edible species is vital when foraging on potentially toxic food, such as fungi. The short lifespan of very small mammals implies that individuals will continue to encounter new truffle species during one season or two seasons at most. Most likely, recognition of edible fungal species is based on innate and possibly generic olfactory clues and individual experience. The lifespan of sciurids can extend over several years; therefore, learning, experimentation and imitation of conspecifics might be more relevant.

21.9 Seasonality in Mycophagists and Truffles

Small mammal populations are limited by food supply and predation (Prevedello et al. 2013). Cyclic overpopulation of small mammals and poor seed crops likely enhance the importance of mycophagy in reducing periodical food stress (Fogel and Trappe 1978). Seasonal build-up of rodent populations may coincide with fungal fruiting peaks in late summer and autumn. Individual species of truffles may be available for restricted periods only, but collectively, the production of hypogeous fruiting bodies throughout the year provides a more reliable food resource for small mammals (Maser and Maser 1988).

The periods of the year which are most critical for survival differ according to overwintering strategies. In hibernating species such as *G. glis* and certain sciurids, food availability in autumn and spring is most important. Small non-hibernating species such as shrews (*Sorex* spp.) and red-backed voles (*Myodes* spp.) depend on a continuous food supply during winter. Hoarders like many sciurids and murids can use the available resources more flexibly.

The diversity of hibernation strategies implies that different species of small mammals are available as vectors of truffles at different seasons. In periods of food shortage, fungal food may be most important for survival and most readily eaten and vectored. Seasonal variation in the availability and activity of small mammal vectors likely is a factor of selection acting on truffle phenology.

21.10 Mycophagy, Truffles and Plant Community Succession

Mycophagy may contribute to plant establishment and vegetation succession, whenever supply of mycorrhizal fungal inoculum is limiting, thereby influencing forest ecology, diversity and productivity (Johnson 1996; Maser et al. 2008; Schickmann et al. 2012). The provision of mycorrhizal inoculum can be critical for plant establishment during succession after major disturbances (Allen et al. 1992). Terwilliger and Pastor (1999) demonstrated that spore dispersal of hypogeous fungi by red-backed vole (*M. gapperi*) promotes the invasion of conifers into meadows formed in abandoned, drained beaver ponds. Cázares and Trappe (1994) showed that various mammal species are vectors of ECM hypogeous fungi (incl. *Tuber*) in successional vegetation close to a receding glacier forefront. Schickmann et al. (2012) found specific communities of small mammals and truffles including true truffles in early successional forest regeneration after stand-replacing windthrow. *Tuber whetstonense* J.L. Frank, D. Southw. and Trappe was frequently found in fecal pellets of small rodents trapped in *Quercus garryana* Dougl. savanna at up to 35 m distance from mature oak trees. *Tuber candidum* Harkn. was present as mycorrhiza in mature trees and experimental seedlings and occasionally in fecal pellets, too (Frank et al. 2009). *Tuber californicum* Harkn. and *Tuber* sp. were part of the ECM community developed on the roots of bioassay seedlings grown in soil collected immediately after stand-replacing wildfire, indicating that propagules of true truffle species survived the fire in the soil spore bank (Taylor and Bruns 1999). However, it was not assessed whether these propagules were animal-dispersed spores.

Various truffle species are found in different types and successional states of vegetation; certain species grow best in earlier stages of forest development and are less competitive in mature undisturbed forests with closed canopy (e.g. Schickmann et al. 2012). If early successional and certain post-disturbance states of woodland development are optimal for the growth of certain species of true truffles (e.g. Schickmann et al. 2012), efficient colonization of these optimal habitats is critical for the species' reproduction. Different truffle species may be prevalent in the ECM state and in the spore population dispersed by mycophagists, at least within a given time frame (Frank et al. 2009).

Typical preferential mycophagists are adapted to living in closed forest with abundant shelter. Small mammals have relatively small home ranges, which may limit their dispersal potential. Therefore, their capacity to colonize and distribute spores to massively disturbed habitats may be limited. More opportunistic species with flexible habitat use (Maser et al. 2008; Schickmann et al. 2012) and larger mammals (Piattoni et al. 2012) are probably more effective in colonizing such types of habitat.

21.11 Small Mammal Mycophagy, Truffle Cultivation and the Truffle Life Cycle

As major consumers of plant, fungal and small animal biomass, small mammals have multiple direct and indirect effects on true truffles, their host trees and habitats, in natural truffle grounds as in plantations. The burrowing activity of many small mammal species disturbs the soil and is likely to be beneficial for pioneer species, including certain species of true truffles. Soil disturbance reduces herbal vegetation cover, potentially assisting in the formation of brûlés, and can improve water retention, by improved infiltration and reduced capillarity (Maser et al. 2008).

Voles like the common vole (*Microtus arvalis*) can cause damage to host trees in young truffle plantations, particularly when population sizes are high (Urban et al. 2012). Voles, like other small mammals, are important prey of many different predators, such as various species of birds of prey, owls, red fox (*Vulpes vulpes*), wildcat (*Felis silvestris*) and mustelids such as weasel (*Mustela nivalis*), European polecat (*Mustela putorius*) and European badger (*Meles meles*). Providing habitat for the predators is a natural way of keeping rodent populations at an acceptable level.

Insectivorous (*Sorex* spp.) and omnivorous (*Apodemus* spp., *M. glareolus*) small mammals may improve the fitness of the truffles' host trees by feeding on invertebrates, thereby reducing the pressure of herbivorous and seed-predating insects, e.g. the curculionid *Otiorhynchus* spp.

In natural truffle populations, the importance of small mammal vectoring of truffle spores for the formation of new mycelia is obvious. Gut passage might promote germination of truffle spores, as it was observed in spores having passed through the gut of *Sus scrofa* (Piattoni et al. 2014). In managed plantations, however, the seedlings are already mycorrhized with truffles before planting. When truffles start to grow, small mammals can consume a part of the harvest, redistributing the spores by endozoochory. Is the continuous supply of truffle spores necessary to maintain the productivity of a truffle plantation? This question is linked to the life cycle and population ecology of true truffles. Murat et al. (2013) found relatively small genets of *T. melanosporum* mycelia (less than 5 m distance between ramets of the same genet) in productive truffle plantations. Genet turnover between two consecutive years was found considerable, suggesting that the population structure in truffle plantations is highly dynamic. More surprisingly, diverse genets associated with one host tree were of the same mating type, reducing the probability that genets of opposing mating type get into contact (Linde and Selmes 2012; Murat et al. 2013). This pattern raised the question of the pathways of mating in true truffles, which were only recently reconfirmed as sexual (Paolocci et al. 2006; Murat and Martin 2008). Small, short-lived mycelia originating from meiospores are a possible solution. If this mechanism is true, ascospore dispersal by animal vectors or, alternatively, by orchard management practices is essential for truffle orchard productivity and might be compared to pollination in fruit orchards. Alternatively or additionally, microconidia might act as spermatia. However, this type of spores has been observed in few species of Puberulum group only, thus far (Urban et al. 2004; Healy et al. 2013).

A potential economic drawback of mycophagy in truffle orchards is the dispersal of non-marketable or low-value competing fungi, since mycophagists use to feed on a variety of species of hypogeous fungi (Urban et al. 2012). It can be predicted that following intentional and unintentional (Murat et al. 2008) introductions of true truffles into new habitats, particularly in the southern hemisphere, many new mycophagous mammal species (both marsupials and placentaria) will start to feed on and disperse true truffles, and native ECM host species are likely to get mycorrhized.

21.12 Conclusions

True truffles, like most other hypogeous fungi, are involved in two different mutualistic networks, mycorrhiza and mycophagy. These relationships are a key component of forest food webs, which structure forest biodiversity and sustain forest productivity and resilience. They existed throughout the evolution of the genus *Tuber* and can be considered the ultimate causes of characteristic traits of truffles, such as truffle odours and the absence of toxicity.

The synthesis of available literature on the feeding of small mammals on true truffles supports a series of conclusions and hypotheses: (1) Mycophagy is widespread among small mammals, involving a phylogenetically diverse array of species, with diverse nutritional habits, life cycle traits and foraging behaviours. Fidelity to a fungal diet is high among a subset of species and populations. In the natural distribution range of true truffles, red-backed voles and certain species of the squirrel family (Sciuridae) are recognized as preferential mycophagists. (2) Small mammal mycophagy likely accounts for a large proportion of animal-vectored spore dispersal, at least at the local level. (3) The proportion of spores of the genus *Tuber* is low in most forest habitats studied thus far, dominated by basidiomycetes. (4) The nutritional value of true truffles is relatively high, compared to other food items of plant and fungal origin. (5) Nutrient assimilation from hypogeous fungi is variable among small mammal mycophagists and appears to be phylogenetically conserved. (6) Allometric constraints on acceptable food quality potentially limit mycophagy in extremely small mammal species such as *S. minutus* and may be at the origin of more selective mycophagy, which could be a driving force in the evolution of nutritional quality of truffles. Information on food choice among different species of hypogeous fungi is still limited. (7) Small mammal mycophagy is essential for short-distance dispersal and, possibly, mating of true truffles. (8) Progress in DNA metabarcoding of fungal communities in environmental samples offers new opportunities for assessing the diversity of fungi consumed by mycophagists with unprecedented taxonomic resolution.

References

Allen MF, Crisafulli C, Friese CF, Jeakins SL (1992) Re-formation of mycorrhizal symbioses on Mount St Helens, 1980–1990: interactions of rodents and mycorrhizal fungi. Mycol Res 96 (6):447–453. doi:10.1016/S0953-7562(09)81089-7

Andary C, Privat G, Bourrier MJ (1985) Variations of monomethylhydrazine content in *Gyromitra esculenta*. Mycologia 77(2):259–264. doi:10.2307/3793077

Bertolino S, Vizzini A, Wauters LA, Tosi G (2004) Consumption of hypogeous and epigeous fungi by the red squirrel (*Sciurus vulgaris*) in subalpine conifer forests. For Ecol Manage 202 (1):227–233. doi:10.1016/j.foreco.2004.07.024

Bielby J, Mace GM, Bininda-Emonds ORP, Cardillo M, Gittleman JL, Jones KE, Orme CDL, Purvis L (2007) The fast-slow continuum in mammalian life history: an empirical reevaluation. Am Nat 169(6):748–757. doi:10.1086/516847

Bozinovic F, Muñoz-Pedreros A (1995) Nutritional ecology and digestive responses of an omnivorous-insectivorous rodent (*Abrothrix longipilis*) feeding on fungus. Physiol Zool 68 (3):474–489

Butet A, Delettre YR (2011) Diet differentiation between European arvicoline and murine rodents. Acta Theriol 56(4):297–304. doi:10.1007/s13364-011-0049-6

Carey AB (1995) Sciurids in Pacific Northwest managed and old-growth forests. Ecol Appl 5 (3):648–661

Carey AB, Colgan W, Trappe JM, Molina R (2002) Effects of forest management on truffle abundance and squirrel diets. Northwest Sci, WSU Press. http://hdl.handle.net/2376/949

Cázares E, Trappe JM (1994) Spore dispersal of ectomycorrhizal fungi on a glacier forefront by mammal mycophagy. Mycologia 86(4):507–510. doi:10.2307/3760743

Claridge A, Cork S (1994) Nutritional-value of hypogeal fungal sporocarps for the long-nosed Potoroo (Potorous-Tridactylus), a forest-dwelling mycophagous marsupial. Aust J Zool 42 (6):701–710. doi:10.1071/ZO9940701

Claridge AW, Trappe JM (2005) Sporocarp mycophagy: nutritional, behavioural, evolutionary and physiological aspects. In: Dighton J, White JF, Oudemans P (eds) The fungal community: its organization and role in the ecosystem, 3rd edn, Mycology series vol 23. Taylor & Francis, Boca Raton, pp 599–611

Claridge AW, Trappe JM, Cork SJ, Claridge DL (1999) Mycophagy by small mammals in the coniferous forests of North America: nutritional value of sporocarps of *Rhizopogon vinicolor*, a common hypogeous fungus. J Comp Physiol B 169(3):172–178. doi:10.1007/s003600050208

Coli R, Maurizi Coli A, Granetti B, Damiani P (1990) Composizione chimica e valore nutritivo del tartufo nero (*T. melanosporum* Vitt.) e del tartufo bianco (*T. magnatum* Pico) raccolti in Umbria. In: Bencivenga M, Granetti B (eds) Atti del 2° congresso internazionale sul tartufo. Comunitá Montana dei Monti Martani e del Serano, Spoleto, 24–27 Nov 1988, pp 511–516

Cork SJ (1994) Digestive constraints on dietary scope in small and moderately-small mammals: how much do we really understand? In: The digestive system in mammals—Food form and function. Cambridge University Press, Cambridge, pp 337–369

Cork SJ, Kenagy GJ (1989a) Nutritional value of hypogeous fungus for a forest-dwelling ground squirrel. Ecology 70(3):577–586. doi:10.2307/1940209

Cork SJ, Kenagy GJ (1989b) Rates of gut passage and retention of hypogeous fungal spores in two forest-dwelling rodents. J Mammal 70(3):512–519. doi:10.2307/1381423

Currah RS, Smreciu EA, Lehesvirta T, Niemi M, Larsen KW (2000) Fungi in the winter diets of northern flying squirrels and red squirrels in the boreal mixedwood forest of northeastern Alberta. Can J Bot 78(12):1514–1520. doi:10.1139/b00-123

Danks MA (2012) Gut-retention time in mycophagous mammals: a review and a study of truffle-like fungal spore retention in the swamp wallaby. Fungal Ecol 5(2):200–210. doi:10.1016/j.funeco.2011.08.005

Dodd NL, States JS, Rosenstock SS (2003) Tassel-eared squirrel population, habitat condition, and dietary relationships in North-Central Arizona. J Wildl Manage 67(3):622–633. doi:10.2307/3802719

Donaldson R, Stoddart M (1994) Detection of hypogeous fungi by Tasmanian bettong (*Bettongia gaimardi*: Marsupialia; Macropodoidea). J Chem Ecol 20(5):1201–1207. doi:10.1007/BF02059754

Dróżdż D (1966) Food habits and food supply of rodents in the beech forest. Acta Theriol 11:363–384. doi:10.4098/AT.arch.66-15

Dubay SA, Hayward GD, Martínez del Rio C (2008) Nutritional value and diet preference of arboreal lichens and hypogeous fungi for small mammals in the Rocky Mountains. Can J Zool 86(8):851–862. doi:10.1139/Z08-054

Fogel R, Trappe JM (1978) Fungus consumption (mycophagy) by small animals. Northwest Sci 52 (1):1–31

Frank J, Barry S, Madden J, Southworth D (2008) Oaks belowground: mycorrhizas, truffles, and small mammals. In: Proceedings of the sixth Californian oak symposium: today's challenges,

tomorrow's opportunities. USDA Forest Service Pacific Southwest Research Station, General Technical Report GTR-PSW-217, Albany, pp 131–138
Frank JL, Anglin S, Carrington EM, Taylor DS, Viratos B, Southworth D (2009) Rodent dispersal of fungal spores promotes seedling establishment away from mycorrhizal networks on *Quercus garryana*. Botany 87(9):821–829. doi:10.1139/B09-044
Gliwicz J, Taylor JRE (2002) Comparing life histories of shrews and rodents. Acta Theriol 47 (1):185–208. doi:10.1007/BF03192487
Gomez DM, Anthony RG, Hayes JP (2005) Influence of thinning of Douglas-fir forests on population parameters and diet of northern flying squirrels. J Wild Manage 69 (4):1670–1682. doi:10.2193/0022-541X(2005)69[1670:IOTODF]2.0.CO;2
Grönwall O, Pehrson Å (1984) Nutrient content in fungi as a primary food of the red squirrel *Sciurus vulgaris* L. Oecologia 64(2):230–231. doi:10.1007/BF00376875
Hanski I (1984) Food consumption, assimilation and metabolic rate in six species of shrew (Sorex and Neomys). Ann Zool Fenn 21(2):157–165
Hansson L (1985) Clethrionomys food: generic, specific and regional characteristics. Ann Zool Fenn 22(3):315–318
Healy RA, Smith ME, Bonito GM, Pfister DH, Ge ZW, Guevara GG, Williams G, Stafford K, Kumar L, Lee T, Hobart C, Trappe J, Vilgalys R, McLaughlin DJ (2013) High diversity and widespread occurrence of mitotic spore mats in ectomycorrhizal Pezizales. Mol Ecol 22 (6):1717–1732. doi:10.1111/mec.12135
Holmes DJ, Austad SN (1994) Fly now, die later: life-history correlates of gliding and flying in mammals. J Mammal 75(1):206–224. doi:10.2307/1382255
Jacobs KM, Luoma DL (2008) Small mammal mycophagy response to variations in green-tree retention. J Wildl Manage 72(8):1747–1755. doi:10.2193/2007-341
Johnson CN (1994) Nutritional ecology of a mycophagous marsupial in relation to production of hypogeous fungi. Ecology 75(7):2015–2021. doi:10.2307/1941606
Johnson CN (1996) Interactions between mammals and ectomycorrhizal fungi. Trends Ecol Evol 11(12):503–507. doi:10.1016/S0169-5347(96)10053-7
Kataržytė M, Kutorga E (2011) Small mammal mycophagy in hemiboreal forest communities of Lithuania. Cent Eur J Biol 6(3):446–456. doi:10.2478/s11535-011-0006-z
Kotter MM, Farentinos RC (1984) Tassel-eared squirrels as spore dispersal agents of hypogeous mycorrhizal fungi. J Mammal 65(4):684–687. doi:10.2307/1380853
Kumerloeve H (1968) Über die Pilznahrung des Eichhörnchens. Veröff Naturwiss Vereins Osnabrück 32:161–164
Læssøe T, Hansen K (2007) Truffle trouble: what happened to the Tuberales? Mycol Res 111 (9):1075–1099. doi:10.1016/j.mycres.2007.08.004
Lehmkuhl JF, Gould LE, Cázares E, Hosford DR (2004) Truffle abundance and mycophagy by northern flying squirrels in eastern Washington forests. For Ecol Manage 200(1):49–65. doi:10.1016/j.foreco.2004.06.006
Linde CC, Selmes H (2012) Genetic diversity and mating type distribution of *Tuber melanosporum* and their significance to truffle cultivation in artificially planted truffiéres in Australia. Appl Environ Microbiol 78(18):6534–6539. doi:10.1128/AEM.01558-12
Linsdale JM, Tevis LP (1951) The dusky-footed wood rat. University of California Press, Berkeley
List PH, Luft P (1968) Gyromitrin, das gift der frühjahrslorchel. 16. Mitt. über Pilzinhaltsstoffe. Arch Pharm 301(4):294–305. doi:10.1002/ardp.19683010410
Lovegrove BG (2000) The zoogeography of mammalian basal metabolic rate. Am Nat 156 (2):201–219. doi:10.1086/303383
Lurz PWW, South AB (1998) Cached fungi in non-native conifer forests and their importance for red squirrels (*Sciurus vulgaris* L.). J Zool 246(4):468–471. doi:10.1111/j.1469-7998.1998.tb00184.x
MacArthur R, Wilson EO (1967) The theory of island biogeography. Princeton University Press, Princeton

Maser C, Maser Z (1987) Notes on mycophagy in four species of mice in the genus Peromyscus. Great Basin Nat 47(2):308–313
Maser C, Maser Z (1988) Mycophagy of red-backed voles, Clethrionomys californicus and C. gapperi. Great Basin Nat 48(2):269–273
Maser C, Trappe JM, Nussbaum RA (1978) Fungal-small mammal interrelationships with emphasis on Oregon coniferous forests. Ecology 59:799–809. doi:10.2307/1938784
Maser C, Maser Z, Molina R (1988) Small-mammal mycophagy in rangelands of central and southeastern Oregon. J Range Manage 41(4):309–312
Maser C, Claridge A, Trappe J (2008) Trees, truffles, and beasts: how forests function. Rutgers University Press, Piscataway
McIlwee AP, Johnson CN (1998) The contribution of fungus to the diets of three mycophagous marsupials in eucalyptus forests, revealed by stable isotope analysis. Funct Ecol 12 (2):223–231. doi:10.1046/j.1365-2435.1998.00181.x
Meyer MD, North MP, Kelt DA (2005) Fungi in the diets of northern flying squirrels and lodgepole chipmunks in the Sierra Nevada. Can J Zool 83(12):1581–1589. doi:10.1139/z05-156
Murat C, Martin F (2008) Sex and truffles: first evidence of Perigord black truffle outcrosses. New Phytol 180(2):260–263. doi:10.1111/j.1469-8137.2008.02634.x
Murat C, Zampieri E, Vizzini A, Bonfante P (2008) Is the Perigord black truffle threatened by an invasive species? We dreaded it and it has happened! New Phytol 178(4):699–702. doi:10.1111/j.1469-8137.2008.02449.x
Murat C, Rubini A, Riccioni C, De la Varga H, Akroume E, Belfiori B, Guaragno M, Le Tacon F, Robin C, Halkett F, Martin F, Paolocci F (2013) Fine-scale spatial genetic structure of the black truffle (*Tuber melanosporum*) investigated with neutral microsatellites and functional mating type genes. New Phytol 199(1):176–187. doi:10.1111/nph.12264
Nieminen P, Hyvärinen H (2000) Seasonality of leptin levels in the BAT of the common shrew (*Sorex araneus*). Z Naturforsch C 55(5–6):455–460. doi:10.1515/znc-2000-5-623
Orczán ÁK, Vetter J, Merényi Z, Bonifert E, Bratek Z (2012) Mineral composition of hypogeous fungi in Hungary. J Appl Bot Food Qual 85(1):100–104
Paolocci F, Rubini A, Riccioni C, Arcioni S (2006) Reevaluation of the life cycle of *Tuber magnatum*. Appl Environ Microbiol 72(4):2390–2393. doi:10.1128/AEM.72.4.2390-2393.2006
Pastor J, Dewey B, Christian DP (1996) Carbon and nutrient mineralization and fungal spore composition of fecal pellets from voles in Minnesota. Ecography 19(1):52–61. doi:10.1111/j.1600-0587.1996.tb00154.x
Piattoni F, Ori F, Morara M, Iotti M, Zambonelli A (2012) The role of wild boars in spore dispersal of hypogeous fungi. Acta Mycol 47(2):145–153. doi:10.5586/am.2012.017
Piattoni F, Amicucci A, Iotti M, Ori F, Stocchi V, Zambonelli A (2014) Viability and morphology of *Tuber aestivum* spores after passage through the gut of *Sus scrofa*. Fungal Ecol 9:52–60. doi:10.1016/j.funeco.2014.03.002
Prevedello JA, Dickman CR, Vieira MV, Vieira EM (2013) Population responses of small mammals to food supply and predators: a global meta-analysis. J Anim Ecol 82(5):927–936. doi:10.1111/1365-2656.12072
Rosentreter R, Hayward GD, WicklowHoward M (1997) Northern flying squirrel seasonal food habits in the interior conifer forests of Central Idaho, USA. Northwest Sci 71(2):97–102
Ruf T, Geiser F (2015) Daily torpor and hibernation in birds and mammals. Biol Rev 90 (3):891–926. doi:10.1111/brv.12137
Sakaguchi E (2003) Digestive strategies of small hindgut fermenters. Anim Sci J 74(5):327–337. doi:10.1046/j.1344-3941.2003.00124.x
Saltarelli R, Ceccaroli P, Cesari P, Barbieri E, Stocchi V (2008) Effect of storage on biochemical and microbiological parameters of edible truffle species. Food Chem 109(1):8–16. doi:10.1016/j.foodchem.2007.11.075

Schickmann S, Kräutler K, Kohl G, Nopp-Mayr U, Krisai-Greilhuber I, Hackländer K, Urban A (2011) Comparison of extraction methods applicable to fungal spores in faecal samples from small mammals. Sydowia 63(2):237–247

Schickmann S, Urban A, Kräutler K, Nopp-Mayr U, Hackländer K (2012) The interrelationship of mycophagous small mammals and ectomycorrhizal fungi in primeval, disturbed and managed Central European mountainous forests. Oecologia 170(2):395–409. doi:10.1007/s00442-012-2303-2

Shattuck MR, Williams SA (2010) Arboreality has allowed for the evolution of increased longevity in mammals. Proc Natl Acad Sci USA 107(10):4635–4639. doi:10.1073/pnas.0911439107

Sidlar K (2012) The role of sciurids and murids in the dispersal of truffle-forming ectomycorrhizal fungi in the Interior Cedar-Hemlock biogeoclimatic zone of British Columbia. PhD thesis. University of British Columbia

Smith WP (2007) Ecology of *Glaucomys sabrinus*: habitat, demography, and community relations. J Mammal 88(4):862–881. doi:10.1644/06-MAMM-S-371R1.1

Talou T, Gaset A, Delmas M, Kulifaj M, Montant C (1990) Dimethyl sulphide: the secret for black truffle hunting by animals? Mycol Res 94(2):277–278. doi:10.1016/S0953-7562(09)80630-8

Taylor DL, Bruns TD (1999) Community structure of ectomycorrhizal fungi in a *Pinus muricata* forest: minimal overlap between the mature forest and resistant propagule communities. Mol Ecol 8(11):1837–1850. doi:10.1046/j.1365-294x.1999.00773.x

Taylor DS, Frank J, Southworth D (2009) Mycophagy in Botta's Pocket Gopher (*Thomomys bottae*) in Southern Oregon. Northwest Sci 83(4):367–370. doi:10.3955/046.083.0408

Terwilliger J, Pastor J (1999) Small mammals, ectomycorrhizae, and conifer succession in Beaver Meadows. Oikos 85(1):83–94. doi:10.2307/3546794

Tevis L Jr (1952) Autumn foods of chipmunks and golden-mantled ground squirrels in the northern Sierra Nevada. J Mammal 33(2):198–205. doi:10.2307/1375929

Trappe JM, Molina R, Luoma DL, Cázares E, Pilz, D, Smith JE, Castellano MA, Miller SL, Trappe MJ (2009) Diversity, ecology, and conservation of truffle fungi in forests of the Pacific Northwest. General Technical Report PNW-GTR-772. USDA, Pacific Northwest Research Station, Oregon

Urban A, Neuner-Plattner I, Krisai-Greilhuber I, Haselwandter K (2004) Molecular studies on terricolous microfungi reveal novel anamorphs of two *Tuber* species. Mycol Res 108 (7):749–758. doi:10.1017/S0953756204000553

Urban A, Kataržytė M, Schickman S, Kräutler K, Pla T (2012) Is small mammal mycophagy relevant for truffle cultivation? Acta Mycol 47(2):139–143. doi:10.5586/am.2012.016

Vernes K, Poirier N (2007) Use of a Robin's nest as a cache site for truffles by a red squirrel. Northeast Nat 14(1):145–149. doi:10.1656/1092-6194(2007)14[145:UOARNA]2.0.CO;2

Vernes K, Blois S, Bärlocher F (2004) Seasonal and yearly changes in consumption of hypogeous fungi by northern flying squirrels and red squirrels in old-growth forest, New Brunswick. Can J Zool 82:110–117. doi:10.1139/z03-224

Villares A, García-Lafuente A, Guillamón E, Ramos Á (2012) Identification and quantification of ergosterol and phenolic compounds occurring in *Tuber* spp. truffles. J Food Compos Anal 26 (1):177–182. doi:10.1016/j.jfca.2011.12.003

Wallace AR (1894) The palæarctic and nearctic regions compared as regards the families and genera of their mammalia and birds. Nat Sci 4:435–445

Wallis IR, Claridge AW, Trappe JM (2012) Nitrogen content, amino acid composition and digestibility of fungi from a nutritional perspective in animal mycophagy. Fungal Biol 116 (5):590–602. doi:10.1016/j.funbio.2012.02.007

Whitaker JO, Maser C (1976) Food habits of five western Oregon shrews. Northwest Sci 50 (2):102–107

Zabel CJ, Waters JR (1997) Food preferences of captive northern flying squirrels from the Lassen National Forest in northeastern California. Northwest Sci 71(2):103–107

Chapter 22
Interrelationships Between Wild Boars (*Sus scrofa*) and Truffles

Federica Piattoni, Francesca Ori, Antonella Amicucci, Elena Salerni, and Alessandra Zambonelli

22.1 Introduction

The wild boar (*Sus scrofa*) has its origins on islands in Southeast Asia such as Indonesia and the Philippines (Chen et al. 2007). Fossil records show that it then migrated to Asia and finally Europe in the early Pleistocene (Groves et al. 1997). It is well adapted to a broad range of environments from semidesert to wetlands, mountains, and forests which has allowed it to colonize the far west of Europe to the far east of Asia as well as Oceania, the Americas, and Africa. It is also found on many islands such as Japan and New Zealand and small remote islands like the Channel Islands of California and Mexico and the subantarctic islands south of New Zealand, having been introduced as a source of food (Challies 1975; Spitz 1999; Schley and Roper 2003; Krajick 2005; New Zealand Department of Conservation 2015). Consequently, excepting man, the pig has the widest geographical distribution of all large mammals (d'Huart 1991). The European population of wild boars has increased considerably since the 1960s (Sáez Royuela and Tellería 1986), with a consequential increased damage to agroecosystems (Herrero et al. 2006).

F. Piattoni (✉) • A. Zambonelli
Department of Agricultural Sciences, University of Bologna, Viale Fanin 46, 40127 Bologna, Italy
e-mail: federica.piattoni@unibo.it

F. Ori
Department of Life, Health and Environmental Sciences, University of L'Aquila, Via Vetoio (Coppito 1), 67100 L'Aquila, Italy

A. Amicucci
Department of Biomolecular Sciences, University of Urbino "Carlo Bo", Via Saffi 2, 61029 Urbino, PU, Italy

E. Salerni
Department of Life Sciences, University of Siena, Via P.A. Mattioli 4, 53100 Siena, Italy

A. Zambonelli et al. (eds.), *True Truffle (*Tuber *spp.) in the World*, Soil Biology 47,
DOI 10.1007/978-3-319-31436-5_22

The distribution of wild boars in Europe and Asia coincides with the natural distribution of the most important commercial *Tuber* species including *Tuber magnatum* Pico, *Tuber melanosporum* Vittad., and *Tuber aestivum* Vittad. and the most important east Asian species such as the *Tuber indicum* Cooke & Masse and *Tuber sinoaestivum* J.P. Zhang & P.G. Liu (Hall et al. 2007; Chen et al. 2011; Zhang et al. 2012; Fekete et al. 2014; see Chap. 2). Wild boars are also common in those parts of Oceania and the Americas where the European truffles, *T. melanosporum*, *T. aestivum*, and *Tuber borchii* Vittad., are now cultivated (Hall and Haslam 2012; Reyna and Garcia Barreda 2014).

In this chapter, the interrelationships between truffles and wild boars are discussed and include the importance of truffles in their diet, their role in spore dispersal, and their impact on cultivated truffières.

22.2 The Diet of the Wild Boar

Wild boars are opportunistic omnivores and eat a wide variety of foods including plant matter, animals, and fungi. Ballari and Barrios-Garcia (2013) reported that about 90 % of the diet is plant material and includes bulbs, roots, aerial parts of plants, fruits, and seeds. When one food is not available, they will switch to another so that in warmer months, aerial parts of plants are eaten, while in winter, bulbs, roots, and above all tubers are more important (Genov 1981; Herrero et al. 2005). If cereals are available, they will browse these in preference to other foods (Hahn and Eisfeld 1998; Schley and Roper 2003). Animal matter represents a relatively small part of the diet but can be as high as 16 % and may include mammals, birds, insects, reptiles, amphibians, invertebrates, and crustaceans (Schley and Roper 2003; Ballari and Barrios-Garcia 2013). Fungi represent a small part of the diet but may be as high as 7 % (Ballari and Barrios-Garcia 2013). Depending on where a wild boar is living, algae and garbage may also be important parts of the diet although inorganic items such as plastic and stones found in stomach contents (Ballari and Barrios-Garcia 2013) are unlikely to have much nutritional value! Significant differences in diet were found in native and in introduced ranges, particularly where natural sources of food are unavailable (Ballari and Barrios-Garcia 2013).

22.3 Damage Caused by Wild Boars

Conflicts between humans and wild animals range from damage to forests (Reimoser and Gossow 1996), transmission of infectious diseases to livestock and humans (Meng et al. 2009), and especially damage to agricultural crops (Schley and Roper 2003). Because of their wide ranging diet, profligate reproduction (Barrios-García and Ballari 2012), and typical rooting behavior, wild boars cause extensive soil disturbance and extensive damage both to agriculture and forestry. Rooting primarily reduces plant cover and diversity, affects the first 15–70 cm of the litter layer, and affects up to 80 % of the forest soil surface. Several studies have been carried out to estimate how rooting physically, chemically, and biologically affects forest soil. These studies showed alterations in soil carbon content and other nutrient element concentration in addition to decomposition and mineralization rate changes (Wirthner et al. 2011).

Damage to agricultural crops by wild boars in Europe was considered an important issue as early as the 1940s (Klemm 1948), and since then, the problem has worsened as wild boar populations have increased. The most important source of crop loss, which can be as high as 90–95 %, is crop destruction through trampling, even though real consumption constitutes only 5–10 % of crop loss (Schley and Roper 2003). A study conducted by Herrero et al. (2006) in the northeastern part of Iberia showed that the highest damage occurs in maize, wheat, and alfalfa fields where these foods formed more than 75 % of the diet. In that study, maize was clearly the favorite food. In another study, Schley and Roper (2003) also concluded that maize was the wild boar's preferred food and could be used to entice the animals so that they could be hunted. There are clear seasonal patterns in wild boar rooting activity on agricultural land. In England, it was worst during the first 3 months of the year, in Poland during summer, and in Germany during the ripening period from June to October (Wilson 2004). In the USA, annual economic losses have been estimated at $ 800 million per year (Pimentel et al. 2005).

Wild boars also transmit viruses, bacteria, and parasites to livestock (when they eat contaminated feed or by direct contact with wild boars) and may cause relevant management costs for eradication programs, in addition to economic losses for livestock mortality (Gortázar et al. 2007; Ruiz-Fons et al. 2008; Barrios-Garcia 2012).

During the rooting activity, wild boars feed on several varieties of arthropods, especially earthworms. The earthworms represent a significant part of wild boar diet: as revealed by a study run in the French Alps, their frequency in the diet is around 92 % (Baubet et al. 2003). Moreover, wild boar may affect earthworms' population not only by direct feeding but also by indirect alteration of their habitat. As previously reported, wild boar behavior modifies the properties of soils (Lacki and Lancia 1983), and so it subsequently affects the earthworms' communities, degrading their habitat (Bueno and Jiménez 2014). One more ecological damage is constituted by eating avian material, mainly from ground-nesting birds like woodcocks (Schley and Roper 2003).

Pigs have a great sense of smell and can be trained for truffle research (Sourzat 1989). Anyway, in nature wild boars occasionally eat truffles (see Sect. 22.4.2). However, wild boars may cause great economic losses to cultivated truffières, not only in terms of truffle predation but also of soil disturbance caused by excavation (Moreno-Arroyo et al. 2005; Ricci 2008; Salerni et al. 2013) and destruction of truffle-inoculated seedlings (Samils et al. 2008). However, in some situations, wild boar rooting activity seems to stimulate the fructification process improving soil moisture retention (Ławrynowicz et al. 2006).

An example of the damage to truffle production caused by wild boars is the study carried out in Tuscany by Salerni et al. 2013, between 2006 and 2008. This study aimed at quantifying the damage caused by wild boars to *T. aestivum* production in a natural *Quercus cerris* L. truffière. The area was studied for 3 years using the BACI (Before-After-Control-Impact) sampling design; after the identification of ten plots (1000 m^2 each), one-half of them was fenced in the 2007 spring, in order to prevent wild boar access (Fig. 22.1). This study showed that before the introduction of fences, the area was commonly frequented by wild boars, as the estimated soil damage in almost all plots was, on average, close to 50 % (as percentage of surface area turned over). A significant increase in production was observed in all plots over the 3-year study period, in comparison to the situation at the outset. The number and the yield of truffles found in the fenced plots were significantly higher than in the non-fenced plots (Fig. 22.2), confirming the big damages on truffle production caused by wild boars.

22.4 Mycophagy by Wild Boars

22.4.1 Methods of Study

Stomach contents or feces can be used to study the diet of wild boars and in particular their ingestion of epigeous and hypogeous fungi (Schley and Roper 2003). The major advantage of analyzing stomach contents is that the undigested parts of fungal fruiting bodies, lichens, arthropods, invertebrates, mollusks, small mammals, and indigestible vascular plants can be identified especially when ingestion is relatively recent (Schley and Roper 2003).

Hohmann and Huckschlag (2005) assessed stomach contents by sealing off the stomach immediately after slaughter with cable toes adjacent to the cardiac and pyloric sphincters. The stomach was then cut longitudinally and the contents, including that adhering to the lining of the stomach, emptied into a container for weighing. Stomachs that could not be assessed immediately were frozen until needed. After weighing, the contents are spread into a container and broadly divided into green matter—plants, mosses, or leaves; brownish-grainy matter containing fungi, roots, and earthworms; cereals from artificial bait sites; and miscellaneous items including arthropods, vertebrates, invertebrates, acorns, and nuts. Using this method, the relative proportion of each category provides a good indication of the diet prior to slaughter. For a more precise quantitative measure, a 50 g sample representative of the stomach contents is washed with tap water and filtered through a 2 mm sieve with or without a Buchner pump (Hohmann and Huckschlag 2005). Material retained on the sieve is then observed under a binocular microscope where pieces of truffle can be seen. Further examination under high power microscope can allow identification to species level.

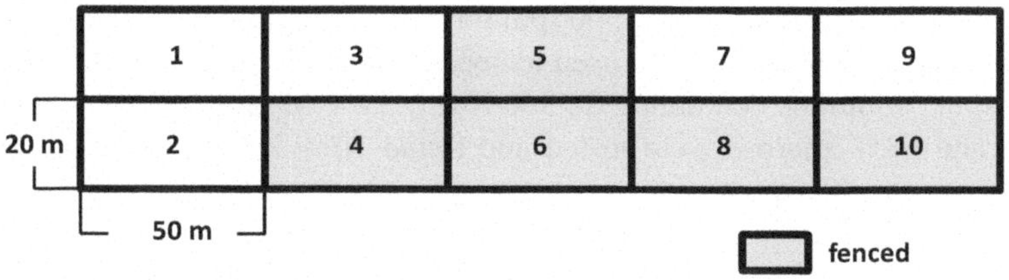

Fig. 22.1 Experimental design for the natural *Q. cerris* truffière

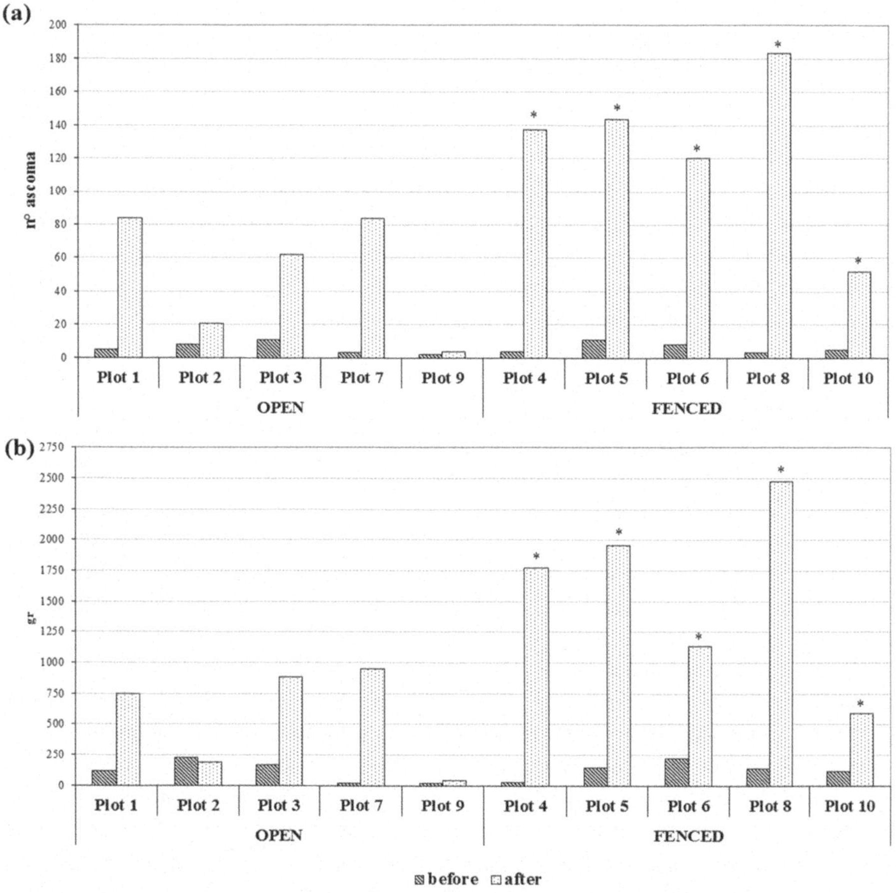

Fig. 22.2 Number of truffles (**a**) and total weight (**b**) counted in the ten plots before and after the fencing-off. Asterisk indicates significant differences at $p < 0.05$ by Tukey's test

An alternative to observing stomach contents is to dilute the contents of the rectum with distilled water, blend, allow to settle, and then the solid matter passed through a graded series of sieves with mesh diameter 800, 400, 150, 60, and 20 μm (Piattoni et al. 2012). The material collected between 150 and 20 μm is then washed in distilled water, centrifuged at 1500 rpm for 3 min, and the supernatant discarded. Differential flotation can then be used to separate one fraction from the rest of the solid matter. Solutions containing 26.5 % NaCl, 35 % $MgSO_4$, 33 % $ZnSO_4$, 70 % $ZnSO_4$, and 68 % sucrose were tested and found 70 % $ZnSO_4$ to be the best for separating out truffle spores.

22.4.2 Consumption of Fungi

Studies analyzing feces and stomach contents of wild boars suggest that they are best considered casual or opportunistic mycophagists, with consumption simply dependent on availability (Wood and Roark 1980; Genov 1981; Baron 1982; Génard et al. 1986; Fournier Chambrillon et al. 1995; Baubet et al. 2003; Herrero et al. 2004, 2005; Skewes et al. 2007).

In the native range, mushrooms are present in the wild boar diet with a frequency not higher than 30 % and a volume lower than 2 % (Genov 1981; Baubet et al. 2003). Only a study about wild boars inhabiting the Montpellier garrigue showed that mushrooms were found frequently in their stomach (60 %) although in low quantity (2 %) (Fournier Chambrillon et al. 1996).

In the introduced range, there are just a few records of the consumption of fungi by wild boars, but their intake generally occurs more frequently than in the native range. In Chile, Skewes et al. (2007) found fungi in 65 % of stomachs mostly collected in winter and autumn and made up 16 % of the volume. Similarly, Wood and Roark (1980) in North Carolina (USA) found fungi during all seasons with a frequency varying from 57 % in spring to 78 % in winter. In studies carried out at the Auckland Islands (New Zealand), wild boars were also found to ingest mushrooms, although with a low frequency (Challies 1975; Chimera et al. 1995); however, it must be considered that this survey was made during the summer of 1972–1973.

There are few studies reporting the consumption of hypogeous fungi by wild boars (Genov 1981; Génard et al. 1986; Skewes et al. 2007; Piattoni et al. 2012) (Table 22.1) although they seem to be eaten more frequently than epigeous mushrooms (Génard et al. 1986; Skewes et al. 2007). In these studies, most of the hypogeous fruiting bodies found in stomachs or feces were not identified or identified only at genus level (Skewes et al. 2007) (Table 22.1). Several studies carried out in the eastern part of Europe showed that the only fungus eaten by wild boars during the whole year is the deer truffle (*Elaphomyces granulatus* Fr.), which is considered the dominant source of radiocaesium contamination in wild boars (Genov 1981; Hohmann and Huckschlag 2005; Steiner and Fielitz 2009).

A study run by Piattoni et al. (2012) in the Regional Park of Gessi and Calanchi dell' Abbadessa (Bologna, Northern Italy), an area where numerous hypogeous fungi were found in 4 years of surveys (15 species, 13 Ascomycetes including *T. aestivum*, *T. magnatum*, *T. borchii*, *Tuber brumale* Vittad., *Tuber dryophilum* Tul. & C. Tul., *Tuber excavatum* Vittad., *Tuber macrosporum* Vittad., and *Tuber*

Table 22.1 Taxa of fungi reported in wild boar diet

Fungal species[a]	Frequency of occurrence %[b]	Type of analysis[c]	Country	Reference
Agaricus spp.	nd	S	Chile	Skewes et al. (2007)
Agrocybe pediades (Fr.) Fayod	nd	S	Spain	Abáigar (1993)
Aphyllophorales	nd	S	Chile	Skewes et al. (2007)
Boletus sp.	5	F	France	Génard et al. (1986)
Boletales	4	F	France	Génard et al. (1986)
Ciliaria sp.	nd	S	Chile	Skewes et al. (2007)
Suillus bellinii (Inzenga) Kuntze	nd	S	Spain	Abáigar (1993)
Suillus granulatus (L.) Roussel	nd	S	Spain	Abáigar (1993)
Suillellus queletii (Schulzer) Vizzini, Simonini & Gelardi	nd	S, F	France	Fournier Chambrillon et al. (1996)
Coprinus sp.	nd	S	Spain	Abáigar (1993)
Cortinarius sp. 24	nd	S	Chile	Skewes et al. (2007)
***Elaphomyces granulatus* Fr.**	6	S	Poland	Steiner and Fielitz (2009)
	nd	F, S	Poland	Genov (1981)
Geastrum fimbriatum Fr. (=*Geastrum sessile* Pouz)	nd	S, F	France	Fournier Chambrillon et al. (1996)
***Genea verrucosa* Vittad.**	7	F	Italy	Piattoni et al. (2012)
***Genea* sp.**	1	F	France	Génard et al. (1986)
Gomphidius sp.	1	F	France	Génard et al. (1986)
***Hymenogaster lycoperdineus* Vittad.**	21	F	Italy	Piattoni et al. (2012)
***Hymenogaster* sp.**	14	F	Italy	Piattoni et al. (2012)
	nd	S	Chile	Skewes et al. (2007)
	5	F	France	Génard et al. (1986)

(continued)

Table 22.1 (continued)

Fungal species[a]	Frequency of occurrence %[b]	Type of analysis[c]	Country	Reference
***Hysterangium* sp.**	19	F	France	Génard et al. (1986)
***Hallingea purpurea* (Zeller & C.W. Dodge) Castellano (=*Hysterangium purpureum* Zeller & C.W. Dodge)**	nd	S	Chile	Skewes et al. (2007)
Leccinum sp.	1	F	France	Génard et al. (1986)
Lepiota sp.	nd	S	Spain	Abáigar (1993)
Lepista sp.	nd	S	Spain	Abáigar (1993)
Lycoperdon lividum Pers.	nd	S	Spain	Abáigar (1993)
Lycoperdon perlatum Pers.	nd	S	Spain	Abáigar (1993)
***Melanogaster* sp.**	2	F	France	Génard et al. (1986)
Paxillus spp.	nd	S	Chile	Skewes et al. (2007)
***Octavianina asterosperma* Vittad.**	4	F	France	Génard et al. (1986)
***Picoa carthusiana* Tul. & C. Tul.**	16	F	France	Génard et al. (1986)
Russulales	3	F	France	Génard et al. (1986)
	nd	S	Spain	Abáigar (1993)
Suillus granulatus (L.) Roussel	nd	S	Spain	Abáigar (1993)
Suillus bellinii (Inzenga) Kuntze	nd	S	Spain	Abáigar (1993)
***Stephensia bombycina* (Vittad.) Tul.**	7	F	Italy	Piattoni et al. (2012)
Tricholoma terreum (Schaeff.) P. Kumm.	nd	S	Spain	Abáigar (1993)
***Tuber aestivum* Vittad.**	14	F	Italy	Piattoni et al. (2012)
	14	F	France	Génard et al. (1986)
	nd	S, F	France	Fournier Chambrillon et al. (1996)

(continued)

Table 22.1 (continued)

Fungal species[a]	Frequency of occurrence %[b]	Type of analysis[c]	Country	Reference
***Tuber ferrugineum* Vittad.**	1	F	France	Génard et al. (1986)
***Tuber magnatum* Pico**	14	F	Italy	Piattoni et al. (2012)
***Tuber rufum* Pico**	7	F	Italy	Piattoni et al. (2012)
***Tuber* sp.**	14	F	Italy	Piattoni et al. (2012)
	6	F	France	Génard et al. (1986)
	nd	S, F	France	Fournier Chambrillon et al. (1996)

[a]In bold hypogeous fungi; [b]*nd*, not determined; [c]*S*, stomach analysis; *F*, fecal analysis

rufum Pico and 2 Basidiomycetes), showed the presence of many of these hypogeous fungi in the feces of 10 of the 14 wild boars examined. Moreover, also the spores of *T. aestivum* and *T. magnatum*, which represent the most widespread and economically important species of that area, were found in their feces. Also Génard et al. (1986), in France, analyzing 105 fecal samples from wild boars, found *Tuber* spp. spores in 21 % of the feces examined, and in 14 % on the cases, they belonged to *T. aestivum* (Table 22.1).

In the Czech Republic and Slovakia, truffle harvesting is forbidden because they are considered endangered species (Gryndler et al. 2011). Thus, the detection of truffle spores in the feces of wild boars may provide a first rough measure of the abundance of hypogeous fungi in an area and might be used as an alternative method to map the presence of truffles.

22.5 The Role of Wild Boars in the Dispersal of Truffle Spores

After air, the main vector for agarics, animals are the most important vehicle for spore dispersal of coprophilous and hypogeous ectomycorrhizal (ECM) fungi. The latter group, having evolved underground fruiting bodies, developed strategies for spore dispersal, such as production of volatile chemical attractants (Trappe and Claridge 2005). These attract animals, including wild boars, which eat the fruiting bodies and disperse the spores in the feces (Piattoni et al. 2014). In the domestic pig, the time the spores remain in the gut ranges between 36 and 48 h (unpublished data). Because wild boars can move 15 km a day, they are very likely to play a major role in the dispersal of mushrooms they have ingested (Andrzejewski and Jezierski 1978; Singer et al. 1981), thus promoting the long-distance spore dispersal in both natural and introduced ranges. This means that their linear dispersal potential is one of the most efficient among mycophagous animals (Halbwachs and Bässler 2015). In particular, in Europe spore dispersal by boars may explain the movement of truffles from the southern refugia to their present distribution after the last glaciation (Murat et al. 2004; Hall and Zambonelli 2012; Piattoni et al. 2012). This movement presumably would continue if there was a warming of northern reaches of Europe (Büntgen et al. 2012). In countries where European truffles have been introduced such as Australia, Chile, China, and New Zealand, wild boars may spread in natural forests the spores of the eaten truffle and in particular those of the European truffles *T. melanosporum*, *T. aestivum*, and *T. borchii*. Although this may appear an economical advantage, the risk that these truffles might compete with the local ECM fungi or favor the invasion of exotic trees must not be underestimated (Nuñez et al. 2013). Moreover, spore dispersal may favor truffle sexual reproductions overcoming the problem of finding a sexual partner in the field where large patches of ECMs of the same mating type are commonly present (Murat et al. 2013).

22.5.1 Effect on Formation of Mycorrhizas

A recent study investigated the effects of digestive enzymes on the viability of *T. aestivum* truffle spores that had been mixed with boiled potatoes and fed to a pot-bellied piglet (Piattoni et al. 2014). Over the following 48 h, all the feces were collected, and the spores were separated by sieving and washing. Fluorescein diacetate stain confirmed that the spores were still viable despite passing through the gut. The digested spores and an equal number of fresh spores were then used to inoculate sterile *Quercus robur* L. seedlings. Ten months later, it was found that the mean percentage of *T. aestivum* colonization was significantly higher in the plants inoculated with the digested spores (66 %) than in the plants inoculated with the undigested spores (49 %). These results differed from an earlier study that used a rodent instead of a pig and different species of truffle *Tuber shearii* Harkn. and *Tuber canaliculatum* Gilkey (Miller 1985). Piattoni et al. (2014) explained these differences by different gut retention times, animal body temperatures, anatomy of the digestive tract, and gut microbial composition, all factors that may have a different effect on spores. To better investigate the effects of the passage through the digestive tract of pig, which may have promoted spore germination, microscopic analyses were carried out by Piattoni et al. (2014). Analyses at light microscope reveal that the majority of the spores digested by the pig (87 %) were free of asci, in contrast to the spores from the fresh fruiting bodies (only 7 %), suggesting that the pig digestive enzymes are able to degrade the ascus spore wall and thus may promote spore germination (Fig. 22.3). Moreover, during the passage through the digestive tract, spores undergo cell wall degradation by digestive

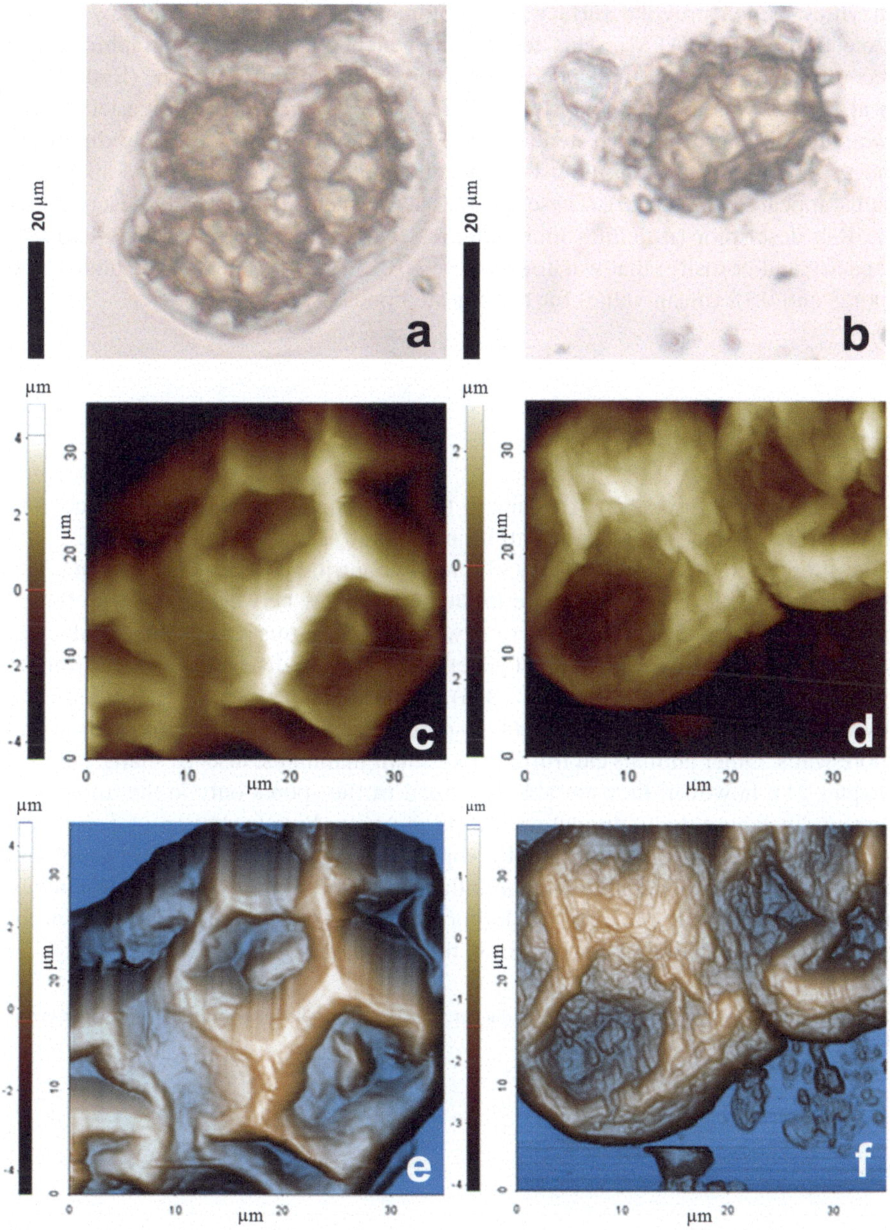

Fig. 22.3 *Tuber aestivum* uneaten spores (**a**, **c**, **e**) and *T. aestivum* pig-digested (**b**, **d**, **f**) spores observed by light microscope (**a**, **b**) and by AFM in Force Modulation Mode (**c**, **d**, **e**, **f**), scan size 35 × 35 µm at 0.2 Hz. The images E and F have been magnified with enhanced colors

enzymes, which provoke surface morphological modifications visible by atomic force microscope (AFM). This technique, which provides a topographic image accompanied also by mechanical and biochemical information (Radmacher et al. 1992), was used to study the wall surface of *T. aestivum* undigested spores compared to that of pig-digested spores (Piattoni et al. 2014). After pig digestion, most spore ornamentations almost disappeared, and several small peaks/valleys and blubs appeared (Fig. 22.3). These changes have been quantitatively described by the Rsk descriptor (it defines morphological variations in terms of load carrying capacity and porosity) that was found to be −0.15 (mean value) for the undigested spores and 0.18 (mean value) for the digested ones.

22.6 Conclusions

Wild boars can cause severe damage to truffle production through predation, soil disturbance, and destruction of truffle-inoculated seedlings (Samils et al. 2008; Salerni et al. 2013). However, they efficiently contribute to dispersal of fungi and have probably played a pivotal role in the genetic flow between and within fungal populations and in the colonization of new habitats. For this reason, animal obligate spore dispersal decreases the risk of genetic isolation by distance which is common by air dispersed-spores (Taschen 2015). Moreover, they seem to facilitate the mycorrhiza formation for the effects of spore digestion on degradation of asci and spore walls. Other animals eat truffles (like small mammals, insects, snails, etc., see chapter 21); however, they are able to spread truffle spores only in the immediate surroundings, and their stimulating effects of mycorrhiza formation of *Tuber* spp. are not still proven. Considering all complex interactions occurring between wild boars and truffle soil ecology and the influence of the former on truffle production and colonization, any project considering the impact of wild boars on the environment has also to take into consideration the role of this animal.

Acknowledgments We would like to thank Dr. Ian Hall for helpful comments and suggestions during the writing of this manuscript.

References

Abáigar T (1993) Régimen alimentario del jabalí (*Sus scrofa* L. 1758) en el sureste Ibérico. Doñana. Acta Vertebrata 20(1):35–48

Andrzejewski R, Jezierski W (1978) Management of a wild boar population and its effects on commercial land. Acta Theriol 23(19):309–339

Ballari SA, Barrios-Garcia MN (2013) A review of wild boar *Sus scrofa* diet and factors affecting food selection in native and introduced ranges. Mammal Rev 44(2):124–134. doi:10.1111/mam.12015

Baron J (1982) Effects of feral hogs (*Sus scrofa*) on the vegetation of Horn Island, Mississippi. Am Midl Nat 107(1):202–205

Barrios-Garcia MN (2012) Multi-level impacts of introduced wild boar on Patagonian ecosystems. PhD thesis, University of Tennessee

Barrios-García MN, Ballari SA (2012) Impact of wild boar (*Sus scrofa*) in its introduced and native range: a review. Biol Invasions 14(11):2283–2300. doi:10.1007/s10530-012-0229-6

Baubet E, Ropert-Coudert Y, Brandt S (2003) Seasonal and annual variations in earthworm consumption by wild boar (*Sus scrofa* L.). Wildl Res 30(2):179–186. doi:10.1071/WR00113

Bueno C, Jiménez JJ (2014) Livestock grazing activities and wild boar rooting affect alpine earthworm communities in the Central Pyrenees (Spain). Appl Soil Ecol 83:71–78. doi:10.1016/j.apsoil.2014.04.013

Büntgen U, Egli S, Camarero JJ, Fischer EM, Stobbe U, Kauserud H, Tegel W, Sproll L, Stenseth NC (2012) Drought-induced Périgord black truffle decline. Nat Clim Change 2(12):827–829. doi:10.1038/nclimate1733

Challies CN (1975) Feral pigs (*Sus scrofa*) on Auckland Island: status, effects on vegetation and nesting sea birds. New Zeal J Zool 2(4):479–490. doi:10.1080/03014223.1975.9517889

Chen K, Baxter T, Muir W, Groenen M, Schook L (2007) Genetic resources, genome mapping and evolutionary genomics of the pig (*Sus scrofa*). Int J Biol Sci 3(3):153–165. doi:10.7150/ijbs.3.153

Chen J, Guo SX, Liu PG (2011) Species recognition and cryptic species in the *Tuber indicum* complex. PLoS One 6(1), e14625. doi:10.1371/journal.pone.0014625

Chimera C, Coleman MC, Parkes JP (1995) Diet of feral goats and feral pigs on Auckland Island, New Zealand. New Zeal J Ecol 19(2):203–207

d'Huart JP (1991) Habitat utilization of old world wild pigs. In: Barrett RG, Spitz F (eds) Biology of Suidae. IRGM, Grenoble, pp 30–48

Fekete A O, Bagi I, Paz Conde I, Iotti M, Zambonelli A (2014) First report on the truffles of Azerbaijan. In: Abstract of the 5th conference of *Tuber aestivum/uncinatum* (TAUESG5), Rabat, 9–13 Apr 2014, p 56

Fournier Chambrillon C, Maillard D, Fournier P (1995) Diet of wild boar (*Sus scrofa* L.) inhabiting the Monpellier garrigue. J Mt Ecol 3:174–179

Fournier Chambrillon C, Maillard D, Fournier P (1996) Variabilité du régime alimentaire du sanglier (*Sus scrofa* L.) dans les garrigues de Montpellier (Hérault). Gibier Faune Sauvage 13 (4):1457–1476

Génard M, Lescourret F, Durrieu G (1986) Mycophagie chez le sanglier et dissemination des spores de champignons hypogés. Gaussenia 2:17–23

Genov P (1981) Food composition of wild boar in North-eastern and Western Poland. Acta Theriol 26(10):185–205

Gortázar C, Ferroglio E, Höfle U, Frölich K, Vicente J (2007) Diseases shared between wildlife and livestock: a European perspective. Eur J Wildl Res 53(4):241–256. doi:10.1007/s10344-007-0098-y

Groves CP, Shaller GB, Amato G, Khounboline K (1997) Rediscovery of the wild pig *Sus bucculentus*. Nature 386:335. doi:10.1038/386335a0

Gryndler M, Hršelová H, Soukupová L, Streiblová E, Valda S, Borovička J, Gryndlerová H, Gažo J, Miko M (2011) Detection of summer truffle (*Tuber aestivum* Vittad.) in ectomycorrhizae and in soil using specific primers. FEMS Microbiol Lett 318(1):84–91. doi:10.1111/j.1574-6968.2011.02243.x

Hahn N, Eisfeld D (1998) Diet and habitat use of wild boar (*Sus scrofa*) in SW-Germany. Gibier Faune Sauvage 15:595–606

Halbwachs H, Bässler C (2015) Gone with the wind—a review on basidiospores of lamellate agarics. Mycosphere 6(1):78–112. doi:10.5943/mycosphere/6/1/10

Hall IR, Haslam W (2012) Truffle cultivation in the Southern Hemisphere. In: Zambonelli A, Bonito GM (eds) Edible ectomycorrhizal mushrooms: current knowledge and future prospects, Soil biology, vol 34. Springer, Berlin, pp 191–208. doi:10.1007/978-3-642-33823-6_11

Hall IR, Zambonelli A (2012) The cultivation of mycorrhizal mushrooms—still the next frontier! In: Zhang J, Wang H, Chen M (eds) Proceedings 18th congress of the international society for Mushroom Science, Beijing, pp 26–30

Hall IR, Brown GT, Zambonelli A (2007) Taming the truffle. The history, lore, and science of the ultimate mushroom. Timber Press, Portland

Herrero J, Couto S, Rosell C, Arias P (2004) Preliminary data on the diet of wild boar living in a mediterranean coastal wetland. Galemis 16(1):115–123

Herrero J, Irizar I, Laskurain NA, García-Serrano A, García-González R (2005) Fruits and roots: wild boar foods during the cold season in the southwestern Pyrenees. Ital J Zool 72(1):49–52. doi:10.1080/11250000509356652

Herrero J, García Serrano A, Couto S, Ortuño VM, García González R (2006) Diet of wild boar (*Sus scrofa* L.) and crop damage in an intensive agroecosystem. Eur J Wildl Res 52 (4):245–250. doi:10.1007/s10344-006-0045-3

Hohmann U, Huckschlag D (2005) Investigation on the radiocaesium contamination of wild boar (*Sus scrofa*) meat in Rhineland-palatinate: a stomach content analysis. Eur J Wildl Res 51 (4):263–270. doi:10.1007/10344-05-0108-x

Klemm M (1948) Schwarzwild und Schwarzwildschäden in Deutschland im Jahre 1946. Nachrichtenblatt für den Deutschen Pflanzenschutzdienst 2:74–77

Krajick K (2005) Winning the war against island invades. http://tonytonello.blog.com/files/2011/02/ARTIGO-1.pdf. Accessed 3 Apr 2015

Lacki MJ, Lancia RA (1983) Changes in soil properties of forests rooted by wildboar. Proceedings of the Annual Conference Southeastern Association of Fish and Wildlife Agencies 37:228–236

Ławrynowicz M, Faliński JB, Bober J (2006) Interactions among hypogeous fungi and wild boars in the subcontinental pine forest. Biodivers Res Conserv 1–2:102–106

Meng XJ, Lindsay DS, Sriranganathan N (2009) Wild boars as sources for infectious diseases in livestock and humans. Philos Trans R Soc B 364(1530):2697–2707. doi:10.1098/rstb.2009.0086

Miller SL (1985) Rodent pellet as ectomycorrhizal inoculum for *Tuber* spp. In: Molina R (ed) Proceedings of the 6th North American conference on mycorrhizae. Forestry Research Laboratory, Corvallis, p 273

Moreno-Arroyo B, Gòmez J, Putido E (2005) Tesoros de muestro montes. Trufas de Andalucía. Consejria de Medio Ambiente, Junta de Andalucía, Cordoba

Murat C, Díez J, Luis P, Delaruelle C, Dupré C, Chevalier G, Bonfante P, Martin F (2004) Polymorphism at the ribosomal DNA ITS and its relation to postglacial re-colonization routes of the Perigord truffle *Tuber melanosporum*. New Phytol 164(2):401–411. doi:10.1111/j.1469-8137.2004.01189.x

Murat C, Rubini A, Riccioni C, De la Varga H, Akroume E, Belfiori B, Guaragno M, Le Tacon F, Robin C, Halkett F, Martin F, Paolocci F (2013) Fine-scale spatial genetic structure of the black truffle (*Tuber melanosporum*) investigated with neutral microsatellites and functional mating type genes. New Phytol 199(1):176–187. doi:10.1111/nph.12264

New Zealand Department of Conservation (2015) Feral pigs. http://www.doc.govt.nz/parks-and-recreation/things-to-do/hunting/what-to-hunt/feral-pigs/. Accessed 3 Apr 2015

Nuñez MA, Hayward J, Horton TR, Amico GC, Dimarco RD, Barrios-Garcia MN, Simberloff D (2013) Exotic mammals disperse exotic fungi that promote invasion by exotic trees. PLoS One 8(6), e66832. doi:10.1371/journal.pone.0066832

Piattoni F, Ori F, Morara M, Iotti M, Zambonelli A (2012) The role of wild boars in spore dispersal of hypogeous fungi. Acta Mycol 47(2):145–153. doi:10.5586/am.2012.017

Piattoni F, Amicucci A, Ori F, Iotti M, Stocchi V, Zambonelli A (2014) Viability and morphology of *Tuber aestivum* spores after passage through the gut of *Sus scrofa*. Fungal Ecol 9:52–60. doi:10.1016/j.funeco.2014.03.002

Pimentel D, Zuniga R, Morrison D (2005) Update on the environmental and economic costs associated with alien-invasive species in the United States. Ecol Econ 52(3):273–288. doi:10.1016/j.ecolecon.2004.10.002

Radmacher M, Tillmann RW, Fritz M, Gaub HE (1992) From molecules to cells: imaging soft samples with the AFM. Science 257(5078):1900–1905. doi:10.1126/science.1411505

Reimoser F, Gossow H (1996) Impact of ungulates of forest vegetation and its dependence on the silvicultural system. For Ecol Manage 88(1):107–119. doi:10.1016/S0378-1127(96)03816-9

Reyna S, Garcia Barreda S (2014) Black truffle cultivation: a global reality. For Syst 23 (2):317–328. doi:10.5424/fs/2014232-04771

Ricci G (2008) Indagine sull'impatto ambientale ed economico degli ungulati selvatici nell'Appennino Bolognese. Fondazione Itaca, Pianoro

Ruiz-Fons F, Segales J, Gortazar C (2008) A review of viral diseases of the European wild boar: effects of population dynamics and reservoir role. Vet J 176(2):158–169. doi:10.1016/j.tvjl.2007.02.017

Sáez Royuela C, Tellería JL (1986) The increased population of wild boar (*Sus scrofa* L.) in Europe. Mammal Rev 16(2):97–101. doi:10.1111/j.1365-2907.1986.tb00027.x

Salerni E, Gardin L, Baglioni F, Perini C (2013) Effects of wild boar grazing on the yield of summer truffle (Tuscany, Italy). Acta Mycol 48(1):73–80. doi:10.5586/am.2013.009

Samils N, Olivera A, Danell E, Alexander SJ, Fischer C, Colinas C (2008) The socioeconomic impact of truffle cultivation in rural Spain. Econ Bot 62(3):331–340. doi:10.1007/s12231-008-9030-y

Schley L, Roper TJ (2003) Diet of wild boar *Sus scrofa* in Western Europe, with particular reference to consumption of agricultural crops. Mammal Rev 33(1):43–56. doi:10.1046/j.1365-2907.2003.00010.x

Singer FJ, Otto DK, Tipton AR, Hable CP (1981) Home ranges, movements, and habitat use of European wild boar in Tennessee. J Wildl Manage 45(2):343–353

Skewes O, Rodriguez R, Jaksic FM (2007) Trophic ecology of the wild boar (*Sus scrofa*) in Chile. Rev Chil Hist Nat 80(3):295–307

Sourzat P (1989) Guide practique de trufficulture. 2nd ed. Délégation Départementale des Services d'Agronomie du Lot, Le Montat, Cahors

Spitz F (1999) *Sus scrofa* (Linnaeus, 1758). In: Mitchell-Jones AJ, Amori G, Boganowicz W, Kryštufek B, Reijnders PJH, Spitzenberger F, Stubbe M, Thissen JBM, Vohralík V, Zima J (eds) The atlas of European mammals, Poyser natural history. Academic, London, pp 380–381

Steiner M, Fielitz U (2009) Deer truffles the dominant source of radiocaesium contamination of wild boar. Radioprotection 44(5):585–588. doi:10.1051/radiopro/20095108

Taschen E (2015) Biotic interactions and reproductive biology of the Black Truffle (*Tuber melanosporum* Vittad.), from spontaneous to planted truffle grounds. PhD thesis, Université Montpellier II, France

Trappe JM, Claridge AW (2005) Hypogeous fungi: evolution of reproductive and dispersal strategies through interactions with animals and mycorrhizal plants. In: Dighton J, White JF, Oudemans P (eds) The fungal community: its organization and role in the ecosystem. Taylor & Francis, Boca Raton, pp 613–623

Wilson CJ (2004) Rooting damage to farmland in Dorset, southern England, caused by feral wild boar *Sus scrofa*. Mammal Rev 34(4):331–335. doi:10.1111/j.1365-2907.2004.00050.x

Wirthner S, Frey B, Busse MD, Schütz M, Risch AC (2011) Effects of wild boar (*Sus scrofa* L.) rooting on the bacterial community structure in mixed–hardwood forest soils in Switzerland. Eur J Soil Biol 47(5):296–302. doi:10.1016/j.ejsobi.2011.07.003

Wood GW, Roark DN (1980) Food habits of feral hogs in coastal South Carolina. J Wildl Manage 44(2):506–511. doi:10.2307/3807990

Zhang JP, Liu PG, Chen J (2012) *Tuber sinoaestivum* sp. nov., an edible truffle from southwestern China. Mycotaxon 122:73–82. doi:10.5248/122.73

Part V
Biochemistry

Chapter 23
The Smell of Truffles: From Aroma Biosynthesis to Product Quality

Richard Splivallo and Laura Culleré

23.1 Introduction

Truffle fungi (*Tuber* spp.) along with caviar and champagne stand on top of the list of luxury food items. Indeed, in Europe, the retail price of the Périgord truffle *Tuber melanosporum* Vittad. can reach up to 2,000 € per kg, while the Piedmont truffle, *Tuber magnatum* Pico, can reach up to 5,000 € per kg. Additionally, especially large specimens of the latter species have even been sold at auctions for 50,000 € per kg, rendering truffles as expansive as gold. In the USA, local truffle species (i.e., *Tuber lyonii* Butters, *Tuber oregonense* Trappe, Bonito, & P. Rawl.) also sell for thousands of € per kg. These high prices are the result of an undersupplied market and the limited seasonal availability of fresh truffles. Indeed, in Europe and the USA, the most requested truffle sorts are harvested during a few months in autumn and in winter.

What makes truffles so famous, besides their high prices and rarity, is their unique smell. *Tuber melanosporum* can be considered as one of the most aromatic species and it has been defined as a "black diamond of cuisine" due to its potent and complex aroma. It is endemic to Spain, France, and Italy and has been introduced in numerous countries of the southern and northern hemispheres (i.e., in some

R. Splivallo (✉)
Institute of Molecular Biosciences, University of Frankfurt, Max-von-Laue-Str. 9, 60438 Frankfurt, Germany

Integrative Fungal Research Cluster (IPF), Georg-Voigt-Str. 14-16, 60325 Frankfurt, Germany
e-mail: richard.splivallo@a3.epfl.ch

L. Culleré
Laboratory for Aroma Analysis and Enology, Instituto Agroalimentario de Aragón, (IA2), Department of Analytical Chemistry, Faculty of Sciences, University of Zaragoza, 50009 Zaragoza, Spain

A. Zambonelli et al. (eds.), *True Truffle (*Tuber *spp.) in the World*, Soil Biology 47,
DOI 10.1007/978-3-319-31436-5_23

European countries as well the USA, New Zealand, Australia, and Chile (Jeandroz et al. 2008; Bonito et al. 2010; Hall and Haslam 2012; Lefevre 2012; see also Chap. 2). This species is very vulnerable to fraud given that the Chinese truffle *Tuber indicum* Cooke & Massee looks very similar (dark gleba and black peridium) but has a much lower market value. Indeed it is difficult to tell both species apart by traditional morphological observations. Additionally, the aroma of the Chinese species is less intense and complex than the characteristic aroma of *T. melanosporum* (Culleré et al. 2013a). *Tuber brumale* Vittad. is another black truffle with a characteristic musky odor, with "earthy" notes. *Tuber aestivum* Vittad. is a black truffle which can be collected all over Europe (Jeandroz et al. 2008; Splivallo et al. 2012a; see also Chap. 4). It is less aromatic than *T. melanosporum*, although it has a good aroma quality and it is very appreciated by end consumers. Lastly, *T. magnatum*, the most expensive truffle on the market, is often considered as the finest species and has a complex aroma reminiscent of garlic and cheese. It is essentially collected in Italy and in the Carpathian Basin of Central Eastern Europe. Besides the five species briefly described above, about 5–10 additional ones are traded commercially throughout the world.

Truffle aromas are made of a mixture of hydrocarbons containing alcohol, ketone, aldehyde, and ester functional groups but also sulfur atoms. To date, more than 300 volatiles have been described from about eleven species (Splivallo et al. 2011). The aromatic profile of a single species typically contains 30–60 volatile constituents (Bellesia et al. 1998; Díaz et al. 2003; Mauriello et al. 2004; Gioacchini et al. 2005; March et al. 2006; Splivallo et al. 2007). Some of these compounds are not only widespread among truffles but also in other fungi and even plants, while others are much more specific and occur in a few or one truffle sort only. Common truffle volatiles include, for example, 2-methyl-1-butanol and 3-methyl-1-butanol and their respective aldehydes, as well as 1-octen-3-ol (Fig. 23.1), an eight carbon-containing volatile with typical fungal flavor (Combet et al. 2006). Sulfur-containing volatiles such as dimethyl sulfide (Fig. 23.1) and dimethyl disulfide also commonly occur in truffles (Splivallo et al. 2011). By contrast to common truffle volatiles, other compounds are restricted to a few truffle species. This is the case, for example, for the thiophene derivatives 3-methyl-4,5-dihydrothiophene (Fig. 23.1) and 2-methyl-4,5-dihydrothiophene which are only found in *Tuber borchii* Vittad. (Bellesia et al. 2001; Splivallo et al. 2011, 2014). Similarly, the other sulfur volatile 2,4-dithiapentane (Fig. 23.1) occurs in a few species only: it is the dominant constituent of *T. magnatum* and has been reported once in *Tuber panniferum* Tul. & C. Tul. (Mauriello et al. 2004; Splivallo et al. 2011). Yet other volatiles such as 1-methoxy-3-methylbenzene have been only described in the black truffle species *T. aestivum*, *T. brumale*, *Tuber mesentericum* Vittad., and *T. melanosporum* (Díaz et al. 2003; Mauriello et al. 2004; March et al. 2006).

Overall truffles are stand-alone gourmet products on their own rights. This is essentially due to their rarity and to their unique fragrances. Considering that only 10–15 truffle sorts are traded commercially out of the 180 existing species (Bonito

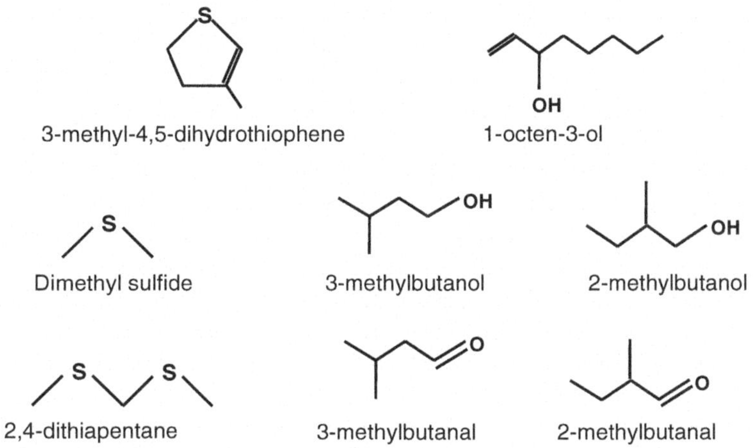

Fig. 23.1 Characteristic volatiles from various truffle species

et al. 2010), truffles represent a valuable and largely unexplored source of aroma compounds.

23.2 Do Truffles or Microbes Produce Truffle Aromas?

Truffles form underground following sexual reproduction between two individuals of opposite mating type (Martin et al. 2010; Rubini et al. 2011). During this process, numerous microbes might end up trapped inside fruiting bodies. As a matter of fact, truffle-fruiting bodies are a hotspot for microbial diversity and host bacteria (Barbieri et al. 2005, 2007; Antony-Babu et al. 2013; Splivallo et al. 2014; see also Chap. 18), yeasts (Buzzini et al. 2005), filamentous fungi (Pacioni et al. 2007; see also Chap. 17), and possibly even viruses (Stielow and Menzel 2010; see also Chap. 20). Bacteria are the dominant group of truffle's microbiomes, the community composition of which have been described in details for *T. melanosporum*, *T. magnatum*, *T. aestivum*, and *T. borchii* in a recent review (Vahdatzadeh et al. 2015). The role of this microbiome in truffle aroma formation has for long been speculative. A historical perspective reveals that Buzzini et al. (2005) demonstrated that yeasts isolated from *T. magnatum* and *T. melanosporum* and grown in pure culture in the presence of L-methionine had the ability to produce some key aroma constituents characteristic of the latter species. The authors therefore hypothesized that yeasts present at the surface of fruiting bodies might fulfill the same role. Indirect evidence further supporting the role of microbes in aroma formation came from the observation that some volatiles characteristic of *T. borchii* fruiting bodies were not found in mycelium grown in pure culture (Tirillini et al. 2000; Splivallo et al. 2007). By contrast, the genome of *T. melanosporum* suggested that truffles might possess all the molecular toolkits necessary to synthesize their volatiles on their own (Martin et al. 2010). This observation was nevertheless speculative since it was based on the presence of genes which function had not been demonstrated (Martin et al. 2010; Splivallo et al. 2011). In 2014, using *T. borchii* as a model organism, the role of the microbiome in truffle aroma formation was finally demonstrated for a group of volatiles (thiophene derivatives) characteristic of the latter species (Splivallo et al. 2014) (Fig. 23.2). The results indicated that inside *T. borchii* fruiting bodies, an unidentified precursor is metabolized into volatile thiophene derivatives only by bacteria and not by truffle mycelium or yeasts. Whether other volatiles characteristic of truffle aromas are produced by truffles or their microbiomes remains to be demonstrated. Based on literature data, common truffle volatiles (i.e., as dimethyl sulfide, dimethyl disulfide) are emitted by mycelial pure cultures (Splivallo and Maier 2011; Liu et al. 2013) but also by numerous yeasts and bacteria (Lemfack et al. 2014). Hence they might hypothetically be produced by both truffles and their microbiomes inside fruiting bodies (Vahdatzadeh et al. 2015). Demonstrating their exact origins will be a challenge considering that microbe-free truffles cannot be currently obtained. A recent meta-analysis about truffle aroma and their microbiomes nevertheless suggests that microbes might play a central role in aroma formation for most, if not all truffle species (Vahdatzadeh et al. 2015).

Regardless of the organismic origin of truffle volatiles, their biosynthesis might follow a scheme which is rather conserved in microbes. Sulfur volatiles such as dimethyl sulfide and dimethyl disulfide and dimethyl trisulfide could be derived from the catabolism of L-methionine (Martin et al. 2010; Splivallo et al. 2011). Two routes might possibly exist in truffles (Splivallo et al. 2011). Methionine could either be converted to methanethiol (MTL) by a C-S lyase or transaminated to 4-methylthio-2-oxobutyric acid, which can be transformed in a series of intermediates eventually leading to MTL. The latter volatile can spontaneously decompose into dimethyl sulfide and dimethyl disulfide. The biosynthetic pathways leading to more complex truffle volatiles (i.e., thiophene derivatives or 2,4-dithiapentane, respectively, characteristic of *T. borchii* and *T. magnatum*) have not been elucidated. L-Methionine remains their hypothetical precursor; however feeding this amino acid to bacteria either grown alone or in the presence of *T. borchii* mycelium did not induce the thiophene derivatives typical of fruiting bodies but only derivatives with related structures (Splivallo et al. 2014). This indicates that a piece of the puzzle is missing and that the biosynthesis of thiophene derivatives might possibly proceed through more complex organismic interactions (i.e., yeasts, bacteria, and truffle).

Other volatiles common to numerous truffle species include branched chain and aromatic hydrocarbons such as 2-methylbutanal, 3-methylbutanal, 2-methylpropanal,, and 2-phenylethanol. Similarly to sulfur volatiles, these might be derived from the catabolism of the specific amino acids through the Ehrlich pathway (Hazelwood et al. 2008; Martin et al. 2010; Splivallo et al. 2011). Indeed feeding isoleucine to *T. borchii* and *T. melanosporum* mycelia induces the production of 2-methylbutanal and 2-methylbutanol; similarly feeding leucine induces 3-methylbutanal and 3-methylbutanol (Splivallo and Maier 2011). Clearly the

Fig. 23.2 Illustration of bacteria eating truffles and emitting thiophene derivatives

identity of the precursors and intermediates should be confirmed by feeding labeled compounds.

The last group of volatiles which are extremely important in truffles and fungi in general is eight carbon-containing volatiles. The most well known is 1-octen-3-ol which occurs as two stereoisomers (R and S forms). Fungi predominantly produce the R form which has also a typical fungal smell compared to the S form which had a moldy, grassy smell (Combet et al. 2006). It is likely that truffles similarly produce the R form even if this has not yet been demonstrated. In mushrooms 1-octen-3-ol is synthesized from linoleic acid which is also the dominant fatty acid in *T. melanosporum* (Harki et al. 2006). Ascomycete fungi might synthesize these volatiles from linoleic acid through a single enzyme known as Ppo (Brodhun et al. 2010). Indeed in the ascomycete mold *Aspergillus nidulans* (Eidam) G. Winter, linoleic acid is transformed through the dioxygenase activity of the Ppo enzyme to 10-hydroperoxyoctadecadienoic acid which might decompose nonenzymatically to 1-octen-3-ol (Garscha and Oliw 2009; Brodhun et al. 2010). The genome of the Périgord truffle *T. melanosporum* suggests that two copies of the Ppo enzymes might exist in truffles. Characterizing the exact pathway will require combining molecular and chemical techniques to fully identify enzymes and precursors.

23.3 Which Factors Influence Truffle Aroma?

As for any other food products of vegetal or fungal origin, truffle aroma is influenced by a series of biotic and abiotic factors. These might, for example, not only include fruiting body maturity (Zeppa et al. 2004), species identity (Mauriello et al. 2004), genotype (Splivallo et al. 2012b; Molinier et al. 2015), and microbial community (Splivallo et al. 2014; Splivallo and Ebeler 2015; Vahdatzadeh et al. 2015) but also soil nutritional content or the identity of the host tree. Unlike with culture intensive crops for which clones are available and for which the influence of these factors can be systematically determined, it is much more difficult to address these questions with truffles. Indeed, trees infested with truffle clones are not yet available commercially, and even within the same orchard, the same genotypes (clones) rarely yield fruiting bodies from one season to the next (Murat et al. 2013; Molinier et al. 2015). This greatly complicates the study of the influence of a single factor on the aroma, since these investigations should be conducted under controlled conditions with the guarantee to collect sufficient truffles for sound statistical analysis. Despite these limitations, studies have been conducted by various groups on truffles collected in natural and artificial truffle orchards and highlighted the possible influence of specific factors. For example, fruiting body maturation has been reported to influence the aroma of *T. borchii*. In fresh fruiting bodies, specific thiophene derivatives were suggested to occur only in fully mature truffles (Zeppa et al. 2004). Recently it has been demonstrated that thiophene derivatives were produced by bacteria colonizing truffle-fruiting bodies

(Splivallo et al. 2014). The observation that the structure of the bacterial communities evolves during maturation in *T. melanosporum* (Antony-Babu et al. 2013) indicates that maturity and microbial community are linked. Furthermore in *T. borchii*, a correlation between the concentration of thiophene derivatives and microbial density has been observed during storage at room temperature (Splivallo et al. 2014). Therefore telling apart the contribution of maturity and bacterial community will only be achieved by studying both factors at the same time in fresh and stored truffles.

Other factors which possibly influence truffle aroma are geographical origin and genotype. Indeed fruiting bodies of *T. magnatum* collected from distinct geographical locations might differ in their aroma profile specifically for terpenoid volatiles (Gioacchini et al. 2008; Federico et al. 2015). Considering that genotypes of truffle-fruiting bodies might not only differ between geographical regions but also from one season to the next, it will be necessary to sample truffles over many seasons to know which factor (i.e., genotype, soil, bacterial community, maturation) determines the production of terpenoids in *T. magnatum*. Indeed in *T. aestivum*, genotype determines the concentration of eight carbon-containing volatiles (Splivallo et al. 2012b; Molinier et al. 2015); therefore this might also be the case in *T. magnatum*. Investigating many specimens for telling apart the contribution of each factor influencing truffle aroma formation will be necessary to understand whether specific aroma constituents can be used as markers of geographical origin.

23.4 Human-Sensed Truffle Aroma

Volatiles responsible for truffle characteristic aromatic blend are alcohols, ketones, aldehydes, aromatic, and sulfur compounds and were first described in the 1980s (Ney and Freitag 1980; Claus et al. 1981). Later, Talou and colleagues (Talou et al. 1987) identified a total of 14 volatiles as black Périgord truffle (*T. melanosporum*) aroma constituents, but none showed the same characteristic as bis(methylthio)methane, the single volatile responsible of the smell of the white truffle *T. magnatum* (Fiecchi et al. 1967). Moreover, a synthetic truffle aroma made up of less than ten volatiles was formulated in 1989 (Talou et al. 1989). According to both sniffing gas chromatography (GC) and sensory analyses carried out in the study with synthetic model mixtures, they concluded that (a) none of the identified compounds was uniquely responsible for the typical aroma of the Périgord truffle and (b) that dimethyl sulfide and 2-methylbutanal, respectively, responsible for sulfurous and pungent notes, had the greatest importance in the final aroma impression.

Volatile sulfur compounds can be naturally generated in oceans, marshes, grounds, and vegetation areas. They can also have an anthropogenic origin in processes such as residual water treatments, composting installations, and recovery plants of organic matter (Smet et al. 1999). For example, the Kraft process used to degrade lignin and produce wood pulp emits as wastes methanethiol, dimethyl sulfide, carbonyl sulfide, and hydrogen sulfide (de Zwart and Kuenen 1992). These sulfur compounds are relevant to food science, since they are present in products such as cheese (Burbank and Qian 2005), beer (Gerbersmann et al. 1995), and wine (Segurel et al. 2004; López et al. 2007; Franco-Luesma and Ferreira 2014). Volatile sulfur compounds can communicate unpleasant smells and tastes often described as rotten eggs (hydrogen sulfide), putrefaction (methanethiol), asparagus (dimethyl sulfide), garlic (diethyl sulfide), cauliflower (dimethyl disulfide), or onion (diethyl disulfide). At small concentrations, however, they are normal components of several food and beverages and can even exert a decisive role on sensory descriptors (Escudero et al. 2007; Mottram and Madruga 1994).

In recent years, some studies have analyzed the volatile compounds of different species of truffles using mainly headspace–solid-phase microextraction (HS–SPME) technique (Pelusio et al. 1995; Bellesia et al. 1998; Díaz et al. 2003, 2009; Mauriello et al. 2004; Gioacchini et al. 2005) and direct headspace analysis (March et al. 2006). The information obtained by gas chromatography–mass spectrometry (GC–MS) analysis is interesting but incomplete since it does not reveal if specific constituents play a role in human-sensed aroma. For this purpose it is indeed necessary to evaluate which of the compounds are in fact important odorants. This can be achieved through sensory analysis using a gas chromatography–olfactometry (GC–O) approach, which essentially relies on the human nose as a detector. On one side GC–O is more time consuming than GC–MS since it requires that a single sample be analyzed by numerous panelists to generate statistically sound results. On the other side GC–O can actually be even more sensitive than an MS detector essentially due to the irreplaceable ability of the human nose to detect specific odorants (i.e., sulfur compounds). Hence GC–MS and GC–O should be regarded as complementary methods.

Techniques in sensory science (i.e., GC–O) have been applied essentially to commercially relevant truffle species. Jansen et al. (2003) characterized and compared the odor profile of a by-product of the black truffle industry to the one of a reference product (truffle, cooking juice). The chemical nature of the compounds responsible for these odors has nevertheless not been identified. Piloni et al. (2005) characterized white truffle aroma by GC–MS and by GC–O analysis and demonstrated that besides bis(methylthio)methane, other sulfur volatiles contributed to the smell of that species. The aromatic compositions of two truffle species (*T. melanosporum* and *T. aestivum*) were compared by GC–O (Culleré et al. 2010). As expected, the olfactogram of *T. melanosporum* was more intense and complex than the one of *T. aestivum*. The aroma emitted by *T. melanosporum* was due to at least 17 different aroma molecules, six of which were reported for the first time: 1-hexen-3-one, 2-methyl-3-furanthiol, furaneol, 3-ethylphenol, 3-propylphenol, and 5-methyl-2-propylphenol. The most important aroma compounds of *T. melanosporum* were 2,3-butanedione, dimethyl disulfide, ethyl butyrate, dimethyl sulfide, 3-methyl-1-butanol, and 3-ethyl-5-ethylphenol. In addition, relevant differences between the aroma profiles from both species were observed. While dimethyl sulfide and dimethyl disulfide and 3-methyl-1-butanol were in both cases among the five most important aroma compounds, the species differed in their content of 2,3-butanedione and ethyl butyrate (higher in *T. melanosporum*) and of methional (higher in *T. aestivum*).

An olfactometric comparison (GC–O) of *T. indicum* with *T. melanosporum* was carried out for the first time by Culleré et al. (2013a). *Tuber melanosporum* aroma was much more intense than the one of *T. indicum*. This might partially explain the higher gastronomical value of *T. melanosporum*. Nevertheless, eight important odorants were identified in *T. indicum*. In order of aromatic significance, these were 1-octen-3-one and 1-octen-3-ol, followed by two ethyl esters (ethyl isobutyrate and ethyl 2-methylbutyrate), 3-methyl-1-butanol, isopropyl acetate, and finally the two sulfides dimethyl disulfide and dimethyl sulfide. Some of these odorants have also been reported in another study on *T. indicum* and on two other Asian truffle species (Liu et al. 2012). Interestingly 1-octen-3-one and 1-octen-3-ol contributed more to the aroma of *T. indicum* than of *T. melanosporum*. By contrast the opposite was true for dimethyl sulfide and dimethyl disulfide. The volatile profiles of both species were also studied by means of headspace–solid-phase microextraction (HS–SPME/GC–MS). This confirmed that the family of C8 compounds (3-octanone, octanal, 1-octen-3-one, 3-octanol, and 1-octen-3-ol) was present at much higher levels in *T. indicum* than in *T. melanosporum*. Therefore either of the two chromatographic methods (GC–O or HS–SPME/GC–MS) could be used as a screening technique to distinguish between *T. indicum* and *T. melanosporum* and thus avoid possible fraud. However, as genotype might influence truffle aroma (Molinier et al. 2015; Splivallo et al. 2012b) and since at least two genetic groups exist for *T. indicum* (see Chap. 2), these results should be validated with *T. indicum* samples belonging to defined genetic groups.

Besides the truffle species described above, sensory techniques were also applied to the whitish truffle *T. borchii*. Interestingly, some of the most important contributors of *T. borchii* aroma were thiophene derivatives and sulfur cyclical compounds, which are derived from truffle's microbiome (Splivallo et al. 2014; Splivallo and Ebeler 2015). This highlights the central role of microbes in human-sensed truffle aroma.

23.5 Truffle Conservation Methods and Substitutes

Distinct preservation methods might affect aroma differently and it is therefore important to evaluate their impact on product quality. Some studies have been carried out in order to examine the influence of the freezing process on the volatile composition of several products, but only a few of them investigated mushrooms (Al-Ruqaie 2006; Jaworska and Bernaś 2009). Al-Ruqaie revealed that freezing might be more effective than drying to conserve the desert truffles *Terfezia claveryi* Chatin and *Terfezia hafizi* Chatin and concluded that blanching in a 4 % NaCl solution and storage at −18 °C proved to be the best preservation method in terms of quality (Al-Ruqaie 2006). Jaworska and Bernaś (2009) proposed a maximum storage period of 4 months for the frozen product in the case of unblanched *Boletus edulis* Bull.

Freezing and frozen storage might affect truffle aroma as well, and this has indeed been recently demonstrated for *T. melanosporum* (Culleré et al. 2013b). Volatile composition data revealed that *T. melanosporum* aromatic profile was deeply modified as a consequence of freezing. These aromatic changes could explain the loss of freshness observed in all frozen truffles. Methional and some phenols appeared as markers of freezing time. Interestingly, 1-octen-3-one was as a general marker of freezing process. After only 24 h of freezing, there was a significant loss of the "characteristic truffle aroma," although intensity was unaffected. Indeed all panelists agreed that none of the frozen samples retained the aroma of fresh truffles. However, the descriptive term "sulfur," usually attributed to the characteristic aroma of truffles, was unaffected by freezing (irrespective of the duration or intensity, −20 or −80 °C). As a matter of fact, the concentration of sulfur volatiles such as dimethyl sulfide and dimethyl disulfide and dimethyl sulfoxide was not affected by freezing, corroborating the sensory results. Overall besides the sulfur attribute, other notes contributed substantially to the typical truffle aroma.

One downside of freezing is that it often destroys food texture. Other "softer" methods have consequently been developed for preservation. Irradiation is recognized as a safe and effective preservation method and it is used to extend the shelf life of raw and processed foods worldwide. The main benefit of irradiation is widely accepted as eliminating microorganisms, insects, or parasites capable of inflicting food spoilage and toxicity, thus replacing chemical fumigants. One downside of food irradiation is that it might decompose fatty acids and subsequently generate off-flavors. Irradiation is nevertheless interesting with truffles since their fruiting bodies are colonized by many microbes (bacteria, yeasts, filamentous fungi, and viruses) and occasionally by insects. Some of these microbes might produce part of truffle aroma; however their uncontrolled growth might also result in off-flavors and spoilage. Indeed the average shelf life of untreated fresh truffles is 10–15 days. In the case of *T. melanosporum*, two irradiation methods were tested for shelf life extension. An electron beam treatment induced important changes in aromatic profile but γ-irradiation at 2.5 kGy did not result in any significant changes. γ-Irradiation therefore constitutes a technique which could potentially be used to increase the shelf life of fresh truffles (Culleré et al. 2012). An alternative method has recently been proposed for *T. magnatum*. It consists of a transparent edible thin film described in a patent application and which is said to double shelf life of truffle-fruiting bodies (Pacioni et al. 2011). But interestingly, traditional methods for conserving fresh truffles are also not equivalent. Indeed simply warping truffles in paper seems more efficient than storing them in rice or under vacuum (Costa et al. 2015).

Overall understanding how truffle aroma is formed through the intimate interactions of truffles with their microbiome will help in developing preservation techniques for shelf-life extension. These techniques should consider the fragile balance between the truffle microbiome and microbes responsible of spoilage/off-flavors and possibly preserve truffle's microbiome but limit the spread of foreign microbes.

23.6 Conclusions

Truffle-fruiting bodies offer a unique niche to investigate from simple to rather complex pathways leading to aroma formation. Techniques in sensory science have complemented traditional profiling approaches by revealing the most important truffle odorants. These studies established the central role of sulfur volatiles in truffle aroma.

Elucidating the pathways leading to characteristic sulfur volatiles as well as the involvement of the different players of truffles' microbiome will require creative thinking. Characterizing these pathways might eventually lead to the biotechnological production of truffle aroma.

References

Al-Ruqaie I (2006) Effect of different treatment processes and preservation methods on the quality of truffles: I. Conventional methods (drying/freezing). J Food Process Preserv 30(3):335–351. doi:10.1111/j.1745-4549.2006.00069.x

Antony-Babu S, Deveau A, Van Nostrand JD, Zhou J, Le Tacon F, Robin C, Frey-Klett P, Uroz S (2013) Black truffle—associated bacterial communities during the development and maturation of *Tuber melanosporum* ascocarps and putative functional roles. Environ Microbiol 16 (9):2831–2847. doi:10.1111/1462-2920.12294

Barbieri E, Bertini L, Rossi I, Ceccaroli P, Saltarelli R, Guidi C, Zambonelli A, Stocchi V (2005) New evidence for bacterial diversity in the ascoma of the ectomycorrhizal fungus *Tuber borchii* Vittad. FEMS Microbiol Lett 247(1):23–35. doi:10.1016/j.femsle.2005.04.027

Barbieri E, Guidi C, Bertaux J, Frey-Klett P, Garbaye J, Ceccaroli P, Saltarelli R, Zambonelli A, Stocchi V (2007) Occurrence and diversity of bacterial communities in *Tuber magnatum* during truffle maturation. Environ Microbiol 9(9):2234–2246. doi:10.1111/j.1462-2920.2007.01338.x

Bellesia F, Pinetti A, Bianchi A, Tirillini B (1998) The volatile organic compounds of black truffle (*Tuber melanosporum* Vitt.) from Middle Italy. Flavour Fragr J 13(1):56–58, doi:10.1002/(SICI)1099-1026(199801/02)13:1<56::AID-FFJ692>3.0.CO;2-X

Bellesia F, Pinetti A, Tirillini B, Bianchi A (2001) Temperature-dependent evolution of volatile organic compounds in *Tuber borchii* from Italy. Flavour Fragr J 16(1):1–6. doi:10.1002/1099-1026(200101/02)16:1<1::AID-FFJ936>3.0.CO;2-Y

Bonito GM, Gryganskyi AP, Trappe JM, Vilgalys R (2010) A global meta-analysis of *Tuber* ITS rDNA sequences: species diversity, host associations and long-distance dispersal. Mol Ecol 19 (22):4994–5008. doi:10.1111/j.1365-294X.2010.04855.x

Brodhun F, Schneider S, Göbel C, Hornung E, Feussner I (2010) PpoC from Aspergillus nidulans is a fusion protein with only one active haem. Biochem J 425:553–565. doi:10.1042/BJ20091096

Burbank HM, Qian MC (2005) Volatile sulfur compounds in Cheddar cheese determined by headspace solid-phase microextraction and gas chromatograph-pulsed flame photometric detection. J Chromatogr A 1066(1–2):149–157. doi:10.1016/j.chroma.2005.01.027
Buzzini P, Gasparetti C, Turchetti B, Cramarossa MR, Vaughan-Martini A, Martini A, Pagnoni UM, Forti L (2005) Production of volatile organic compounds (VOCs) by yeasts isolated from the ascocarps of black (*Tuber melanosporum* Vitt.) and white (*Tuber magnatum* Pico) truffles. Arch Microbiol 184(3):187–193. doi:10.1007/s00203-005-0043-y
Claus R, Hoppen HO, Karg H (1981) The secret of truffles: a steroidal pheromone? Cell Mol Life Sci 37(11):1178–1179. doi:10.1007/BF01989905
Combet E, Henderson J, Eastwood DC, Burton KS (2006) Eight-carbon volatiles in mushrooms and fungi: properties, analysis, and biosynthesis. Mycoscience 47(6):317–326. doi:10.1007/s10267-006-0318-4
Costa R, Fanali C, Pennazza G, Tedone L, Dugo L, Santonico M, Sciarrone D, Cacciola F, Cucchiarini L, Dachà M, Mondello L (2015) Screening of volatile compounds composition of white truffle during storage by GCxGC-(FID/MS) and gas sensor array analyses. LWT-Food Sci Technol 60(2):905–913. doi:10.1016/j.lwt.2014.09.054
Culleré L, Ferreira V, Chevret B, Venturini ME, Sánchez-Gimeno AC, Blanco D (2010) Characterisation of aroma active compounds in black truffles (*Tuber melanosporum*) and summer truffles (*Tuber aestivum*) by gas chromatography–olfactometry. Food Chem 122(1):300–306. doi:10.1016/j.foodchem.2010.02.024
Culleré L, Ferreira V, Venturini ME, Marco P, Blanco D (2012) Evaluation of gamma and electron-beam irradiation on the aromatic profile of black truffle (*Tuber melanosporum*) and summer truffle (*Tuber aestivum*). Innov Food Sci Emerg 13:151–157. doi:10.1016/j.ifset.2011.09.003
Culleré L, Ferreira V, Venturini ME, Marco P, Blanco D (2013a) Potential aromatic compounds as markers to differentiate between *Tuber melanosporum* and *Tuber indicum* truffles. Food Chem 141(1):105–110. doi:10.1016/j.foodchem.2013.03.027
Culleré L, Ferreira V, Venturini ME, Marco P, Blanco D (2013b) Chemical and sensory effects of the freezing process on the aroma profile of black truffles (*Tuber melanosporum*). Food Chem 136(2):518–525. doi:10.1016/j.foodchem.2012.08.030
de Zwart JMM, Kuenen JG (1992) C1-cycle of sulfur compounds. Biodegradation 3(1):37–59. doi:10.1007/BF00189634
Díaz P, Ibáñez E, Señoráns FJ, Reglero G (2003) Truffle aroma characterization by headspace solid-phase microextraction. J Chromatogr A 1017(1–2):207–214. doi:10.1016/j.chroma.2003.08.016
Díaz P, Ibáñez E, Reglero G, Señoráns FJ (2009) Optimization of summer truffle aroma analysis by SPME: comparison of extraction with different polarity fibres. LWT- Food Sci Technol 42(7):1253–1259. doi:10.1016/j.lwt.2009.02.017
Escudero A, Campo E, Fariña L, Cacho J, Ferreira V (2007) Analytical characterization of the aroma of five premium red wines. Insights into the role of odor families and the concept of fruitiness of wines. J Agric Food Chem 55(11):4501–4510. doi:10.1021/jf0636418
Federico V, Cosimo T, Antonio P, Nadia B, Valentina L, Stefano M, Amedeo A (2015) Volatile organic compounds in truffle (*Tuber magnatum* Pico): comparison of samples from different regions of Italy and from different seasons. Sci Rep 5:12629. doi:10.1038/srep12629
Fiecchi A, Kienle MG, Scala A, Cabella P (1967) Bis-methylthiomethane, an odorous substance from white truffle, *Tuber magnatum* Pico. Tetrahedron Lett 8(18):1681–1682
Franco-Luesma E, Ferreira V (2014) Quantitative analysis of free and bonded forms of volatile sulfur compounds in wine. Basic methodologies and evidences showing the existence of reversible cation-complexed forms. J Chromatogr A 1359:8–15. doi:10.1016/j.chroma.2014.07.011
Garscha U, Oliw EH (2009) Leucine/valine residues direct oxygenation of linoleic acid by (10R)- and (8R)-dioxygenases. Expression and site-directed mutagenesis of (10R)-dioxygenase with epoxyalcohol synthase activity. J Biol Chem 284:13755–13765. doi:10.1074/jbc.M808665200

Gerbersmann C, Lobinski R, Adams FC (1995) Determination of volatile sulfur compounds in water samples, beer and coffee with purge and trap gas chromatography—microwave-induced plasma atomic emission spectrometry. Anal Chim Acta 316(1):93–104. doi:10.1016/0003-2670(95)00344-Y

Gioacchini AM, Menotta M, Bertini L, Rossi I, Zeppa S, Zambonelli A, Piccoli G, Stocchi V (2005) Solid-phase microextraction gas chromatography/mass spectrometry: a new method for species identification of truffles. Rapid Commun Mass Spectrom 19(17):2365–2370. doi:10.1002/rcm.2031

Gioacchini AM, Menotta M, Guescini M, Saltarelli R, Ceccaroli P, Amicucci A, Barbieri E, Giomaro G, Stocchi V (2008) Geographical traceability of Italian white truffle (Tuber magnatum Pico) by the analysis of volatile organic compounds. Rapid Commun Mass Spectrom 22(20):3147–3153. doi:10.1002/rcm.3714

Hall IR, Haslam W (2012) Truffle cultivation in the southern hemisphere. In: Zambonelli A, Bonito GM (eds) Edible ectomycorrhizal mushrooms, soil biology, vol 34. Springer, Berlin, pp 191–208

Harki E, Bouya D, Dargent R (2006) Maturation-associated alterations of the biochemical characteristics of the black truffle *Tuber melanosporum* Vitt. Food Chem 99(2):394–400. doi:10.1016/j.foodchem.2005.08.030

Hazelwood LA, Daran JM, van Maris AJA, Pronk JT, Dickinson JR (2008) The Ehrlich pathway for fusel alcohol production: a century of research on *Saccharomyces cerevisiae* metabolism. Appl Environ Microbiol 74(8):2259–2266. doi:10.1128/AEM.02625-07

Jansen O, Raynaud C, Talou T, Gaset A (2003) Flavour profiling of protected black truffle flavourings issued from black truffle industry by-products. In: Le Quéré JL, Etiévant PX (eds) Flavour research at the dawn of the twenty-first century. Proceedings of the 10th Weurman flavour research symposium, Editions Tec & Doc, Beaune, June 2002, p 638–641

Jaworska G, Bernaś E (2009) The effect of preliminary processing and period of storage on the quality of frozen *Boletus edulis* (Bull: Fr.) mushrooms. Food Chem 113(4):936–943. doi:10.1016/j.foodchem.2008.08.023

Jeandroz S, Murat C, Wang Y, Bonfante P, Tacon FL (2008) Molecular phylogeny and historical biogeography of the genus *Tuber*, the "true truffles". J Biogeogr 35(5):815–829. doi:10.1111/j.1365-2699.2007.01851.x

Lefevre C (2012) Native and cultivated truffles of North America. In: Zambonelli A, Bonito GM (eds) Edible Ectomycorrhizal Mushrooms, Current knowledge and future prospects, vol 34, Soil Biology. Springer, Berlin, pp 209–226. doi:10.1007/978-3-642-33823-6_12

Lemfack MC, Nickel J, Dunkel M, Preissner R, Piechulla B (2014) mVOC: a database of microbial volatiles. Nucleic Acids Res 42(D1):D744–D748. doi:10.1093/nar/gkt1250

Liu RS, Li DC, Li HM, Tang YJ (2012) Evaluation of aroma active compounds in Tuber fruiting bodies by gas chromatography-olfactometry in combination with aroma reconstitution and omission test. Appl Microbiol Biotechnol 94(2):353–363. doi:10.1007/s00253-011-3837-7

Liu RS, Zhou H, Li HM, Yuan ZP, Chen T, Tang YJ (2013) Metabolism of L-methionine linked to the biosynthesis of volatile organic sulfur-containing compounds during the submerged fermentation of *Tuber melanosporum*. Appl Microbiol Biotechnol 97(23):9981–9992. doi:10.1007/s00253-013-5224-z

López R, Lapeña AC, Cacho J, Ferreira V (2007) Quantitative determination of wine highly volatile sulfur compounds by using automated headspace solid-phase microextraction and gas chromatography-pulsed flame photometric detection: critical study and optimization of a new procedure. J Chromatogr A 1143(1–2):8–15. doi:10.1016/j.chroma.2006.12.053

March RE, Richards DS, Ryan RW (2006) Volatile compounds from six species of truffle-head-space analysis and vapor analysis at high mass resolution. Int J Mass Spectrom 249:60–67. doi:10.1016/j.ijms.2005.12.038

Martin F, Kohler A, Murat C, Balestrini R, Coutinho PM, Jaillon O, Montanini B, Morin E, Noel B, Percudani R, Porcel B, Rubini A, Amicucci A, Amselem J, Anthouard V, Arcioni S, Artiguenave F, Aury JM, Ballario P, Bolchi A, Brenna A, Brun A, Buée M, Cantarel B,

Chevalier G, Couloux A, Da Silva C, Denoeud F, Duplessis S, Ghignone S, Hilselberger B, Iotti M, Mello M, Miranda M, Pacioni G, Quesneville H, Riccioni C, Ruotolo R, Splivallo R, Stocchi V, Tisserant E, Viscomi AR, Zambonelli A, Zampieri E, Henrissat B, Lebrun MH, Paolocci F, Bonfante P, Ottonello S, Wincker P (2010) Perigord black truffle genome uncovers evolutionary origins and mechanisms of symbiosis. Nature 464(7291):1033–1038. doi:10.1038/nature08867

Mauriello G, Marino R, DAuria M, Cerone G, Rana GL (2004) Determination of volatile organic compounds from truffles via SPME-GC-MS. J Chromatogr Sci 42(6):299–305. doi:10.1093/chromsci/42.6.299

Molinier V, Murat C, Frochot H, Wipf D, Splivallo R (2015) Fine-scale spatial genetic structure analysis of the black truffle *Tuber aestivum* and its link to aroma variability. Environ Microbiol 17(8):3039–3050. doi:10.1111/1462-2920.12910

Mottram DS, Madruga MS (1994) Important sulfur-containing aroma volatiles in meat. In: Sulfur compounds in foods, ACS symposium series. American Chemical Society, pp 180–187

Murat C, Rubini A, Riccioni C, De la Varga H, Akroume E, Belfiori B, Guaragno M, Le Tacon F, Robin C, Halkett F, Martin F, Paolocci F (2013) Fine-scale spatial genetic structure of the black truffle (*Tuber melanosporum*) investigated with neutral microsatellites and functional mating type genes. New Phytol 199(1):176–187. doi:10.1111/nph.12264

Ney K, Freitag W (1980) Trüffel-aroma. Gordian 9:214

Pacioni G, Leonardi M, Aimola P, Ragnelli AM, Rubini A, Paolocci F (2007) Isolation and characterization of some mycelia inhabiting *Tuber* ascomata. Mycol Res 111(12):1450–1460. doi:10.1016/j.mycres.2007.08.016

Pacioni G, Ragnelli A, Stagnini E (2011) Un film edibile e realizzato per preservare la vitalita e le caratteristiche organolettiche dei tartufi freschi. Italian Patent Application AQ2011000001

Pelusio F, Nilsson T, Montanarella L, Tilio R, Larsen B, Facchetti S, Madsen J (1995) Headspace solid-phase microextraction analysis of volatile organic sulfur compounds in black and white truffle aroma. J Agric Food Chem 43(8):2138–2143. doi:10.1021/jf00056a034

Piloni M, Tat L, Tonizzo A, Battistutta F (2005) Aroma characterisation of white truffle by GC-MS and GC-O. Ital J Food Sci 17(4):463–468

Rubini A, Belfiori B, Riccioni C, Arcioni S, Martin F, Paolocci F (2011) *Tuber melanosporum*: mating type distribution in a natural plantation and dynamics of strains of different mating types on the roots of nursery inoculated host plants. New Phytol 189(3):723–735. doi:10.1111/j.1469-8137.2010.03493.x

Segurel MA, Razungles AJ, Riou C, Salles M, Baumes RL (2004) Contribution of dimethyl sulfide to the aroma of Syrah and Grenache Noir wines and estimation of its potential in grapes of these varieties. J Agric Food Chem 52(23):7084–7093. doi:10.1021/jf049160a

Smet E, Van Langenhove H, De Bo I (1999) The emission of volatile compounds during the aerobic and the combined anaerobic/aerobic composting of biowaste. Atmos Environ 33(8):1295–1303. doi:10.1016/S1352-2310(98)00260-X

Splivallo R, Ebeler SE (2015) Sulfur volatiles of microbial origin are key contributors to human-sensed truffle aroma. Appl Microbiol Biotechnol 99(6):2583–2592. doi:10.1007/s00253-014-6360-9

Splivallo R, Maier C (2011) Production of natural truffle flavours from truffle mycelium. US Patent Application No PCTIB2010052913

Splivallo R, Bossi S, Maffei M, Bonfante P (2007) Discrimination of truffle fruiting body versus mycelial aromas by stir bar sorptive extraction. Phytochemistry 68(20):2584–2598. doi:10.1016/j.phytochem.2007.03.030

Splivallo R, Ottonello S, Mello A, Karlovsky P (2011) Truffle volatiles: from chemical ecology to aroma biosynthesis. New Phytol 189(3):688–699. doi:10.1111/j.1469-8137.2010.03523.x

Splivallo R, Rittersma R, Valdez N, Chevalier G, Molinier V, Wipf D, Karlovsky P (2012a) Is climate change altering the geographical distribution of truffles? Front Ecol Environ 10(9):461–462. doi:10.1890/12.WB.020

Splivallo R, Valdez N, Kirchhoff N, Ona MC, Schmidt JP, Feussner I, Karlovsky P (2012b) Intraspecific genotypic variability determines concentrations of key truffle volatiles. New Phytol 194(3):823–835. doi:10.1111/j.1469-8137.2012.04077.x

Splivallo R, Deveau A, Valdez N, Kirchhoff N, Frey-Klett P, Karlovsky P (2014) Bacteria associated with truffle-fruiting bodies contribute to truffle aroma. Environ Microbiol 17 (8):2647–2660. doi:10.1111/1462-2920.12521

Stielow B, Menzel W (2010) Complete nucleotide sequence of TaV1, a novel totivirus isolated from a black truffle ascocarp (*Tuber aestivum* Vittad.). Arch Virol 155(12):2075–2078. doi:10.1007/s00705-010-0824-8

Talou T, Delmas M, Gaset A (1987) Principal constituents of black truffle (*Tuber melanosporum*) aroma. J Agric Food Chem 35(5):774–777. doi:10.1021/jf00077a031

Talou T, Delmas M, Gaset A (1989) Black Perigord truffle: from aroma analysis to aromatizer formulation. In: Flavors and off-flavors, Proceedings of the 6th international flavor conference, Rethymnon, pp 715–728

Tirillini B, Verdelli G, Paolocci F, Ciccioli P, Frattoni M (2000) The volatile organic compounds from the mycelium of Tuber borchii Vitt. Phytochemistry 55(8):983–985. doi:10.1016/S0031-9422(00)00308-3

Vahdatzadeh M, Deveau A, Splivallo R (2015) The role of the microbiome of truffles in aroma formation: a meta-analysis approach. Appl Environ Microbiol 81(20):6946–6952. doi:10.1128/AEM.01098-15

Zeppa S, Gioacchini AM, Guidi C, Guescini M, Pierleoni R, Zambonelli A, Stocchi V (2004) Determination of specific volatile organic compounds synthesised during *Tuber borchii* fruit body development by solid-phase microextraction and gas chromatography/mass spectrometry. Rapid Commun Mass Spectrom 18(2):199–205. doi:10.1002/rcm.1313

Chapter 24
A Proteomic View of Truffles: Aspects of Primary Metabolism and Molecular Processes During Their Life Cycle

Antonella Amicucci, Marselina Arshakyan, Paola Ceccaroli, Francesco Palma, Giovanni Piccoli, Roberta Saltarelli, Vilberto Stocchi, and Luciana Vallorani

24.1 Introduction

Proteins are responsible for innumerable tasks within the cell. The complete set of proteins in a cell can be referred to as its proteome, and the study of protein structure and function and what every protein in the cell does is known as proteomics. The proteome is highly dynamic and it changes periodically in response to different environmental stimuli.

Proteomics, as the analysis of the protein components of organisms, is a well-known tool providing a global perspective (whole-organism approach) of cellular physiology that allows understanding the cellular protein expression alterations in response to various biotic and abiotic stresses. By virtue of its capacity to yield definitive information on protein identity, localization, posttranslational modification, and the accuracy of in silico gene model prediction in fungi, proteomics has become an integral component of all large-scale "omic" and systems approaches to understand the rich complexity of fungal biochemistry (Doyle 2011).

The life cycle of truffles (*Tuber* spp.) is very complex and can be divided into distinct stages: a growth as filamentous vegetative mycelium, an association of the fungal hyphae with the host roots to form the ectomycorrhizas (ECMs), and the organization of hypogeous fruiting bodies with asci and ascospore.

The proteomic assets of the *Tuber* spp. have been thoroughly studied for the past 20 years using various methods such as the purification of proteins from biological sources and heterologous expression in prokaryotes and eukaryotic cells. Several proteins have been characterized in the different phases of the life cycle of the most

A. Amicucci • M. Arshakyan • P. Ceccaroli • F. Palma • G. Piccoli • R. Saltarelli • V. Stocchi • L. Vallorani (✉)
Dipartimento di Scienze Biomolecolari, Università of Urbino Carlo Bo, Via Saffi 2, 61029, Urbino, PU, Italy
e-mail: luciana.vallorani@uniurb.it

A. Zambonelli et al. (eds.), *True Truffle (*Tuber *spp.) in the World*, Soil Biology 47,
DOI 10.1007/978-3-319-31436-5_24

Table 24.1 Summary of the proteins described in this study (PS, protein symbol; GS, gene symbol; PE, phase of expression)

Species[a]	Protein	PS	GS	PE[b]	Activity and function
Tbo	Enolase	ENO-1	*eno-1*	Myc	Lyase (interconversion of 2-phosphoglycerate in phosphoenolpyruvate)
Tbo	Glutamate dehydrogenase	GDH	*gdh*	Myc	Reductive amination 2-oxoglutarate to glutamate (primary nitrogen metabolism in response to different nitrogen sources)
Tbo	High-affinity ammonium transporter	TbAMT1	*TbAMT1*	Myc	Ammonium transporter (transport ammonium onto the cell; expression modulated upon perturbation of the N status)
Tbo	Cell division control protein 42	TbCdc42	*Tbcdc42*	Myc	Small GTPase (Rho family) (polarized growth and actin cytoskeleton organization; required for polarity establishment and maintenance, in bud formation and shmooing)
Tbo	Fruiting body lectin 1	TBF-1	*tbf-1*	Asc	Lectin (sorting bacteria during the ascoma formation)
Tbo	Glutamine synthetase	TbGS	*TbGS*	Myc ECM Asc	Condensation of glutamate and ammonia to form glutamine (respond positively to N deprivation and return rapidly to basal levels upon resupplementation of various forms of nitrogen, especially nitrate)
Tbo	Hexokinase	TBHXK1	*hxk-1*	Myc	Phosphorylase (transfer of a high-energy phosphate group of a donor to glucose, producing glucose-6-phosphate)
Tbo	Hexose transporter	TBHXT1	*Tbhxt1*	Myc ECM	High-affinity glucose transporter (glucose and fructose uptake)
Tbo	Mannitol dehydrogenase	TbMDH	*Tbmdh*	Myc	Dehydrogenase reductases (fructose reduction and mannitol oxidation)
Tbo	Nitrite reductase	TBNR1	*tbnr1*	Myc ECM	Catalyze the nitrite reduction (inducible gene responding mainly to nitrate; upregulated after symbiosis establishment)
Tbo	Transporter	TbNrt2	*TbNrt2*	Myc ECM	Nitrate and nitrite transporter (high-affinity transporter bispecific for nitrate and nitrite, expressed in free-living mycelia and ECMs; upregulated following Myc transfer to a nitrogen-free medium)

(continued)

Table 24.1 (continued)

Species[a]	Protein	PS	GS	PE[b]	Activity and function
Tbo	Rho GTPase dissociation inhibitor	TbRhoGDI	*Tbgdi*	Myc	Small GTP-binding proteins (Rho family) (important modulator of cytoskeleton reorganization during polarized apical growth that antecedes symbiosis instauration)
Tbo	Homologue of the blue-light photoreceptor	TbWC-1	*Tbwc-1*	Myc	Photoreceptor (prevent mycelia from growing toward the light)
Tdr	Fruiting body lectin 1 (Tbf-1 orthologues)	TDF-1	*tdf-1*	Asc	Lectin (sorting bacteria during the ascoma formation)
Tme	Invertase	TmelINV	*tmelinv*	Myc EC Asc	β-Fructofuranosidase (catalyzes the hydrolysis of sucrose in fructose and glucose)

[a]Tbo, *T. borchii*; Tdr, *T. dryophilum*; Tme, *T. melanosporum*
[b]M, Myc; ECM, ectomycorrhiza; Asc, Ascoma

important *Tuber* species in terms of agronomics and economics. In particular, the availability of a reliable *Tuber borchii* Vittad. in vitro system has allowed researchers to investigate its complex life cycle (Sisti et al. 1998), while the complete genome of *Tuber melanosporum* Vittad. provides a wealth of information for in-depth studies. On the contrary, *Tuber magnatum* Pico is more difficult to cultivate in the laboratory, and therefore studies have been focused only on its ascoma.

Herein we review some of the most interesting investigations regarding the characterization of specific proteins and metabolic pathways that can give us insights into these intriguing organisms belonging to the genus *Tuber*. The proteins described in this review are summarized in Table 24.1.

24.2 Proteomic Studies on Truffles

Modern molecular biology benefits greatly from the growing availability of genomic sequences; however, a genome sequence only represents a picture of an organism's physiological cell development, and adaptive potential. Hence, to gain an in-depth understanding of metabolic processes as well as physiological adaptation to environmental changes, it is necessary to bring transcriptomics and proteomics into play. While genomics may provide a comprehensive picture of everything related to nucleic acids, transcriptomics sheds light about the set of RNA transcripts that are produced by the genome under specific circumstances and proteomics provides an actual description of specific proteins or complexes of proteins in their biological milieu. Furthermore, proteomes are much more multiform and dynamic than related upstream genomes, hence requiring complex experimental pipelines, including separation, detection, and data processing to highlight protein content and/or modifications during the complex life cycle of an organism. Proteins are very challenging compounds; their amino acid composition determines their physicochemical properties (e.g., hydrophobicity, polarity, acid/base character) and affects the formation of specific tertiary and quaternary structures (e.g., globular or fibrillar proteins, mono- or multimers, supercomplexes). In addition, proteins can be post-translationally modified (e.g., phosphorylation, glycosylation) to alter their catalytic, regulatory, or structural properties. Considering the enormous structural and functional diversity of proteins, only rational combinations of various analytical approaches allow us to gain an effective view of the overall state of the cell.

Protein complements are separated by standard methods, including gel electrophoresis and liquid chromatography (Baggerman et al. 2005; Rabilloud et al. 2010), and individual protein species are usually identified by mass spectrometry (Domon and Aebersold 2006). The successful coupling of multidimensional separations with mass spectrometry (MS) for protein and peptide analyses has been achieved thanks to the advent of the matrix-assisted laser desorption ionization (MALDI) and electrospray ionization (ESI) techniques and the evolution of new powerful MS instruments. While MALDI is usually combined with gel-based separations, ESI is frequently coupled with "online" high-pressure liquid chromatography (HPLC) or reversed-phased liquid chromatography (RPLC) separations. Two main analytical approaches are commonly used for protein analyses for proteome separation and characterization: "top-down" and "bottom-up." The top-down method is used for the analysis of intact proteins separated by HPLC and identified by MS. In the bottom-up approach, samples require a gel-based purification step, followed by an enzymatic or chemical digestion that provides peptides separated by multidimensional chromatography such as 2D-HPLC, which are finally identified by tandem MS (Cui et al. 2011; Wu et al. 2012).

The majority of proteomic studies on truffles have been focused on cultured mycelia and fruiting bodies. One of the first comparisons of electrophoretic protein patterns from fruiting bodies of various truffle species using one-dimensional sodium dodecyl sulfate-polyacrylamide gel electrophoresis (1D-SDS-PAGE) showed significant differences in protein expression among *T. magnatum*, *T. borchii*, *Tuber dryophilum* Tul. & C. Tul., and *Tuber puberulum* Berk. & Broome (De Bellis et al. 1998). In particular, a predominant protein with a molecular weight of 11.9 kDa was found in *T. borchii* and *T. dryophilum* (Agostini et al. 2000), but not in the other species. This protein was purified, characterized, and named TBF-1, whose function will be described below.

The variability of protein profiles across different *Tuber* species was also highlighted in a study on the effect of different storage treatments on *T. magnatum*, *T. borchii*, *T. melanosporum*, and *Tuber aestivum* Vittad. fruiting bodies (Saltarelli et al. 2008). One-dimensional SDS-PAGE revealed species-specific bands even in species that were very similar morphologically such as *T. magnatum* and *T. borchii*. Further differences in protein profiles were emerged during the storage of fresh fruiting bodies at 4 °C; in particular, during the time course of 30 days, one protein of about 45 kDa progressively disappeared, while two proteins of about 25 kDa and 20 kDa increased in *T. borchii* fruiting bodies. During the storage at 4 °C, changes in proteins were observed also in *T. melanosporum* and *T. aestivum*, while the fruiting bodies of *T. magnatum* remained quite stable. In addition to the changes which have been observed in *T. borchii* fruiting bodies during the storage, changes in protein expression during maturation were reported by Pierleoni et al. (2004). The comparison of two-dimensional electrophoresis (2-DE) maps obtained from immature and mature *T. borchii* fruiting bodies showed that in the mature fruiting bodies the expression level of many 30–150 kDa proteins decreases. This is not surprising because in the fructification initial phase, a higher energy supply is necessary to meet the cell demand since biosynthetic pathways are strongly induced. However, some proteins are more expressed in the mature fruiting bodies. To understand the processes leading to ascoma development in *T. borchii*, the identification of proteins specific to one or both differentiation phases was attempted. Of the 19 spots excised from 2-DE gels, four were identified, one of which was found to be more expressed in the immature fruiting body (ATP synthetase A chain), one upregulated in the mature ascoma (endoglucanase), while metallothionein and ferredoxin present in equal measure in each maturation phase. Endoglucanase is able to degrade crystalline cellulose to glucose and must be involved in the establishment of symbiosis. Moreover, this enzyme allows fungi to grow and fruit on particular substrates. The decrease in ATP synthetase A chain during truffle maturation is in agreement with the higher energy supply required in the initial phase of fructification for the formation of membrane, cell walls, and other processes. The lack of information on the *T. borchii* genome has limited our ability to identify a greater number of proteins both in the fruiting body and in mycelium.

The only available study on the proteome of the *T. borchii* mycelium was carried out by Vallorani et al. (2000). The authors provide a 2D-PAGE map of 30-day-old *T. borchii* mycelium grown in a liquid medium. The map reveals the presence of about 800 spots, 23 of which were analyzed using both amino acid composition and amino terminal sequence tagging. However, the majority of the spots did not provide information when subjected to N-terminal sequencing, probably because the N-terminus was blocked in some way. Only five spots yielded positive results, but no similarity with sequences in databases was found except in one case. The amino acid composition data obtained for all the spots did not match with any protein in the databases except for the spot corresponding to ubiquitin, which was also confirmed by an N-terminal tag. To date, these sequences do not show any similarities in databases, including *T. melanosporum* database. In the 2-DE map, spots corresponding to glutamate dehydrogenase were also identified by coelectrophoresis with the purified enzyme (Vallorani et al. 2002). The proteomic results on *T. borchii* presented in this investigation, although limited in the number of proteins identified, may be considered useful for subsequent studies designed to better understand the events responsible for the formation and maturation of the *T. borchii* fruiting body. Proteomic analysis has recently been used to characterize the differences occurring in *T. magnatum* fruiting bodies naturally grown in four different regions of central and northern Italy (Vita et al. 2013). The area where truffles are grown significantly influences the quality of fruiting bodies and their protein profile. This is due to differences in environmental growth conditions, the particular developmental stage of the fungus, and genetic variations. The comparison of 2D gels of *T. magnatum* from different geographic areas revealed quantitative differences that remained constant over different collecting years. Using a precise bioinformatics method of analysis, the authors identified 17 spots suitable to discriminate the area of origin, 15 of which were sequenced by mass spectrometer. One of the proteins identified was NADP-dependent mannitol dehydrogenase, the enzyme that catalyzes the reversible conversion of mannitol to fructose. Another enzyme affected by environment was malate dehydrogenase, which is involved in the Krebs cycle, the glyoxylate cycle, and the gluconeogenesis pathway. The differential expression of these proteins is not surprising as truffles are not autotrophic organisms and rely on plant-originated carbohydrate breakdown for their metabolism. However, for most of the proteins identified, it was difficult to extrapolate specific physiological roles related to their different environments of origin. The results obtained by Vita and colleagues indicate that proteomic analysis is a suitable method for characterizing the differences in *T. magnatum* grown in various geographical areas, even if the genome sequencing of this species is not available. This finding may also have important economic implications, as the fruiting bodies from different areas have different organoleptic characteristics that affect the price; thus, geographic characterization might allow us to unmask fraudulent products (Gioacchini et al. 2008).

Regarding *T. melanosporum*, the 2010 publication of its genome (Martin et al. 2010) provided useful information to gain a better understanding, at the genomic level, of its carbohydrate metabolism, transcription, mating behavior, volatiles that produce aroma, and other aspects of transcriptional and genomic control. However, to fully understand the biology of this fungus, a combined bioinformatics and proteomic study to characterize the *T. melanosporum* proteome was carried out (Islam et al. 2013). The authors applied a sequential BLAST method and annotated the truffle proteome from the 2010 *T. melanosporum* genome comprising 12771 putative proteins. They provided high-quality annotation for 2587 sequences and proteomic evidence of 836 truffle proteins. Approximately 20 % of all of the putative proteins were assigned putative biological functionality. In particular, nine proteins responsible for a part of the aroma profile were identified. Furthermore, the authors suggested a potential enzymatic pathway for the production of the primary volatiles in black truffles. These studies, along with new findings which identified and characterized novel proteins (Payen et al. 2015), have begun to shed light on truffle biological pathways paving the way for the identification of potential biomarkers of authenticity, freshness, and aroma.

24.3 The Major Pathways of Primary Metabolism in Truffles

24.3.1 Carbon Metabolism

Ectomycorrhizal (ECM) symbiosis between certain soil fungi and plant roots is one way that these respective partners overcome their nutritional limitations. Fungi contribute to tree nutrition by means of mineral weathering and mobilization of nutrients from organic matter and obtain plant-derived carbohydrates as a response (Nehls 2008). Primary carbon metabolism, related to carbon/nitrogen exchange at the plant-fungus interface, as well as metabolic support of mycelium and fruiting body development, plays a fundamental role in the *Tuber* sp. Under axenic growth conditions, which do not necessarily reflect the situation in soil, the best carbon sources for *T. borchii* growth are simple sugars, such as glucose, fructose, mannose, and mannitol (Saltarelli et al. 1998; Ceccaroli et al. 2001). On the contrary, *T. melanosporum* also grows in liquid media containing sucrose as the sole carbon source and a sucrose-based fed-batch fermentation process has been developed for the efficient production of mycelia and bioactive polysaccharides from this fungus (Liu et al. 2009). In fact, *T. melanosporum* possesses an acid invertase, the first enzyme of this kind to be identified in a plant symbiotic filamentous fungus (Ceccaroli et al. 2011), and Hacquard et al. (2013) report the accumulation of the invertase transcripts in the Hartig net. This evidence suggests that *T. melanosporum* may have the ability to hydrolyze plant-derived sucrose to provide the substrate for fungal uptake at the symbiotic interface. In fact, Le Tacon et al. (2013), using carbon flux assessment with labeled $^{13}CO_2$ by the host to *T. melanosporum* mycorrhizas and ascomata, demonstrated that the mycorrhizas first form a carbon sink and accumulate ^{13}C prior to ascoma development. Then, the mycorrhizas transfer ^{13}C to the ascomata to provide constitutive carbon. The ascomata accumulate host carbon until reaching complete maturity (Le Tacon et al. 2013).

Carbohydrate uptake by fungal cells is the prerequisite for both mycelium growth and the development of ECM tissues. The first hexose transporter to be identified and completely characterized in *Tuber* sp. is TBHXT1 from *T. borchii* (Polidori et al. 2007). TBHXT1 has a strong preference for D-glucose (Km 38 ± 10 μM) over D-fructose (Km 16 ± 5 mM), and uncoupling experiments indicate that TBHXT1 catalyzes the transport via a proton-symport mechanism. The gene *Tbhxt1* is not regulated by fructose, but it reaches its highest level of expression at 3 mM glucose and it is repressed by very high glucose concentration. Prolonged carbon starvation conditions upregulate *Tbhxt1*, while its expression remains at a basal level in the ECM tissue. The mode of regulation of *Tbhxt1* is consistent with its role as a high-affinity D-glucose transporter; hence, it has been concluded that TBHXT1 is mainly responsible for carbohydrate uptake by soil-growing hyphae and the reduction of sugar leakage from hyphae (Polidori et al. 2007). The sequenced genome of *T. melanosporum* contains about 20 gene models, which are homologous to sugar transporters identified in other basidiomycetes or ascomycetes but not functionally characterized (Ceccaroli et al. 2011). Five potential candidate genes are upregulated in ECM tissue including orthologous to *Tbhxt1* (Hacquard et al. 2013).

The sugars imported by fungal hyphae are utilized for ATP generation, amino acid biosynthesis, and the formation of (potential) carbohydrate storage compounds. Genome analysis and biochemical studies have allowed the reconstruction of the central carbohydrate metabolism in *Tuber* sp. (Ceccaroli et al. 2011). All the genes of Embden–Meyerhof (EM) pathway, which converts glucose into pyruvate, and oxidative pentose phosphate (PP) pathway, which generates NADPH and pentose sugars, have been identified (also in multiple copies of some genes) in the *T. melanosporum* genome (Ceccaroli et al. 2011). The evaluation of enzymatic activities in *T. borchii* has shown that most of the enzymes of these pathways are more active in mycelium than in fruiting bodies (Saltarelli et al. 1998).

One of the major regulatory steps in the EM pathway is represented by the ATP-dependent conversion of hexose to hexose 6-phosphate catalyzed by hexokinase. In *T. borchii*, three hexokinase activities (HKM1, HKM2, HKM3) have been identified and purified by anion exchange chromatography. These enzymatic forms are detectable during the different phases of fungus growth: HKM1 and HKM2 are present in the young mycelium, whereas HKM3 appears only in the final phase of growth and it is also the unique form found in the *T. borchii* fruiting body (Ceccaroli et al. 1999). Despite the presence of three distinct enzymatic forms, only one gene from *T. borchii* coding for hexokinase (*hxk-1*) has been isolated and characterized (Agostini et al. 2001). The *T. melanosporum* genome harbors two genes codifying for hexokinase: one gene codes for a protein showing the highest similarity with *T. borchii hxk-1* (AF309088, 94 % identity) (Agostini et al. 2001) and the second gene shares maximum identity (39 %) with a *Neurospora crassa* Shear and B.O. Dodge hypothetical protein (NCU06996) and with several putative hexokinase family proteins XprF, which are purely regulatory in terms of function and play a role in the response to carbon starvation (Ceccaroli et al. 2011). Among the enzymes of the glycolytic pathway, enolase has also been purified and the corresponding gene (*eno-1*) has been cloned and characterized in *T. borchii* (Polidori et al. 2004). This enzyme seems to be an unregulated constitutively expressed protein in most environmental conditions. Moreover, in the case of *T. borchii* mycelium, the *eno-1* transcriptional level and enzymatic activity did not change when different carbon sources were used in the medium, but *eno-1* mRNA transcript was overexpressed during the mycelial growth and the enzymatic activity similarly increased (Polidori et al. 2004).

In addition to rapid carbohydrate uptake and metabolism, storage carbohydrates are of particular interest in ECM fungi. Although trehalose, mannitol, and glycogen are the most common carbohydrates in fungi, the accumulation of these compounds varies greatly among ECM species. HPLC analyses of *T. melanosporum* mycelia and fruiting bodies have revealed the presence of both trehalose and mannitol, and all genes involved in these pathways have been identified in the genome (Ceccaroli et al. 2011). RT-qPCR analyses show that mannitol dehydrogenases and mannitol 1-phosphate dehydrogenase mRNA transcripts are more highly expressed in mycelium than in fruiting bodies, suggesting that mannitol metabolism mainly operates in mycelial tissue. On the contrary, trehalose metabolism is upregulated in fruiting bodies and all glycogen metabolism genes are expressed at the highest level in ECMs (Ceccaroli et al. 2011). In *T. borchii* mycelium, the specific activities of all enzymes of mannitol metabolism have been determined and nuclear magnetic resonance spectroscopy (NMR) experiments have shown that mannitol synthesis occurs, whereas trehalose and its metabolic pathway have not been detected (Ceccaroli et al. 2003). In this fungus, D-mannitol is produced from D-fructose by a novel $NADP^+$-dependent D-mannitol dehydrogenase, named TbMDH, which has been purified and characterized and whose corresponding gene has been cloned. The enzyme is a homotetramer with two zinc atoms per subunit and catalyzes both D-fructose reduction and D-mannitol oxidation, although it shows the highest substrate specificity and catalytic efficiency for D-fructose. The gene *Tbmdh* is upregulated when mycelia are transferred to a culture medium containing D-mannitol or D-fructose. The enzyme identified a new group of proteins, most of them annotated in databases as hypothetical zinc-dependent dehydrogenases, forming a distinct subfamily among the polyol dehydrogenase family (Ceccaroli et al. 2007).

24.3.2 Nitrogen Metabolism

Nitrogen retrieval and assimilation by symbiotic ECM fungi is thought to play a central role in the mutualistic interaction between these organisms and their plant hosts. Urea, ammonium, and nitrate are the major nitrogenous compounds of forest soils and can be used by *Tuber* sp. as a source of nitrogen through the activation of dedicated transporters at the soil/mantle interface (Hacquard et al. 2013). In *T. borchii* mycelium, an ammonium transporter called TbAMT1 has been isolated and functionally characterized. TbAMT1 protein is a high-affinity ammonium transporter of 52 kDa with a Mep/Amt-like membrane topology, and the *TbAMT1* mRNA transcript in *T. borchii* mycelia is modulated upon perturbation of the N status (Montanini et al. 2002). The nitrate transporter TbNrt2 protein of *T. borchii* is a high-affinity transporter ($Km = 4.7$ μM nitrate) that is bispecific for nitrate and nitrite. It is expressed in free-living mycelia and in mycorrhizas, where it preferentially accumulates in the plasma membrane of root-contacting hyphae (Montanini et al. 2006). The *TbNrt2* mRNA was specifically upregulated following transfer of mycelia to nitrate (or nitrite)-containing medium. TbNrt2 was also upregulated (at both mRNA and protein level) following transfer to a nitrogen-free medium. The functional and expression properties delineate TbNrt2 as a versatile transporter that may be especially suited to cope with the fluctuating (and often low) mineral nitrogen concentrations found in most natural, especially forest soils.

After assimilation, the transfer of nitrogen from the fungal hyphae to the plant is the result of various metabolic processes made possible by the action of several enzymes, such as glutamate dehydrogenase (GDH), glutamine synthetase (GS), nitrate reductase (NR), nitrite reductase (NiR), and amino transferases (ATs). In *T. borchii*, some of these proteins have been identified and biochemically characterized and their genes have been cloned. NADP-GDH appears to be a hexameric protein consisting of identical subunits and its Km value for glutamate is much lower than that of other fungi (Km = 2.79 mM), which points to the greater affinity of *T. borchii* GDH for this organic N source. Biochemical and Northern blot analyses carried out with mycelia grown on different nitrogen sources clearly show an upregulation in response to different nitrogen sources such as urea, nitrate, alanine, and glutamate and in the symbiotic phase (Vallorani et al. 2002).

Until now, only one gene that codes for a GS polypeptide (TbGS) has been identified in *T. borchii*. This polypeptide autonomously assembles into catalytically active homooligomers. TbGS displays a strong positive cooperativity ($n = 1.7 \pm 0.29$) and an unusually high Km value (54 ± 16 mM) for glutamate and a correspondingly low sensitivity toward inhibition by the glutamate analogue herbicide phosphinothricin. The gene *TbGS* is upregulated in N-starved mycelia and returns to basal levels upon resupplementation with various forms of N, the most effective of which is nitrate. Both responses are accompanied by parallel variations in the amount of TbGS protein and glutamine synthetase activity, thus indicating that *TbGS* mRNA levels are primarily controlled at the pre-translational level (Montanini et al. 2003).

The nitrate reductase gene (*tbnr1*) from *T. borchii* encodes a protein of 929 amino acids. Biochemical and Northern blot analyses reveal that nitrate assimilation in *T. borchii* is an inducible system that responds mainly to nitrate. Enzymatic assays show that NADPH-dependent nitrite formation markedly increases in ECMs, suggesting that the fungal partner plays a fundamental role in nitrate assimilation.

The nitrite reductase gene of *T. borchii* (*tbnir1*) forms a gene cluster with the nitrate transporter gene and is downregulated in the presence of primary nitrogen sources and upregulated following transfer to either nitrate- or nitrogen-limited conditions. Real-time PCR assays reveal that both nitrate transporter and nitrite reductase expression levels are about 15-fold and 10-fold higher in ECM tissues than in control mycelia, respectively, suggesting that the symbiotic fungus *T. borchii* contributes to improving the host plant's ability to make use of nitrate/nitrite in its nitrogen nutrition (Guescini et al. 2007).

24.4 Proteins Involved in the Changes Occurring During the Truffle Life Cycle

Truffles have a complex life cycle involving the reorganization of mycelium into specialized structures (mycorrhiza and fruiting body). This process is affected by numerous morpho-functional modifications linked to spatiotemporal developmental changes in genes and related protein expression in the fungus. Moreover, endogenous factors as well as biotic and abiotic features affect the truffle life cycle. Stimuli such as the presence of the plant, volatile compounds, root extracts, different carbohydrate sources, and bacteria as well as climate dynamic can induce morphological and functional changes (Amicucci et al. 2010; Potenza et al. 2012).

Among the genes differentially expressed during its life cycle, an 11.9-kDa protein specifically expressed in the *T. borchii* fruiting body is of particular interest. The protein, called TBF-1 (Swiss-Prot accession no. P80708), represents the most abundant protein in aqueous extract of truffle ascoma (De Bellis et al. 1998). This protein has been purified and sequenced and the related gene cloned (De Bellis et al. 1998). More recently, the TBF-1 has been produced in *Escherichia coli* (Palma et al. 2005). Structural characterization has shown that TBF-1 is a homodimeric protein, not-glycosylated, consisting two monomers of 108 amino acids. Furthermore, the TBF-1 has no orthologue in *T. melanosporum* genome, whereas it has significant homology with the TDF-1 in *T. dryophilum* (Agostini et al. 2000); in fact, these species are phylogenetically closely related and belonging to Puberulum group (see Chap. 7). A comparison of the amino acid sequence deduced from the *tbf-1* gene and the protein sequence has revealed the presence of an N-terminal 12-amino acid peptide. Further studies, using heterologous expression of several engineered variants of TBF-1 on yeast, showed that the peptide is a noncanonical signal peptide, which might enable the TBF-1 to enter the secretory pathway, bypassing interaction with a signal recognition particle (SRP). On the other hand, it leads the TBF-1 to the classical pathway via a nonconventional route (Nombela et al. 2006; Palma et al. 2007).

Localization of TBF-1 within the fruiting body tissue has allowed the localization of the protein in the cell walls of hyphae and asci in a co-localization with chitin and glucans. On the contrary, the presence of TBF-1 was not detected on the spore surface (De Bellis et al. 1998). Moreover, in the yeast cells expressing TBF-1, the protein was localized both loosely bound to the cell wall and free in the medium (Palma et al. 2007).

Hemagglutination assay, using rabbit erythrocytes, has shown that TBF-1 is a lectin able to bind the exopolysaccharides extracted by *Rhizobium* strains from *T. borchii* ascoma and does not react with either the specific *Rhizobium* of Leguminosae or with *Rhizobia* isolated from the *T. magnatum* (Cerigini et al. 2008). The high binding specificity of TBF-1 towards rhizobial surface polysaccharides seems to suggest an active role of the protein in sorting these bacteria during the fructification phase. Hence, the presence of nitrogen-fixing bacteria in the fruiting bodies of *Tuber* spp. and in association with other fungi (Frey-Klett et al. 2007) suggests that these associated microorganisms could play an important role in fungal growth and nutrition during ascoma formation (see Chap. 18). In addition to their interaction with the microbial community, in the soil, truffles also respond to abiotic factors.

It has been reported that organoleptic properties of truffle are equilibrium among genetic, microorganism inside ascomata, environment, and climate (Molinier et al. 2015). Thus, the influence of light, on *T. borchii* mycelial growth, has been investigated. Ballario and her team reported for the first time a phenotypic response to light in a "fully" hypogeous organism (Ambra et al. 2004). *Tuber borchii* mycelia were found to respond to blue light by inhibiting apical growth and producing in vitro mycelial colonies of reduced size. Furthermore, to study light transduction, the gene *Tbwc-1* was cloned, resulting in the homologue of the blue-light photoreceptor of *N. crassa*, NcWC-1. The structural characterization of the protein expressed by the *T. borchii* gene showed the presence of the light-oxygen-voltage (LOV)-sensing (LOV) domain that is present in signal transduction proteins, a nuclear localization signal (NLS) domain, and a Per/Arnt/Sim (PAS) domain, which function as a signal sensor, but not the glutamine-rich regions (activator domain) present in *N. crassa*. These findings support the hypothesis that the TbWC-1 is a blue-light photoreceptor, but its role as transcriptional activator remains unclear. However, why should a hypogenous organism need a photoreceptor? TbWC-1 might prevent mycelia from growing toward the light, thus favoring growth toward plant roots in order to form mycorrhizas. At the level of spore dispersal from the fruiting body, the perception of light might inhibit the growth of the germinating mycelium in this direction and therefore promote the underground growth.

Filamentous fungi grow by the polar expansion of hyphae, and polarized growth is the dominant growth of filamentous fungi, while, in single cell yeasts, such as budding yeast and fission yeast, polarized growth is restricted to certain times during the cell cycle (i.e., in starvation) (Banuett et al. 2008; Harris 2006; Steinberg 2007). In fact, the continuous polarized growth of filamentous fungi at their tips is fundamental to their expansion in the soil. Such growth aims to get nutrients and to meet the opposite mating-type hypha, to move toward the plant roots, and to develop the mycorrhiza and fruiting body.

Studies on *T. borchii* have shown that Cdc42 and Rho-Gdi have a fundamental role in hyphal growth. *Tuber borchii* Cdc42 is involved in polarized growth and has a fundamental role in the organization of the actin cytoskeleton (Menotta et al. 2007, 2008). In particular, heterologous expression assays with a constitutively active TbCdc42 variant (Q63L mutation) show that in yeast Cdc42 is able to switch on a series of evident morphological modifications, including apical growth and branching (Menotta et al. 2007). Cdc42 together with other small GTP-binding proteins belonging to Rho GTPases is involved primarily in the reorganization of the actin cytoskeleton and, hence, in all cell processes linked to morphological modifications. The processes include cytokinesis, cell motility, vacuole trafficking, secretion, and apoptosis in all eukaryotes, and more recently, it has been shown that they are also upstream of complex signaling pathways that modulate gene expression and cell growth (Etienne-Manneville and Hall 2002; Park and Bi 2007). Their functions are tightly regulated by a molecular switch modality. The cycling between the GTP-bound (active) and the GDP-bound (inactive) state is regulated by three classes of proteins: guanine nucleotide exchange factors (GEFs), GTPase-activating proteins (GAPs), and GDP-dissociation inhibitors (Gdis). By interacting with Rho-Gdi-binding proteins, by phosphorylation or upon alterations in the lipid composition of membranes, they can actively contribute to the delivery of Rho proteins to specific subcellular membranes and signaling pathways (Bassilana et al. 2005).

Regarding the features of the deduced amino acidic sequence of TbCdc42, comparison of the amino acid sequence with those of other fungal species and representative animal Cdc42 proteins reveals the five typical functional domains of this protein family associated with binding and hydrolysis of guanine nucleotides. These domains are highly homologous with those in mammals and yeast. The deduced amino acid sequence of TbCdc42 also contains the typical carboxyl-terminal consensus sequence, CAAX (C, cysteine; A, any aliphatic amino acid; and X, any amino acid) and a lysine-rich polybasic region upstream of the CAAX domain important for intracellular localization. The residue Thr37, which is involved in the coordination of Mg^{2+} cations in the interaction with Rho-Gdi in competition with GEF, is present inside the typical domain I (switch I), and the residue Gln63, involved in GTP hydrolysis, is present in switch II. In addition, the amino acid sequence of TbGdi is inferred from the gene sequence. The amino acid sequence comparison with diverse Rho-GDI proteins reveals the two functional domain characteristic of this protein family as shown by biochemical and structural studies: a short N-terminal regulatory domain (residues 5–55) linked by residues 56–64 to a C-terminal domain with barrel structure (residues 65–202) (Hoffman et al. 2000). The C-terminal domain has an immunoglobulin-like fold that contains a hydrophobic pocket for insertion of the isoprenyl moiety of the Rho GTPase. The nonpolar residues that form this pocket are conserved in TbRhoGdi, particularly a "hydrophobic triad." This triad is critical for the formation of the binding site for the distal isoprene unit. These residues are fundamental because they interact with the polybasic region at the C-terminus of the Rho GTPases.

Further information regarding TbGdi and TbCdc42 proteins has been obtained using deduced virtual three-dimensional structures, which can be created thanks to the high degree of conservation and the numerous crystallographic structures available in the Swiss-Prot database, allowing the verification of the three-dimensional shape of the predicted domains. The predicted three-dimensional structure reveals the presence of active sites matching those present in other eukaryotes, mainly fungi. Knowledge obtained through the analyses of the deduced amino acidic sequences and the construction of three-dimensional structures has been deepened through an interaction study. In yeast two-hybrid systems and translocation assays, TbRhoGdi has been found to interact with TbCdc42 (Menotta et al. 2008), confirming the regulatory role of TbGdi vs the GTPase TbCdc42. Furthermore, most interestingly, it was found that the *Tbgdi* and *Tbcdc42* genes are expressed at higher levels in the presence of the symbiotic host, root exudates, or glucose (Menotta et al. 2007). An increased expression of both genes in the presence of the host plant and/or root exudates and/or glucose (a component of root exudates) is necessary to the fungus. It determines apical growth and branching activation and consequently it induces the fungus to prepare itself for the development of mycorrhizal symbiosis, making it get physically closer to the plant roots through the prime of branching events and polar growth.

Regarding other genes involved in these mechanisms, thanks to the information made available by the recent sequencing of the *T. melanosporum* genome, a large number of proteins involved in cytoskeletal reorganization, hyphal growth, and branching have been found. Briefly, 149 genes have been identified and functionally grouped according to the deduced amino acid sequences (Amicucci et al. 2011), permitting the description of a hypothetical metabolic pathway. A detailed gene annotation shows that most components of the machinery for cell polarity, morphogenesis, and cytoskeleton, as characterized in yeasts and filamentous fungi, are conserved, albeit the degree of similarity varies from strong to weak.

These findings raise many questions regarding fungal morphogenesis and should spur us on to gain a better understanding of the role of these hyphal growth genes in the context of plant/fungal interactions. Long-standing questions around issues such as the larger hyphal diameter shown by truffles at the moment of the contact with the plant or the abundance of incomplete transverse walls during the mantle formation probably depend on a differential regulation of genes controlling the axiality of the cell growth.

24.5 Conclusions

Truffle research has several different goals: understanding the fungus' complex life cycle, developing tools for its identification, and improving the production of mycorrhized seedlings and their cultivation. Proteomic approaches represent a good starting point to broaden our knowledge of the *Tuber* species and to investigate their biology. Gaining a better understanding of the species' primary metabolism is a necessary step toward understanding its physiology. Indeed, primary metabolism affects all the phenotypical traits of filamentous fungi. In particular, it affects the fungi's reaction to extracellular stimuli and its production of precursor molecules required for cell division, morphological changes, and provision of monomer building blocks for the production of secondary metabolites and extracellular polypeptides.

Our newfound knowledge of the genome of *T. melanosporum* has also given a boost to the discovery of new proteins and the characterization of their genes. Proteins can also be characterized using heterologous systems to produce recombinant variants or to complement yeast strains. However, the genome by itself is not completely representative of the functional aspects of an organism. In fact, it is the proteomic network that explains and describes i) the biology of an organism, ii) its functional organization, and iii) its ability to respond to endogenous programming in the realization of its life cycle and to external biotic and abiotic stimuli. Moreover, interestingly, the results mentioned herein highlight the potential of methodologies based on proteomic investigations for ecological studies, in particular in the identification of the geographical origin of truffles (where an ascoma is harvested). Indeed, protein profiles, either detected by 2D profiles or by transcriptome analysis, are greatly influenced by environmental factors, which comprise the pedo-climatic conditions: soil composition and temperature, precipitation, as well as the fauna and microfauna and flora.

References

Agostini D, De Bellis R, Polidori E, Piccoli G, Palma F, Stocchi V (2000) Identification, purification and gene cloning of a protein from *Tuber dryophilum* fruitbodies homologous to TBF-1 protein. Mycol Res 104(05):533–536. doi:10.1017/S0953756299001951

Agostini D, Polidori E, Palma F, Ceccaroli P, Saltarelli L, Tonelli D, Stocchi V (2001) Cloning, expression, and characterization of the hxk-1 gene from the white truffle *Tuber borchii* Vittad.: a first step toward understanding sugar metabolism. Fungal Genet Biol 33(1):15–23. doi:10.1006/fgbi.2001.1268

Ambra R, Grimaldi B, Zamboni S, Filetici P, Macino G, Ballario P (2004) Photomorphogenesis in the hypogeous fungus *Tuber borchii*: isolation and characterization of *Tbwc-1*, the homologue of the blue-light photoreceptor of *Neurospora crassa*. Fungal Genet Biol 41(7):688–697. doi:10.1016/j.fgb.2004.02.004

Amicucci A, Zambonelli A, Iotti M, Polidori E, Menotta M, Saltarelli R, Potenza L, Stocchi V (2010) Morphological and molecular modifications induced by different carbohydrate sources in *Tuber borchii*. J Mol Microbiol Biotechnol 18(2):120–128. doi:10.1159/000297915

Amicucci A, Balestrini R, Kohler A, Barbieri E, Saltarelli R, Faccio A, Roberson RW, Bonfante P, Stocchi V (2011) Hyphal and cytoskeleton polarization in *Tuber melanosporum*: a genomic and cellular analysis. Fungal Genet Biol 48(6):561–572. doi:10.1016/j.fgb.2010.12.002

Baggerman G, Vierstraete E, de Loof A, Schoofs L (2005) Gel-based versus gel-free proteomics: a review. Com Chem High T Scr 8(8):669–677. doi:10.2174/138620705774962490

Banuett F, Quintanilla RH, Reynaga-Peña CG (2008) The machinery for cell polarity, cell morphogenesis, and the cytoskeleton in the Basidiomycete fungus *Ustilago maydis*. A survey of the genome sequence. Fungal Genet Biol 45(Suppl 1):S3–S14. doi:10.1016/j.fgb.2008.05.012

Bassilana M, Hopkins J, Arkowitz RA (2005) Regulation of the *Cdc42*/*Cdc24* GTPase module during *Candida albicans* hyphal growth. Eukaryot Cell 4(3):588–603. doi:10.1128/EC.4.3.588-603.2005

Ceccaroli P, Saltarelli R, Buffalini M, Piccoli G, Stocchi V (1999) Three different forms of hexokinase are identified during *Tuber borchii* mycelium growth. Mol Cell Biochem 194 (1–2):71–77. doi:10.1023/A:1006908501788

Ceccaroli P, Saltarelli R, Cesari P, Zambonelli A, Stocchi V (2001) Effects of different carbohydrate sources on the growth of *Tuber borchii* Vittad. mycelium strains in pure culture. Mol Cell Biochem 218(1):65–70. doi:10.1023/A:1007265423786

Ceccaroli P, Saltarelli R, Cesari P, Pierleoni R, Sacconi C, Vallorani L, Rubini P, Stocchi V, Martin F (2003) Carbohydrate and amino acid metabolism in *Tuber borchii* mycelium during glucose utilization: a $^{(13)}$C NMR study. Fungal Genet Biol 39(2):168–175. doi:10.1016/S1087-1845(03)00006-9

Ceccaroli P, Saltarelli R, Guescini M, Polidori E, Buffalini M, Menotta M, Pierleoni R, Barbieri E, Stocchi V (2007) Identification and characterization of the *Tuber borchii* D-mannitol dehydrogenase, which defines a new subfamily within the polyol-specific medium chain dehydrogenases. Fungal Genet Biol 44(10):965–978. doi:10.1016/j.fgb.2007.01.002

Ceccaroli P, Buffalini M, Saltarelli R, Barbieri E, Polidori E, Ottonello S, Kohler A, Tisserant E, Martin F, Stocchi V (2011) Genomic profiling of carbohydrate metabolism in the ectomycorrhizal fungus *Tuber melanosporum*. New Phytol 189(3):751–764. doi:10.1111/j.1469-8137.2010.03520.x

Cerigini E, Palma F, Barbieri E, Buffalini M, Stocchi V (2008) The *Tuber borchii* fruiting body-specific protein TBF-1, a novel lectin which interacts with associated *Rhizobium* species. FEMS Microbiol Lett 284(2):197–203. doi:10.1111/j.1574-6968.2008.01197.x

Cui W, Rohrs HW, Gross ML (2011) Top-down mass spectrometry: recent developments, applications and perspectives. Analyst 136(19):3854–3864. doi:10.1039/c1an15286f

De Bellis R, Agostini D, Piccoli G, Vallorani L, Potenza L, Polidori E, Sisti D, Amoresano A, Pucci P, Arpaia G, Macino G, Balestrini R, Bonfante P, Stocchi V (1998) The *tbf-1* gene from the white truffle *Tuber borchii* codes for a structural cell wall protein specifically expressed in fruitbody. Fungal Genet Biol 25(2):87–99. doi:10.1006/fgbi.1998.1092

Domon B, Aebersold R (2006) Mass spectrometry and protein analysis. Science 312 (5771):212–217. doi:10.1126/science.1124619

Doyle S (2011) Fungal proteomics: from identification to function. FEMS Microbiol Lett 321 (1):1–9. doi:10.1111/j.1574-6968.2011.02292.x

Etienne-Manneville S, Hall A (2002) Rho GTPases in cell biology. Nature 420(6916):629–635. doi:10.1038/nature01148

Frey-Klett P, Garbaye JA, Tarkka M (2007) The mycorrhiza helper bacteria revisited. New Phytol 176(1):22–36. doi:10.1111/j.1469-8137.2007.02191.x

Gioacchini AM, Menotta M, Guescini M, Saltarelli L, Ceccaroli P, Amicucci A, Barbieri E, Giomaro G, Stocchi V (2008) Geographical traceability of Italian white truffle (*Tuber magnatum* Pico) by the analysis of volatile organic compounds. Rapid Commun Mass Spectrom 22(20):3147–3153. doi:10.1002/rcm.3714

Guescini M, Zeppa S, Pierleoni R, Sisti D, Stocchi L, Stocchi V (2007) The expression profile of the *Tuber borchii* nitrite reductase suggests its positive contribution to host plant nitrogen nutrition. Curr Genet 51(1):31–41. doi:10.1007/s00294-006-0105-y

Hacquard S, Tisserant E, Brun A, Legué V, Martin F, Kholer A (2013) Laser microdissection and microarray analysis of *Tuber melanosporum* ectomycorrhizas reveal functional heterogeneity between mantle and Hartig net compartments. Environ Microbiol 15(6):1853–1869. doi:10.1111/1462-2920.12080

Harris SD (2006) Cell polarity in filamentous fungi: shaping the mold. Int Rev Cytol 251:41–77. doi:10.1016/S0074-7696(06)51002-2

Hoffman GR, Nassar N, Cerione RA (2000) Cell structure of the Rho family GTP-binding protein Cdc42 in complex with the multifunctional regulator RhoGDI. Cell 100(3):345–356. doi:10.1016/S0092-8674(00)80670-4

Islam MT, Mohamedali A, Garg G, Khan JM, Gorse AD, Parsons J, Marshall P, Ranganathan S, Baker MS (2013) Unlocking the puzzling biology of the Black Périgord Truffle *Tuber melanosporum*. J Proteome Res 12(12):5349–5356. doi:10.1021/pr400650c

Le Tacon F, Zeller B, Plain C, Hossann C, Bréchet C, Robin C (2013) Carbon transfer from the host to *Tuber melanosporum* mycorrhizas and ascocarps followed using a ^{13}C pulse-labeling technique. PLoS One 8(5), e64626. doi:10.1371/journal.pone.0064626

Liu QN, Liu RS, Wang YH, Mi ZY, Li DS, Zhong JJ, Tang YJ (2009) Fed-batch fermentation of *Tuber melanosporum* for the hyperproduction of mycelia and bioactive *Tuber* polysaccharides. Bioresour Technol 100(14):3644–3649. doi:10.1016/j.biortech.2009.02.037

Martin F, Kohler A, Murat C, Balestrini R, Coutinho PM, Jaillon O, Montanini B, Morin E, Noel B, Percudani R, Porcel B, Rubini A, Amicucci A, Amselem J, Anthouard V, Arcioni S, Artiguenave F, Aury JM, Ballario P, Bolchi A, Brenna A, Brun A, Buée M, Cantarel B, Chevalier G, Couloux A, Da Silva C, Denoeud F, Duplessis S, Ghignone S, Hilselberger B, Iotti M, Mello M, Miranda M, Pacioni G, Quesneville H, Riccioni C, Ruotolo R, Splivallo R, Stocchi V, Tisserant E, Viscomi AR, Zambonelli A, Zampieri E, Henrissat B, Lebrun MH, Paolocci F, Bonfante P, Ottonello S, Wincker P (2010) Perigord black truffle genome uncovers evolutionary origins and mechanisms of symbiosis. Nature 464(7291):1033–1038. doi:10.1038/nature08867

Menotta M, Amicucci A, Basili G, Rivero F, Polidori E, Sisti D, Stocchi V (2007) Molecular characterisation of the small GTPase CDC42 in the ectomycorrhizal fungus *Tuber borchii* Vittad. Protoplasma 231(3–4):227–237. doi:10.1007/s00709-007-0254-y

Menotta M, Amicucci A, Basili G, Polidori E, Stocchi V, Rivero F (2008) Molecular and functional characterization of a Rho GDP dissociation inhibitor in the filamentous fungus *Tuber borchii*. BMC Microbiol 8(1):57. doi:10.1186/1471-2180-8-57

Molinier V, Murat C, Frochot H, Wipf D, Splivallo R (2015) Fine scale spatial genetic structure analysis of the black truffle *Tuber aestivum* and its link to aroma variability. Environ Microbiol 17(8):3039–3050. doi:10.1111/1462-2920.12910

Montanini B, Moretto N, Soragni E, Percudani R, Ottonello S (2002) A high-affinity ammonium transporter from the mycorrhizal ascomycete *Tuber borchii*. Fungal Genet Biol 36(1):22–34. doi:10.1016/S1087-1845(02)00001-4

Montanini B, Betti M, Márquez AJ, Balestrini R, Bonfante P, Ottonello S (2003) Distinctive properties and expression profiles of glutamine synthetase from a plant symbiotic fungus. Biochem J 373:357–368. doi:10.1042/bj20030152

Montanini B, Viscomi AR, Bolchi A, Martin Y, Siverio J, Balestrini R, Bonfante P, Ottonello S (2006) Functional properties and differential mode of regulation of the nitrate transporter from a plant symbiotic ascomycete. Biochem J 394:125–134. doi:10.1042/BJ20051199

Nehls U (2008) Mastering ectomycorrhizal symbiosis: the impact of carbohydrates. J Exp Bot 59 (5):1097–1108. doi:10.1093/jxb/erm334

Nombela C, Gil C, Chaffin WL (2006) Non-conventional protein secretion in yeast. Trends Microbiol 14(1):15–21. doi:10.1016/j.tim.2005.11.009

Palma F, Agostini D, Cerigini E, Polidori E, Stocchi V (2005) Expression and purification of a *Tuber borchii* fruit body-specific protein, TBF-1, from *Escherichia coli*: generation of polyclonal antibodies. Prep Biochem Biotechnol 35(2):145–153. doi:10.1081/PB-200054736

Palma F, Cerigini E, Stocchi V (2007) Yeast expression of the *Tuber borchii* fruiting body specific protein, TBF-1: identification of a noncanonical signal peptide. FEMS Microbiol Lett 272 (1):114–119. doi:10.1111/j.1574-6968.2007.00748.x

Park HO, Bi E (2007) Central roles of small GTPases in the development of cell polarity in yeast and beyond. Microbiol Mol Biol Rev 71(1):48–96. doi:10.1128/MMBR.00028-06

Payen T, Murat C, Gigant A, Morin E, De Mita S, Martin F (2015) A survey of genome-wide single nucleotide polymorphisms through genome re-sequencing in the Périgord black truffle (*Tuber melanosporum* Vittad.). Mol Ecol Resour 15(5):1243–1255. doi:10.1111/1755-0998.12391

Pierleoni R, Buffalini M, Vallorani L, Guidi C, Zeppa S, Sacconi C, Pucci P, Amoresano A, Casbarra A, Stocchi V (2004) *Tuber borchii* fruit body: 2-dimensional profile and protein identification. Phytochemistry 65(7):813–820. doi:10.1016/j.phytochem.2004.02.012

Polidori E, Saltarelli R, Ceccaroli P, Buffalini M, Pierleoni R, Palma F, Bonfante P, Stocchi V (2004) Enolase from the ectomycorrhizal fungus *Tuber borchii* Vittad.: biochemical characterization, molecular cloning, and localization. Fungal Genet Biol 41(2):157–167. doi:10.1016/j.fgb.2003.10.008

Polidori E, Ceccaroli P, Saltarelli R, Guescini M, Menotta M, Agostini D, Palma F, Stocchi V (2007) Hexose uptake in the plant symbiotic ascomycete *Tuber borchii* Vittadini: biochemical features and expression pattern of the transporter TBHXT1. Fungal Genet Biol 44(3):187–198. doi:10.1016/j.fgb.2006.08.001

Potenza L, Saltarelli R, Polidori E, Ceccaroli P, Amicucci A, Zeppa S, Zambonelli A, Stocchi V (2012) Effect of 300 mT static and 50 Hz 0.1 mT extremely low frequency magnetic fields on *Tuber borchii* mycelium. Can J Microbiol 58(10):1174–1182. doi:10.1139/w2012-093

Rabilloud T, Chevallet M, Luche S, Lelong C (2010) Two-dimensional gel electrophoresis in proteomics: past, present and future. J Proteomics 73(11):2064–2077. doi:10.1016/j.jprot.2010.05.016

Saltarelli R, Ceccaroli P, Vallorani L, Zambonelli A, Citterio B, Malatesta M, Stocchi V (1998) Biochemical and morphological modifications during the growth of *Tuber borchii* mycelium. Mycol Res 102(4):403–409. doi:10.1017/S0953756297004875

Saltarelli R, Ceccaroli P, Cesari P, Barbieri E, Stocchi V (2008) Effect of storage on biochemical and microbiological parameters of edible truffle species. Food Chem 109(1):8–16. doi:10.1016/j.foodchem.2007.11.075

Sisti D, Giomaro G, Rossi I, Ceccaroli P, Citterio B, Stocchi V, Zambonelli A, Benedetti PA (1998) In vitro mycorrhizal synthesis of micropropagated *Tilia platyphyllos* Scop. plantlets with *Tuber borchii* Vittad. mycelium in pure culture. Acta Hortic 457:379–387. doi:10.17660/ActaHortic.1998.457.47

Steinberg G (2007) Hyphal growth: a tale of motors, lipids, and the Spitzenkörper. Eukaryot Cell 6(3):351–360. doi:10.1128/EC.00381-06

Vallorani L, Bernardini F, Sacconi C, Pierleoni R, Pieretti B, Piccoli G, Buffalini M, Stocchi V (2000) Identification of *Tuber borchii* Vittad. mycelium proteins separated by two-dimensional polyacrylamide gel electrophoresis using amino acid analysis and sequence tagging. Electrophoresis 21(17):3710–3716. doi:10.1002/1522-2683(200011)21:17<3710::AID-ELPS3710>3.0.CO;2-9

Vallorani L, Polidori E, Sacconi C, Agostini D, Pierleoni R, Piccoli G, Zeppa S, Stocchi V (2002) Biochemical and molecular characterization of NADP-glutamate dehydrogenase from the ectomycorrhizal fungus *Tuber borchii*. New Phytol 154(3):779–790. doi:10.1046/j.1469-8137.2002.00409.x

Vita F, Lucarotti V, Alpi E, Balestrini R, Mello A, Bachi A, Alessio M, Alpi A (2013) Proteins from *Tuber magnatum* Pico fruiting bodies naturally grown in different areas of Italy. Proteome Sci 11(1):7. doi:10.1186/1477-5956-11-7

Wu Q, Yuan H, Zhang L, Zhang Y (2012) Recent advances on multidimensional liquid chromatography-mass spectrometry for proteomics: from qualitative to quantitative analysis-a review. Anal Chim Acta 731:1–10. doi:10.1016/j.aca.2012.04.010

Index

A. Zambonelli et al. (eds.), *True Truffle (*Tuber *spp.) in the World*, Soil Biology 47,
DOI 10.1007/978-3-319-31436-5

The manufacturer's authorised representative in the EU is Springer Nature Customer Service Centre GmbH, Europaplatz 3, 69115 Heidelberg, Germany. If you have any concerns regarding our products, please contact ProductSafety@springernature.com

Printed and bound by CPI Group (UK) Ltd, Croydon, CR0 4YY
30/09/2026
02225652-0018